S. I. Adian

The Burnside Problem and Identities in Groups

Translated from the Russian
by John Lennox and James Wiegold

Springer-Verlag
Berlin Heidelberg New York 1979

Sergei I. Adian
Steklov Mathematical Institute, Moscow, USSR

John Lennox, James Wiegold
University College, Cardiff, Wales, U.K.

Title of the Russian Original Edition:
Problema Bernsaida i tozhdestva v gruppakh
Publisher: Nauka, Moscow, 1975

AMS Subject Classification (1970): 20 E 10, 20 F 50

ISBN-13: 978-3-642-66934-7 e-ISBN-13: 978-3-642-66932-3
DOI: 10.1007/978-3-642-66932-3

Library of Congress Cataloging in Publication Data. Adian, S. I. The Burnside problem and identities in groups (Ergebnisse der Mathematik und ihrer Grenzgebiete; 95). Translation of Problema Bernsaida i tozhdestva v gruppakh. Bibliography: p. Includes indexes. 1. Burnside problem. 2. Groups, Theory of – 3. Groups, Theory of – Generators. I. Title. II. Title: Identities in groups. III. Series. QA171.A313. 512'.22. 78-16896

Softcover reprint of the hardcover 1st edition 1979

2141/3140-543210

Preface to the English Edition

Three years have passed since the publication of the Russian edition of this book, during which time the method described has found new applications.

In [26], the author has introduced the concept of the periodic product of two groups. For any two groups G_1 and G_2 without elements of order 2 and for any odd $n \geqslant 665$, a group $G_1 ⓝ G_2$ may be constructed which possesses several interesting properties. In $G_1 ⓝ G_2$ there are subgroups $\bar{G}_1$ and $\bar{G}_2$ isomorphic to G_1 and G_2 respectively, such that $\bar{G}_1$ and $\bar{G}_2$ generate $G_1 ⓝ G_2$ and intersect in the identity. This operation "ⓝ" is commutative, associative and satisfies Mal'cev's postulate (see [27], p. 474), i.e., it has a certain hereditary property for subgroups. For any element x which is not conjugate to an element of either $\bar{G}_1$ or $\bar{G}_2$, the relation $x^n = 1$ holds in $G_1 ⓝ G_2$. From this it follows that when G_1 and G_2 are periodic groups of exponent n, so is $G_1 ⓝ G_2$. In addition, if G_1 and G_2 are free periodic groups of exponent n the group $G_1 ⓝ G_2$ is also free periodic with rank equal to the sum of the ranks of G_1 and G_2. I believe that groups having many interesting properties can be constructed using this notion of periodic product. For example, it has been proved recently that a periodic product $G_1 ⓝ G_2$ is a simple group if and only if each of the groups G_1 and G_2 coincides with the subgroup generated by its n-th powers.

Using a modification of our methods, one can prove that the word problem and the conjugacy problem are solvable for any finitely presented group which has only defining relations of the form $A^n = 1$ with elementary periods A and given odd $n \geqslant 665$.

A contradiction within the system of parameter conditions, on which Britton based his argument in [25], was demonstrated in the introduction to the Russian edition. It was clear to me at that time that this mistake was connected with the principal difficulties inherent in the Burnside Problem, and would prevent Britton from completing his proof with any ease. In fact, Britton has not yet published a correction of the mistake in his proof.

In conclusion I would like to express my sincere thanks to Professors James Wiegold and John Lennox who took on themselves the onerous task of translating this book.

Moscow, May 15, 1978 S. Adian

Translators' Preface

We would like to thank Professor Adian for willingly giving us so much of his time while the translation was being prepared. His help and patience over difficult points in the Russian were invaluable, and we are very grateful.

Thanks, too, to the University of Bielefeld, in particular to Professor Jens Mennicke, for inviting one of us (J.W.) to the Burnside meeting of July 1977. Not only was this an enjoyable visit, but it also made it possible to have many long discussions with Professor Adian concerning the translation.

Finally, our thanks to Springer-Verlag for their unfailing courtesy at all times.

Preface

To the memory of
P. S. Novikov

This book is based on a special course that the author delivered to the Faculty of Mechanics and Mathematics at Moscow University in the academic years 1971/72 and 1972/73. It presents a new and improved version of the method of investigating groups with an identical relation of the form $x^n = 1$ evolved by P.S. Novikov and the author for solving Burnside's problem on periodic groups, first published in the joint paper [5]. The distinguishing feature of that method is the proof of a large number of assertions (more than a hundred) by simultaneous induction over a natural parameter. Comparing now with [5], certain new concepts are introduced here, the definitions of a number of the old concepts are altered somewhat, and a large number of new lemmas are added. These changes have made it possible to simplify the proof significantly and to reduce the bound for n from $n \geqslant 4381$ to $n \geqslant 665$. I have succeeded in giving the definitions of all the main concepts, which also go by induction on a natural parameter, in the first chapter. Undoubtedly, this facilitates the reading of the book, since the reader has the opportunity of grasping the definitions of the concepts before beginning to take the proof apart. No special knowledge is required of the reader. If any difficulties arise in a first reading, it is recommended that cumbersome proofs of individual lemmas be omitted, and appeal made to the subject index.

After giving the solution of the Burnside problem, we shall prove the results contained in the author's lecture to the International Congress of Mathematicians in Nice [9]. Moreover, a construction is given of finitely generated torsion-free groups such that every pair of cyclic subgroups have non-trivial intersection. This is a non-commutative analogue of the additive group of rational numbers. There is reason to suppose that the methods of the book will find application to the solution of other problems in the theory of infinite groups.

A recent unsuccessful attempt was made in [25] to give a simpler solution of the Burnside problem. In that paper the author follows, in the main, the original scheme proposed by P. S. Novikov in 1959 (see [4]), which was based on a use of transformations of cyclic words and the method of V. A. Tartakovskiĭ (see [24]). As in [5], the proof of a large number of the assertions proceeds by simultaneous induction over a natural parameter. However, several of the concepts that were essential in [5] are omitted from [25]. For example, it does not contain the concept of mutual normalisability in given rank, which was central in [5], nor that of cascade of rank α. In [25] the proof rests on so-called parametric conditions, which involve a system of some hundreds of equalities and inequalities in 302 parameters. The consistency of this system is not proved in [25]. More than that, an analysis carried out jointly by Ju. I. Hmelevskiĭ, the editor of this book, and myself, shows that the system of parametric conditions used in [25] is contradictory*.

[25] is therefore erroneous.

I am heartily grateful to Ju. I. Hmelevskiĭ, who read the manuscript with great care, checked all the proofs, made a number of useful remarks, and was of essential help to me in the analysis of [25].

*For example, the conditions

$u_4 = u_1 + r_{25}$ (p. 145, line 10 from below),

$r_{25} \geqslant u_{37} + 54/e$ (p.283, line 4 from below),

$u_{37} > 14\alpha + 214/e$, where $\alpha = \varepsilon_{30} + u_{13} + 6u_4$ (p.221, lines 11 and 12 from below)

give an obvious contradiction: $u_4 > r_{25} > u_{37} > u_4$.

Contents

Introduction

In 1902, Burnside [1] formulated the following problem:

"Is every group with a finite number of generators and satisfying an identical relation $x^n = 1$ finite?"

This problem is known as the *Burnside problem for groups of finite exponent*, and it remained open for a considerable time. The negative solution was obtained in a joint paper of P. S. Novikov and the author [5], where it was shown that, for every odd $n \geqslant 4381$ and every $m > 1$, there exists an infinite group $\Gamma(m, n)$ on m generators and satisfying the identical relation $x^n = 1$.

Up to then, a positive answer to the Burnside problem had been obtained for $n \leqslant 3$ (see [1]), $n = 4$ (see [2]) and $n = 6$ (see [3]). The existence of a periodic group on two generators having no bound on the orders of its elements was established in [12].

In order to describe the group $\Gamma(m, n)$, a classification of periodic words in a group alphabet was introduced in [5], and a theory constructed of transformation of words corresponding to an identical relation $x^n = 1$ for fixed odd $n \geqslant 4381$.

In the first five chapters of this book we present an improved version of this theory for odd exponent $n \geqslant 665$. On the basis of this theory we prove in Chapter VI the existence of infinite groups of odd exponent $n \geqslant 665$. In that Chapter we also give the proofs of all the results about the properties of free groups of odd exponent $n \geqslant 665$ that were published in [6, 7, 8, 10]. Chapter VII contains an account of the applications of our method to questions not connected with periodic groups. These applications were published in [10, 11].

The Chapters of the book are divided into sections, and the sections into subsections. Reference to the assertion or definition contained in subsection 16 of §5 of Chapter II, for example, will be made in the form II.5.16. Reference to the Chapter will not be made within the Chapter itself.

In order to avoid numerous repetitions in the execution of the complicated simultaneous induction to which Chapters II-V are devoted, we shall not cite the separate formulations of the inductive assumptions (and we shall not even number them separately, as was done in [5]). In referring to one or other inductive assumption, we shall directly indicate the corresponding assertion that is formulated and proved for larger values of the inductive parameter in the succeeding chapters and sections. For example, if assertion *IV.1.7* is encountered in the text at a stage before it is proved in rank α, this means that the assertion obtained from IV.1.7 on replacing the inductive parameter α by a suitable $\beta \leqslant \alpha - 1$ is assumed to be proven at that stage, in accordance with the inductive assumption. We shall distinguish such references from all others by writing them in *italics*.

All assertions considered are trivial when $\alpha = 0$. This saves us from special considerations in verifying the first steps of the induction.

Chapter I. Basic Concepts and Notation

We shall use the following notation for logical connectives:

$\exists$, $\forall$ the existence and universal quantifiers,
& conjunction ("and"),
$\vee$ disjunction ("or"),
$\neg$ negation ("not"),
$\Rightarrow$ implication ("if ..., then, ..."),
$\Longleftrightarrow$ logical equivalence ("if and only if").

The symbol $\rightleftharpoons$ denotes equality by definition, and it is used to introduce notation for certain expressions.

If X and Y are elements of a set $\mathscr{M}$, then for brevity we write $(X \in \mathscr{M})$ & $(Y \in \mathscr{M})$ as $X, Y \in \mathscr{M}$.

The symbol $\subset$ is used to denote inclusion of one set in another, $\cup$ for union of sets and $\cap$ for intersection of sets. The empty set is denoted by $\varnothing$.

We fix an integer $m > 1$ and an odd number $n \geqslant 665$, which will be unchanged for the duration of Chapters I-VI. In addition, we shall use the fixed values of the following numerical parameters:

$$p = 9, \quad p_1 = 17,$$
$$q_1 = 2p_1 + 3 = 37, \quad q_2 = q_1 + p_1 = 54, \quad q = q_2 + 2p_1 + 2 = 90\,.$$

§ 1. Words and Occurrences

We shall consider words in a group alphabet

$$a_1, a_2, \ldots, a_m, a_1^{-1}, a_2^{-1}, \ldots, a_m^{-1}\,. \tag{1}$$

The empty word is denoted by Λ. Letter-for-letter equality of two words X and Y is denoted by $X \overset{\circ}{=} Y$.

1.1. Two letters a_i and a_i^{-1} for given i are said to be *mutually inverse*. If

$$A \overset{\circ}{=} b_1 b_2 b_3 \ldots b_j \ldots b_{r-1} b_r\,,$$

where each b_j is one of the letters in (1), then the word

$$b_r^{-1}b_{r-1}^{-1}\dots b_j^{-1}\dots b_3^{-1}b_2^{-1}b_1^{-1},$$

with $(a_i^{-1})^{-1} \rightleftharpoons a_i$, is called the *inverse* of A and is denoted by A^{-1}. Clearly, $A^{-1} \equiv E^{-1}\, D^{-1}$ whenever $A \equiv DE$.

We denote the length of the word A by $\partial(A)$, that is, $\partial(A)$ is the number of letters comprising A. In particular, $\partial(\Lambda) = 0$.

1.2. A word E is said to *occur* in a word X if there exist words R and Q such that $X \equiv REQ$. If the word R (the word Q) is empty, then E is said to be a *start* (an *end*) of X.

Clearly, one of any two starts of a given word X is a start of the other, and similarly one of two ends of A is an end of the other.

One and the same word E may occur in a given word X in different places. In order to distinguish between two different occurrences of E in X we use an extra symbol $*$. If X is a word in an alphabet not containing the letter $*$, and $X \equiv REQ$, then we shall call the word $R * E * Q$ an *occurrence of E in X*.

If $V \rightleftharpoons R * E * Q$, we denote $Q^{-1} * E^{-1} * R^{-1}$ by V^{-1}.

The word E is said to be the *base* of the occurrence $R * E * Q$. If V denotes the occurrence $R * E * Q$, we shall write $E = \text{Bas}\,(V)$.

We shall consider only occurrences with non-empty bases. An occurrence of the form $* E *$ will be identified with the word E.

Upper-case Latin letters U, V, W, with or without suffices, will be used to denote occurrences in words in the alphabet (1).

1.3. Let $P * E * Q$ and $R * D * S$ denote two occurrences in one and the same word, *i.e.*, $PEQ \equiv RDS$. All the relations defined in this subsection are meaningful only for occurrences in one and the same word.

We shall say that the occurrence $P * E * Q$ is *contained in* $R * D * S$ if $\partial(R) \leqslant \partial(P)$ and $\partial(S) \leqslant \partial(Q)$, that is, if R is a start of P and S is an end of Q. If in addition $P \equiv R$ (or $Q \equiv S$), we say that $P * E * Q$ is a *start* (or an *end*) of $R * D * S$, or else that $R * D * S$ *starts* (or *ends*) with the occurrence $P * E * Q$.

We say that the occurrences $P * E * Q$ and $R * D * S$ *intersect* if there is an occurrence V with nonempty base that is contained in $P * E * Q$ and in $R * D * S$. We shall call the maximal such occurrence V (with respect to length of base) the *common part* of $P * E * Q$ and $R * D * S$, or their *intersection.* If $P * E * Q$ is contained in $R * D * S$, their common part is $P * E * Q$.

If $\partial(P) < \partial(R)$ and $\partial(S) < \partial(Q)$, we say that the occurrence $P * E * Q$ *lies to the left* of $R * D * S$, and write $P * E * Q < R * D * S$. If PE is a start of R, we shall say that $P * E * Q$ *lies strictly to the left* of $R * D * S$, and write $P * E * Q \ll R * D * S$.

If $P * E * Q < R * D * S$, then neither of the occurrences $P * E * Q$ and $R * D * S$ is contained in the other; conversely, if neither is contained in the other, then $P * E * Q < R * D * S$ or $R * D * S < P * E * Q$.

If $P * E * Q \ll R * D * S$, then $P * E * Q$ and $R * D * S$ do not intersect; conversely, if they do not intersect, then either $P * E * Q \ll R * D * S$ or $R * D * S \ll P * E * Q$.

The *union* of occurrences $P * E * Q$ and $R * D * S$ is the occurrence containing both of them that has base of shortest length. Suppose that U is the union of $P*E*Q$ and $R*D*S$. If $P*E*Q$ is contained in $R*D*S$, then $U = R*D*S$. If $P*E*Q < R*D*S$, then $P*E*Q$ is a start of U, and $R*D*S$ is an end of it.

1.4. Let $P*E*Q$ and P_1*E*Q_1 be occurrences in words X and Y and $V \rightleftharpoons R*C*S$ an occurrence in X contained in $P*E*Q$. Then there exist words A and B such that $R \eqcirc PA$, $S \eqcirc BQ$ and $E \eqcirc ACB$. In such a case the occurrence $P_1A_1*C*BQ_1$ in Y is denoted by

$$\phi(V; P*E*Q, P_1*E*Q_1).$$

For any two occurrences W and W_1 with the same base, the function $V_1 \rightleftharpoons \phi(V; W, W_1)$ sets up a one-to-one mapping of the set of occurrences in W onto the set of all occurrences in W_1.

If $V_1 = \phi(V; W, W_1)$, then $V = \phi(V_1; W_1, W)$.

Let W_1, W_2, W_3 be occurrences with the same base and suppose that V_1 is contained in W_1. If $V_2 = \phi(V_1; W_1, W_2)$ and $V_3 = \phi(V_2; W_2, W_3)$, then $V_3 = \phi(V_1; W_1, W_3)$.

Clearly, the function $V_1 = \phi(V; W, W_1)$ preserves the relations $<$, $\ll$ and carries the common part (union) of two occurrences contained in W to the common part (union) of their images in W_1.

§ 2. Periodic Words

For any integer $t > 0$, let A^t stand for the word $AA\ldots A$, with A repeated t times. For $t < 0$ we set $A^t \rightleftharpoons (A^{-1})^{-t}$. Finally, for any word A we set by definition

$$A^0 = \Lambda.$$

We shall call a word of the form $A_1A^tA_2$, where A_2 is a start of A, A_1 is an end of A and $\partial(A_1A^tA_2) > 2\partial(A)$, a *periodic word with period* A. The set of all periodic words with period A is denoted by Per(A). For empty A_1 (or A_2), we call A the *left* (or *right*) *period* of $A_1A^tA_2$.

Clearly, if $A \eqcirc B^r$, then $\mathrm{Per}(A) \subset \mathrm{Per}(B)$.

2.1. We say that a word B is a *cyclic shift* of a word A if $A \eqcirc PQ$ and $B \eqcirc QP$ for some P and Q.

Clearly, if B is a cyclic shift of A, then $\mathrm{Per}(A) = \mathrm{Per}(B)$. If $X \in \mathrm{Per}(A)$, B occurs in X and $\partial(B) = \partial(A)$, then B is a cyclic shift of A.

2.2. *If $AB \eqcirc BA$, then there is a word D such that $A \eqcirc D^t$ and $B \eqcirc D^r$ for some $t, r \geqslant 0$.*

We may assume that $\partial(A) \geqslant \partial(B)$. If B is empty, then $B \eqcirc A^0$. So we assume that B is nonempty and that the assertion is true whenever $\partial(AB) < j$, and prove it for $\partial(AB) = j$. Suppose that $AB \mathrel{\overline{\underline{\circ}}} BA$. Then for some C we have $A \mathrel{\overline{\underline{\circ}}} BC$ and $CB \mathrel{\overline{\underline{\circ}}} BC$. Since $\partial(CB) = j - \partial(B)$, the inductive assumption gives that $C \mathrel{\overline{\underline{\circ}}} D^k$ and $B \mathrel{\overline{\underline{\circ}}} D^r$ for some D, k and r. In that case $A \mathrel{\overline{\underline{\circ}}} D^{k+r}$.

2.3. *If $A^t A' \mathrel{\overline{\underline{\circ}}} B^r B'$, where A' is a start of A, B' is a start of B and $\partial(A^t A') \geqslant \partial(AB)$, there exists a word D such that $A \mathrel{\overline{\underline{\circ}}} D^k$ and $B \mathrel{\overline{\underline{\circ}}} D^s$ for some k and s.*

Suppose that $\partial(A) \geqslant \partial(B)$. Since A is a start of $B^r B'$, we have $A \mathrel{\overline{\underline{\circ}}} B^{r_1} B_1$, where $r_1 > 0$ and $B \mathrel{\overline{\underline{\circ}}} B_1 B_2$. Cancelling A on the left of the original equality, we get $A^{t-1} A' \mathrel{\overline{\underline{\circ}}} B_2 B^{r-r_1-1} B'$, where now $B_1 B_2$ is a start of the left hand side and $B_2 B_1$ is a start of the right hand side. Thus $B_1 B_2 \mathrel{\overline{\underline{\circ}}} B_2 B_1$. By 2.2, there exists a word D such that $B_1 \mathrel{\overline{\underline{\circ}}} D^{s_1}$ and $B_2 \mathrel{\overline{\underline{\circ}}} D^{s_2}$ for some s_1, s_2. We can take $s \rightleftharpoons s_1 + s_2$ and $k \rightleftharpoons r_1(s_1 + s_2) + s_1$.

2.4. Suppose that $X \in \operatorname{Per}(A)$. The occurrence of $P * E * Q$ in X is said to be *interior relative to the period A* if $\partial(P) \geqslant 8\,\partial(A)$ and $\partial(Q) \geqslant 8\,\partial(A)$. We denote the set of all occurrences of this sort in X by $\operatorname{Inn}(X, A)$.

2.5. Suppose that $X \in \operatorname{Per}(A)$ again. Two occurrences $V \rightleftharpoons FP * E * QG$ and $W \rightleftharpoons FR * E * SG$ in one and the same word FXG, where F and G are arbitrary words, are said to *correspond in phase relative to the period A* if there is an integer r such that

$$\partial(R) - \partial(P) = r\,\partial(A).$$

The words F and G here may be empty. For $r = 0$ we have that $W \mathrel{\overline{\underline{\circ}}} V$. If $r > 0$ (or $r < 0$) we shall say that W is *the result of shifting V to the right (to the left) by r periods A.*

We denote by $\operatorname{Corr}_A(V, W)$ the predicate that is valid if and only if V and W are occurrences in some word FXG that correspond in phase relative to period A, where now $X \in \operatorname{Per}(A)$. It is clear that the relation $\operatorname{Corr}_A(V, W)$ is symmetric and transitive.

2.6. We extend the concept of correspondence in phase defined in 2.5 to occurrences in different periodic words with given period. Two occurrences $P * E * Q$ and $R * E * S$ in the words $X \in \operatorname{Per}(A)$ and $Y \in \operatorname{Per}(A)$ are said to *correspond in phase* if one of the words P and R is an end of the other and one of Q and S is a start of the other.

We remark that if U and V are two occurrences in a given word $X \in \operatorname{Per}(A)$, then $\operatorname{Corr}_A(U, V)$ implies that they correspond in phase in the sense just defined if we take $Y \rightleftharpoons X$.

2.7. A word A is said to be *simple* if it cannot be represented in the form D^r for $r > 1$.

If A is a non-empty word, then there is a simple word B such that $A \mathrel{\overline{\underline{\circ}}} B^t$ for some $t \geqslant 1$.

This is proved by induction on $\partial(A)$. If A is not simple, then $A \mathrel{\overline{\underline{\circ}}} D^r$ for some D and some $r > 1$. Since $\partial(D) < \partial(A)$, the inductive hypothesis gives the existence of a simple B such that $D \mathrel{\overline{\underline{\circ}}} B^k$, where $k \geqslant 1$. Thus $A \mathrel{\overline{\underline{\circ}}} B^{rk}$.

2.8. *If AB is a simple word, then BA is simple.*

It is enough to prove this for the case where B consists of a single letter a. Assume that $aA \overset{\circ}{=} D^r$, where $r > 1$. Then for some E, $D \overset{\circ}{=} aE$, so that $Aa \overset{\circ}{=} (Ea)^r$.

2.9. *Suppose that* $A'A_1 \overset{\circ}{=} B'B_1$, *where* $\partial(A'A_1) \geqslant \partial(AB)$, A_1 *is a start of* A *and* B_1 *is a start of* B. *If* A *is simple, then* $B \overset{\circ}{=} A^k$ *for some* k.

Indeed, by 2.3 there is a D such that $A \overset{\circ}{=} D^s$ and $B \overset{\circ}{=} D^k$. Clearly, we may take $s > 0$. Since A is a simple word, $s = 1$, that is, $D \overset{\circ}{=} A$.

§ 3. Aperiodic Words

In what follows we shall need an infinite sequence

$$t_1, t_2, t_3, \ldots, t_i, t_{i+1}, \ldots \tag{2}$$

whose terms are either 1 or 2, such that for each i the word $t_1 t_2 \ldots t_i$ does not have any occurrence of a non-empty word of the form E^3. We mention now a method of constructing such a sequence, suggested by Aršon [13].

3.1. Consider all permutations of three symbols, 1,2,3:

$$\begin{array}{ccc} 1\ 2\ 3, & & 3\ 2\ 1, \\ 2\ 3\ 1, & & 1\ 3\ 2, \\ 3\ 1\ 2, & & 2\ 1\ 3. \end{array}$$

We call the permutations in the left-hand column *odd*, and those in the right-hand column *even*. The odd permutations are numbered by their first elements, and the even ones by their last elements. Every even permutation is the mirror image of the odd permutation with the same numeral.

We shall construct words A_i by induction on i, for $i \geqslant 1$. Set

$$A_1 = 1 .$$

If the word A_i has been constructed already, and $A_i = h_1\ h_2 \ldots h_r$, then we let A_{i+1} stand for the result of replacing every symbol h_j in A_i by the even or odd permutation with numeral h_j, according to the parity of j. Let us write down the first few A_i:

$$\begin{aligned} A_2 &= 123\,, \\ A_3 &= 123\ 132\ 312\,, \\ A_4 &= 123\ 132\ 312\ 321\ 312\ 132\ 312\ 321\ 231\,. \end{aligned}$$

We have introduced a *triple* of symbols here so as to be able to survey the whole word more easily. It is clear that A_i is a start of A_{i+1} in all cases.

Ergebnisse der Mathematik und ihrer Grenzgebiete 95

A Series of Modern Surveys in Mathematics

3.2. *There is no occurrence of a nonempty word of the form EE in any* A_i, $i \geqslant 1$.

Proceed by induction on i. Suppose that it is true for A_i. Decompose A_{i+1} into triples: the number of them is the length of A_i. We shall call the occurrences of these triples in A_{i+1} the *constituent triples* of A_{i+1}.

By definition of A_i, there is no sequence in it of two triples of the same parity. Moreover, it follows from the inductive assumption that there is no sequence of two triples with one and the same numeral.

Suppose that $A_{i+1} \overline{\underline{\circ}}\, PEEQ$, where E is non-empty. We show by an analysis of cases that this assumption leads to a contradiction.

Let $V \rightleftarrows R * abc * S$ be the last triple intersecting the occurrence $U \rightleftarrows P * E * EQ$.

If $\partial(E) = 1$, then E is different from a and b, and since $E \overline{\underline{\circ}}\, c$, we get that there is a triple to the right of V and having the same numeral. Consequently $\partial(E) > 1$.

If $\partial(E) = 2$, then $E \overline{\underline{\circ}}\, bc$, and then to the right of V there stands a triple of the same parity. Thus $\partial(E) \geqslant 3$.

Assume that V is an end of the occurrence $P * E * EQ$, that is, $E \overline{\underline{\circ}}\, E_1\, abc$ and $R \overline{\underline{\circ}}\, PE_1$.

Suppose that $\partial(E) = 3j$. If j is odd, then the triple abc occurs in the decomposition of A_i both as an even and as an odd triple. If j is even, then A_i has a subword of the form D^2, where $\partial(D) = j$, which contradicts the inductive assumption.

Suppose that $\partial(E) = 3j + 1$. Then $\partial(E_1 ab)$ is a multiple of three, that is, there is a triple $V_1 \rightleftarrows PEE_2 * cab * cQ$ in the decomposition of A_{i+1}. It is easy to convince oneself in this case that c is a start of E. In fact, if E_2 is not empty, the triple immediately to the left of V is of the form $PE_3 * bac * abcEQ$, that is, $E_2 \overline{\underline{\circ}}\, E_3 ba$, whence it follows by analogy with the preceding case that the triple immediately to the left of V_1 is of the form $PEE_4 * cba * cabcQ$, *etc.* But c cannot be a start of E, since then the word c^2 would occur in A_{i+1}.

Finally, suppose that $\partial(E) = 3j + 2$. Then $\partial\,(E_1 a)$ is a multiple of 3, that is, the decomposition of A_{i+1} contains a triple of the form $PEE_2 * cba * bcQ$, where $E_2 cb \overline{\underline{\circ}}\, E_1$. If E_2 is not empty, then a is an end of E_2. As in the preceding case, we can convince ourselves that c must be a start of E, which is impossible.

Thus we have shown that V cannot be an end of the occurrence $P * E * EQ$. It remains to consider the following two cases:

$$E \overline{\underline{\circ}}\, E_1 a \overline{\underline{\circ}}\, bcE_2 \overline{\underline{\circ}}\, bcE_3 a \tag{3}$$

and

$$E \overline{\underline{\circ}}\, E_1 ab \overline{\underline{\circ}}\, cE_2 \overline{\underline{\circ}}\, cE_3 ab\,,$$

where $PE_1 \overline{\underline{\circ}}\, R$. Since these cases are analogous, we may restrict attention to the first one. Suppose that (3) is satisfied.

If $\partial(E) = 3j$, then $P * bc * E_2 EQ$ is an end of some constituent triple, that is, $P \overline{\underline{\circ}}\, P_1 a$. Then $P_1 * abcE_3 * abcE_3 aQ$ ends with some constituent triple, which is impossible, as was proved above.

If $\partial(E) = 3j + 1$, then by (3), $\partial(bcE_3)$ is a multiple of three. Consequently, the occurrence $P * bcE_3 * aEQ$ begins with the triple bca, that is, $E_3 \overline{\underline{\circ}}\, aE_4$. Then to the right of V there stands the triple $PEbc * acb * E_5 Q$, that is, $E_3 \overline{\underline{\circ}}\, abcE_5$. Thus the second triple contained in $P * bcE_3 * aEQ$ also ends with a. In exactly the same way it can be proved that all the triples in this occurrence end on letter a. But this is impossible, since a^2 does not occur in A_{i+1}.

If $\partial(E) = 3j + 2$, then by (3), $\partial(cE_3)$ is a multiple of three. Then the occurrence $Pb * cE_3 * abcE_3aQ$ begins with one of the triples *cab*, *cba*, from which it follows that the occurrence $PE_1abc * E_3a * Q$ starts with one of the triples *abc*, *bac*. By considering the following triples in these same occurrences, we can convince ourselves in a similar way that E_3a must end in one of the triples *abc* or *bac*, that is, again we get a contradiction.

3.3. Set

$$\nu(1) = 12, \ \nu(2) = 121, \ \nu(3) = 1221 .$$

For an arbitrary word X in the alphabet $\{1,2,3\}$, we denote by $\nu(X)$ the result of replacing each symbol i in X by $\nu(i)$. We shall call the words $\nu(i)$ the *components* of $\nu(X)$.

Clearly, every component starts with the word 12. On the other hand, 12 can occur in $\nu(X)$ only as a start of some component. Thus the number of occurrences of 12 in $\nu(X)$ is equal to $\partial(X)$, and the mapping $Y = \nu(X)$ is one-to-one, that is, $X \overset{\circ}{\equiv} Z$ whenever $\nu(X) \overset{\circ}{\equiv} \nu(Z)$.

It also follows from what we have said that for any words P and Q,

$$P12Q \overset{\circ}{\equiv} \nu(X) \Rightarrow X \overset{\circ}{\equiv} \nu^{-1}(P)\,\nu^{-1}(12Q) . \tag{4}$$

3.4. *There is no occurrence of a word of the form EEE in* $\nu(A_i)$.

Assume that for some P and Q,

$$\nu(A_i) = PEEEQ ,$$

where E is not empty. Then both symbols 1 and 2 must occur in E. Thus $\partial(E^3) \geqslant 6$. Suppose that $PR * 12 * SQ$ is the first occurrence of 12 in $P * EEE * Q$. Then $\partial(R) < \partial(E)$ and $E \overset{\circ}{\equiv} R1G$ for some G. Suppose that $1GR \overset{\circ}{\equiv} 12T$. Then

$$\nu(A_i) = PR12T12T1GQ ,$$

whence by (4) it follows that the following equalities hold for some F, D and H:

$$A_i \overset{\circ}{\equiv} FDH,\ \nu(F) = PR,\ \nu(D) = 12T \text{ and } \nu(H) = 12T1GQ .$$

If G is not empty, then it begins with the symbol 2. Then D is a start of H, which contradicts 3.2. Consequently $R \overset{\circ}{\equiv} 2T$ and $\nu(F) = P2T$. If T is empty, then F ends with 1 and $D \overset{\circ}{\equiv} 1$, which is impossible by 3.2. If T ends with 1, then D is a start of H and the word DD occurs in A_i. Finally, if T ends with 2, then 12 must be an end of $\nu(D)$, that is, 12 is an end of T. Suppose that $T \overset{\circ}{\equiv} B12$. Then by (4) we have

$$D = \nu^{-1}(12B)1,\ F = \nu^{-1}(P2B)1,\ H = \nu^{-1}(12B)\,\nu^{-1}(121GQ) ,$$

that is, the square of $1\nu^{-1}(12B)$ occurs in A_i, and this contradicts 3.2. This is what we wanted to prove.

3.5. *Since for every $i \geqslant 1$ the word $\nu(A_i)$ is a start of $\nu(A_{i+1})$, by completing the construction of the sequence of words $\nu(A_i)$ to $\nu(A_{i+1})$ we get the desired sequence (2) in which for all i the word $t_1 t_2 \ldots t_i$ has no occurrence of a non-empty word of the form E^3.*

§ 4. Inductive Definitions

We shall define certain concepts by simultaneous induction on a natural parameter α, called the rank, that relate to words in the alphabet (1) and occurrences in such words. Of these concepts we first of all mention the following.

4a. $\mathscr{N}_\alpha$, $\mathscr{P}_\alpha$, $\mathscr{R}_\alpha$, $\mathscr{K}_\alpha$, $\mathscr{L}_\alpha$ $\mathscr{M}_\alpha$, $\bar{\mathscr{M}}_\alpha$, are certain sets of words in the alphabet (1) such that

$$\bar{\mathscr{M}}_\alpha \subset \mathscr{M}_\alpha \subset \mathscr{L}_\alpha \subset \mathscr{K}_\alpha \subset \mathscr{R}_\alpha \subset \mathscr{P}_\alpha \subset \mathscr{N}_\alpha .$$

4b. Certain occurrences in a word $X \in \mathscr{N}_\alpha$ are said to be *kernels of rank α* in X. We denote by Ker(α,X) the set of all kernels of rank α in the word X. The set Ker(α, X) is fully ordered under $<$, that is, for two different kernels U and V of rank α of a given word X, either $U < V$ or $V < U$. In general, the kernels from Ker(α,X) may intersect. For any occurrence W in X, we denote by $\partial_\alpha(W)$ the number of kernels of rank α of X contained in W.

4c. *A regular occurrence of rank α* in a word $X \in \mathscr{N}_\alpha$ is an occurrence in X which starts and ends in a kernel of rank α. We denote by Reg(α,X) the set of all regular occurrences of rank α in X. Clearly

$$\mathrm{Ker}(\alpha, X) \subset \mathrm{Reg}(\alpha, X) .$$

4d. There is defined a relation of *equivalence in rank α* on the set $\mathscr{P}_\alpha$. It is reflexive, symmetric and transitive. The notation $X \overset{\alpha}{\sim} Y$ means that $X, Y \in \mathscr{R}_\alpha$ and that X is equivalent to Y in rank α.

4e. The function $W = f_\alpha(V; X,Y)$ is defined when $X \overset{\alpha}{\sim} Y$ and maps Reg (α, X) one-one onto Reg(α, Y). In addition $f_\alpha(V; X,Y) \in$ Ker(α, Y) if $V \in$ Ker(α, X).

$$\text{If } W = f_\alpha(V;\ X,Y),\ \text{then } V = f_\alpha(W;\ Y,X) .$$
$$\text{If } X \overset{0}{=} Y,\ \text{then } f_\alpha(V;\ X,Y) = V .$$

4f. There is defined a reflexive, symmetric and transitive relation on $\bigcup_{X \in \mathscr{R}_\alpha} \mathrm{Reg}(\alpha,X)$, denoted by MutNorm$_\alpha$($V$,$W$). If MutNorm$_\alpha$($V$,$W$), we shall say that the occurrences V and W are *mutually normalised in rank α.*

4.1. We give the definitions of the concepts mentioned in 4a-4f for $\alpha = 0$.

A word in the alphabet (1) is said to be *uncancellable* if there is no occurrence in it of a word of the form $a_i a_i^{-1}$ or $a_i^{-1} a_i$. We denote by $\mathcal{R}_0$ the set of all uncancellable words in (1). We set by definition

$$\overline{\mathcal{M}}_0 = \mathcal{M}_0 = \mathcal{K}_0 = \mathcal{L}_0 = \mathcal{R}_0 = \mathcal{P}_0 = \mathcal{N}_0 .$$

A kernel of rank 0 of a word $X \in \mathcal{N}_0$ is any occurrence $P * E * Q$ in X such that E is a letter in the alphabet (1). Thus Ker(0,X) is the set of all occurrences in X with single-letter bases. Thus for every occurrence V in a word $X \in \mathcal{N}_0$ we have $\partial_0(V) = \partial(\mathrm{Bas}(V))$.

The set Reg(0, X) is the set of all occurrences in X with non-empty base.

Two words $X, Y \in \mathcal{R}_0$ are said to be *equivalent in rank* 0 (and we write $X \overset{0}{\sim} Y$) if $X \stackrel{\circ}{=} Y$.

When $X \overset{0}{\sim} Y$ and $V \in \mathrm{Reg}(0,X)$, we set by definition $f_0\ (V; X,Y) \rightleftharpoons V$.

Occurrences V and W in words $X \in \mathcal{R}_0$ and $Y \in \mathcal{R}_0$ are said to be *mutually normalised in rank* 0 if $\mathrm{Bas}(V) \stackrel{\circ}{=} \mathrm{Bas}(W)$.

4.2. Suppose that $\alpha > 0$ and that all the concepts in 4a–4f have been defined for ranks $< \alpha$. We define them for rank α; the remaining part of the section will be devoted to this task. In addition to these concepts, we shall define certain others for ranks α and $\alpha - 1$. We did not include these last in the original list simply because either they are meaningful only for $\alpha > 0$ or else they are easily defined in terms of the concepts enumerated in 4a–4f.

Incidentally we shall mention without proof the individual properties of the concepts that we define, since advance familiarity with them may facilitate understanding of the text. The proofs of all these properties are contained in the subsequent Chapters II–V.

We recommend the reader to interpret at the outset all the concepts defined below for $\alpha = 1$. As a help towards this we shall give some illustrative examples in rank 1.

4.3. A word $X \in \mathcal{R}_{\alpha-1}$ is said to be a *periodic word of rank α with period A* if $X \in \mathrm{Per}(A)$, $\partial(X) \geqslant 27\partial(A)$, and there is a kernel V of rank $\alpha - 1$ of X such that $V \in \mathrm{Inn}(X,A)$ and $\partial(\mathrm{Bas}(V)) < 2\partial(A)$.

We denote the set of all periodic words of rank α with period A by Per(α,A). If Per(α,A) is non-empty, A is said to be a *period of rank α*.

By definition 4.1 every cyclically uncancellable* word is a period of rank 1.

4.4. Let $X \in \mathrm{Per}(\alpha,A)$. If $V \in \mathrm{Ker}(\alpha - 1,X)$ and $V \in \mathrm{Inn}(X,A)$, we shall call V a *supporting kernel of rank $\alpha - 1$ of X*. We shall call an occurrence W in a word $X \in \mathrm{Per}(\alpha,A)$ a *normal generating occurrence of rank α in X* if it begins and ends with supporting kernels of rank $\alpha - 1$.

It will be shown in II.1.10 that, if V is a supporting kernel of rank $\alpha - 1$ of $X \in$ Per (α,A), $W \in \mathrm{Inn}(X,A)$ and $\mathrm{Corr}_A(V,W)$, then W is also a supporting kernel of rank $\alpha - 1$.

*A word A is said to be *cyclically uncancellable* if AA is uncancellable.

4.5. A word Y is said to be an *integral word of rank α with period A* if there is a word $X \in \mathrm{Per}(\alpha, A)$ such that $X \overset{\alpha-1}{\simeq} Y$. If $X \in \mathrm{Per}(\alpha, A)$, we denote by $\mathrm{Int}(X, \alpha, A)$ the set of all words Y such that $X \overset{\alpha-1}{\simeq} Y$.

It follows from 4.1 that $\mathrm{Int}(X, \alpha, A)$ consists of the single element X when $\alpha = 1$.

4.6. Let $Y \in \mathrm{Int}(X, \alpha, A)$. If V is a supporting kernel of rank $\alpha - 1$ of X, we say that the occurrence $f_{\alpha-1}(V; X, Y)$ is a *supporting kernel of rank $\alpha - 1$ of the word Y*. If W is a normal generating occurrence of rank α in the word X, then $f_{\alpha-1}(W; X, Y)$ is said to be a *normal generating occurrence of rank α in the word Y*.

It follows from 4e and 4b in rank $\alpha - 1$ that every supporting kernel of rank $\alpha - 1$ of Y is a kernel of Y of rank $\alpha - 1$ and, for every two supporting kernels U and V of rank $\alpha - 1$ of Y, either $U < V$ or $V < U$.

It follows easily from 4.1 that for $\alpha = 1$ every occurrence $V \in \mathrm{Inn}(X, A)$ with non-empty base is a normal generating occurrence of rank 1.

From the properties of the mapping $f_{\alpha-1}(V; X, Y)$ it follows that an occurrence of W in a word $Y \in \mathrm{Int}(X, \alpha, A)$ is a normal generating occurrence of rank α if and only if it starts and ends in supporting kernels of rank $\alpha - 1$ (see II.2.1).

We extend the relation $\mathrm{Corr}_A(V, W)$ defined in 2.5 for occurrences in a word $X \in \mathrm{Per}(\alpha, A)$ to normal generating occurrences of rank α in a word $Y \in \mathrm{Int}(X, \alpha, A)$ as follows:

$$\begin{aligned}\mathrm{Corr}_{\alpha,A}(V, W) \rightleftharpoons \exists V_1 W_1 (&V_1, W_1 \in \mathrm{Reg}(\alpha - 1, X)\ \& \\ &V_1, W_1 \in \mathrm{Inn}(X, A)\ \&\ \mathrm{Corr}_A(V_1, W_1)\ \& \\ &V = f_{\alpha-1}(V_1; X, Y)\ \&\ W = f_{\alpha-1}(W_1; X, Y)).\end{aligned}$$

Occurrences V and W such that $\mathrm{Corr}_{\alpha,A}(V, W)$ holds will be said to *correspond in phase in rank α relative to the period A*.

4.7. Suppose that $Y \in \mathrm{Int}(X, \alpha, A)$. We say that an occurrence W in a word Y with non-empty base is a *generating occurrence of rank α* if it is contained in some normal generating occurrence of rank α. We shall call a supporting kernel of rank $\alpha - 1$ of Y contained in W a *supporting kernel of rank $\alpha - 1$* of the generating occurrence W.

It follows easily from 4.1 that every generating occurrence of rank 1 is a normal generating occurrence of rank 1.

Let U be the leftmost supporting kernel of rank $\alpha - 1$ of a generating occurrence W (in general it may not be a start of W). By *left segments of rank α of the occurrence W* we mean supporting kernels V of rank $\alpha - 1$ of W such that

$$\mathrm{Corr}_{\alpha,A}(U, V).$$

The left segments of rank α of W are fully ordered. The first one is U itself.

Taking for U a rightmost supporting kernel U' of rank $\alpha - 1$, we get the definition of *right segments of rank α of the occurrence W*, which in contrast to left segments we enumerate from right to left.

Although right and left segments coincide only when $\mathrm{Corr}_{\alpha,A}(U, U')$, it follows from II.2.4 that the number of left segments of an arbitrary generating occurrence W of rank α is the same as the number of right segments. We shall denote this number by $l_{\alpha,A}(W)$ and call it the *number of segments of rank* α of W. If W has no supporting kernels of rank $\alpha - 1$, we set $l_{\alpha,A}(W) = 0$.

Any base E of a generating occurrence $P * E * Q$ of rank α is said to be a *semi-integral word of rank* α. If $l_{\alpha,A}(P * E * Q) \geqslant r$, then we say that E is a *semi-integral r-power of rank* α *with period* A.

The word $(a_1a_2)^{2q}a_1$ is an example of a semi-integral q-power of rank 1 with period $(a_1a_2)^2$. It is also a semi-integral $(2q + 1)$-power of rank 1 with period a_1a_2.

4.8. Let U be a supporting kernel of rank $\alpha - 1$ of a word $X \in \mathrm{Per}(\alpha, A)$, V the result of shifting U to the right by the period A, let $V \in \mathrm{Inn}(X, A)$, and let W be the union of the occurrences U and V. We call $\partial_{\alpha-1}(W) - 1$ *the density of* X *in rank* α *relative to the period* A and denote it by $\rho_{\alpha,A}(X)$.

If W is a generating occurrence in a word $Y \in \mathrm{Int}(X, \alpha, A)$, we set

$$\rho_{\alpha,A}(W) \rightleftharpoons \rho_{\alpha,A}(Y) \rightleftharpoons \rho_{\alpha,A}(X) .$$

It is easy to see that $\rho_{\alpha,A}(X) = \partial(A)$ when $\alpha = 1$.

It will be proved in Chapter II that $\rho_{\alpha,A}(X)$ does not depend on the the choice of the original kernel U and, for any supporting kernels Y and V of rank $\alpha - 1$ of the word $Y \in \mathrm{Int}(X, \alpha, A)$,

$$\mathrm{Corr}_{\alpha,A}(U, V) \Longleftrightarrow \partial_{\alpha-1}(W) - 1 = (l_{\alpha,A}(W) - 1)\rho_{\alpha,A}(Y) ,$$

where W is the union of U and V (see II.2.9).

4.9. Suppose that $\mathrm{Per}(\alpha, A)$ is not empty. We shall say that A is a *minimal period of rank* α (or that the period A is *minimal in rank* α), if for any normal generating occurrences W and W_1 of rank α in any integral words $Y \in \mathrm{Int}(X, \alpha, A)$ and $Y_1 \in \mathrm{Int}(X_1, \alpha, B)$ respectively, such that $\mathrm{MutNorm}_{\alpha-1}(W, W_1)$ and $l_{\alpha,A}(W) \geqslant 3$, we have $\rho_{\alpha,A}(X) \leqslant \rho_{\alpha,B}(X_1)$.

We note that IV.3.4 gives that $\mathrm{MutNorm}_{\alpha-1}(W, W_1)$ implies $\partial_{\alpha-1}(W) = \partial_{\alpha-1}(W_1)$.

Since $\rho_{1,A}(X) = \partial(A)$ whenever $X \in \mathrm{Per}(1, A)$, it follows that every minimal period of rank 1 is a simple word. On the other hand, it follows easily from 2.9 that every cyclically uncancellable simple word A is a minimal period of rank 1.

Running ahead a little, we remark that the simple word

$$a_1a_2{}^{3q}\ a_1a_2{}^{-n+3q}$$

is an example of a period of rank 2 that is not minimal in rank 2 (see (14) in §4.27).

4.10. We call a minimal period A of rank α an *elementary period of rank* α if, for every minimal period B of rank α and all generating occurrences V and W in integral words $Y \in \mathrm{Int}(X, \alpha, A)$ and $Y_1 \in \mathrm{Int}\ (X_1, \alpha, B)$ such that $\mathrm{MutNorm}_{\alpha-1}(V, W)$ and $l_{\alpha,B}(W) \geqslant 9$, we have $\rho_{\alpha,A}(X) = \rho_{\alpha,B}(X_1)$.

It follows easily from 2.9 that a minimal period A of rank 1 is elementary in rank 1 if and only if no word $E \in \text{Per}(B)$ such that $\partial(E) > 8\partial(B)$ occurs in A^4. Thus $(a_1a_2)^8a_3$ is an elementary period of rank 1, while the period $(a_1a_2)^8a_1$ is not elementary in rank 1. The word $(a_1a_2)^8a_1$ is the simplest example of an elementary period of rank 2.

If A is an elementary period of rank α, a base E of any generating (normal generating) occurrence $P * E * Q$ of rank α in a word $Y \in \text{Int}(X, \alpha, A)$ is said to be an *elementary (normal elementary) word of rank* α. We shall say that the word E is *generated by the occurrence* $P * E * Q$. If in addition $l_{\alpha,A}(P * E * Q) \geqslant r$, then E is said to be an *elementary r-power of rank* α.

By 4.1, every elementary word of rank 1 is a normal elementary word of rank 1.

4.11. Let $W_1 \rightleftharpoons R * E * S$ be an occurrence in an arbitrary word Z of an elementary word E generated by an occurrence $W \rightleftharpoons P * E * Q$ in a word $Y \in \text{Int}(X, \alpha, A)$, where the period A is elementary in rank α. If V is a supporting kernel of rank $\alpha - 1$ of the occurrence W, we shall call $\phi(V; W, W_1)$ a *pseudokernel of rank* $\alpha - 1$ *of the occurrence* W_1. If R and S are empty, we also call the occurrence $\phi(V; W, * E *)$ a *pseudokernel of rank* $\alpha - 1$ *of the word* E. For any pseudokernels U and V of rank $\alpha - 1$ of the occurrence $W_1 = R * E * S$ we set

$$\text{Corr}_{\alpha,A}(U, V) \Longleftrightarrow \text{Corr}_{\alpha,A}(\phi(U; W_1, W), \phi(V; W_1, W)).$$

If $\text{Corr}_{\alpha,A}(U, V)$, the pseudokernels U and V of rank $\alpha - 1$ of the occurrence W_1 are said to *correspond in phase in rank* α.

We call a pseudokernel of rank $\alpha - 1$ of an occurrence W_1 a *left (right) segment of rank* α of W_1 if $\phi(V; W_1, W)$ is a left (right) segment of rank α of the generating occurrence W. Left segments are counted from left to right, and right segments from right to left. The number $l_{\alpha,A}(W)$ of segments, which by 4.7 is equal to the number of pseudokernels of rank $\alpha - 1$ of W_1 that correspond in phase in rank α with the leftmost pseudokernel of rank $\alpha - 1$, will be called the *number of segments of the occurrence* W_1 *(of the elementary word E)*, and we shall denote it by $l_{\alpha,A}(W_1)$ or $l_{\alpha,A}(E)$, that is,

$$l_{\alpha,A}(W_1) \rightleftharpoons l_{\alpha,A}(E) \rightleftharpoons l_{\alpha,A}(W).$$

Let W_2 be an arbitrary occurrence in the same word Z. We say that W_2 *contains r segments of the occurrence* W_1 if they have non-empty intersection W' and $l_{\alpha,A}(W') = r$, that is, $l_{\alpha,A}(\phi(W'; W_1, W)) = r$.

The *density* $\rho_\alpha(W_1)$ *of the occurrence* W_1 *in rank* α is the density $\rho_{\alpha,A}(W)$ of the generating occurrence W.

In those cases where it is clear from the text which period A is the relevant one, we shall omit the symbol A from the symbols $\text{Corr}_{\alpha,A}(U, V)$, $l_{\alpha,A}(W)$ and $\rho_{\alpha,A}(W)$.

It will be proved in Chapter II that for an elementary 9-power E of rank α the concepts of pseudokernel of rank $\alpha - 1$, the relation $\text{Corr}_\alpha(U, V)$, the density $\rho_\alpha(W_1)$, and the number $l_\alpha(E)$ of segments do not depend on the choice of the elementary period A of rank α nor the generating occurrence W (see II.5.1).

4.12. Let $W \rightleftharpoons R * E * S$ be an occurrence in an arbitrary word Z of a normal elementary word E of rank α generated by an occurrence $P * E * Q$ in a word $Y \in \text{Int}(X, \alpha, A)$ with some elementary period A. We call an occurrence $W_1 \rightleftarrows R_1 * uEv * S_1$, where $R \mathrel{\overline{\overline{\circ}}} R_1 u$ and $S \mathrel{\overline{\overline{\circ}}} vS_1$, a *continuation of the occurrence W with respect to the generating occurrence $P * E * Q$* if there is a normal generating occurrence $F * uEv * G$ in some word $Y_1 \in \text{Int}(X_1, \alpha, B)$ with elementary period B such that

$$\text{MutNorm}_{\alpha-1}(Fu * E * vG, P * E * Q)$$

and

$$l_{\alpha,B}(Fu * E * vG) = l_{\alpha,A}(P * E * Q).$$

If in addition the word u (the word v) is empty, we call W_1 a *continuation of W to the right* (or to the *left*). If u (or v) is not empty, we say that W can be *continued to the left* (or *to the right*) *relative to the generating occurrence $P * E * Q$*.

An occurrence W is said to be *maximal* if it cannot be continued either to the left or to the right relative to any generating occurrence. A *maximal continuation of an occurrence W* is any continuation W_1 of it which is a maximal occurrence. By a *maximal continuation to the right* (*to the left*) of an occurrence W we mean a continuation W_1 to the right (to the left respectively) which cannot be continued any further to the right (to the left respectively).

For example, the occurrence $a_2^{-10}a_1 * (a_1^2a_2)^9 * a_1^2a_2a_1a_2$ can be continued to the right but not to the left. A maximal continuation of it is the occurrence $a_2^{-10}a_1 * (a_1^2a_2)^{10}a_1 * a_2$.

It will be proved in Chapter II that, for every occurrence of a normal elementary 9-power of rank α, the maximal continuations as well as the maximal continuations to the left (and right) are uniquely determined (II.5.13).

4.13. An occurrence $R * E * S$ of a normal elementary word E of rank α in a word $Z \in \mathcal{R}_{\alpha-1}$ is said to be *normalised* if for some generating occurrence $P * E * Q$ we have

$$\text{MutNorm}_{\alpha-1}(P * E * Q, R * E * S).$$

The set of all normalised occurrences of elementary r-powers of rank α in a word $Z \in \mathcal{R}_{\alpha-1}$, $r > 0$, is denoted by

$$\text{Norm}(\alpha, Z, r).$$

By assumption 4f in rank $\alpha - 1$, every normal generating occurrence of an elementary word of rank α is normalised, and also for any word $Z \in \mathcal{R}_{\alpha-1}$ and any $r > 0$,

$$\text{Norm}(\alpha, Z, 9) \subset \text{Reg}(\alpha - 1, Z).$$

By 4.1 every occurrence of an elementary word of rank 1 is normalised.

If we require in all the definitions in 4.12 that W and W_1 be normalised, we get the concepts of *normalised occurrence*, normalised continuation to the left (or to the right), *normalised continuability of an occurrence to the left* or *to the right*, *maximal normalised occurrence*, *maximal normalised continuation* W_1 *of* W and maximal normalised continuation to the left (to the right).

We denote the set of all maximal normalised occurrences of elementary r-powers of rank α in a word $Z \in \mathcal{R}_{\alpha-1}$ by

$$\mathrm{MaxNorm}\,(\alpha, Z, r)\,.$$

It will be proved in Chapter II that the maximal normalised continuation, the maximal normalised continuation to the right (and to the left) of every occurrence $W \in \mathrm{Norm}\,(\alpha, Z, 9)$ are uniquely determined (II.7.3).

4.14. Two occurrences U and V of elementary words of rank α in one and the same word Z are said to be *compatible* if their union is a continuation of both of them with respect to some generating occurrences of their bases. We shall denote this reflexive symmetric relation between occurrences of elementary words in a given word by

$$\mathrm{Comp}(U, V)\,.$$

We shall prove in Chapter II that this relation does not depend on the choice of generating occurrences for elementary 9-powers, and that it is transitive.

For example, the occurrence

$$a_1^{-2}a_2^3 * (a_1^2a_2)^q a_1^2 * (a_2a_1^2)^9 a_2a_1a_2^{-3}(a_1^2a_2)^{10}a_1^{-2}$$

of an elementary q-power of rank 1 with period $a_1^2a_2$ is compatible with the occurrence

$$a_1^{-2}a_2^3(a_1^2a_2)^{15}a_1 * (a_1a_2a_1)^{20}a_1a_2 * (a_1^2a_2)^{q-26}a_1a_2^{-3}(a_1^2a_2)^{10}a_1^{-2}\,,$$

but not with the occurrence

$$a_1^{-2}a_2^3(a_1^2a_2)^{q+10}a_1a_2^{-3} * (a_1^2a_2)^9a_1^2 * a_2a_1^{-2}$$

in the same word. The occurrence

$$a_1^{-2}a_2^2 * a_2(a_1^2a_2)^{q+10}a_1 * a_2^{-3}(a_1^2a_2)^{10}a_1^{-2}$$

is the common maximal continuation of the first two of these occurrences.

4.15. If A is an elementary period of rank α, we denote the set of all elementary 3-powers of rank α generated by occurrences in a word $Y \in \mathrm{Int}\,(X, \alpha, A)$ by $\mathrm{El}(\alpha, A)$. It follows easily from 4.1 that

$$\mathrm{El}(1, A) \subset \mathrm{Per}(A)\,.$$

We say that two elementary words E and D are *related* if they lie in one and the same set El (α, A). In addition, we also call arbitrary occurrences of the form $P * E * Q$ and $R * D * S$ related. We write these relations as

$$\mathrm{Rel}(E, D) \text{ or } \mathrm{Rel}(P * E * Q,\ R * D * S)\,.$$

It is clear that this relation is reflexive and symmetric. We shall establish transitivity in Chapter II (II.5.20).

Two elementary words $E \in \mathrm{El}(1, A)$ and $D \in \mathrm{El}(1, B)$ of rank 1 are related if and only if B is a cyclic shift of A. This follows easily from the fact that Int $(X, 1, A)$ contains only the word $X \in$ Per $(1, A)$ (see 4.5).

4.16.* Suppose that $r \geqslant 9$, $t > 0$, $h > 0$, $X, Y \in \mathscr{K}_{\alpha-1}$ and that occurrences

$$RT * A^t A_1 * T_1 Q \in \mathrm{MaxNorm}\,(\alpha, X, r)\,,$$
$$RH * D^h D_1 * H_1 Q \in \mathrm{MaxNorm}\,(\alpha, Y, r)\,,$$
$$P * A^t A_1 * S \in \mathrm{MaxNorm}\,(\alpha, Z, 9)\,,$$

are given, where $Z \in \mathscr{M}_{\alpha-1}$, $A \overset{\circ}{=} A_1 A_2$ is an elementary period of rank α, $D \overset{\circ}{=} D_1 D_2$ is a cyclic shift of the word A^{-1}, and the following relations are satisfied:

$$A^{n+20} \overset{\alpha-1}{\simeq} A^{t+10} A_1 T_1 H_1^{-1} D_1^{-1} D^{-h} H^{-1} T A^{10}\,, \tag{5}$$

$$D^{n+20} \overset{\alpha-1}{\simeq} D^{h+10} D_1 H_1 T_1^{-1} A_1^{-1} A^{-t} T^{-1} H D^{10}\,, \tag{6}$$

the left and right-hand sides of which belong to $\mathscr{K}_{\alpha-1}$.

Then the transition

$$RTA^t A_1 T_1 Q \to RHD^h D_1 H_1 Q \tag{7}$$

is said to be a *simple r-reversal of rank* α, or a *simple r-reversal*, of an arbitrary *occurrence* $W \in$ Norm $(\alpha, X, 9)$ compatible with $RT * A^t A_1 * T_1 Q$.

If $\alpha = 1$, it follows from (5) that $TA^t A_1 T_1 \in$ Per (A) and $HD^h D_1 H_1 \in \mathrm{Per}(A^{-1})$. Thus, by the remark in 4.13 regarding rank 1, we get that, for $\alpha = 1$, the words T, T_1, H and H_1 in the simple reversal (7) of rank 1 must be empty. That is, every simple r-reversal of rank 1 has the form

$$RA^t A_1 Q \to RD^h D_1 Q\,,$$

where $A^t A_1$ and $D^h D_1$ are elementary r-powers of rank 1, $A^n \overset{\circ}{=} A^t A_1 D_1^{-1} D^{-h}$ and, consequently, $D \overset{\circ}{=} A^{-1}$.

As an example, we may carry out a simple q-reversal of rank 1 of the occurrence $a_1^{-2} * (a_2^3 a_1^{-1})^q * a_2 a_1^{-7} a_2^5$:

---

*See 4.18.

$$a_1^{-2}(a_2^3a_1^{-1})^q a_2 a_1^{-7} a_2^5 \to a_1^{-1}(a_2^{-3}a_1)^{n-q-1} a_2^{-2} a_1^{-7} a_2^5 ,$$

but it is impossible to carry out even a simple 9-reversal of rank 1 of the occurrence

$$a_1^{-2} * (a_2^3a_1^{-1})^{n-5} * a_2 a_1^{-7} a_2^5 .$$

4.17. A regular occurrence $F * E * G$ of rank $\alpha - 1$ in the word $RTA^tA_1T_1Q$ is said to be *stable in the simple reversal* (7) of rank α if either $R \mathrel{\overline{\underline{\circ}}} FER_1$ and

$$\mathrm{MutNorm}_{\alpha-1}(F * E * R_1TA^tA_1T_1Q\,, \quad F * E * R_1HD^hD_1H_1Q)\,,$$

or $Q \mathrel{\overline{\underline{\circ}}} Q_1EG$ and

$$\mathrm{MutNorm}_{\alpha-1}(RTA^tA_1T_1Q_1 * E * G\,, \quad RHD^hD_1H_1Q_1 * E * G)\,.$$

In such a case the occurrence $F * E * R_1HD^hD_1H_1Q$ ($RHD^hD_1H_1Q_1 * E * G$ respectively) will be called an *image of the occurrence* $F * E * G$ in the reversal (7) and will be denoted by $f_{X \to Y}(F * E * G)$.

We note that if (7) is a simple r-reversal of rank 1, then, by 4.1, every occurrence with non-empty base contained in $* R * A^tA_1Q$ or in $RA^tA_1 * Q *$ is stable in that reversal and, conversely, every occurrence that is stable in this reversal must be contained in one of the occurrences $* R * A^tA_1Q$ or $RA^tA_1 * Q *$.

We shall say that an arbitrary occurrence $U \in \mathrm{Norm}(\alpha, X, 9)$ compatible with $RH * D^hD_1 * H_1Q$ is an *image of the occurrence* $W \in \mathrm{Norm}(\alpha, X, 9)$ *in the proper reversal* (7).

4.18. The conditions given in 4.16 for simple r-reversals of rank α are not independent. Thus, for example, the relation (6) can be derived from (5). However, a redundant system of conditions of this sort turns out to be convenient for proving the symmetry of r-reversals (see III.1.2).

If the question of simplifying the definition of simple r-reversals arises, it should be noted that for $n \geqslant 1003$ the following easier definition of simple r-reversal of rank α may be used.

The transition

$$RA^tA_1Q \to R(A^{-1})^{n-t-1}A_2^{-1}Q$$

is said to be a *simple r-reversal of the occurrence W* if $W \in \mathrm{Norm}(\alpha, RA^tA_1Q, 9)$, RA^tA_1Q, $RA^{-n+t+1}A_2^{-1}Q \in \mathscr{K}_{\alpha-1}$, $t < n - 1$, $A \mathrel{\overline{\underline{\circ}}} A_1A_2$ is an elementary period of rank α, W is compatible with $R * A^tA_1 * Q$, the word A^n occurs in some word $Z \in \mathscr{M}_{\alpha-1}$, and the occurrences $R * A^tA_1 * Q$ and $R * (A^{-1})^{n-1-t}A_2^{-1} * Q$ are compatible with some normalised occurrences of elementary r-powers of rank α*.

4.19. Suppose that $X \in \mathscr{R}_{\alpha-1}$ and that $W \in \mathrm{Norm}(\alpha, X, 9)$. A transition $X \to Y$

---

*The proof of the existence of such a reversal depends on the existence of a left (right) bounded local coupling of rank α (see VII. 2.6).

is said to be an *r-reversal of the occurrence* W (an *r-reversal of rank* α), if words $X_1 \in \mathscr{K}_{\alpha-1}$ and $Y_1 \in \mathscr{K}_{\alpha-1}$ can be found such that $X \overset{\alpha-1}{\simeq} X_1$, $Y \overset{\alpha-1}{\simeq} Y_1$ and the transition $X_1 \to Y_1$ is a simple r-reversal of the occurrence $W_1 \rightleftharpoons f_{\alpha-1}(W; X, X_1)$. We shall call an arbitrary 9-reversal of an occurrence W a *reversal* of the occurrence W.

It readily follows from 4.1 that every r-reversal of rank 1 is a simple r-reversal of rank 1.

We shall say that a regular occurrence V of rank $\alpha - 1$ in a word X is *stable in the reversal* $X \to Y$ if the occurrence $V_1 \rightleftharpoons f_{\alpha-1}(V; X, X_1)$ is stable in the simple reversal $X_1 \to Y_1$.

If an occurrence V_2 is an image of V_1 (or of W_1) in the reversal $X_1 \to Y_1$, then the occurrence $f_{\alpha-1}(V_2; Y_1, Y)$ is said to be an *image of* V (*of* W) *in the reversal* $X \to Y$.

4.20. Since the function $f_{\alpha-1}(V; X, Y)$ is single-valued, it follows from definitions 4.17 and 4.19 that, if the occurrence $V \in \operatorname{Reg}(\alpha - 1, X)$ is stable in a given reversal $X \to Y$ of rank α, then its image in this reversal is uniquely defined. We shall denote the image of a stable occurrence V in a given reversal $X \to Y$ by

$$f_{X \to Y}(V).$$

The image of an occurrence $W \in \operatorname{Norm}(\alpha, X, 9)$ in a reversal of itself is not uniquely defined. However, as we shall convince ourselves in III.1.3, all images of an occurrence W in a reversal of itself are pairwise compatible.

A *maximal image of an occurrence W in a reversal $X \to Y$ of it* of rank α is an image which belongs to the set MaxNorm $(\alpha, Y, 9)$.

It is clear that if $R * A^t A_1 * Q \in \operatorname{MaxNorm}(1, X, 9)$ and

$$RA^t A_1 Q \to Z$$

is a 9-reversal of rank 1 of the occurrence $R * A^t A_1 * Q$ of the elementary 9-power $A^t A_1$ of rank 1 with period $A \overset{\circ}{=} A_1 A_2$, then $Z \overset{\circ}{=} RA^{-n+t+1} A_2^{-1} Q$, and the occurrence $R * A^{-n+t+1} A_2^{-1} * Q$ is a maximal image of the occurrence $R * A^t A_1 * Q$ in this reversal.

4.21. We denote by $\mathscr{N}_\alpha$ the set of all words $X \in \mathscr{R}_{\alpha-1}$ for which it is possible to carry out a q_1-reversal of an arbitrary occurrence $V \in \operatorname{Norm}(\alpha, X, q_1)$ and a q_1-reversal of an arbitrary occurrence $W \in \operatorname{Norm}(\alpha, Y, q_1)$, where Y is the result of an arbitrary q_1-reversal $X \to Y$ of the occurrence $V \in \operatorname{Norm}(\alpha, X, q_1)$.

As an example we have $(a_1^8 a_2)^q a_1^{-n+q_1} a_2^q \notin \mathscr{N}_1$, since carrying out a q_1-reversal on the occurrence $*\,(a_1^8 a_2)^q * a_1^{-n+q_1} a_2^q$ yields the word $(a_2^{-1} a_1^{-8})^{n-q} a_1^{-n+q_1} a_2^q$, on which it is impossible to carry out a q_2-reversal of the occurrence

$$(a_2^{-1} a_1^{-8})^{n-q-1} a_2^{-1} * a_1^{-n-8+q_1} * a_2^q .$$

Let $X \in \mathscr{N}_\alpha$, $U, V \in \operatorname{Norm}(\alpha, X, q_1)$ and $\neg\operatorname{Comp}(U, V)$. We shall say that occurrences U and V *do not adjoin one another* if a maximal normalised continuation of each of them is stable in a reversal of the other, and the same thing is true both for maximal normalised continuations of their images U_1 and V_1 in a reversal $X \to Y$

of either of them and for maximal normalised continuations of images of the occurrences U_1 and V_1 in a q_1-reversal of either of the latter. In the contrary case we shall say that the occurrences U and V *adjoin one another*.

To take an example, we have

$$(a_1^8 a_2)^q a_1^{-n+2q_1} a_2^q \in \mathscr{N}_1,$$

where the occurrences

$$(a_1^8 a_2)^q * a_1^{-n+2q_1} * a_2^q \text{ and } (a_1^8 a_2)^q a_1^{-n+2q_1} * a_2^q *$$

do not adjoin one another, but the occurrences

$$(a_1^8 a_2)^q * a_1^{-n+2q_1} * a_2^q \text{ and } * (a_1^8 a_2)^q * a_1^{-n+2q_1} a_2^q$$

do adjoin one another.

4.22. Suppose that $X \in \mathscr{N}_\alpha$, $P * A_0 * Q \in \text{Norm}\ (\alpha, X, q)$ and that $V_0 \rightleftharpoons P * A_0 * Q$. We shall call the set of occurrences

$$V_i \rightleftharpoons PA_0 u_1 A_1 u_2 \ldots u_i * A_i * u_{i+1} \ldots A_{r-1} u_r A_r Q_1 \tag{8}$$

in the word X, where $i = 1, 2, \ldots, r$ and $r \geqslant 1$, a *right cascade of rank* α of the occurrence V_0 if the following conditions are satisfied:

(a) For $0 < i \leqslant r$, $V_i \in \text{Norm}\ (\alpha, X, q_{t_i})$, where t_i is the i-th term of the sequence (2) constructed in § 3; for $0 < i < r$, a maximal normalised continuation of the occurrence V_i to the right contains less than $q_{t_i} + 17$ segments, the occurrence V_r has no normalised continuation to the right, and $l_\alpha(V_r) \geqslant q_{t_r} + 17$.

(b) For $0 \leqslant i < r$, the occurrences V_i and V_{i+1} adjoin one another.

(c) Each V_i, $0 < i \leqslant r$, is stable in reversals of V_{i-1} and V_{i+1}, but no proper normalised continuation of it has this property.

We shall call V_1 the *first element of the cascade* (8) and V_0 its *head.*

The concept of a *left cascade of rank* α of an occurrence V_0 is symmetrical to that of a right cascade and is defined analogously.

For example, the occurrence

$$a_3^q * (a_1 a_2)^{2q-1} a_1 * a_2 (a_1^2 a_2^6)^{q_1+2} a_1^{q_2+5} (a_1 a_2^{-1})^q a_2^{-2q} \tag{9}$$

has a right cascade of rank 1, consisting of the following three occurrences:

$$a_3^q (a_1 a_2)^{2q} a_1 * a_1 a_2^6 (a_1^2 a_2^6)^{q_1+1} * a_1^{q_2+5} (a_1 a_2^{-1})^q a_2^{-2q},$$
$$a_3^q (a_1 a_2)^{2q} (a_1^2 a_2^6)^{q_1+2} a_1^2 * a_1^{q_2+3} * (a_1 a_2^{-1})^q a_2^{-2q},$$
$$a_3^q (a_1 a_2)^{2q} (a_1^2 a_2^6)^{q_1+2} a_1^{q_2+6} * a_2^{-1} (a_1 a_2^{-1})^{q-1} * a_2^{-2q}$$

(of course we are taking into account the facts that $t_1 = 1$ and $t_2 = 2$ in the sequence (2)). The occurrence (9) does not have a left cascade of rank 1, since the neighbouring occurrence of a_3^q does not adjoin it. If, however, we replaced a_3 by a_2, then (9) would have a left cascade consisting of a single occurrence.

4.23. We shall say that an occurrence $V \in \text{Norm}(\alpha, X, q)$ in a word $X \in \mathcal{N}_\alpha$ is *completely stable in rank α from the right (from the left)* if either V does not have a right (left) cascade of rank α, or it is stable in a reversal of the first element of a right (left) cascade of it.

Those occurrences which are completely stable both from the left and from the right in rank α are said to be *completely stable occurrences in rank α*.

The occurrence (9) is an example of a completely stable occurrence of rank 1. It is easy to convince oneself that the occurrence $(a_1a_2)^{2q} * a_1^{q+3} * a_2^{-q}$ is completely stable from the right but not from the left. It is compatible with the completely stable occurrence

$$(a_1a_2)^{2q}a_1 * a_1^{q+2} * a_2^{-q}. \tag{10}$$

Suppose that $X \to Y$ is a q-reversal of an occurrence $V \in \text{Norm}(\alpha, X, 9)$, where $X, Y \in \mathcal{N}_\alpha$. We shall say that this reversal is a *real reversal of rank α* if completely stable occurrences V_1 and W_1 of rank α can be found in the words X and Y respectively such that V_1 is compatible with V and W_1 is an image of V in the reversal $X \to Y$. We shall call an occurrence V an *active occurrence of rank α* if $V \in \text{Norm}(\alpha, X, 9)$ and it is possible to find some real reversal of it of rank α. The set of all active occurrences of rank α in a given word $X \in \mathcal{N}_\alpha$ is denoted by

$$\text{Act}(\alpha, X).$$

The occurrence (10) is an example of an active occurrence of rank 1. Indeed, we have a q-reversal of rank 1,

$$(a_1a_2)^{2q}a_1^{q+3}a_2^{-q} \to (a_1a_2)^{2q}\, a_1^{-n+q+3}a_2^{-q},$$

where $(a_1a_2)^{2q} * a_1^{-n+q+3} * a_2^q$ is a completely stable occurrence of rank 1. It is easy to convince oneself that the occurrences

$$* (a_1a_2)^{2q} * a_1^{q+3}a_2^{-q}, \quad (a_1a_2)^{2q}a_1^{q+3} * a_2^{-q} * \tag{11}$$

in the same word are also active.

We shall prove in Chapter III that, if $X \to Y$ is a real reversal of rank α, then $Y \to X$ is also a real reversal of rank α (see III.3.25).

4.24. We shall call a normalised occurrence V of some elementary 9-power of rank α in a word $X \in \mathcal{N}_\alpha$ a *kernel of rank α in the word X* if it is stable in a reversal of an arbitrary active occurrence of rank α in X that is not compatible with it and if no proper normalised continuation of it has this property. Furthermore, we shall call a kernel V an *active kernel of rank α* if $V \in \text{Act}(\alpha, X)$. The set of all kernels of rank α in a word $X \in \mathcal{N}_\alpha$ is denoted by

$$\text{Ker}(\alpha, X).$$

Occurrences (10) and (11) are kernels of rank 1 in $(a_1a_2)^{2q}a_1^{q+3}a_2^{-q}$. The word

$$(a_1^{10}a_2^{2q}a_1^{10}a_2^{-n+2q})^n \tag{12}$$

has $2n$ inactive kernels of rank 1 with base a_1^{10} and n active kernels of rank 1 with each of the bases a_2^{2q} and a_2^{-n+2q}.

4.25. We now list some properties of real reversals and kernels of rank α that will be proved later on.

(a) If U and V are distinct kernels of rank α of a word $X \in \mathscr{N}_\alpha$, then they are not compatible and neither of them is contained in the other, that is either $U < V$ or $V < U$ (see IV.1.2).

(b) If a kernel V of rank α of a word X is active, then a maximal normalised continuation of it does not intersect other kernels of rank α (see IV.1.3).

(c) If $X \in \mathscr{N}_\alpha$ and $W \in \text{Norm}(\alpha, X, 43)$, then W is compatible with some kernel of rank α of X (see IV.1.4).

(d) If $V \in \text{Ker}(\alpha, X)$ and $X \overset{\alpha-1}{\simeq} Y$, then $f_{\alpha-1}(V; X, Y) \in \text{Ker}(\alpha, Y)$ (see IV.1.8).

(e) Suppose that the transition $X \to Y$ is a real reversal of rank α. For each kernel $V \in \text{Ker}(\alpha, X)$ there exists one and only one kernel $W \in \text{Ker}(\alpha, Y)$ that is the image of V in the reversal $X \to Y$. This one-to-one correspondence of the set $\text{Ker}(\alpha, X)$ with $\text{Ker}(\alpha, Y)$ will be denoted by $f_\alpha(V; X, Y)$. It preserves the relations $<$ and $\ll$ (see IV.1.10 and IV.1.11).

4.26. We define the sets $\mathscr{P}_\alpha$, $\mathscr{R}_\alpha$, $\mathscr{K}_\alpha$, $\mathscr{L}_\alpha$ and $\mathscr{M}_\alpha$ as follows:

$$X \in \mathscr{P}_\alpha \Longleftrightarrow X \in \mathscr{R}_{\alpha-1} \;\&\; \text{Norm}(\alpha, X, n-88) = \varnothing,$$

$$X \in \mathscr{R}_\alpha \Longleftrightarrow X \in \mathscr{P}_\alpha \;\&\; \forall W\, (W \in \text{Ker}(\alpha, X) \Rightarrow l_\alpha(W) \leqslant n - 176),$$

$$X \in \mathscr{K}_\alpha \Longleftrightarrow X \in \mathscr{P}_\alpha \cap \mathscr{K}_{\alpha-1} \;\&\; \forall W\, (W \in \text{Ker}(\alpha, X) \Rightarrow l_\alpha(W) \leqslant n - 218),$$

$$X \in \mathscr{L}_\alpha \Longleftrightarrow X \in \mathscr{P}_\alpha \cap \mathscr{L}_{\alpha-1} \;\&$$

$$\&\; \forall W\, (W \in \text{Ker}(\alpha, X) \Rightarrow l_\alpha(W) \leqslant \frac{n+1}{2} + 21),$$

$$X \in \mathscr{M}_\alpha \Longleftrightarrow X \in \mathscr{P}_\alpha \cap \mathscr{M}_{\alpha-1} \;\&\; \forall W\, (W \in \text{Ker}(\alpha, X) \Rightarrow l_\alpha(W) \leqslant \frac{n+1}{2}).$$

Since $n \geqslant 665$ the following relations follow directly from the definitions:

$$\mathscr{M}_\alpha \subset \mathscr{L}_\alpha \subset \mathscr{K}_\alpha \subset \mathscr{R}_\alpha \subset \mathscr{P}_\alpha \subset \mathscr{R}_{\alpha-1}.$$

It is easy to see that the word (12) belongs to $\mathscr{R}_1$, but not to $\mathscr{M}_1$.

4.27. We shall say that two words X, Y in $\mathscr{P}_\alpha$ are *equivalent in rank* α if either $X \overset{\alpha-1}{\simeq} Y$ or there exists a sequence of real reversals of rank α of the form:

$$X \to X_1 \to X_2 \to \cdots \to X_t \to X_{t+1} \to \cdots \to X_\lambda \to Y. \tag{13}$$

By $X \overset{\alpha}{\sim} Y$ we shall denote the combined assertion: $X \in \mathscr{R}_\alpha$, $Y \in \mathscr{R}_\alpha$ and X is equivalent to Y in rank α.

The relation $\overset{\alpha}{\sim}$ is reflexive, symmetric and transitive (see IV.2.7).

On carrying out on the word (12) a sequence of n real reversals of all kernels of rank 1 whose numerals from the left are divisible by 4, we get

$$(a_1^{10} a_2^{2q} a_1^{10} a_2^{-n+2q})^n \overset{\alpha}{\sim} (a_1^{10} a_2^{2q} a_1^{10} a_2^{2q})^n. \tag{14}$$

Clearly, $(a_1^{10} a_2^{2q})^{2n} \in \mathscr{M}_1$.

4.28. Suppose that $X \overset{\alpha}{\sim} Y$. We define the function $f_\alpha(V; X, Y)$ mapping $\mathrm{Ker}(\alpha, X)$ onto $\mathrm{Ker}(\alpha, Y)$ as follows:

If $X \overset{\alpha-1}{\sim} Y$, *then we set* $f_\alpha(V; X, Y) \rightleftharpoons f_{\alpha-1}(V; X, Y)$.

If there exists a sequence (13) *of real reversals of rank* α, *then we define* $f_\alpha(V; X, Y)$ *to be the superposition of the mappings for the reversals in the sequence* (13) *indicated in* 4.25(e) (see IV.2.8).

It will be proved in Chapter IV that this mapping preserves the linear ordering relation between kernels of rank α and that it is symmetric, that is, $V = f_\alpha(W; Y, X)$ whenever $W = f_\alpha(V; X, Y)$ (IV. 2.9).

4.29. We shall call an occurrence W in a word $X \in \mathscr{N}_\alpha$ a *regular occurrence of rank* α if some start and some end of it lies in $\mathrm{Ker}(\alpha, X)$. We shall denote the set of all regular occurrences of rank α in a word X by

$$\mathrm{Reg}(\alpha, X).$$

4.30. We make a natural extension of the mapping $f_\alpha(V; X, Y)$ for $X \overset{\alpha}{\sim} Y$ introduced in 4.28 to the set $\mathrm{Reg}(\alpha, X)$. Suppose that $X \overset{\alpha}{\sim} Y$ and $W \in \mathrm{Reg}(\alpha, X)$. We look at kernels $U, V \in \mathrm{Ker}(\alpha, X)$, where U is a start of W and V is an end of W. By 4.28 there are uniquely defined kernels $f_\alpha(U; X, Y)$ and $f_\alpha(V; X, Y)$ of Y whose union belongs to $\mathrm{Reg}(\alpha, Y)$. We denote the union by $f_\alpha(W; X, Y)$. In this way we obtain a function $W_1 = f_\alpha(W; X, Y)$ that maps the set $\mathrm{Reg}(\alpha, X)$ one-to-one onto the set $\mathrm{Reg}(\alpha, Y)$.

4.31. Suppose that $X, Y \in \mathscr{R}_\alpha$. We say that occurrences $V \in \mathrm{Reg}(\alpha, X)$ and $W \in \mathrm{Reg}(\alpha, Y)$ are *mutually normalised in rank* α if the following conditions are fulfilled:

(a) If Z is an arbitrary word such that $X \overset{\alpha}{\sim} Z$, then a word Z_1 with $Y \overset{\alpha}{\sim} Z_1$ can be found such that

$$\mathrm{Bas}(f_\alpha(V; X, Z)) \mathrel{\underset{\circ}{=}} \mathrm{Bas}(f_\alpha(W; Y, Z_1)), \tag{15}$$

and in addition an arbitrary occurrence U contained in $f_\alpha(V; X, Z)$ is an active $W \in$ (inactive) kernel of rank α if and only if the occurrence $\phi(U; f_\alpha(V; X, Z), f_\alpha(W; Y, Z_1))$ is an active (inactive respectively) kernel of rank α.

(b) Conversely, if Z_1 is an arbitrary word such that $Y \overset{\alpha}{\sim} Z_1$, then a word Z with $X \overset{\alpha}{\sim} Z$ can be found satisfying (15) together with the condition mentioned in (a) above on kernels of rank α contained in $f_\alpha(V; X, Z)$.

To denote the fact that occurrences V and W are mutually normalised in rank α, we shall write

$$\mathrm{MutNorm}_\alpha(V, W).$$

4.32. We list now some properties of the relation $\mathrm{MutNorm}_\alpha(V, W)$ which will be proved in §3, Chapter IV:

(a) The relation $\mathrm{MutNorm}_\alpha(V, W)$ is reflexive, symmetric and transitive,
(b) $W = f_\alpha(V; X, Y) \Rightarrow \mathrm{MutNorm}_\alpha(V; W)$,
(c) $\mathrm{MutNorm}_\alpha(P * E * Q, R * E * S) \Rightarrow \mathrm{MutNorm}_{\alpha-1}(P * E * Q, R * E * S)$.

4.33. If W is an occurrence in a word $X \in \mathscr{N}_\alpha$ then every kernel of rank α of X contained in W is called a *kernel of rank α of the occurrence W*. We use $\partial_\alpha(W)$ to denote the number of kernels of rank α of an occurrence W.

4.34. Put

$$\mathscr{A} \rightleftharpoons \bigcup_{\alpha=1}^{\infty} \mathscr{A}_\alpha,$$

where

$$X \in \mathscr{A}_\alpha \Longleftrightarrow X \in \mathscr{R}_{\alpha-1} \ \&\ \mathrm{Norm}(\alpha, X, 9) = \varnothing.$$

We shall call the elements of $\mathscr{A}$ *absolutely reduced words*.

4.35. By 4.1 we have $\bar{\mathscr{M}}_0 = \mathscr{R}_0$. We define the set $\bar{\mathscr{M}}_\alpha$ as follows:

$$X \in \bar{\mathscr{M}}_\alpha \Longleftrightarrow X \in \mathscr{M}_\alpha \ \&\ X \in \bar{\mathscr{M}}_{\alpha-1} \ \&$$
$$\&\ \forall\, UV((U, V \in \mathrm{Ker}(\alpha, X) \ \&\ \mathrm{MutNorm}_\alpha(U, V)) \Rightarrow \mathrm{Rel}(U, V)).$$

We shall prove in Chapter IV that, for an arbitrary word $X \in \mathscr{R}_\alpha$, it is possible to find a $Y \in \bar{\mathscr{M}}_\alpha$ such that $X \overset{\alpha}{\sim} Y$ (see IV.3.12).

The words in $\bar{\mathscr{M}}_{\alpha-1}$ are characterised in the following way:

If an occurrence $P * E * Q$ in a word $X \in \mathscr{R}_{\alpha-1}$ is mutually normalised in rank $\alpha - 1$ with some occurrence $R * E * S$ in some word $Y \in \bar{\mathscr{M}}_{\alpha-1}$, then a base for each occurrence $V \in$ Norm $(\alpha, X, 9)$ *contained in $P * E * Q$ is a periodic 9-power with elementary period of rank α*(see II.4.5).

We shall therefore say that an occurrence $P * E * Q$ satisfying

$$\mathrm{MutNorm}_{\alpha-1}(P * E * Q, R * E * S)$$

for some RES in $\bar{\mathscr{M}}_{\alpha-1}$ is *periodised in rank $\alpha - 1$*.

It is clear that every occurrence $P * E * Q$ in a word $X \in R_0$ is periodised in rank 0.

4.36. For every $\alpha \geqslant 0$ and arbitrary words $X, Y \in \mathscr{R}_\alpha$ we shall define a binary operation $[X, Y]_\alpha = Z$, called *coupling of rank α*, as follows:

$$[X, Y]_\alpha = PQ \Longleftrightarrow \exists T(X \overset{\alpha}{\sim} PT \ \&\ Y \overset{\alpha}{\sim} T^{-1}Q \ \&\ PQ \in \mathscr{R}_\alpha).$$

In Chapter V we shall prove that, for an arbitrary pair of words $X, Y \in \mathcal{R}_\alpha$, the result of this operation is uniquely defined to within equivalence in rank α. The coupling operation is associative. All of these properties are obvious for $\alpha = 0$.

4.37. Using the set $\mathcal{A}$, the relation $\overset{\alpha}{\sim}$ and the operation $[X, Y]_\alpha = Z$ as ingredients, we shall define in Chapter VI a group $\boldsymbol{\Gamma}(m, n)$ (see VI.1.3 and VI.1.4) and shall prove that it is infinite, has m generators and satisfies the identical relation $X^n = 1$ (see VI.1.5).

§ 5. Symmetry and Effectiveness

The object of this paragraph is to draw attention to two general properties of the concepts that we are considering. We formulate these properties as two principles that are essentially metatheorems concerning all the assertions about words and occurrences that will be proved in what follows.

5.1. We shall say that two words X and Y in some alphabet are *mutually symmetric* if one of them is obtained on reading the other from right to left. In this paragraph we shall denote by X' the word symmetric to a given word X. If

$$X \overset{\circ}{=} d_1 d_2 \dots d_{r-1} d_r ,$$

where the d_i are letters in the alphabet under consideration, then

$$X' \overset{\circ}{=} d_r d_{r-1} \dots d_2 d_1 .$$

Clearly, $\partial(X') = \partial(X)$ and $(X')' \overset{\circ}{=} X$. The empty word and all single letter words are self-symmetric.

If R is a start (end) of a word Q, then R' is an end (start) of Q'.

If V is an occurrence in a word X in the alphabet (1), then V' is an occurrence in X'. Furthermore, if E is a base of V, then E' is a base of V'. If E occurs in X, then E' occurs in X'. If X is uncancellable then so is X'.

If $V < W$ ($V \ll W$), then $W' < V'$ ($W' \ll V'$ respectively).

If V is the i-th occurrence of the word E in X from the left (right), then V' is the i-th occurrence of E' in X' from the right (left).

It is clear that the list of assertions of this sort could be extended. Indeed, all the concepts we consider, their properties and the relationships between them, turn out to be invariant in the sense defined under substitution of words and occurrences by words and occurrences symmetric to them.

Suppose that

$$\mathfrak{F}(X_1, X_2, \dots, X_r, V_1, V_2, \dots, V_t) \tag{16}$$

is an arbitrary assertion about words X_i and occurrences V_j. We denote by

$$\mathfrak{F}'(X_1', X_2', \ldots, X_r', V_1', V_2', \ldots, V_t') \tag{17}$$

the result of replacing X_i by X_i' and, V_j by V_j' and the terms "start", "end", "left" "right" by "end", "start", "right", "left", respectively in (16). The assertions (16), (17) are said to be mutually *symmetric*.

5.2. The Principle of Symmetry. *Any assertion* (16) *connected with any of the concepts in this book is true if and only if the corresponding symmetric assertion* (17) *is true.*

In the first place, observe that it is sufficient to convince ourselves that the principle of symmetry is true for elementary assertions, that is, for predicates which correspond to all the sets of words and occurrences and to all the relations between words and occurrences which were defined in § 4 or will be defined in Chapter II-V. But for all of these predicates, which, in general, depend on a natural parameter α, the desired assertion is proved by simultaneous induction on α in the same order in which the corresponding concepts were defined in § 4. As examples, we write down the elementary assertions which correspond to the concepts enumerated under 4a-4f:

(a)
$$X \in \mathscr{R}_\alpha \Longleftrightarrow X' \in \mathscr{R}_\alpha,$$
$$X \in \mathscr{M}_\alpha \Longleftrightarrow X' \in \mathscr{M}_\alpha$$

and the same for the sets $\mathscr{K}_\alpha$ and $\mathscr{L}_\alpha$:

(b) $$W \in \operatorname{Ker}(\alpha, X) \Longleftrightarrow W' \in \operatorname{Ker}(\alpha, X');$$

(c) $$W \in \operatorname{Reg}(\alpha, X) \Longleftrightarrow W' \in \operatorname{Reg}(\alpha, X');$$

(d) $$X \overset{\alpha}{\sim} Y \Longleftrightarrow X' \overset{\alpha}{\sim} Y';$$

(e) $$W = f_\alpha(V; X, Y) \Longleftrightarrow W' = f_\alpha(V'; X', Y');$$

(f) $$\operatorname{MutNorm}_\alpha(V, W) \Longleftrightarrow \operatorname{MutNorm}_\alpha(V', W').$$

For $\alpha = 0$ each of these assertions is obvious. Furthermore, on the assumption that they are all true for rank $< \alpha$, we can successively prove point for point that all the concepts defined in § 4 and the relations between words and occurrences for rank α satisfy the principle of symmetry as well. Moreover, the verification for each of the concepts involves no difficulty, since the definition of the concept itself reduces the problem to an analogous problem for the preceding concepts or for the same concepts in lower rank.

The above remarks furnish a fully adequate proof of the principle of symmetry. The principle means that we need only formulate and prove one of two mutually symmetric assertions. However, for convenience of reference when considering unwieldy assertions, we shall sometimes formulate both of the symmetric assertions, although we shall of course prove only one of them.

5.3. The principle of symmetry is essentially a corollary to a peculiarity of all the concepts that we are considering, namely, that when we formulate an arbitrary assertion (16) all the words and occurrences different from X_i and V_j figure as variables restricted by a universal or existential quantifier.

We note that in all of our definitions the letters a_i and a_i^{-1} occur in an equivalent way, that is, all of the definitions are invariant with respect to the transformation of letters replacing a_i by a_i^{-1} and conversely. Such a transformation takes an arbitrary word X^{-1} in the alphabet (1) into the word X' symmetric to X and conversely. Therefore the following variant of the principle of symmetry holds:

An arbitrary assertion (16) *about any of the concepts in this book is true if and only if the assertion*

$$\mathfrak{F}'(X_1^{-1}, X_2^{-1}, \ldots, X_r^{-1}, V_1^{-1}, V_2^{-1}, \ldots, V_t^{-1})$$

obtained by replacing words X_i by X_i^{-1}, occurrences V_j by V_j^{-1} and the terms "start", "end", "left" and "right" by the terms "end", "start", "right" and left" respectively in (16), *is true.*

In fact this variant of the principle of symmetry was used in [5] under the name "*principle of duality*".

Here are some examples of assertions which are obtained by means of the above variant of the principle of symmetry:

$$X \in \mathscr{R}_\alpha \Longleftrightarrow X^{-1} \in \mathscr{R}_\alpha,$$

$$X \overset{\alpha}{\sim} Y \Longleftrightarrow X^{-1} \overset{\alpha}{\sim} Y^{-1},$$

$$V \in \operatorname{Ker}(\alpha, X) \Longleftrightarrow V^{-1} \in \operatorname{Ker}(\alpha, X^{-1}),$$

$$W = f_\alpha(V; X, Y) \Longleftrightarrow W^{-1} = f_\alpha(V^{-1}; X^{-1}, Y^{-1}),$$

$$\operatorname{MutNorm}_\alpha(V, W) \Longleftrightarrow \operatorname{MutNorm}_\alpha(V^{-1}, W^{-1}),$$

$$Y \in \operatorname{Int}(X, \alpha, A) \Longleftrightarrow Y^{-1} \in \operatorname{Int}(X^{-1}, \alpha, A^{-1}),$$

$$V < W \Longleftrightarrow W^{-1} < V^{-1}.$$

5.4. The principle of effectiveness. *All sets, functions and relations between words and occurrences considered in this book are algorithmically effective, that is, for each of the relations we encounter there exists an algorithm to decide whether a given relation on an arbitrary set of words or occurrences holds or not. Furthermore, for each assertion in the succeeding chapters about the existence of some object or other, an algorithm may be found that produces the desired object in terms of prescribed values of the parameters on which the object depends.*

We prove this principle for all the relevant concepts and assertions by simultaneous induction on the rank α in the order in which these concepts and assertions appear in the text of this book. Each step of the proof is a simple consequence of the inductive hypothesis concerning the algorithmic effectiveness of previously defined concepts and the relative effectiveness of the definition of the concept under consideration or of the proof of the assertion under consideration, adduced at the corresponding place. Successive verification of the relative effectiveness of all of these definitions and proofs will not demand any special effort on the part of the reader, since all quantifiers encountered in the definitions in § 4 may be assumed effectively bounded.

This is proved for 4.27 in V.2.6. The effective boundedness of all quantifiers in definitions 4.31 and 4.36 follows from this.

By II.3.1, we may assume in definition 4.9 that $l_{\alpha,A}(W) = 3$. Then, by V.2.10, V.2.8 and V.2.6, we may assume that the quantifiers on X, Y, B, X_1, Y_1 and W_1 in 4.9 are all effectively bounded. By II.3.7 we may assume that $l_{\alpha,A}(V) < 3$ in definition 4.10, from which it follows that all quantifiers here may be taken to be effectively bounded. This is true for definitions 4.12, 4.13 and 4.14, in view of V.2.9 and V.2.6.

The effective boundedness of all quantifiers in the remaining definitions in § 4 is clear.

Hence the recognition of an arbitrary relation among the definitions of § 4 for a prescribed choice of words or occurrences reduces to the verification of some of the previous relations for an effectively bounded finite number of choices of words or occurrences.

The algorithms obtained in this way have something of the nature of case-by-case verification, and in many cases simpler algorithms may be found. Thus, for example, for recognising whether a given period A of rank α is minimal in rank α or not, it is natural, having taken a word $X \in \mathrm{Per}(\alpha, A)$, to use IV.3.12 at the outset to find a word $Y \in \mathcal{M}_{\alpha-1}$ such that $X \overset{\alpha}{\sim} Y$. Then, for some normal generating occurrence $P * E * Q$ of rank α in X, one finds an occurrence

$$f_{\alpha-1}(P * E * Q; X, Y) = R * E_1 * S.$$

Then by II.4.1 the word E_1 can be represented in the form $D^t D_1$, where D is an image of the period A in the occurrence $R * E_1 * S$. By II.4.3 and II.3.2 the period A is minimal in rank α if and only if D is a simple word; and this is easy to check.

However, in proving the principle of effectiveness, the complexity of the algorithms is of no real importance.

Chapter II. Periodic and Elementary Words of Rank α

§ 1. Periodic Words of Rank α

The definition of periodic words of rank α was given in I.4.3. In this section we shall prove some properties of such words on the basis of the inductive assumptions.

1.1. *If* $X \in \mathrm{Per}(\alpha, B)$ *and* A *is a cyclic shift of* B, *then* $X \in \mathrm{Per}(\alpha, A)$.
This follows at once from I.2.1 and I.4.3.

1.2. *If* $\alpha > 1$, *then* $\mathrm{Per}(\alpha, A) \subset \mathrm{Per}(\alpha - 1, A)$.
Suppose that $X \in \mathrm{Per}(\alpha, A)$. Since $X \in \mathscr{R}_{\alpha-1}$, we have by definition *I.4.26* that $X \in \mathscr{R}_{\alpha-2}$. Suppose that $W \in \mathrm{Ker}(\alpha - 1, X)$, $W \in \mathrm{Inn}(X, A)$ and $\partial(\mathrm{Bas}(W)) < 2\partial(A)$. Then by inductive assumption *IV.1.1** we have $W \in \mathrm{Reg}(\alpha - 2, X)$. But, by I.4.29, this means that some start V of the occurrence W is a kernel of rank $\alpha - 2$ of X. Clearly, $V \in \mathrm{Inn}(X, A)$ and $\partial(\mathrm{Bas}(V)) < 2\partial(A)$. Hence $X \in \mathrm{Per}(\alpha - 1, A)$.

1.3. *If* $\alpha > 1$, $X \in \mathrm{Per}(\alpha, A)$ *and* $W \in \mathrm{Ker}(\alpha - 1, X)$, *where* $W = P * E * Q$, $\partial(E) < 2\partial(A)$ *and* $\partial(P), \partial(Q) \geqslant 8\partial(A)$, *then* W *is not compatible with the result of shifting it to the left (to the right) by one period* A (see I.2.5).
It is clearly sufficient to look at a shift W_1 of the occurrence W to the right. By I.4.24, $W \in \mathrm{Norm}(\alpha - 1, X, 9)$, that is, $l_{\alpha-1}(E) \geqslant 9$ (see *II.5.1*). By 1.1. the period A can be chosen in such a way that P ends with some power of A and E is a start of A^2. Then for some u and G we have $A^2 \overset{\circ}{=} Eu$, $Q \overset{\circ}{=} uA^6G$ and $W_1 = PA * E * uA^5G$. Let us suppose that W_1 is compatible with W. Then by *II.5.8* and *II.5.12*, W is also compatible with the occurrence $W_6 \rightleftharpoons PEuEuEu * E * uG$, which is the result of shifting W to the right by 6 periods A. By I.4.14 this means that $P * EuEuEuE * uG$ is a continuation of the kernel W to the right. But this is impossible, since by *II.5.2* we have $l_{\alpha-1}(EuEuE) \geqslant 3l_{\alpha-1}(E) - 2 \geqslant 25$, which contradicts *IV.1.7*. Hence the occurrences W and W_1 are not compatible.

1.4. *If* $X \in \mathrm{Per}(\alpha, A)$ *and* $X \overset{\circ}{=} RDS$, *where* D *is an elementary word of rank* $\alpha - 1$, *then* $\partial(D) < 9\partial(A)$.
Suppose that $\partial(D) \geqslant 9\partial(A)$. Retaining the notation of 1.3, we have $D \overset{\circ}{=} vA^8v_1$, where $A^2 \overset{\circ}{=} Eu$. We look at a generating occurrence $V \rightleftharpoons R_1 * D * S_1$ of the elementary word D in the corresponding integral word Y of rank $\alpha - 1$. Since $P * E * Q \in \mathrm{Norm}(\alpha - 1, X, 9)$ by hypothesis, it follows from *II.6.14* that the

*We recall that we agreed in the introduction to write references to inductive assumptions in italics (see p. XI).

occurrences $V_1 \rightleftharpoons R_1vEu * E * uEuA^2v_1S_1$ and $V_2 \rightleftharpoons R_1vAEu * E * uEuAv_1S$ are normalised. It follows by assumption *II.5.7* that V_1 and V_2 are compatible with V. By *II.5.12* they are then compatible with one another. But this is impossible by 1.3 and II.5.8. Therefore, $\partial(D) < 9\partial(A)$.

1.5. *If* $X \in \mathrm{Per}(\alpha, A)$ *and* $X \overset{\circ}{=} PDQ$, *where* D *is an elementary 9-power of rank* $\alpha - 1$, *then the occurrence* $V \rightleftharpoons P * D * Q$ *is not compatible with the result of shifting it to the right (or to the left) by one period* A.

Furthermore, if $l_{\alpha-1}(D) > 20$, *then* $\partial(D) < 2\partial(A)$. *If* V *is normalised and* $\partial(P)$, $\partial(Q) \geqslant 3\partial(A)$, *then* $\partial(D) < 2\partial(A)$.

We suppose that V is compatible with the result of shifting it to the right (or left) by one period A. Then by *II.5.8* and *II.5.12*, all of its shifts to the right or left by an integral number of periods A are pairwise compatible. Let W be the union of the extreme right and the extreme left of these shifts. Since $\partial(X) \geqslant 27\partial(A)$, $\partial(\mathrm{Bas}(W)) > 25\partial(A)$. By definitions I.4.14 and I.4.12, $\mathrm{Bas}(W)$ is an elementary word of rank $\alpha - 1$. But this is impossible, by I.4.

Suppose that $l_{\alpha-1}(D) > 20$. We assume that $\partial(D) \geqslant 2\partial(A)$. By 1.1 we may assume that A is a start of D, that is, $D \overset{\circ}{=} AR \overset{\circ}{=} RB$, where R is the base for the part common to V and the shift of it to the right by the period A. Since $\mathrm{Rel}(D, D)$, it then follows by *II.6.7* that R contains at least 10 segments of the elementary word D. Since A is a start of R, by *II.5.2* we have $l_{\alpha-1}(D) \leqslant 2l_{\alpha-1}(R) + 2 \leqslant 20$. We have reached a contradiction.

Suppose now that the occurrence V is normalised and that $\partial(P), \partial(Q) \geqslant 3\partial(A)$. It follows from 1.4 that the length of one of the words P, Q is greater than $4\partial(A)$. Suppose that $\partial(Q) < 4\partial(A)$. According to definition I.4.3, there is an occurrence in X of some elementary 9-power E of rank $\alpha - 1$, where $\partial(W) < 2\partial(A)$. Since V is normalised it follows by *II.6.14* that its shift V_1 to the right by the period A is also normalised. Since $\mathrm{Rel}(V, V_1)$, it follows from *II.5.7* that the base R of the intersection of V and V_1 contains less than two segments of D, and hence so also does A. It follows from this by *II.5.2* that if $\partial(D) \geqslant 2\partial(A)$, then $l_{\alpha-1}(D) \leqslant l_{\alpha-1}(R) + l_{\alpha-1}(A) + 2 < 9$. Hence $\partial(D) < 2\partial(A)$.

1.6. *Suppose that* $X \in \mathrm{Per}(\alpha, A)$ *and that* $W \in \mathrm{Norm}(\alpha - 1, X, 9)$, *where* $W = P * E * Q$ *and* $\partial(P), \partial(Q) \geqslant 3\partial(A)$. *Then*

(a) *each occurrence*

$$R * E * S \tag{1}$$

in the word X *that corresponds to* W *in phase relative to the period* A *and satisfies*

$$\partial(R), \partial(S) \geqslant 2\partial(A)\,, \tag{2}$$

is normalised;

(b) *all these occurrences are pairwise incompatible;*

(c) *if* $\partial(P), \partial(Q) \geqslant 4\partial(A)$ *and the occurrence*

$$W_1 \rightleftharpoons P_1 * uEv * Q_1 \tag{3}$$

is a maximal normalised continuation of the occurrence W, *where* $P \overset{\circ}{\equiv} P_1 u$ *and* $Q \overset{\circ}{\equiv} vQ_1$, *then* $\partial(u) < \partial(A)$, $\partial(v) < \partial(A)$ *and* $\partial(uEv) < 2\partial(A)$. *If, in addition,* $\partial(R)$, $\partial(S) \geqslant 4\partial(A)$ *then a maximal normalised continuation of the occurrence* (1) *has the form*

$$R_1 * uEv * S_1 , \tag{4}$$

where $R_1 u \overset{\circ}{\equiv} R$, $vS_1 \overset{\circ}{\equiv} S$.

By 1.5 and 1.1 we may assume that $A^2 \overset{\circ}{\equiv} EG$ for some G. Then for some P', Q', R', S' we have $P \overset{\circ}{\equiv} P'EG$, $R \overset{\circ}{\equiv} R'EG$, $Q \overset{\circ}{\equiv} GEQ'$, $S \overset{\circ}{\equiv} GES'$, and (a) follows from *II.6.14*.

We suppose that two different occurrences V_1 and V_2 of type (1) satisfying condition (2) are incompatible. Clearly one of them is a shift of the other to the right by r periods A, where $r > 0$. But, by 1.5 and I.1.14, this is impossible.

Suppose that (3) is a maximal normalised continuation of the occurrence W, where $\partial(P)$, $\partial(Q) \geqslant 4\partial(A)$. If $\partial(u) \geqslant \partial(A)$, then the shift W' of W to the left by period A is contained in (3) and, by (a), it is normalised. Then by *II.5.7*, W' is compatible with W_1, and thus with W, which is impossible by (b). Hence $\partial(u) < \partial(A)$. The inequality $\partial(v) < \partial(A)$ is obtained in an analogous way. The inequality $\partial(uEv) < 2\partial(A)$ follows from 1.5.

Suppose that $\partial(R)$, $\partial(S) \geqslant 4\partial(A)$. Since (1) corresponds in phase to W, (4) corresponds in phase to the occurrence (3). Since (3) is a continuation of W, by I.4.12 the occurrence (4) is a continuation of the occurrence (1) and this is normalised, by (a). If (4) had a proper normalised continuation to the left (or right), then, by analogy to the foregoing, we could prove that (3) has a proper normalised continuation to the left (respectively to the right).

1.7. It is easy to convince oneself that *all assertions in section* 1.6 *will remain valid if we keep the assumption that* $X \in \mathrm{Per}(\alpha, A)$, *but instead of occurrence* (1) *we consider occurrences in an arbitrary word* $Y \in \mathrm{Per}(A)$, *where* $Y \in \mathscr{R}_{\alpha-1}$ *and* $\partial(Y) \geqslant 27\partial(A)$, *which correspond to the occurrence* W *in phase in the sense of definition* I.2.6.

1.8. *Suppose that* $X \in \mathrm{Per}(\alpha, A)$, $Z \in \mathrm{Per}(A) \cap \mathscr{R}_{\alpha-1}$, $\partial(Z) \geqslant 27\partial(A)$, *and that* U *and* V *are interior relative to the period* A *and are occurrences in* X *and* Z *corresponding to one another in phase in the sense of* I.2.6. *If* $U \in \mathrm{Reg}(\alpha - 1, X)$, *then* $V \in \mathrm{Reg}(\alpha - 1, Z)$ *and* $\mathrm{MutNorm}_{\alpha-1}(U, V)$.

The truth of the assertion for $\alpha = 1$ follows trivially from I.4.1.

Let $\alpha > 1$. We suppose that some active kernel $R * E * Q \in \mathrm{Ker}(\alpha - 1, X)$ satisfies the condition

$$\partial(P), \partial(Q) \geqslant 6\partial(A) . \tag{5}$$

By *IV.1.3*, $P * E * Q$ is normalised and $l_{\alpha-1}(E) \geqslant q$. From 1.5 it follows that $\partial(E) < 2\partial(A)$. By 1.7 and 1.6(a), all the occurrences $R * E * S$ in X or Z which correspond to $P * E * Q$ in phase relative to the period A, with $\partial(R)$, $\partial(S) \geqslant 2\partial(A)$, are also normalised. By *III.3.30*, each such occurrence with $\partial(R)$, $\partial(S) \geqslant 4\partial(A)$ is active, whence it follows by *IV.1.15* that each such occurrence with $\partial(R), \partial(S) \geqslant 6\partial(A)$ is a kernel of rank $\alpha - 1$. Then, by *IV.1.3*, we have $\partial(E) \leqslant \partial(A)$.

Set $U = P_1 * D * Q_1$, $V = P_2 * D * Q_2$. Since $\partial(P_i), \partial(Q_i) \geqslant 8\partial(A)$, we have that $P_i \overset{\circ}{=} F_iEu$ and $Q_i \overset{\circ}{=} vEG_i$, where $\partial(F_i), \partial(G_i) \geqslant 6\partial(A)$ and the occurrences $F_i * E * uDvEG_i$ and $F_iEuDv * E * G_i$ are active kernels of rank $\alpha - 1$ for $i = 1, 2$. Then by *IV.3.24* and the fact that $U \in \mathrm{Reg}(\alpha - 1, X)$, we have $\mathrm{MutNorm}_{\alpha-1}(U, V)$. Hence $V \in \mathrm{Reg}(\alpha - 1, Z)$.

We are now left with the case where, for every active kernel $P * E * Q$ of rank $\alpha - 1$ of each of the words X and Z, either $\partial(P) < 6\partial(A)$ or $\partial(Q) < 6\partial(A)$.

We consider an arbitrary $R * D * S \in \mathrm{MaxNorm}(\alpha - 1, X, 9)$, where $\partial(R)$, $\partial(S) \geqslant 7\partial(A)$. If the word X has no active kernel of rank $\alpha - 1$, then by I.4.24, $R * D * S \in \mathrm{Ker}(\alpha - 1, X)$. Let $W \rightleftarrows R * E * Q$ be the first active kernel to the right of rank $\alpha - 1$ of the word X that satisfies $\partial(P) < 2\partial(A)$. Since $l_{\alpha-1}(E) \geqslant q$, by 1.5 we have $\partial(E) < 2\partial(A)$. Let W_1 be a shift of W to the right by period A. By 1.5, W_1 is not compatible with W. Let W_2 be a maximal end of the occurrence W_1 that is normalised and stable in a reversal of the kernel W. By *II.6.12* and *III.2.10*, W_2 contains not less than $q - 21$ segments. By the remarks in *II.7.3*, the occurrences $V_1 \leftrightarrows R * D * S$ and W_1 are not compatible. Then by *II.6.7*, V_1 contains less then 17 segments of W_1, that is, $W_2 < V_1$. From this it follows by *II.7.1* and *III.2.4* that V_1 is stable in a reversal of the kernel W. Then by *III.2.11*, V_1 is stable in a reversal of an arbitrary active kernel of rank $\alpha - 1$, situated to its left. In a similar manner it can be proved that V_1 is stable in a reversal of an arbitrary kernel of rank $\alpha - 1$, situated to its right. By I.4.24, this means that V_1 is a kernel of rank $\alpha - 1$ of X.

On the other hand, if $U_1 \rightleftarrows R_1 * D_1 * S_1$ is a kernel of rank $\alpha - 1$ of X, interior with respect to the period A, and $U_2 \rightleftarrows R_2 * uD_1v * S_2$ is a maximal normalised continuation of it, then by 1.6(c) we have $\partial(u), \partial(v) < \partial(A)$ and, consequently, $\partial(R_2)$, $\partial(S_2) > 7\partial(A)$. Then by what has been proved, U_2 is a kernel of rank $\alpha - 1$ and so $U_2 = U_1$.

Thus we have proved that for each occurrence U_1 in the word X which is interior relative to the period A,

$$U_1 \in \mathrm{Ker}(\alpha - 1, X) \Longleftrightarrow U_1 \in \mathrm{MaxNorm}(\alpha - 1, X, 9)\,. \qquad (6)$$

Clearly the same is true for the word Z. Since by hypothesis the words U and V correspond in phase to one another, it follows by (6), 1.7 and definition I.4.29 that $V \in \mathrm{Reg}(\alpha - 1, Z)$, since $U \in \mathrm{Reg}(\alpha - 1, X)$. Then by *IV.1.1*, *II.5.5* and *II.7.3* we have $\mathrm{MutNorm}_{\alpha-2}(U,V)$. Since U and V do not contain active kernels of rank $\alpha - 1$, it then follows from this by *IV.3.9* that $\mathrm{MutNorm}_{\alpha-1}(U, V)$.

1.9. *Suppose that* $X \in \mathrm{Per}(\alpha, A)$, $U \in \mathrm{Inn}(X, A)$, $V \in \mathrm{Inn}(X, A)$ *and* $\mathrm{Corr}_A(U, V)$. *If* $U \in \mathrm{Reg}(\alpha - 1, X)$, *then* $\mathrm{MutNorm}_{\alpha-1}(U, V)$.

This is the special case of lemma 1.8 with $Z \overset{\circ}{=} X$, since it follows from $\mathrm{Corr}_A(U, V)$ that U and V correspond to each other in phase in the sense of definition I.2.6.

1.10. *Suppose that* $X \in \mathrm{Per}(\alpha, A)$, *$V$ is a supporting kernel of rank $\alpha - 1$ of the word X,* $W \in \mathrm{Inn}(X, A)$ *and* $\mathrm{Corr}_A(V, W)$. *Then W is also a supporting kernel of rank*

$\alpha - 1$ *and* $\mathrm{MutNorm}_{\alpha-1}(V, W)$. *If* V *is active, so also is* W.

This follows from 1.9 and *IV.3.3*.

1.11. *If* V *is a supporting kernel of rank* $\alpha - 1$ *of the word* $X \in Per(\alpha, A)$, *then* $\partial(Bas(V)) < 2\partial(A)$. *If* V *is active, then* $\partial(\mathrm{Bas}(V)) \leqslant \partial(A)$.

The first assertion follows from 1.5 and the second from 1.10 and *IV.1.3*.

1.12. Suppose that $X \in \mathrm{Per}(\alpha, A)$. If every supporting kernel of rank $\alpha - 1$ of the word X is inactive, we shall say that X is of *type I*. In the contrary case X will be said to be of *type II*.

It follows from 1.8 and *IV.3.3* that, for a given period A, any two words in the set $\mathrm{Per}(\alpha, A)$ are of one and the same type. Hence we can speak of a period A of rank α of type I or type II.

1.13. *If the word* $X \in \mathrm{Per}(\alpha, A)$ *is of type I, then for an arbitrary occurrence* $U \in \mathrm{Inn}(X, A)$,

$$U \in \mathrm{Ker}(\alpha - 1, X) \Longleftrightarrow U \in \mathrm{MaxNorm}(\alpha - 1, X, 9)\,.$$

In fact, we proved in 1.8 that, if some active kernel $P * E * Q$ of rank $\alpha - 1$ of a word X satisfies condition (5), then X is of type II, and that in the contrary case (6) holds.

1.14. *If* $X \in \mathrm{Per}(\alpha, A)$ *and* $A \mathrel{\underline{\overline{\circ}}} B'$, *then* $X \in \mathrm{Per}(\alpha, B)$.

Suppose that $X \in \mathrm{Per}(\alpha, A)$, that is, $X \in \mathrm{Per}(A)$, $X \in \mathscr{R}_{\alpha-1}$ and for some $V \in \mathrm{Ker}(\alpha - 1, X)$, $V \in \mathrm{Inn}(X, A)$ and $\partial(\mathrm{Bas}(V)) < 2\partial(A)$. Clearly then $X \in \mathrm{Per}(B)$ and $V \in \mathrm{Inn}(X, B)$. It remains to prove that $\partial(\mathrm{Bas}(V)) < 2\partial(B)$. Let V_1 and V_2 be right shifts of V by the periods B and A respectively. Since $V \in \mathrm{Norm}(\alpha - 1, X, 9)$ it follows by 1.7 that $V_1 \in \mathrm{Norm}(\alpha - 1, X, 9)$. Since $\neg\,\mathrm{Comp}(V, V_2)$ by 1.3, we have $\neg\,\mathrm{Comp}(V, V_1)$ from *II.5.8* and *II.5.12*. Then from *II.5.7* and *II.5.2* it follows that $\partial(\mathrm{Bas}(V)) < 2\partial(B)$.

1.15. *Suppose that* $X \in \mathscr{R}_{\alpha-1}$, $X \in \mathrm{Per}(A)$ *and* $\partial(X) \geqslant 27\partial(A)$. *If there is an occurrence* $W \in \mathrm{Inn}(X, A)$ *such that* $\partial(\mathrm{Bas}(W)) < 4\partial(A)$ *and* $W \in \mathrm{MaxNorm}(\alpha - 1, X, 9)$, *then* $X \in \mathrm{Per}(\alpha, A)$.

Let W_1 be the result of applying a shift to the right by the period A to W. By *II.6.14* and *I.4.14*, W_1 is normalised and is incompatible with W. It then follows from *II.5.7* and *II.5.2* that $\partial(\mathrm{Bas}(W)) < 2\partial(A)$. If $\mathrm{Bas}(W) \mathrel{\underline{\overline{\circ}}} E$, we may assume without loss of generality that $A^2 \mathrel{\underline{\overline{\circ}}} Eu$.

If the word X has no active kernels of rank $\alpha - 1$, then by 1.13, $W \in \mathrm{Ker}(\alpha - 1, X)$. By I.4.3, $X \in \mathrm{Per}(\alpha, A)$.

Suppose some kernel $U \in \mathrm{Ker}(\alpha - 1, X)$ is active. By *IV.1.3*, $U \in \mathrm{Norm}(\alpha - 1, X, q)$. By repeating the arguments of 1.4, it is easy to convince oneself that $\partial(D) < 9\partial(A)$ if the elementary word D occurs in X. From this it follows by *II.5.8* and *II.5.12* that U is incompatible with its shifts in the left or right by period A. Then by *II.6.7* and *II.5.2* we have $\partial(\mathrm{Bas}(U)) < 2\partial(A)$. Let $U_1 \rightleftarrows P * F * Q$ be a shift of U such that $\partial(P) \geqslant 9\partial(A)$ and $\partial(Q) \geqslant 9\partial(A)$. By *II.6.6*, U_1 contains some occurrence $U_2 \in \mathrm{Norm}(\alpha - 1, X, q - 8)$. Let U_3 be a maximal normalised continuation of U_2. In a way analogous to 1.6(c), it can be proved that $U_3 \in \mathrm{Inn}(X, A)$

and $\partial(\mathrm{Bas}(U_3)) < 2\partial(A)$. Finally, by *IV.1.4* there is some kernel $V \in \mathrm{Ker}(\alpha - 1, X)$ compatible with U_3 and therefore contained in it. Thus $V \in \mathrm{Inn}(X, A)$.

1.16. *Suppose that* $X \in \mathrm{Per}(\alpha - 1, A) \cap \mathscr{R}_{\alpha-1}$. *If* A *is a minimal period which is not elementary in rank* $\alpha - 1$, *then* $X \in \mathrm{Per}(\alpha, A)$.

By *II.3.7*, an occurrence $V \in \mathrm{Inn}(X, A)$ and a generating occurrence U in some word $Y' \in \mathrm{Int}(X', \alpha - 1, B)$ with elementary period B, can be found such that

$$\mathrm{Bas}(V) \mathrel{\underline{\underline{\circ}}} \mathrm{Bas}(U),\ \mathrm{MutNorm}_{\alpha-2}(V, U)\,,$$
$$l_{\alpha-1,B}(U) \geqslant 9,\ \ l_{\alpha-1,A}(V) < 3,\ \ \rho_{\alpha-1,B}(X') < \rho_{\alpha-1,A}(X)\,.$$

By *II.1.9* we may assume that $V = P * E * Q$, where $12\partial(A) \leqslant \partial(P) \leqslant 13\partial(A)$. Since $V \in \mathrm{Norm}(\alpha - 1, X, 9)$, then by the remark in *II.7.3*, a maximal normalised continuation W of it is uniquely defined. Let $W = R * F * S$ and let $W_1 \rightleftarrows R_1 * F * S_1$ be a generating occurrence of the word F in the word $Y_1 \in \mathrm{Int}(X_1, \alpha - 1, C)$, with elementary period C. Then by *II.5.5* we have

$$\mathrm{MutNorm}_{\alpha-2}(W, W_1)\,. \tag{7}$$

By definition I.4.12 we have the relations

$$\mathrm{MutNorm}_{\alpha-2}(U, \phi(V;\, W, W_1))$$

and

$$l_{\alpha-1,C}(\phi(V;\, W, W_1)) = l_{\alpha-1,B}(U) \geqslant 9\,,$$

whence $\rho_{\alpha-1,B}(X') = \rho_{\alpha-1,C}(X_1)$ by *II.3.4*.

Let us assume that $\partial(R) < 8\partial(A)$. Suppose then that the occurrence V_1 in X starts with the first kernel of rank $\alpha - 2$ of W from the left that is interior with respect to A and ends with V. Then $V_1 \in \mathrm{Inn}(X, A)$ and $V_1 \in \mathrm{Reg}(\alpha - 2, X)$, that is, V_1 is a normal generating occurrence of rank $\alpha - 1$ in the word $X \in \mathrm{Int}(X, \alpha - 1, A)$. From (7) it follows by *IV.3.6* that $\mathrm{MutNorm}_{\alpha-2}(V_1, \phi(V_1;\, W, W_1))$. Since $\rho_{\alpha-1,C}(X_1) < \rho_{\alpha-1,A}(X)$, it follows that $l_{\alpha-1,A}(V_1) < 3$ because A is minimal in rank $\alpha - 1$, that is, that $\partial(\mathrm{Bas}(V_1)) < 4\partial(A)$. Then $\partial(R) \geqslant 8\partial(A)$.

The inequality $\partial(S) \geqslant 8\partial(A)$ is established in an analogous fashion. Therefore $W \in \mathrm{Inn}(X,A)$. By repeating the argument we carried out above on V_1, we get that $\partial(F) < 4\partial(A)$. We may now refer to 1.15.

1.17. *Suppose that* $X \in \mathrm{Per}(\alpha, A)$, *let* U, V *be supporting kernels of rank* $\alpha - 1$ *of* X *and let* W *be their union. Then* $\partial_{\alpha-1}(W) = \rho_{\alpha,A}(X) + 1$ *if and only if* V *is the result of shifting* U *by one period* A *to the right or left.*

By definition I.4.8, we have for some supporting kernel U_1 of rank $\alpha - 1$, a shift V_1 of it to the right by one period A and their join W_1,

$$\rho_{\alpha,A}(X) = \partial_{\alpha-1}(W_1) - 1\,.$$

Let V be a shift of U to the right by one period A. If $\mathrm{Corr}_A(U, U_1)$, then by 1.8,

MutNorm$_{\alpha-1}(W, W_1)$, whence by *IV.3.4* it follows that $\partial_{\alpha-1}(W) = \partial_{\alpha-1}(W_1) = \rho_{\alpha,A}(X) + 1$.

Suppose that $\neg$ Corr$_A(U, U_1)$. Then there clearly exist supporting kernels U' and V' of rank $\alpha - 1$ such that $U < U' < V$, Corr$_A(U_1, U')$, $U_1 < V' < V_1$ and Corr$_A(V, V')$. Suppose that W' is the union of the kernels U and U', W'' that of U' and V, W_1'' that of U_1 and V', and W_1' that of V' and V_1. Since Corr$_A(W', W_1')$ and Corr$_A(W'', W_1'')$, we have $\partial_{\alpha-1}(W') = \partial_{\alpha-1}(W_1')$ and $\partial_{\alpha-1}(W'') = \partial_{\alpha-1}(W_1'')$. Since U' is the unique kernel of rank $\alpha - 1$ contained in both occurrences W' and W'', $\partial_{\alpha-1}(W) = \partial_{\alpha-1}(W') + \partial_{\alpha-1}(W'') - 1$ and, similarly, $\partial_{\alpha-1}(W_1) = \partial_{\alpha-1}(W_1') + \partial_{\alpha-1}(W_1'') - 1$. Hence $\partial_{\alpha-1}(W) = \partial_{\alpha-1}(W_1)$.

The second assertion is clear.

1.18. *If* $X, X_1 \in$ Per(α, A), *then* $\rho_{\alpha,A}(X) = \rho_{\alpha,A}(X_1)$.

This follows from 1.8 and *IV.3.3*.

1.19. *Let* U_1, U_2, V_1, V_2 *be supporting kernels of rank* $\alpha - 1$ *of the word* $X \in$ Per (α, A), *let* U_1 *be a start of* U, U_2 *an end of* U, V_1 *a start of* V, V_2 *an end of* V. *Then the following holds*:

$$\text{Corr}_A(U, V) \Longleftrightarrow$$
$$\Longleftrightarrow \partial_{\alpha-1}(U) = \partial_{\alpha-1}(V) \;\&\; \text{Corr}_A(U_1, V_1) \;\&\; \text{Corr}_A(U_2, V_2)\,. \tag{8}$$

Suppose that Corr$_A(U, V)$. Then Bas$(U) \overset{\circ}{=}$ Bas(V) and by 1.9 we have MutNorm$_{\alpha-1}(U, V)$, whence by *IV.3.3* it follows that $\partial_{\alpha-1}(U) = \partial_{\alpha-1}(V)$. By 1.10 and *IV.1.2* we have $V_1 = \phi(U_1; U, V)$ and $V_2 = \phi(U_2; U, V)$, that is, all the terms in the conjunction on the right hand side of (8) are valid.

Suppose that the right hand side of (8) holds. Without loss of generality we may assume that V_1 is a shift of U_1 to the right by r periods A and $\partial(\text{Bas}(U)) \leqslant \partial(\text{Bas}(V))$. Suppose that W is a shift of U by r periods A to the right. Assume that Corr$_A(U, V)$ is false. Then W is a proper start of V and, by what has been proved, $\partial_{\alpha-1}(W) = \partial_{\alpha-1}(U) = \partial_{\alpha-1}(V)$. But since $\neg$ Corr$_A(U, V)$, this is impossible, since in this case V_2 is not contained in W. We have obtained a contradiction. Therefore, Corr$_A(U, V)$.

§ 2. Integral Words and Generating Occurrences of Rank α

By definition I.4.5, every integral word Y of rank α is equivalent in rank $\alpha - 1$ to some periodic word of rank α Generally speaking, one and the same Y can be equivalent to different periodic words of rank α. We are not yet interested in the question whether, for a given word $Y \in$ Int(X, α, A), the concept of supporting kernel of rank $\alpha - 1$ and the relation Corr$_{\alpha,A}(U, V)$ depend on the choice of the original word $X \in$ Per(α, A), since from the notation $Y \in$ Int(X, α, A) it will always be clear to us relative to which words $X \in$ Per(α, A) we are considering these concepts.

Since $X \overset{\alpha-1}{\approx} X$ and $f_{\alpha-1}(V; X, X) = V$ for each $V \in \text{Reg}(\alpha - 1, X)$ by *IV.2.18(c)*, it follows that $X \in \text{Int}(X, \alpha, A)$ whenever $X \in \text{Per}(\alpha, A)$ and, furthermore, for arbitrary normal generating occurrences U and V of rank α in X, the relation

$$\text{Corr}_{\alpha,A}(U, V) \Longleftrightarrow \text{Corr}_A(U, V)$$

holds.

2.1. *If V is a supporting kernel of rank $\alpha - 1$ of the word $Y \in \text{Int}(X, \alpha, A)$, then $V \in \text{Ker}(\alpha - 1, Y)$.*

An occurrence U in a word $Y \in \text{Int}(X, \alpha, A)$ is a normal generating occurrence of rank α if and only if it starts and ends with a supporting kernel of rank $\alpha - 1$.

If W is a generating occurrence of rank α in the word Y, then every kernel of rank $\alpha - 1$ contained in it is a supporting kernel of it of rank $\alpha - 1$.

These three assertions follow directly from the corresponding definitions I.4.4–I.4.7 and assumptions *IV.2.18(a), (d), (i)*.

From *IV.1.2* it follows that, for two arbitrary supporting kernels U and V of rank $\alpha - 1$ of a word $Y \in \text{Int}(X, \alpha, A)$, either $U < V$ or $V < U$.

It follows from definition I.4.6 that the relation $\text{Corr}_{\alpha,A}(U, V)$ for normal generating occurrences of rank α in a given word $Y \in \text{Int}(X, \alpha, A)$ is reflexive, symmetric and transitive, since by I.2.5 this is true for occurrences in a word $X \in \text{Per}(\alpha, A)$.

2.2. *If $Y \in \text{Int}(X, \alpha, A)$, $U, V \in \text{Reg}(\alpha - 1, Y)$ and $\text{Corr}_{\alpha,A}(U, V)$, then $\text{MutNorm}_{\alpha-1}(U, V)$.*

By I.4.6 we have $Y \overset{\alpha-1}{\approx} X$, $X \in \text{Per}(\alpha, A)$, $U = f_{\alpha-1}(U_1; X, Y)$, $V = f_{\alpha-1}(V_1; X, Y)$, where $U_1, V_1 \in \text{Inn}(X, A)$, $\text{Corr}_A(U_1, V_1)$ and $U_1, V_1 \in \text{Reg}(\alpha - 1, X)$. From 1.9 it follows that $\text{MutNorm}_{\alpha-1}(U_1, V_1)$, from which we obtain by *IV.3.2* and *IV.3.1* that $\text{MutNorm}_{\alpha-1}(U, V)$.

2.3. *If $Y \in \text{Int}(X, \alpha, A)$, U is a supporting kernel of rank $\alpha - 1$ of Y and $\text{Corr}_{\alpha,A}(U, V)$, then V is a supporting kernel of rank $\alpha - 1$ of Y and $\text{MutNorm}_{\alpha-1}(U, V)$. If U is active, then V is active.*

This follows from 1.10, 2.2 and assumptions *IV.2.17*.

2.4. *Let U, V, W be supporting kernels of rank $\alpha - 1$ of a word $Y \in \text{Int}(X, \alpha, A)$. If $U < V$, $\text{Corr}_{\alpha,A}(U, V)$ and $\neg\, \text{Corr}_{\alpha,A}(U, W)$, then a supporting kernel W_1 can be found such that $U < W_1 < V$ and $\text{Corr}_{\alpha,A}(W, W_1)$.*

This assertion is clear for X, and in the general case it follows from definition I.4.6 and assumption *IV.2.18(f)*.

2.5. Suppose that $Y \in \text{Int}(X, \alpha, A)$, that is, $Y \overset{\alpha-1}{\approx} X$ for some $X \in \text{Per}(\alpha, A)$. Y will be assigned to *type* I or II according to the type assigned by definition 1.12 to X.

From *IV.2.17* it follows that *an integral word Y of rank α has type II if and only if one of its supporting kernels of rank $\alpha - 1$ is active.*

2.6. *Let U, V be supporting kernels of rank $\alpha - 1$ of a word $Y \in \text{Int}(X, \alpha, A)$, $\bar{V}$ a maximal normalised continuation of V and suppose that $\text{Corr}_{\alpha,A}(U, V)$.*

If Y is of type II, then U and $\bar{V}$ do not intersect.

If Y is of type I then $\bar{V}$ (and hence V) contains less than two segments of the occurrence U.

The first assertion follows from 2.5, 2.4, 2.3, *IV.1.3* and *II.5.7*.

Suppose that Y is of type I. By 2.5, U and V are inactive, whence, by definitions I.4.15 and I.4.6 and assumptions *IV.2.14* and *II.5.20*, it follows that $\mathrm{Rel}(U, V)$. Then by *II.5.17*, $\mathrm{Rel}(U, \bar{V})$. Since $\bar{V}$ is incompatible with U it only remains for us to take advantage of *II.5.7*.

2.7. *Let U, V, W be supporting kernels of rank $\alpha - 1$ of a word $Y \in \mathrm{Int}(X, \alpha, A)$ such that $\mathrm{Corr}_{\alpha,A}(U, V)$ and $\mathrm{Corr}_{\alpha,A}(U, W)$. If $U < V < W$, then a maximal normalised continuation $\bar{U}$ of the kernel U does not intersect W.*

This follows from 2.5 and *II.5.2*.

2.8. *Suppose that $W \rightleftharpoons P * E * Q$ is a generating occurrence of rank α in the word $Y \in \mathrm{Int}(X, \alpha, A)$. If $E \mathrel{\bar{\circ}} E_1E_2$, $W_1 \rightleftharpoons P * E_1 * E_2Q$ and $W_2 \rightleftharpoons PE_1 * E_2 * Q$, then*

$$l_{\alpha,A}(W_1) + l_{\alpha,A}(W_2) - 1 \leqslant l_{\alpha,A}(W) \leqslant l_{\alpha,A}(W_1) + l_{\alpha,A}(W_2) + 2\,.$$

Let U and V be the extreme left supporting kernels of rank $\alpha - 1$ of the occurrences W_1 and W_2. If $\mathrm{Corr}_{\alpha,A}(U, V)$, then the first inequality is clear; and if $\neg\,\mathrm{Corr}_{\alpha,A}(U, V)$, it follows from 2.4. The second inequality follows from 2.7 and 2.4.

2.9. *If U_1, U_2, V_1, V_2 are supporting kernels of rank $\alpha - 1$ of a word $Y \in \mathrm{Int}(X, \alpha, A)$, U_1 is a start of the occurrence U, U_2 is an end of U, V_1 is a start of the occurrence V and V_2 is an end of V, then the following relation holds:*

$$\mathrm{Corr}_{\alpha,A}(U, V) \Longleftrightarrow \partial_{\alpha-1}(U) = \partial_{\alpha-1}(V)\ \&$$
$$\&\ \mathrm{Corr}_{\alpha,A}(U_1, V_1)\ \&\ \mathrm{Corr}_{\alpha,A}(U_2, V_2)\,.$$

Let U, V be supporting kernels of rank $\alpha - 1$ of the word $Y \in \mathrm{Int}(X, \alpha, A)$ and W their union. Then

$$\mathrm{Corr}_{\alpha,A}(U, V) \Longleftrightarrow \partial_{\alpha-1}(W) - 1 = \rho_{\alpha,A}(X) \cdot (l_{\alpha,A}(W) - 1)\,.$$

The first assertion follows at once from 1.19 by definition I.4.6 and assumption *IV.2.18(j)*. The second assertion is obtained by induction on $l_{\alpha,A}(W)$ from 1.17, by definitions I.4.6 and I.4.7 and assumption *IV.2.18(j)*.

2.10. *Suppose that W and W_1 are generating occurrences in the words $Y \in \mathrm{Int}(X, \alpha, A)$ and $Y_1 \in \mathrm{Int}(X_1, \alpha, B)$, where $\mathrm{Bas}(W_1) \mathrel{\bar{\circ}} \mathrm{Bas}(W)$ and $\mathrm{MutNorm}_{\alpha-1}(W, W_1)$. If $\rho_{\alpha,A}(X) = \rho_{\alpha,B}(X_1)$, then $l_{\alpha,A}(W) = l_{\alpha,B}(W_1)$ and, for arbitrary occurrences of U and V contained in W,*

$$\mathrm{Corr}_{\alpha,A}(U, V) \Longleftrightarrow \mathrm{Corr}_{\alpha,B}(\phi(U; W, W_1), \phi(V; W, W_1))\,.$$

This follows from 2.9, *IV.3.6* and *IV.3.3*.

2.11. *If $W \rightleftharpoons P * D * Q$ is an occurrence of a semi-integral 3-power of rank α in a word $Z \in \mathscr{R}_{\alpha-1}$, then it is possible to find an occurrence $U \in$ MaxNorm$(\alpha-1, Z, 9)$ which is contained in W and distinct from it.*

Suppose that $W_1 \rightleftharpoons R * D * S$ is a generating occurrence in a word $Y_1 \in$ Int(X_1, α, A). We can find in W_1 supporting kernels V_1, V_2, V_3 of rank $\alpha - 1$ such that $V_1 < V_2 < V_3$ and $\mathrm{Corr}_{\alpha,A}(V_i, V_{i+1})$ for $i = 1,2$. By 2.6, a maximal normalised continuation V of the kernel V_2 contains less than two segments of rank $\alpha - 1$ of V_1 and V_2. Then by *II.6.14*, $\phi(V; W_1, W)$ is the desired occurrence of U.

2.12. *If E is an elementary word of rank $\alpha - 1$, then no semi-integral 3-power of rank α occurs in it.*

Assume that $E \overset{\circ}{=} uDv$, where D is a semi-integral 3-power of rank α. Let $W \rightleftharpoons P * E * Q$ be a generating occurrence of rank $\alpha - 1$ in the word $Y \in$ Int$(X, \alpha - 1, B)$, where the period B is elementary in rank $\alpha - 1$. We may clearly assume that W is normal, that is, $W \in$ Reg$(\alpha - 2, Y)$. By 2.11 there is an occurrence $U \in$ MaxNorm$(\alpha - 1, Y, 9)$ which is contained in $Pu * D * vS$ and is distinct from it. But this is impossible, since in that case W would be a normalised continuation of U, by *II.5.3* and *II.5.7*.

2.13. *Suppose that $Y \overset{\alpha-1}{\approx} Z$. Then $Z \in$ Int(X, α, A) if $Y \in$ Int(X, α, A). If V is a supporting kernel of rank $\alpha - 1$ of Y, then $f_{\alpha-1}(V; Y, Z)$ is a supporting kernel of rank $\alpha - 1$ of Z. For arbitrary supporting kernels of rank $\alpha - 1$ U and V of Y,*

$$\mathrm{Corr}_{\alpha,A}(U, V) \Longleftrightarrow \mathrm{Corr}_{\alpha,A}(f_{\alpha-1}(U; Y, Z), f_{\alpha-1}(V; Y, Z)) .$$

If W is a normal generating occurrence in Y, then $f_{\alpha-1}(W; Y, Z)$ is a normal generating occurrence in Z and $l_{\alpha,A}(f_{\alpha-1}(W; Y, Z)) = l_{\alpha,A}(W)$.

This result follows from definitions I.4.5, I.4.6 and assumptions *IV.2.18(d)*, *(e)*, *(i)*.

2.14. *Let W be a normal generating occurrence of rank α in the word $Y \in$ Int(X, α, A), V an occurrence in some word $Z \in \mathscr{R}_{\alpha-1}$ and suppose that* MutNorm$_{\alpha-1}(W, V)$. *If $Z \overset{\alpha-1}{\approx} Z_1$ and $V_1 = f_{\alpha-1}(V; Z, Z_1)$, then there exists a generating occurrence W_1 of rank α in some word $Y_1 \in$* Int(X, α, A) *such that $Y \overset{\alpha-1}{\approx} Y_1$,* Bas$(W_1) \overset{\circ}{=}$ Bas(V_1), $l_{\alpha,A}(W_1) = l_{\alpha,A}(W)$ *and* MutNorm$_{\alpha-1}(W_1, V_1)$.

By I.4.31 we can find a $Y_1 \overset{\alpha-1}{\approx} Y$ such that Bas$(f_{\alpha-1}(W; Y, Y_1)) \overset{\circ}{=}$ Bas(V_1). Suppose that $W_1 \rightleftharpoons f_{\alpha-1}(W; Y, Y_1)$. From 2.13 it follows that $l_{\alpha,A}(W_1) = l_{\alpha,A}(W)$ and from *IV.3.2* and *IV.3.1* that MutNorm$_{\alpha-1}(V_1, W_1)$.

2.15. *Let W be a normal generating occurrence of rank α in a word $Y \in$ Int(X, α, A), V an occurrence in some word $Z \in \mathscr{R}_{\alpha-2}$, and suppose that* MutNorm$_{\alpha-2}(V, W)$. *If $Z \overset{\alpha-2}{\approx} Z_1$ and $V_1 = f_{\alpha-2}(V; Z, Z_1)$, then there exists a generating occurrence W_1 of rank α in some word $Y_1 \overset{\alpha-1}{\approx} Y$ such that* Bas$(W_1) \overset{\circ}{=}$ Bas(V_1), $l_{\alpha,A}(W_1) = l_{\alpha,A}(W)$ *and* MutNorm$_{\alpha-2}(V_1, W_1)$.

By I.4.31 we can find a $Y_1 \overset{\alpha-2}{\approx} Y$ such that Bas$(V_1) \overset{\circ}{=}$ Bas$(f_{\alpha-2}(W; Y, Y_1))$. Set $W_1 \rightleftharpoons f_{\alpha-2}(W; Y, Y_1)$. Then MutNorm$_{\alpha-1}(V_1, W_1)$ follows from *IV.3.2* and *IV.3.1*, while *IV.2.11* and *IV.2.18(b)* give that $Y \overset{\alpha-1}{\approx} Y_1$ and $W_1 = f_{\alpha-1}(W; Y, Y_1)$. Then 2.13 gives that $l_{\alpha,A}(W_1) = l_{\alpha,A}(W)$.

2.16. *If $\alpha > 1$ and $P * E * Q$ is a normal generating occurrence of rank α in a word $Y \in \mathrm{Int}(X, \alpha, A)$, then there exists a word $Z \in \mathrm{Int}(X, \alpha, A)$ such that $Y \overset{\alpha-1}{\simeq} Z$,*

$$f_{\alpha-1}(P * E * Q; Y, Z) = R * E * S \tag{1}$$

and every kernel $V \in \mathrm{Ker}(\alpha - 1, Z)$ not contained in the occurrence $R * E * S$ is related to the kernel $f_{\alpha-1}(V; Z, X)$.

Let $F * D * H$ be a generating occurrence in the word $X \in \mathrm{Per}(\alpha, A)$, and suppose that $f_{\alpha-1}(F * D * H; X, Y) = P * E * Q$. Then, by *IV.2.26*, there exists a word $Z \in \mathscr{R}_{\alpha-1}$ such that $X \overset{\alpha-1}{\simeq} Z, f_{\alpha-1}(F * D * H; X, Z) = R * E * S$ and $F * D * H$ transforms to $R * E * S$ locally in rank $\alpha - 1$. Relation (1) is satisfied because of *IV.2.18(d)* and (*e*). If some kernel $V \in \mathrm{Ker}(\alpha - 1, Z)$ not contained in $R * E * S$ is not related to $f_{\alpha-1}(V; Z, X)$, then it is active by *IV.2.14* and does not intersect $R * E * S$ by *IV.1.3*. Then by definition *IV.2.25*, we have $\mathrm{Bas}(V) \overset{\circ}{=} \mathrm{Bas}(f_{\alpha-1}(V; Z, X))$, that is, $\mathrm{Rel}(V; f_{\alpha-1}(V; Z, X))$.

2.17. *Let $F * D * H$ be a normal generating occurrence of rank α in a word $X \in \mathrm{Per}(\alpha, A)$, and suppose that $X \overset{\alpha-1}{\simeq} Y$ and $f_{\alpha-1}(F * D * H; X, Y) = P * E * Q$. If $\alpha > 1$ and Y is of type I, then there is a word $Z \in \mathrm{Int}(X, \alpha - 1, A)$ such that $Y \overset{\alpha-1}{\simeq} Z, X \overset{\alpha-2}{\simeq} Z$,*

$$f_{\alpha-2}(F * D * H; X, Z) = f_{\alpha-1}(P * E * Q; Y, Z) = R * E * S \tag{2}$$

and, for every occurrence $V \in \mathrm{Reg}(\alpha - 1, X)$ contained in $F * D * H$,

$$f_{\alpha-1}(V; X, Y) = \phi(f_{\alpha-2}(V; X, Z); R * E * S, P * E * Q)\,. \tag{3}$$

The desired word Z can be found using 2.16. In addition, by 2.5 and *IV.2.14*, we have that $I_{\alpha-1}(X, Z) = 0$, so that $X \overset{\alpha-2}{\simeq} Z$ by *IV.2.13*. Then by *IV.2.18(b)* and (*e*), the desired relation (2) follows from (1). Relation (3) is satisfied because of *IV.2.27*.

2.18. *Let $W \rightleftharpoons P * uEv * Q$ be a normal generating occurrence of rank α in a word $Y \in \mathrm{Int}(X, \alpha, A)$ of type I, where $\alpha > 1$ and the words u, v are not empty. Then for every occurrence V contained in $Pu * E * vQ$, we have*

$$V \in \mathrm{Ker}(\alpha - 1, Y) \Longleftrightarrow V \in \mathrm{MaxNorm}(\alpha - 1, Y, \cdot 9)\,.$$

Set $W_0 \rightleftharpoons f_{\alpha-1}(W; Y, X)$. By 2.17 there is a word $Z \overset{\alpha-2}{\simeq} X$ such that $Z \overset{\alpha-1}{\simeq} Y$ and

$$f_{\alpha-2}(W_0; X, Z) = f_{\alpha-1}(W; Y, Z) = R * uEv * S \tag{4}$$

for some R and S. Set $W_1 \rightleftharpoons R * uEv * S$. By *IV.3.2*, we have $\mathrm{MutNorm}_{\alpha-1}(W, W_1)$, whence $\mathrm{MutNorm}_{\alpha-2}(W, W_1)$ follows because of *IV.3.9*.

Let V be any occurrence contained in $Pu * E * Q$, and suppose that $U \rightleftharpoons \phi(V; W, W_1)$. Then by *IV.3.3*,

$$V \in \mathrm{Ker}(\alpha - 1, Y) \Longleftrightarrow U \in \mathrm{Ker}(\alpha - 1, Z)\,,$$

and by *II.5.15*,

$$V \in \mathrm{MaxNorm}\,(\alpha - 1, Y, 9) \Longleftrightarrow U \in \mathrm{MaxNorm}\,(\alpha - 1, Z, 9)\,.$$

Moreover, it follows from (4) and *II.5.16* (relation (13)) that

$$U \in \mathrm{MaxNorm}\,(\alpha - 1, Z, 9) \Longleftrightarrow f_{\alpha-2}(U; Z, X) \in \mathrm{MaxNorm}\,(\alpha - 1, X, 9),$$

while from *IV.1.8,*

$$U \in \mathrm{Ker}(\alpha - 1, Z) \Longleftrightarrow f_{\alpha-2}(U; Z, X) \in \mathrm{Ker}(\alpha - 1, X)\,.$$

Now all we need do is quote 1.13.

2.19. *Let U, V be supporting kernels of rank $\alpha - 1$ of a word $Y \in \mathrm{Int}(X, \alpha, A)$ of type I. If $\mathrm{Corr}_{\alpha,A}(U, V)$, then $\mathrm{Rel}(U, V)$, $\mathrm{MutNorm}_{\alpha-2}(U, V)$ and $l_{\alpha-1}(U) = l_{\alpha-1}(V)$.*

$\mathrm{MutNorm}_{\alpha-1}(U, V)$ follows from $\mathrm{Corr}_{\alpha,A}(U, V)$, by 2.3. Then $\mathrm{Rel}(U, V)$ by *IV. 3.11*, whence by *IV.3.8* we get that $\mathrm{MutNorm}_{\alpha-2}(U, V)$ and $l_{\alpha-1}(U) = l_{\alpha-1}(V)$.

2.20. *Let $W \leftrightarrows P * uEv * Q$ be a generating occurrence of rank α in a word $Y \in \mathrm{Int}(X, \alpha, A)$ of type II, and suppose that $l_{\alpha,A}(W) \geqslant 7$ and that the occurrence $W_1 \rightleftarrows Pu * E * vQ$ begins with the fourth left and ends with the fourth right segments of rank α of W. Then for every word $RuEvS \in \mathscr{R}_{\alpha-1}$ we have*

$$\mathrm{MutNorm}_{\alpha-1}(Pu * E * vQ, Ru * E * vS)\,, \tag{5}$$

and for every word $RuEvS \in \mathscr{P}_{\alpha-1}$ and every occurrence V contained in $Ru * E * vS$ we have

$$V \in \mathrm{Ker}(\alpha - 1, \quad RuEvS) \Longleftrightarrow$$
$$\Longleftrightarrow \phi(V; Ru * E * vS, Pu * E * vQ) \in \mathrm{Ker}\,(\alpha - 1, Y)\,. \tag{6}$$

Suppose that $RuEvS \in \mathscr{P}_{\alpha-1}$, $W' \rightleftarrows R * uEv * S$, and that V_i is the i-th left segment of rank α of the occurrence W, and $U_i \rightleftarrows \phi(V_i; W, W')$. By 2.6, $V_i \ll V_{i+1}$. If the supporting kernels V_i are not active, then by 2.3 and 2.4 we can find active supporting kernels of rank $\alpha - 1$ situated between them. Thus without loss of generality we may assume that the kernels V_2, V_3 and V_4 are active. Then by *IV.1.3* and *II.6.14* it follows that

$$V_i \in \mathrm{Norm}(\alpha - 1, Y, q) \text{ and } U_i \in \mathrm{Norm}\,(\alpha - 1, Z, q)$$

for $i = 2, 3, 4$, whence $U_3 \in \mathrm{Act}\,(\alpha - 1, RuEvS)$ by *III.3.30* and *IV.1.2*, and moreover $V_3 \ll W_1$. Analogously, we can find an active kernel V_3' of rank $\alpha - 1$ in W such that $W_1 \ll V_3'$ and the occurrence $U_3' \rightleftarrows \phi(V_3'; W, W')$ is also active. Then (6) holds by *IV.1.15*, and (5) by *IV.3.24*.

2.21. *Let $P * uEv * Q$ be a generating occurrence of rank α in a word $Y \in \mathrm{Int}(X, \alpha, A)$ such that the occurrence $Pu * E * vQ$ starts with the second left segment and ends with the second right segment of rank α. Then for all words $RuEvS \in \mathscr{R}_{\alpha-2}$ and $FuEvH \in \mathscr{R}_{\alpha-2}$,*

$$\text{MutNorm}_{\alpha-2}(Ru * E * vS, Fu * E * vH).$$

If in addition $\text{MutNorm}_{\alpha-2}(P * uEv * Q, R * uEv * S)$, *then for all words* $FuEvS$, $RuEvH \in \mathscr{R}_{\alpha-2}$ *we have*

$$\text{MutNorm}_{\alpha-2}(Ru * Ev * S, Fu * Ev * S)$$

and

$$\text{MutNorm}_{\alpha-2}(R * uE * vS, R * uE * vH).$$

The first assertion follows from 2.6 and *II.6.14*. Since $Pu*E*vQ \in \text{Reg}(\alpha - 1, Y)$, we get by *II.7.1* and *II.7.2* that $Ru * E * vS \in \text{ComReg}(\alpha - 2, RuEvS)$. Relations (7) and (8) follow from *II.7.4*.

2.22. *If* $r \geqslant 9$ *and a generating occurrence* $P * E * Q$ *of rank* α *in a word* $Y \in \text{Int}(X, \alpha, A)$ *of type I contains an occurrence* $W \in \text{Norm}(\alpha - 1, Y, r)$, *then* $P * E * Q$ *contains a supporting kernel of rank* $\alpha - 1$ *having not less than* r *segments.*

Since final supporting kernels of rank $\alpha - 1$ of the occurrence $P * E * Q$ are stable in a reversal of any active kernel of rank $\alpha - 1$, it follows from 2.5, *II.7.1* and *III.2.4* that W is also stable in these reversals. If W is compatible with an initial (or final) supporting kernel V of rank $\alpha - 1$ of $P * E * Q$, by definition I.4.24 it is contained in V and thus $l_{\alpha-1}(V) \geqslant r$. In the contrary case, *5.7* gives that $P * E * Q$ contains a maximal normalised continuation U of W, and by 2.18, U is a supporting kernel of rank $\alpha - 1$.

§ 3. Minimal and Elementary Periods of Rank α

In this section we shall consider certain properties of generating occurrences in integral words of rank α with minimal periods in rank α.

3.1. *If the period* A *of a word* $X \in \text{Per}(\alpha, A)$ *is not minimal in rank* α, *then there is a word* $X_1 \in \text{Per}(\alpha, B)$ *such that*

$$\rho_{\alpha,B}(X_1) < \rho_{\alpha,A}(X) \tag{1}$$

and, for some generating occurrences U *and* U_1 *of rank* α *in* X *and* X_1,

$$l_{\alpha,A}(U) = 3 \quad \text{and} \quad \text{MutNorm}_{\alpha-1}(U, U_1). \tag{2}$$

Suppose that the period A is not minimal in rank α. Then by definition I.4.9, there are generating occurrences V_1 and V_2 in some words $Y_1 \in \text{Int}(X_1, \alpha, B)$ and $Y_2 \in \text{Int}(X_2, \alpha, A)$ respectively such that

$$\rho_{\alpha,B}(X_1) < \rho_{\alpha,A}(X_2),\ l_{\alpha,A}(V_2) \geqslant 3,\ \text{MutNorm}_{\alpha-1}(V_1, V_2).$$

Then relation (1) is satisfied, by 1.17. Set

$$W_1 \rightleftharpoons f_{\alpha-1}(V_1;\ Y_1, X_1),\ W_2 \rightleftharpoons f_{\alpha-1}(V_2;\ Y_2, X_2)\,,$$

and let U_2 be a start of W_2 such that $l_{\alpha,A}(U_2) = 3$ and $U_2 \in \mathrm{Reg}(\alpha - 1, X_2)$. Since $\mathrm{MutNorm}_{\alpha-1}(W_1, W_2)$, by *IV.3.7* there exists a start U_1 of the occurrence W_1 such that $\mathrm{MutNorm}_{\alpha-1}(U_1, U_2)$. For the desired occurrence U we take some occurrence $U \in \mathrm{Inn}(X, A)$ corresponding in phase to U_2 in the sense of I.2.6. Then $\mathrm{MutNorm}_{\alpha-1}(U_2, U)$ by 1.8, so that $\mathrm{MutNorm}_{\alpha-1}(U_1, U)$. Clearly, $l_{\alpha,A}(U) = l_{\alpha,A}(U_2) = 3$.

3.2. *If the period A of the word $X \in \mathrm{Per}(\alpha, A)$ is minimal in rank α, then A is a simple word.*

Let us suppose that A is not simple, that is, $A \overset{\circ}{=} D^r$ for some $r > 1$. Then $X \in \mathrm{Per}(\alpha, D)$ by 1.14. Clearly, $\rho_{\alpha,D}(X) < \rho_{\alpha,A}(X)$. Let W be some normal generating occurrence in X relative to the period A, where $l_{\alpha,A}(W) \geqslant 3$. Since W is a generating occurrence relative to D and $\mathrm{MutNorm}_{\alpha-1}(W, W)$, definition I.4.9 gives that the period A is not minimal in rank α.

3.3. *If $\alpha > 1$, the word $X \in \mathrm{Per}(\alpha, A)$ is of type I and the period A is minimal in rank $\alpha - 1$, then it is minimal in rank α.*

Suppose that A is not minimal in rank α. Then, by 3.1, there exist generating occurrences U and U_1 in words X and $X_1 \in \mathrm{Per}\,(\alpha, B)$ such that relations (1) and (2) are satisfied. By I.4.31, there exists a word $Y_1 \overset{\alpha-1}{\sim} X_1$ such that $\mathrm{Bas}(U_2) \overset{\circ}{=} \mathrm{Bas}(U)$ for $U_2 \rightleftharpoons f_{\alpha-1}(U_1; X_1, Y_1)$. By 2.17, there exists a word $Y_2 \overset{\alpha-1}{\sim} Y_1$ such that the word $U_3 \rightleftharpoons f_{\alpha-1}(U_2;\ Y_1, Y_2)$ satisfies the relations

$$Y_2 \overset{\alpha-2}{\sim} X_1,\ \mathrm{Bas}(U_3) \overset{\circ}{=} \mathrm{Bas}(U),\ f_{\alpha-2}(U_1; X_1, Y_2) = U_3\,. \tag{3}$$

Consequently, U_3 and U may be considered both as generating occurrences of rank α in the words $Y_2 \in \mathrm{Int}(X, \alpha, B)$ and $X \in \mathrm{Int}(X, \alpha, A)$ respectively, and as generating occurrences of rank $\alpha - 1$ in the same words $Y_2 \in \mathrm{Int}(X, \alpha - 1, B)$ and $X \in \mathrm{Int}(X, \alpha - 1, A)$.

It follows from (2) that $\mathrm{MutNorm}_{\alpha-1}(U, U_3)$ for $U_3 = f_{\alpha-1}(U_1; X_1, Y_2)$, whence by (3) and *IV.3.9* we have

$$\mathrm{MutNorm}_{\alpha-2}(U, U_3)\,. \tag{4}$$

By (1), (2), *IV.3.3* and 2.9 we have $l_{\alpha,B}(U_3) \geqslant 3$. By 1.17 and 2.9 there exist starts W and W_3 of U and U_3 such that

$$\partial_{\alpha-1}(W) = \rho_{\alpha,A}(X) + 1,\quad \partial_{\alpha-1}(W_3) = \rho_{\alpha,B}\,(X_1) + 1\,, \tag{5}$$

and starts W' and W_3' of these same occurrences U and U_3 such that

$$\partial_{\alpha-2}(W') = \rho_{\alpha-1,A}(X) + 1,\quad \partial_{\alpha-1}(W_3') = \rho_{\alpha-1,B}(X_1) + 1\,. \tag{6}$$

Clearly, the final kernel V' of rank $\alpha - 2$ of W' is a start of the final kernel V of rank

$\alpha - 1$ of W. In a similar way, the final kernel V_3' of rank $\alpha - 2$ of W_3' is a start of the final kernel V_3 of rank $\alpha - 1$ of V_3. In this the transition from the word X_1 to Y_2 requires the use of *IV.2.18(b)* and *(i)*.

Since $\mathrm{MutNorm}_{\alpha-1}(U, U_3)$, it follows from *IV.3.6* and *IV.3.3* that

$$\partial_{\alpha-1}(W_3) = \partial_{\alpha-1}(\phi(W_3;\ U_3,\ U)),$$

and similarly, it follows from (4) that

$$\partial_{\alpha-2}(W_3') = \partial_{\alpha-2}(\phi(W_3';\ U_3,\ U))\,. \tag{7}$$

By (1) and (5),

$$\partial_{\alpha-1}(W) > \partial_{\alpha-1}(W_3) = \partial_{\alpha-1}(\phi(W_3;\ U_3,\ U))\,.$$

Then $\phi(V_3;\ U_3,\ U) < V$, whence by *IV.1.2* in rank $\alpha - 2$ it follows that $\phi(V_3';\ U_3,\ U) < V'$, that is,

$$\partial_{\alpha-2}(W') > \partial_{\alpha-2}(\phi(W_3';\ U_3,\ U))\,. \tag{8}$$

By (6), (7) and (8),

$$\rho_{\alpha-1,B}(X_1) < \rho_{\alpha-1,A}(X)\,. \tag{9}$$

By (2), $l_{\alpha-2,A}(U) \geqslant 3$. It therefore follows from (4) and (9) that the period A is not minimal in rank $\alpha - 1$.

3.4. *For $i = 1, 2$ let V_i be generating occurrences of rank α in words $Y_i \in \mathrm{Int}(X_i, \alpha, A_i)$, where the A_i are minimal periods of rank α and $l_{\alpha,A_i}(V_i) \geqslant 3$. If $\mathrm{MutNorm}_{\alpha-1}(V_1, V_2)$, then $\rho_{\alpha,A_1}(X_1) = \rho_{\alpha,A_2}(X_2)$.*

Indeed, $\rho_{\alpha,A_1}(X_1) \leqslant \rho_{\alpha,A_2}(X_2)$ because of the minimality of the period A_1, while $\rho_{\alpha,A_2}(X_2) \leqslant \rho_{\alpha,A_1}(X_1)$ because of the minimality of A_2.

3.5. *Let A be an elementary period of rank α. Then A is a simple word and all of its cyclic shifts D are elementary periods of rank α such that* $\mathrm{El}(\alpha, D) = \mathrm{El}(\alpha, A)$.

Simplicity of A follows from 3.2. Let D be a cyclic shift of A. Then $\mathrm{Per}\,(\alpha, D) = \mathrm{Per}\,(\alpha, A)$ by 1.1, and $\rho_{\alpha,D}(X) = \rho_{\alpha,A}(X)$ for all $X \in \mathrm{Per}(\alpha, D)$ by 1.17. Since $\partial(D) = \partial(A)$, it follows that $l_{\alpha,D}(W) = l_{\alpha,A}(W)$ for every generating occurrence W of rank α in a word $Y \in \mathrm{Int}(X, \alpha, A)$. Consequently, by definitions I.4.9 and I.4.10, the minimality of the period A is equivalent to that of D, and similarly A is elementary if and only if D is elementary. The equality $\mathrm{El}(\alpha, D) = \mathrm{El}(\alpha, A)$ is also clear.

3.6. *If $P * E * Q$ is a generating occurrence in a word $Y \in \mathrm{Int}\,(X, \alpha, A)$ with minimal but not elementary period A of rank α such that $l_{\alpha,A}(P * E * Q) \geqslant 9$, then there is a generating occurrence $R * D * S$ in some word $Y' \in \mathrm{Int}(X', \alpha, B)$ with minimal period B of rank α such that $E \mathrel{\overset{\circ}{=}} FDG$ for some F and G, and*

$$\mathrm{MutNorm}_{\alpha-1}(R * D * S,\ PF * D * GQ),\ \rho_{\alpha,B}(X') < \rho_{\alpha,A}(X)\,,$$

$$l_{\alpha,B}(R * D * S) \geqslant 9 \text{ and } l_{\alpha,A}(PF * D * GQ) < 3 .$$

Since A is an elementary period in rank α, there exist generating occurrences W_1 and W_2 in some integral words $Y_1 \in \text{Int}(X_1, \alpha, A)$ and $Y_2 \in \text{Int}(X', \alpha, B)$ such that the period B is minimal in rank α,

$$l_{\alpha,B}(W_2) \geqslant 9, \text{ MutNorm}_{\alpha-1}(W_1, W_2) \text{ and } \rho_{\alpha,B}(X') \neq \rho_{\alpha,A}(X_1) .$$

Then $\rho_{\alpha,B}(X') < \rho_{\alpha,A}(X_1)$ since B is a minimal period, and so $l_{\alpha,A}(W_1) < 3$ since A is a minimal period.

Set

$$U \rightleftharpoons f_{\alpha-1}(P * E * Q;\, Y, X) \text{ and } U_1 \rightleftharpoons f_{\alpha-1}(W_1;\, Y_1, X_1) .$$

Since $l_{\alpha,A}(U) \geqslant 9$ and $l_{\alpha,A}(U_1) < 3$ by 2.13, there is an occurrence U_2 contained in U that corresponds in phase to U_1 in the sense of I.2.6. Then by 1.8 we have $\text{MutNorm}_{\alpha-1}(U_2, U_1)$, so that $\text{MutNorm}_{\alpha-1}(U_2, W_2)$. The occurrence $U' \rightleftharpoons f_{\alpha-1}(U_2;\, X, Y)$ is contained in $P * E * Q$, that is, $U' = PF * D * GQ$, where $FDG \mathrel{\overset{\circ}{=}} E$. Further, by 2.14 there exists a generating occurrence W' in some word $Y' \in \text{Int}(X', \alpha, B)$ such that $\text{Bas}(W') \mathrel{\overset{\circ}{=}} \text{Bas}(U') \mathrel{\overset{\circ}{=}} D$, $\text{MutNorm}_{\alpha-1}(W', U')$ and $l_{\alpha,B}(W') = l_{\alpha,B}(W_2) \geqslant 9$. Clearly, W' is the required occurrence $R * D * S$.

3.7. *If $X \in \text{Per}(\alpha, A)$, where the period A is minimal but not elementary in rank α, then there exists an occurrence $V \in \text{Inn}(X, A)$ and a generating occurrence W in some word $Y \in \text{Int}(X', \alpha, B)$ with elementary period B such that*

$$\begin{gathered} \text{Bas}(V) \mathrel{\overset{\circ}{=}} \text{Bas}(W), \text{MutNorm}_{\alpha-1}(V, W) , \\ l_{\alpha,B}(W) \geqslant 9,\, l_{\alpha,A}(V) < 3,\, \rho_{\alpha,B}(X') < \rho_{\alpha,A}(X) . \end{gathered} \tag{10}$$

Since $X \in \text{Int}(X, \alpha, A)$, by 3.6 there exists an occurrence $V \in \text{Inn}(X, A)$ such that relation (10) is satisfied for some generating occurrence W in some word $Y \in \text{Int}(X', \alpha, B)$ with period B that is minimal in rank α. If the period B is not elementary in rank α, then by 3.6 there exist an occurrence V_1 contained in W and a generating occurrence W_1 in some word $Y_1 \in \text{Int}(X_1, \alpha, C)$ with minimal period C such that

$$\begin{gathered} \text{Bas}(W_1) \mathrel{\overset{\circ}{=}} \text{Bas}(V_1),\ \text{MutNorm}_{\alpha-1}(W_1, V_1) , \\ l_{\alpha,C}(W_1) \geqslant 9 \text{ and } \rho_{\alpha,C}(X_1) < \rho_{\alpha,B}(X') . \end{gathered}$$

Then $\text{MutNorm}_{\alpha-1}(W_1, \phi(V_1;\, W, V))$ by *IV.3.6*. Consequently, if we choose a word X' for which the parameter $\rho_{\alpha,B}(X')$ takes the least value, the period B will be elementary in rank α.

3.8. *Let $W \rightleftharpoons P * E * Q$ and $W' \rightleftharpoons R * E * S$ be generating occurrences of rank α in words $Y \in \text{Int}(X, \alpha, A)$ and $Y' \in \text{Int}(X', \alpha, B)$ respectively, such that at least one of them starts with a supporting kernel of rank $\alpha - 1$, at least one ends with a supporting kernel of rank $\alpha - 1$, and*

$$\rho_{\alpha,A}(X) = \rho_{\alpha,B}(X') . \tag{11}$$

If in addition W contains an occurrence $W_1 \in \mathrm{Reg}(\alpha - 1, Y)$ such that $l_{\alpha,A}(W_1) \geqslant 2$ and $\mathrm{MutNorm}_{\alpha-1}(W_1, \phi(W_1; W, W'))$, then

$$\mathrm{MutNorm}_{\alpha-1}(W, W') \quad and \quad l_{\alpha,A}(W) = l_{\alpha,A}(W') . \tag{12}$$

Set $W_1 = PF * G * HQ$, where $FGH \mathrel{\underline{\circ}} E$. The proof goes by induction on $\partial(FH)$. When $\partial(FH) = 0$, we have $W_1 \mathrel{\underline{\circ}} W$. Then the first of the relations in (12) follows by hypothesis, and the second follows from the first in view of 2.10.

Suppose that $\partial(FH) > 0$. Without loss of generality, we may assume that F is non-empty and that W begins with a supporting kernel of rank $\alpha - 1$. Let V_1 be an initial supporting kernel of rank $\alpha - 1$ of the occurrence W_1, V a supporting kernel of rank $\alpha - 1$ of Y that is near as possible to V_1 on its left, and U_1 the union of V and V_1. Since F is non-empty, U_1 is contained in W.

Let V_1' be an initial supporting kernel of rank $\alpha - 1$ of the occurrence $W_2 \rightleftharpoons \phi(W_1; W, W')$, V' a supporting kernel of rank $\alpha - 1$ of Y' that is near as possible to V_1' on its left, and U_1' the union of V' and V_1'. We may not yet assert that U_1' is contained in W', since we do not know whether there is a kernel of rank $\alpha - 1$ contained in W' and situated to the left of W_2.

Since $\partial_{\alpha-1}(U_1) = 2$ and $l_{\alpha,A}(W_1) \geqslant 2$, there exists an occurrence U_2 contained in W_1 such that $\mathrm{Corr}_{\alpha,A}(U_1, U_2)$. We denote by V_2 the final kernel of rank $\alpha - 1$ of U_2. Then $\mathrm{Corr}_{\alpha,A}(V_1, V_2)$. Set

$$V_2' \rightleftharpoons \phi(V_2; W_1, W_2) \text{ and } U_2' \rightleftharpoons \phi(U_2; W_1, W_2) .$$

Since $\mathrm{Corr}_{\alpha,A}(V_1, V_2)$ and relation (11) holds, we have $\mathrm{Corr}_{\alpha,B}(V_1', V_2')$ by 2.10. Then $\mathrm{Corr}_{\alpha,B}(U_1', U_2')$ by 2.9, since $\partial_{\alpha-1}(U_1') = 2 = \partial_{\alpha-1}(U_2')$. $\mathrm{MutNorm}_{\alpha-1}(U_2, U_2')$ follows from $\mathrm{MutNorm}_{\alpha-1}(W_1, W_2)$, because of *IV.3.6*. By 2.2, $\mathrm{MutNorm}_{\alpha-1}(U_1, U_2)$ and $\mathrm{MutNorm}_{\alpha-1}(U_1', U_2')$. Thus $\mathrm{MutNorm}_{\alpha-1}(U_1, U_1')$. Since $\mathrm{Bas}(V_1') \mathrel{\underline{\circ}} \mathrm{Bas}(V_1)$ by *IV.3.3*, one of the words $\mathrm{Bas}(U_1)$ and $\mathrm{Bas}(U_1')$ is an end of the other. But then $\mathrm{Bas}(U_1') \mathrel{\underline{\circ}} \mathrm{Bas}(U_1)$ by *IV.3.10*, that is, $U_1' = \phi(U_1; W, W')$.

Let W_3 be the union of the occurrences U_1 and W_1, and W_4 the union of U_1' and W_2. Since $W_4 = \phi(W_3; W, W')$, we have $\mathrm{MutNorm}_{\alpha-1}(W_3, W_4)$ by *IV.3.25*, whence it follows by the induction hypothesis that (13) is satisfied.

3.9. *Let $W \rightleftharpoons P * E * Q$ and $W' \rightleftharpoons R * E * S$ be generating occurrences in words $Y \in \mathrm{Int}(X, \alpha, A)$ and $Y' \in \mathrm{Int}(X', \alpha, B)$, where the periods A and B are minimal in rank α. If*

$$l_{\alpha,A}(W) \geqslant 9, \; l_{\alpha,B}(W') \geqslant 9 \text{ and } W \in \mathrm{Reg}\,(\alpha - 1, \; Y),$$

then

$$\rho_{\alpha,A}(X) = \rho_{\alpha,B}(X'), \; \mathrm{MutNorm}_{\alpha-1}(W, W') , \tag{13}$$

and, for all occurrences U and V contained in W,

$$U \in \mathrm{Ker}(\alpha - 1, Y) \Longleftrightarrow \phi(U; W, W') \in \mathrm{Ker}(\alpha - 1, Y') \tag{14}$$

$$\mathrm{Corr}_{\alpha,A}(U, V) \Longleftrightarrow \mathrm{Corr}_{\alpha,B}(\phi(U; W, W'), \phi(V; W, W')) . \tag{15}$$

By 3.2, A and B are simple words. If $\alpha = 1$, then $E \in \mathrm{Per}(A)$ and $E \in \mathrm{Per}(B)$, whence by I.2.9 it follows that B is a cyclic shift of A. The desired relations are then clear. Suppose that $\alpha > 1$.

Suppose that

$$\rho_{\alpha,A}(X) \geqslant \rho_{\alpha,B}(X') . \tag{16}$$

Let $E \overset{\circ}{=} uDv$, where the occurrence $Pu * D * vQ$ starts with the fourth left and ends with the fourth right segment of rank α of the occurrence W. Set

$$W_1 \rightleftharpoons Pu * D * vQ \text{ and } W_2 \rightleftharpoons Ru * D * vs .$$

Clearly $l_{\alpha,A}(W_1) \geqslant 3$, and $l_{\alpha,B}(W_2) \geqslant 3$ by (16). Suppose that Y is of type II. Then by 2.20, $\mathrm{MutNorm}_{\alpha-1}(W_1, W_2)$, whence by 2.3 and *IV.3.3* it follows that W_1 contains an active kernel of rank $\alpha - 1$, that is, Y' is also of type II. In a similar manner it is proved that Y is of type II if Y' is of type II.

If Y and Y' are of type I, it follows from 2.18 and *II.6.14* that for every occurrence V contained in W_1,

$$V \in \mathrm{Ker}(\alpha - 1,\ Y) \Longleftrightarrow \phi(V;\ W_1,\ W_2) \in \mathrm{Ker}(\alpha - 1,\ Y') .$$

By *II.5.5*, we have $\mathrm{MutNorm}_{\alpha-2}(V, \phi(V;\ W_1,\ W_2))$ for every kernel of rank $\alpha - 1$ contained in W_1, whence $\mathrm{MutNorm}_{\alpha-1}(V, \phi(V;\ W_1,\ W_2))$ follows from *IV.3.9*. Then by *IV.3.21* we have

$$\mathrm{MutNorm}_{\alpha-1}(W_1, W_2) . \tag{17}$$

If Y and Y' are of type II, relation (17) holds by 2.20.

Since $l_{\alpha,A}(W_1) \geqslant 3$, it follows from (17) and I.4.9 that $\rho_{\alpha,A}(X) \leqslant \rho_{\alpha,B}(X')$. Together with (16), this gives us that $\rho_{\alpha,A}(X) = \rho_{\alpha,B}(X')$.

If together with (16) we assume that $\rho_{\alpha,A}(X) \leqslant \rho_{\alpha,B}(X')$, then by analogy with the foregoing we get (17) and also the relation $\rho_{\alpha,A}(X) = \rho_{\alpha,B}(X')$.

By (17) and 3.8, $\mathrm{MutNorm}_{\alpha-1}(W, W')$. Thus we have proved (13) completely. By *IV.3.3* and 2.10 it follows from this that relations (14) and (15) are satisfied for all occurrences U and V contained in W.

3.10. *Let $W_1 = P * EF * Q$ and $W_2 = R * DG * S$ be normal generating occurrences of rank α in words $Y_1 \in \mathrm{Int}(X_1, \alpha, A)$ and $Y_2 \in \mathrm{Int}(X_2, \alpha, B)$ respectively, where the periods A and B are minimal in rank α and the words E, F, D, G are not empty. Suppose further that $P_1 * EF * Q_1$ and $R_1 * DG * S_1$ are occurrences in words $Z_1, Z_2 \in \mathscr{K}_{\alpha-1}$, where*

$$\mathrm{MutNorm}_{\alpha-1}(W_1, P_1 * EF * Q_1) \textit{ and } \mathrm{MutNorm}_{\alpha-1}(W_2, R_1 * DG * S_1).$$

If $E \overset{\circ}{=} D$, $l_{\alpha,A}(E) \geqslant 9$, $l_{\alpha,B}(D) \geqslant 9$ and neither of the words F and G is a start of the other, then $[F^{-1}, G]_0$ does not occur in any word Z in $\mathscr{R}_{\alpha-1}$.

If $F \overset{\circ}{=} G$, $l_{\alpha,A}(F) \geqslant 9$, $l_{\alpha,B}(G) \geqslant 9$ and neither of the words E and D is an end of the other, then $[D, E^{-1}]_0$ does not occur in any word in $\mathscr{R}_{\alpha-1}$.

In view of the principle of symmetry, we may restrict ourselves to a proof of the first of these assertions.

By *IV.3.14* and 2.13, we may assume that Y_1 and Y_2 also lie in $\mathscr{K}_{\alpha-1}$.

The case $\alpha = 1$ is impossible, since when $\alpha = 1$, it follows by I.2.9 that one of the words F and G is a start of the other since $E \overset{\circ}{=} D$.

Suppose that $\alpha > 1$. Without loss of generality, we may assume that the occurrence $V_1 \rightleftharpoons P * E * FQ$ ends with a kernel of rank $\alpha - 1$, that is, $V_1 \in \operatorname{Reg}(\alpha - 1, Y_1)$. Then we get from 3.9 that

$$\rho_{\alpha,A}(Y_1) = \rho_{\alpha,B}(Y_2) \text{ and } \operatorname{MutNorm}_{\alpha-1}(V_1, V_2)\,, \tag{18}$$

where $V_2 \rightleftharpoons R * E * GS$. Clearly, it is enough to consider the case where $\partial_{\alpha-1}(W_1) = \partial_{\alpha-1}(V_1) + 1$ and $\partial_{\alpha-1}(W_2) = \partial_{\alpha-1}(V_2) + 1$, that is, for $i = 1, 2$ only the final kernel of rank $\alpha - 1$ of the occurrence W_i is not contained in V_i. Let U_i be the union of the final kernels of rank $\alpha - 1$ of W_i and V_i. By (18), it follows from 2.9, 2.2, *IV.3.6* and *IV.3.1* that $\operatorname{MutNorm}_{\alpha-1}(U_1, U_2)$. Set $U_1 = PE_1 * HF * Q$. By *IV. 3.3*, we have $U_2 = RE_1 * HG * S$. By *IV.3.18* there exists a word $Y_3 \in \mathscr{K}_{\alpha-1}$ such that $Y_1 \overset{\alpha-1}{\approx} Y_3$ and

$$f_{\alpha-1}(U_1;\ Y_1, Y_3) = P_1 * HG * Q_1\,. \tag{19}$$

Since $PE_1 * H * FQ \in \operatorname{Ker}(\alpha - 1,\ Y_1)$, it follows from (19) and *IV.2.18(i)*, *IV. 2.27* that

$$f_{\alpha-1}(PE_1 * H * FQ;\ Y_1\ Y_3) = P_1 * H * GQ_1\,.$$

Set $Y \rightleftharpoons PE_1HGQ_1$. By *IV.2.33* we have $Y \in \mathscr{K}_{\alpha-1}$, $Y \overset{\alpha-1}{\approx} Y_1 \overset{\alpha-1}{\approx} Y_3$,

$$f_{\alpha-1}(PE_1 * H * FQ;\ Y_1, Y) = PE_1 * H * GQ_1 \tag{20}$$

and

$$f_{\alpha-1}(P_1 * H * GQ_1;\ Y_3,\ Y) = PE_1 * H * GQ_1\,. \tag{21}$$

By *IV.2.18(i)* and *IV.2.27*, it follows from (20) that

$$f_{\alpha-1}(P * E_1H * FQ;\ Y_1, Y) = P * E_1H * GQ_1\,. \tag{22}$$

Similarly, we get from (21) that

$$f_{\alpha-1}(P_1 * HG * Q_1;\ Y_3, Y) = PE_1 * HG * Q_1,$$

whence, using (19) we find from *IV.2.18(e)* that

$$f_{\alpha-1}(PE_1 * HF * Q; Y_1, Y) = PE_1 * HG * Q_1 . \tag{23}$$

Finally, it follows from (22) and (23) that

$$f_{\alpha-1}(P * EF * Q; Y_1, Y) = P * EG * Q_1 .$$

Then by *V.2.4*, the word

$$[F^{-1}E^{-1}P^{-1}, PEG]_0 = [F^{-1}, G]_0$$

cannot occur in any word in $\mathscr{R}_{\alpha-1}$.

§ 4. Periodisation

In general, integral and semi-integral words of rank α are not periodic. In this section we shall exhibit certain conditions ensuring that the base of a generating occurrence of rank α is a periodic word, and we investigate the properties of such generating occurrences.

4.2. *Let $P * E * Q$ be a normal generating occurrence of rank α in a word $Y \in$ Int (X, α, A) such that* $\mathrm{MutNorm}_{\alpha-1}(P * E * Q, R * E * S)$, *where $R * E * S$ is an occurrence in some word $Z \in \mathscr{M}_{\alpha-1}$.*

Then $E \ \bar{\circeq}\ D^t D_1$ for some D, where D_1 is a start of D, and for all occurrences $U, V \in$ Reg $(\alpha - 1, Y)$ *contained in $P * E * Q$,*

$$\mathrm{Corr}_{\alpha,A}(U, V) \Longleftrightarrow \mathrm{Corr}_D(U, V) . \tag{1}$$

If $l_{\alpha,A}(P * E * Q) \geqslant 2$ and $E \ \bar{\circeq}\ D^t D_1$, we shall call a word D satisfying relation (1) *an image of the period A in the occurrence $P * E * Q$*, and also an *image* of A in every occurrence $R * E * S \in$ Norm $(\alpha, RES, 9)$ whenever A is an elementary period.

We prove first of all that, for all occurrences $U, V \in$ Reg $(\alpha - 1, Y)$ contained in $P * E * Q$, $\mathrm{Bas}(U) \ \bar{\circeq}\ \mathrm{Bas}(V)$ whenever $\mathrm{Corr}_{\alpha,A}(U, V)$. In fact we have from 2.2 that $\mathrm{MutNorm}_{\alpha-1}(U, V)$ for occurrences like this, whence it follows that

$$\mathrm{MutNorm}_{\alpha-1}(\phi(U; P * E * Q, R * E * S),\ \ \phi(V; P * E * Q, R * E * S)) ,$$

by *IV.3.6* and $\mathrm{MutNorm}_{\alpha-1}(P * E * Q, R * E * S)$. By *IV.3.19* we have

$$\mathrm{Bas}(U) \ \bar{\circeq}\ \mathrm{Bas}(\phi(U; P * E * Q, R * E * S)) \ \bar{\circeq}$$
$$\bar{\circeq}\ \mathrm{Bas}(\phi(V; P * E * Q, R * E * S)) \ \bar{\circeq}\ \mathrm{Bas}(V).$$

When $l_{\alpha,A}(P * E * Q) = 1$, we may take $D \rightleftharpoons E$. Suppose that $l_{\alpha,A}(P * E * Q) = r > 1$ and that

$$V_1, V_2, V_3, \ldots, V_r \tag{2}$$

are all the left segments of the occurrence $P * E * Q$, that is, $V_i \in \mathrm{Ker}(\alpha - 1, Y)$, V_1 is a start of $P * E * Q$, $\mathrm{Corr}_{\alpha,A}(V_i, V_{i+1})$ and $V_i < V_{i+1}$. For $1 \leqslant j \leqslant r$, we let W_j denote the union of the kernels V_j and V_{j+1}.

Set $\mathrm{Bas}(W_1) = B$ and $\mathrm{Bas}(V_1) = C$. By what was proved above, we have for $1 \leqslant i \leqslant r$ and $1 \leqslant j < r$ that

$$\mathrm{Bas}(W_j) = B, \mathrm{Bas}(V_i) = C\,. \tag{3}$$

Then C is an end of B, that is, $B \mathrel{\overline{\underline{\circ}}} DC$ for some D. It follows from (3) that $E \mathrel{\overline{\underline{\circ}}} D^{r-1}CF$, where

$$V_r = PD^{r-1} * C * FQ\,.$$

Suppose that $W \rightleftharpoons PD^{r-1} * CF * Q$. Since $W \in \mathrm{Reg}(\alpha - 1, Y)$ and $l_{\alpha,A}(W) = 1$, by 2.4 and 2.9 there exists a proper start W' of the occurrence $W_{r-1} = PD^{r-2} * DC * FQ$ such that $\mathrm{Corr}_{\alpha,A}(W', W)$. Then $\mathrm{Bas}(W') \mathrel{\overline{\underline{\circ}}} CF$, that is, $DC \mathrel{\overline{\underline{\circ}}} CFH$ for some nonempty H, whence it follows easily that $CF \mathrel{\overline{\underline{\circ}}} D^iD_1$ for some $i \geqslant 0$, where D_1 is a start of D. Therefore $E \mathrel{\overline{\underline{\circ}}} D^{r+i-1}D_1$.

We prove (1) first of all for $U, V \in \mathrm{Ker}\,(\alpha - 1, Y)$.

Suppose that $V = V_i$. If $\mathrm{Corr}_{\alpha,A}(U, V)$, then $U = V_j$ for some j, so that $\mathrm{Corr}_D(U, V)$. If $\mathrm{Corr}_D(U, V)$, then $\mathrm{Bas}(U) = \mathrm{Bas}(V_i) \mathrel{\overline{\underline{\circ}}} C$. Since $\partial(CF) < \partial(DC)$, $\neg\,(V_r < U)$. Then $U = V_j$ for some $j \leqslant r$, that is, $\mathrm{Corr}_{\alpha,A}(U, V)$.

Thus we may assume that U and V are different from the kernels (2). Let V_i be the nearest of these kernels to V on the left, and V_j the nearest to U on the left. We denote by $\bar{V}$ and $\bar{U}$ the union of the occurrences V and V_i, and of U and V_j respectively.

Suppose that $\mathrm{Corr}_{\alpha,A}(U, V)$. Since $\mathrm{Corr}_{\alpha,A}(V_i, V_j)$ and $\partial_{\alpha-1}(\bar{U}) = \partial_{\alpha-1}(\bar{V})$ by 2.4, it follows from 2.9 that $\mathrm{Corr}_{\alpha,A}(\bar{U}, \bar{V})$, so that $\mathrm{Bas}(\bar{V}) \mathrel{\overline{\underline{\circ}}} \mathrm{Bas}(\bar{U})$ and $\mathrm{Bas}(V) \mathrel{\overline{\underline{\circ}}} \mathrm{Bas}(U)$ by what was proved above. Thus $\mathrm{Corr}_D(V_i, V_j)$ implies $\mathrm{Corr}_D(\bar{U}, \bar{V})$ and $\mathrm{Corr}_D(U, V)$.

Suppose that $\mathrm{Corr}_D(U, V)$ and $i < j$. Since $\mathrm{Corr}_{\alpha,A}(V_i, V_j)$, by 2.4 there exists a kernel $V' \in \mathrm{Ker}\,(\alpha - 1, Y)$ such that $V_i < V' < V_{i+1}$ and $\mathrm{Corr}_{\alpha,A}(V', U)$. Then $\mathrm{Corr}_D(V', U)$ by what has been proved, so that $\mathrm{Corr}_D(V', V)$, that is, $V' = V$.

It follows from the preceding arguments that

$$U \in \mathrm{Ker}(\alpha - 1, Y)\ \&\ \mathrm{Corr}_D(U, V) \Rightarrow V \in \mathrm{Ker}(\alpha - 1, Y)$$

for all occurrences U, V contained in $P * E * Q$.

We now prove (1) for arbitrary $U, V \in \mathrm{Reg}(\alpha - 1, Y)$. If $\mathrm{Corr}_{\alpha,A}(U, V)$, then $\mathrm{MutNorm}_{\alpha-1}(U, V)$ and $\mathrm{Bas}(U) \mathrel{\overline{\underline{\circ}}} \mathrm{Bas}(V)$. Then, on the basis of relation (1), which has been proved already for kernels, we get $\mathrm{Corr}_D(U, V)$.

Suppose that $\mathrm{Corr}_D(U, V)$. What has been proved already shows that $\mathrm{Corr}_{\alpha,A}(U_0, \phi(U_0; V, U)$ for any supporting kernel U_0 of rank $\alpha - 1$ contained in U, that is, $\phi(U_0; U, V) \in \mathrm{Ker}(\alpha - 1, Y)$. Then $\partial_{\alpha-1}(V) = \partial_{\alpha-1}(U)$ and we get $\mathrm{Corr}_{\alpha,A}(U, V)$ from 2.9.

4.2. *Suppose that the relations* $l_{\alpha,A}(W) \geqslant 2$, $\mathrm{MutNorm}_{\alpha-1}(W, V)$ *and* $\mathrm{Bas}(W) \overset{\circ}{=} \mathrm{Bas}(V)$ *hold for some generating occurrence* W *in a word* $Y \in \mathrm{Int}(X, \alpha, A)$ *and some occurrence* V *in a word* $Z \in \mathscr{M}_{\alpha-1}$. *If* D *is an image of the period* A *in the occurrence* W, *there exists a generating occurrence* W' *in some word* $Y' \in \mathrm{Int}(X, \alpha, A)$ *such that* $Y' \in \mathscr{M}_{\alpha-1}$, $l_{\alpha,A}(W') \geqslant 9$, D *is also an image of* A *in* W', *and for some start* U' *of* W', *we have*

$$\mathrm{MutNorm}_{\alpha-1}(U', W) \textit{ and } \mathrm{Bas}(U') \overset{\circ}{=} \mathrm{Bas}(W)\,.$$

Let $W_1 \rightleftharpoons f_{\alpha-1}(W; Y, X)$. Consider the first left of the occurrences $W_2 \in \mathrm{Inn}(X, A)$ such that $\mathrm{Corr}_A(W_1, W_2)$. Since $\partial(X) \geqslant 27\partial(A)$, W_2 is a start of some generating occurrence W_3, where $l_{\alpha,A}(W_3) \geqslant 9$. By 1.9 we have

$$\mathrm{MutNorm}_{\alpha-1}(W_1, W_2)\,.$$

By I.4.31, there exists a Y_1 such that $Y_1 \overset{\alpha-1}{\sim} X$ and $\mathrm{Bas}(W) \overset{\circ}{=} \mathrm{Bas}(f_{\alpha-1}(W_2; X, Y_1))$. Let $U \rightleftharpoons f_{\alpha-1}(W_2; X, Y_1)$. Since $\mathrm{MutNorm}_{\alpha-1}(U, V)$, it follows from *IV.3.14* that $Y_1 \overset{\alpha-1}{\sim} Y'$ and $\mathrm{Bas}(f_{\alpha-1}(U; Y_1, Y')) \overset{\circ}{=} \mathrm{Bas}(U)$ for some $Y' \in \mathscr{M}_{\alpha-1}$. Let $U \rightleftharpoons f_{\alpha-1}(U; Y_1, Y')$ and $W' \rightleftharpoons f_{\alpha-1}(W_3; X, Y')$. Since D is an image of A in W, it follows by $\mathrm{MutNorm}_{\alpha-1}(W_1, W_2)$ and *IV.3.5* that D is an image of A in U'. Let H be an image of A in W'. Then H is a start of $\mathrm{Bas}(U')$ and $\partial(H) = \partial(D)$, that is, $H \overset{\circ}{=} D$. Clearly, $l_{\alpha,A}(W') \geqslant 9$ and $\mathrm{MutNorm}_{\alpha-1}(U', W)$.

4.3. *Let* $W_1 \rightleftharpoons R * E * S$ *be an occurrence in some word* $Z \in \mathscr{M}_{\alpha-1}$, $W \rightleftharpoons P * E * Q$ *a generating occurrence in a word* $Y \in \mathrm{Int}(X, \alpha, A)$, *such that*

$$l_{\alpha,A}(W) \geqslant 2,\ \mathrm{MutNorm}_{\alpha-1}(W, W_1)\,, \tag{4}$$

and suppose that $E \overset{\circ}{=} D^t D_1$, $D \overset{\circ}{=} D_1 D_2$, *where* D *is an image of the period* A *in the occurrence* W. *Set*

$$X' \rightleftharpoons D_3 D^r D^t D_1 D_2 D^s D_4, \quad W_2 \rightleftharpoons D_3 D^r * D^t D_1 * D_2 D^s D_4\,,$$

where $r, s \geqslant 13$, D_3 *is an end of* D *and* D_4 *is a start of* D.

Then $X' \in \mathrm{Per}(\alpha, D)$, $X' \in \mathscr{K}_{\alpha-1}$,

$$\mathrm{MutNorm}_{\alpha-1}(W, W_2), \tag{5}$$

$$\rho_{\alpha,D}(X') = \rho_{\alpha,A}(X)\,. \tag{6}$$

If the period D *is a simple word, then it is minimal in rank* α.

The period D is minimal (elementary) in rank α if and only if A is minimal (elementary) in rank α.

Since D is an image of A in W, the relation

$$\mathrm{Corr}_{\alpha,A}(U, A) \Longleftrightarrow \mathrm{Corr}_D(U, V) \tag{7}$$

is satisfied by all regular occurrences U and V of rank $\alpha - 1$ contained in W. Since $Z \in \mathscr{M}_{\alpha-1}$, it follows from *IV.1.21* that $E \in \mathscr{K}_{\alpha-1}$. For $\alpha = 1$, $X' \in \mathscr{K}_{\alpha-1}$. If $\alpha > 1$, then since $l_{\alpha,A}(W) \geqslant 2$, it follows that some elementary 9-power of rank $\alpha - 1$ occurs in the word $D^{t-1}D_1$. Then $X' \in \mathscr{K}_{\alpha-2}$ by *II.6.13*. Suppose that $X' \notin \mathscr{K}_{\alpha-1}$. Then by *IV.1.19*, there is an occurrence in X' of some elementary $(n - 217)$-power F of rank $\alpha - 1$. By 4.2 and 2.12, $D^{t+1}D_1$ does not occur in F, that is, $\partial(F) < (t + 3)\partial(D)$. Let V be the occurrence of F that we have found in X', and V' a shift of it to the left or right by the period D. Clearly, V and V' are not compatible. By *II.6.7* and *II.5.2* we have $F \overset{\circ}{=} F_1F_2$, where $\partial(F_1) \leqslant \partial(D)$ and $l_{\alpha-1}(F_1) \geqslant n - 236$. Then F_1 occurs in Z, which is impossible because of *IV.1.18*. Thus $X' \in \mathscr{K}_{\alpha-1}$.

By 4.2 and 2.11, there exists an occurrence $V \in \mathrm{Inn}(X', D)$ such that $V \in \mathrm{MaxNorm}(\alpha - 1, X', 9)$ and $\partial(\mathrm{Bas}(V)) < 4\partial(D)$. Then $X' \in \mathrm{Per}(\alpha, D)$, by 1.15.

We first prove the remaining assertions under the additional condition $l_{\alpha,A}(W) \geqslant 9$.

Set

$$W_3 \rightleftharpoons PD^3 * D^{t-6}D_1 * (D_2D_1)^3Q, \quad W_4 \rightleftharpoons D_3D^{r+3} * D^{t-6}D_1 * (D_2D_1)^{s+3}D_4 .$$

If Y is of type II, then it follows from 2.20 that

$$\mathrm{MutNorm}_{\alpha-1}(W_3,\ W_4)\,. \tag{8}$$

Suppose that Y is of type I. Then 2.18 yields for any occurrence V contained in W_3 that

$$V \in \mathrm{Ker}\,(\alpha - 1,\ Y) \Longleftrightarrow V \in \mathrm{MaxNorm}\,(\alpha - 1,\ Y,\ 9)\,. \tag{9}$$

By 1.11, some elementary 9-power of rank $\alpha - 1$ occurs in D^2. Therefore by II.6.14 we have for such occurrences V,

$$V \in \mathrm{MaxNorm}\,(\alpha - 1, Y, 9) \Longleftrightarrow \phi(V;\, W_3, W_4) \in \mathrm{MaxNorm}(\alpha - 1, X', 9)\,. \tag{10}$$

By 2.20 and *IV.3.3*, the word X' is also of type I. By 1.13 it follows from (9) and (10) that

$$V \in \mathrm{Ker}\,(\alpha - 1,\ Y) \Longleftrightarrow \phi(V;\, W_3,\, W_4) \in \mathrm{Ker}\,(\alpha - 1,\ X').$$

In addition by 2.5, *II.5.5* and *IV.3.9* we get

$$V \in \text{Ker}(\alpha - 1, Y) \Rightarrow \text{MutNorm}_{\alpha-2}(V, \phi(V; W_3, W_4)) \Rightarrow$$
$$\Rightarrow \text{MutNorm}_{\alpha-1}(V, \phi(V; W_3, W_4)).$$

Then relation (8) is satisfied in this case, by *IV.3.21*.

By 2.2 and *IV.3.21*, the desired relation (5) follows from (7) and (8).

Relation (6) follows from (5) and (7), by 2.9, *IV.3.6* and *IV.3.3*.

Let us show that the minimality of the period D in rank α is a consequence both of the simplicity and of the minimality of the period A.

Assume that D is not minimal in rank α. Then by 3.1 there exist generating occurrences U_1 and U_2 in words $X_1 \in \text{Per}(\alpha, B)$ and X' such that

$$\rho_{\alpha,B}(X_1) < \rho_{\alpha,D}(X'), \tag{11}$$

$$l_{\alpha,D}(U_2) = 3 \text{ and MutNorm}_{\alpha-1}(U_1, U_2). \tag{12}$$

Since $l_{\alpha,D}(W_2) \geqslant 9$ by assumption, by 1.9 we may assume that U_2 is contained in W_2.

From (12) and (7) it follows that $l_{\alpha,A}(\phi(U_2; W_2, W) = 3$, and from (12) and (5) we have that $\text{MutNorm}_{\alpha-1}(U_1, \phi(U_2; W_2, W))$. Then A is not minimal in rank α, by (6) and (11). Thus, if A is minimal, then D is also minimal.

Clearly, $\text{Bas}(U_2) \overset{\circ}{=} G^h G_1$, where G is some cyclic shift of the word D, and $h \geqslant 2$. By I.4.31, there is a word Y_1 such that $Y_1 \overset{\alpha-1}{\approx} X_1$ and

$$\text{Bas}(f_{\alpha-1}(U_1; X, Y_1) \overset{\circ}{=} G^h G_1. \tag{13}$$

Set $V_1 \rightleftharpoons f_{\alpha-1}(U_1; X_1, Y_1)$. Since $\text{MutNorm}_{\alpha-1}(V_1, U_1)$, it follows by (12), (4) and (5) that $\text{MutNorm}_{\alpha-1}(V_1, \phi(U_2; W_2, W_1))$. Then $\text{Bas}(V_1) \overset{\circ}{=} H^k H_1$ by 4.1, where H is an image of the period B in V_1. Assume that D is a simple word. Then G is simple, and by I.2.9 it follows from (13) that $\partial(H) \geqslant \partial(G) = \partial(D)$. Then by 2.9, it follows from $\text{MutMorm}_{\alpha-1}(V_1, U_2)$, *IV.3.6* and *IV.3.3* that

$$\rho_{\alpha,B}(X_1) = \partial_{\alpha-1}(V_2) - 1 = \partial_{\alpha-1}(\phi(V_2; V_1, U_2)) - 1 \geqslant \rho_{\alpha,D}(X'),$$

where V_2 is some start of the occurrence V_1. But this contradicts relation (11). Thus if the period D is a simple word, it is minimal in rank α.

Assume that the period A is not minimal in rank α. Then by 3.1, we have

$$\rho_{\alpha,B}(X_1) < \rho_{\alpha,A}(X),\ l_{\alpha,A}(U') = 3 \text{ and MutNorm}_{\alpha-1}(U_1, U')$$

for some generating occurrences U' and U_1 in words X and $X_1 \in \text{Per}(\alpha, B)$ respectively. Set $W' \rightleftharpoons f_{\alpha-1}(W; Y, X)$. Since $l_{\alpha,A}(W') \geqslant 9$, we may assume that U' is contained in W'. Then the occurrence $f_{\alpha-1}(U'; X, Y)$ is contained in W. Set $U_2 \rightleftharpoons \phi(f_{\alpha-1}(U'; X, Y); W, W_2)$. It is clear that $l_{\alpha,D}(U_2) = l_{\alpha,A}(U') = 3$ and $\text{MutNorm}_{\alpha-1}(U_1, U_2)$. Thus the period A is also not minimal in rank α.

Suppose that D is elementary in rank α. Then D and A are minimal in rank α.

If A is not elementary in rank α, by 3.6 there exists an occurrence U contained in W and a generating occurrence U_1 in some word $Y_1 \in \mathrm{Int}(X_1, \alpha, B)$ with period B that is minimal in rank α, such that

$$\rho_{\alpha,B}(X_1) < \rho_{\alpha,A}(X),\ l_{\alpha,B}(U_1) \geqslant 9 \text{ and } \mathrm{MutNorm}_{\alpha-1}(U_1, U)\,.$$

Then $\mathrm{MutNorm}_{\alpha-1}(U_1, \phi(U; W, W_2)$ and D is not elementary in rank α, by (6).

Analogously, D is elementary if A is elementary.

We have therefore considered the case $l_{\alpha,A}(W) \geqslant 9$ completely. For $l_{\alpha,A}(W) < 9$, we first of all find by 4.2 a generating occurrence W' in some word $Y' \in \mathrm{Int}(X, \alpha, A)$ such that $Y' \in \bar{\mathcal{M}}_{\alpha-1}$, $l_{\alpha,A}(W') \geqslant 9$, D is an image of the period A in W', and $\mathrm{MutNorm}_{\alpha-1}(W, W_5)$ and $\mathrm{Bas}(W) \overset{\circ}{=} \mathrm{Bas}(W_5)$ for some start W_5 of the occurrence W'. Suppose that $\mathrm{Bas}(W') \overset{\circ}{=} D^l D'$, where $D \overset{\circ}{=} D'D''$. Then $D^l D' \overset{\circ}{=} D^r D_1 T$ for some T. By what has been proved already, with $X'' \rightleftharpoons D_3 D^r D^l D' D'' D^s$ and $W_6 \rightleftharpoons D_3 D^r * D^l D' * D'' D^s$, we get $X'' \in \mathrm{Per}(\alpha, D)$, $\mathrm{MutNorm}_{\alpha-1}(W', W_6)$ and all the desired assertions except (5). Then by *IV.3.6* we have $\mathrm{MutNorm}_{\alpha-1}(W_5, W_7)$, where $W_7 \rightleftharpoons D_3 D^r * D^r D_1 * T D'' D^s$. Since one of the words $D_2 D^s D_4$ and $TD''D^s$ is a start of the other, we get $\mathrm{MutNorm}_{\alpha-1}(W_7, W_2)$ by 1.8. Thus $\mathrm{MutNorm}_{\alpha-1}(W, W_2)$.

4.4. *Let* $W_1 \rightleftharpoons R_1 * E_1 E * S_1$ *and* $W_2 \rightleftharpoons R_2 * EE_2 * S_2$ *be normal generating occurrences of rank* α *in words* $Y_i \in \mathrm{Int}(X_i, \alpha, A_i)$ *with periods* A_i *that are elementary in rank* α,

$$V_1 \rightleftharpoons R_1 E_1 * E * S_1, \quad V_2 \rightleftharpoons R_2 * E * E_2 S_2$$

and

$$\rho_{\alpha,A_1}(X_1) = \rho_{\alpha,A_2}(X_2).$$

If $\mathrm{MutNorm}_{\alpha-1}(V_1, V_2)$ *and* $l_{\alpha,A_1}(V_1) \geqslant 2$, *there exists a generating occurrence* $R * E_1 E E_2 * S$ *in some word* $Y \in \mathrm{Int}(X, \alpha, D)$ *with period* D *that is elementary in rank* α, *such that*

$$\rho_{\alpha,D}(X) = \rho_{\alpha,A_1}(X_1)\,, \tag{14}$$

$$\mathrm{MutNorm}_{\alpha-1}(W_1, R * E_1 E * E_2 S) \text{ and } \mathrm{MutNorm}_{\alpha-1}(W_2, RE_1 * EE_2 * S). \tag{15}$$

First we consider the case where $Y_2 \in \bar{\mathcal{M}}_{\alpha-1}$. In this case 4.1 gives that $E \overset{\circ}{=} B^r B_1$ and $E \overset{\circ}{=} C^s C_1$, where B and C are images of the periods A_1 and A_2 in occurrences V_1 and V_2 respectively, $B \overset{\circ}{=} B_1 B_2$, $C \overset{\circ}{=} C_1 C_2$. Since $l_{\alpha,A_1}(V_1) \geqslant 2$, we have $\partial(E) > \partial(B)$. Since $\mathrm{MutNorm}_{\alpha-1}(V_1, V_2)$, it follows from 2.10 that $B \overset{\circ}{=} C$, $s = r$ and $B_1 \overset{\circ}{=} C_1$. Clearly, B is the image of period A_2 in the occurrence W_2 as well, that is,

$$EE_2 \overset{\circ}{=} B^k B_3, \text{ where } B \overset{\circ}{=} B_3 B_4\,.$$

By *IV.3.14*, there exists a word $Y_3 \in \mathscr{M}_{\alpha-1}$ such that $Y_3 \overset{\alpha-1}{\approx} Y_1$ and

$$\mathrm{Bas}(f_{\alpha-1}(V_1;\ Y_1,\ Y_3)) \mathrel{\overline{\circ}} \mathrm{Bas}(V_1) \mathrel{\overline{\circ}} E .$$

Set

$$V_3 \rightleftharpoons f_{\alpha-1}(V_1;\ Y_1,\ Y_3) \text{ and } W_3 \rightleftharpoons f_{\alpha-1}(W_1;\ Y_1,\ Y_3) . \tag{16}$$

Then for some R_3, E_3, S_3 we have

$$W_3 = R_3 * E_3E * S_3 \text{ and } V_3 = R_3E_3 * E * S_3 .$$

By 2.13, $Y_3 \in \mathrm{Int}(X_1;\ \alpha,\ A_1)$. It follows from *IV.2.27* that B is an image of the period A_1 in the occurrence V_3. By 4.1, $E_3E \mathrel{\overline{\circ}} D'D_1$, where D is an image of A_1 in W_3 and $D \mathrel{\overline{\circ}} D_1D_2$. Clearly, B is a cyclic shift of D and $E_3EE_2 \in \mathrm{Per}(D)$, where D is a left period of the word E_3EE_2, and B_4B_3 is a right period of it. Set

$$X \rightleftharpoons D^{13}E_3EE_2B_4B^{13},\quad U \rightleftharpoons D^{13} * E_3EE_2 * B_4B^{13} ,$$
$$U_2 \rightleftharpoons D^{13}E_3 * EE_2 * B_4B^{13},\quad U_3 \rightleftharpoons D^{13} * E_3E * E_2B_4B^{13} .$$

By 4.3, U_2 and U_3 are normal generating occurrences of rank α in the word $X \in \mathrm{Per}\ (\alpha, D)$, where the period D is elementary in rank α and relation (14) is satisfied,

$$\mathrm{MutNorm}_{\alpha-1}(W_2,\ U_2) \text{ and } \mathrm{MutNorm}_{\alpha-1}(W_3,\ U_3) . \tag{17}$$

From (16) and (17) it follows that

$$\mathrm{MutNorm}_{\alpha-1}(W_1,\ U_3) . \tag{18}$$

By *IV.3.4*, there exists a word X_0 such that $X \overset{\alpha-1}{\approx} X_0$ and $f_{\alpha-1}(U_3;\ X,\ X_0) = R * E_1E * S'$ for some R and S'. We set

$$Y \rightleftharpoons RE_1EE_2B_4B^{13} \text{ and } U' \rightleftharpoons D^{13}E_3 * E * E_2B_4B^{13}.$$

Since $V_1 \in \mathrm{Reg}\ (\alpha - 1,\ Y_1)$ and (18) holds, it follows from *IV.3.6*, *IV.2.18(i)* and *IV.2.27* that $U' \in \mathrm{Reg}(\alpha - 1, X)$ and $f_{\alpha-1}(U'; X, X_0) = RE_1 * E * S'$. Then by *IV.2.33* we have $X \overset{\alpha-1}{\approx} Y$ and

$$f_{\alpha-1}(U';\ X,\ Y) = RE_1 * E * E_2B_4B^{13},$$

whence, by *IV.2.27*,

$$\begin{aligned} f_{\alpha-1}(U_3; X, Y) &= R * E_1E * E_2B_4B^{13}, \\ f_{\alpha-1}(U_2; X, Y) &= RE_1 * EE_2 * B_4B^{13}, \\ f_{\alpha-1}(U; X, Y) &= R * E_1EE_2 * B_4B^{13}. \end{aligned} \tag{19}$$

Clearly, $R * E_1EE_2 * B_4B^{13}$ is the desired generating occurrence in the word $Y \in \mathrm{Int}(X, \alpha, D)$. In addition relation (15) follows easily from (17), (18) and (19).

We have considered the case where $Y_2 \in \mathscr{M}_{\alpha-1}$. In the general situation, we first of all find by *IV.3.12* a word $Z_2 \in \mathscr{M}_{\alpha-1}$ such that $Z_2 \overset{\alpha-1}{\approx} Y_2$. Set

$$f_{\alpha-1}(V_2;\ Y_2,\ Z_2) = P * F * F_2Q \tag{20}$$

and

$$f_{\alpha-1}(W_2;\ Y_2,\ Z_2) = P * FF_2 * Q. \tag{21}$$

Since $\mathrm{MutNorm}_{\alpha-1}(V_1, V_2)$, by I.4.31 there exists a Z_1 such that $Y_1 \overset{\alpha-1}{\approx} Z_1$,

$$f_{\alpha-1}(V_1;\ Y_1,\ Z_1) = P_1F_1 * F * Q_1 \tag{22}$$

and

$$f_{\alpha-1}(W_1;\ Y_1,\ Z_1) = P_1 * F_1F * Q_1. \tag{23}$$

Since $Z_i \in \mathrm{Int}(X_i, \alpha, A_i)$ and $\mathrm{MutNorm}_{\alpha-1}(P_1F_1 * F * Q_1, P * F * F_2Q)$, what has been proved already shows that there is a generating occurrence

$$R' * F_1FF_2 * S'$$

in some word $Z \in \mathrm{Int}(X, \alpha, D)$ with period D that is elementary in rank α such that relations (14) are satisfied,

$$\mathrm{MutNorm}_{\alpha-1}(P_1 * F_1F * Q_1,\ R' * F_1F * F_2S') \tag{24}$$

and

$$\mathrm{MutNorm}_{\alpha-1}(P * FF_2 * Q,\ R'F_1 * FF_2 * S'). \tag{25}$$

By (23) and (24), there exists a word Z' such that $Z \overset{\alpha-1}{\approx} Z'$ and

$$f_{\alpha-1}(R' * F_1F * F_2S';\ Z,\ Z') = R * E_1E * Q', \tag{26}$$

whence, by (22) and *IV.3.5*,

$$f_{\alpha-1}(R'F_1 * F * F_2S';\ Z,\ Z') = RE_1 * E * Q'. \tag{27}$$

Using (25), (21) and (20), we find in an analogous way a word Z'' such that $Z'' \overset{\alpha-1}{\approx} Z$,

$$f_{\alpha-1}(R'F_1 * FF_2 * S';\ Z,\ Z'') = R'' * EE_2 * S \tag{28}$$

and

$$f_{\alpha-1}(R'F_1 * F * F_2S'; Z, Z'') = R'' * E * E_2S. \tag{29}$$

By *IV.2.18(d)* and (*e*), it follows from (27) and (29) that

$$f_{\alpha-1}(RE_1 * E * Q'; Z', Z'') = R'' * E * E_2S.$$

Thus, setting $Y \rightleftharpoons RE_1EE_2S$, we get from *IV.2.33* that $Z' \overset{\alpha-1}{\approx} Y$ and

$$f_{\alpha-1}(RE_1 * E * Q'; Z', Y) = RE_1 * E * E_2S.$$

Then by *IV.2.18(i)* and *IV.2.7* we have

$$f_{\alpha-1}(R * E_1E * Q'; Z', Y) = R * E_1E * E_2S. \tag{30}$$

Analogously, it follows from the equation

$$f_{\alpha-1}(R'' * E * E_2S; Z'', Y) = RE_1 * E * E_2S$$

that

$$f_{\alpha-1}(R'' * EE_2 * S; Z'', Y) = RE_1 * EE_2 * S. \tag{31}$$

We get from relations (26), (30), (28) and (31) that

$$f_{\alpha-1}(R' * F_1FF_2 * S'; Z, Y) = R * E_1EE_2 * S. \tag{32}$$

Thus (32) is a normal generating occurrence in a word $Y \in \mathrm{Int}(X, \alpha, D)$. The desired relations (15) can be deduced easily from (24) and (25) using *IV.3.2* and *IV.3.1*.

4.5. An occurrence $P * E * Q \in \mathrm{Reg}(\alpha - 1, X)$ is said to be *periodised in rank* $\alpha - 1$ if there is an occurrence $R * E * S$ in some word $Z \in \mathscr{M}_{\alpha-1}$ such that

$$\mathrm{MutNorm}_{\alpha-1}(P * E * Q, R * E * S).$$

Clearly, every occurrence with non-empty base is periodised in rank 0.

If an occurrence is periodised in rank $\alpha - 1$, then by I.4.35 and *IV.3.9* it is periodised in every rank β with $0 \leqslant \beta < \alpha - 1$.

If an occurrence W is periodised in rank $\alpha - 1$, then by *IV.3.6* every regular occurrence of rank α contained in it is also periodised in rank $\alpha - 1$.

If $V \in \mathrm{Norm}(\alpha, X, r)$ and V is periodised in rank $\alpha - 1$, then by 4.1 and 4.3 its base is a periodic r-power of rank α with elementary period.

4.6. *If $P * E * Q$ is a normal generating occurrence of rank α in a word $Y \in$* Int(X, α, A) *such that $l_{\alpha,A}(P * E * Q) \geqslant 3$, then there exists a generating occurrence $P_1 * E * Q_1$ of rank α in some word $Y_1 \in$* Int(X_1, α, B) *such that the period B is* *minimal in rank* α, $\mathrm{MutNorm}_{\alpha-1}(P * E * Q, P_1 * E * Q_1)$ *and* $l_{\alpha,B}(P_1 * E * Q_1) \geqslant l_{\alpha,A}(P * E * Q)$.

By *IV.3.12*, there exists a word $Y' \in \mathscr{M}_{\alpha-1}$ such that $Y \overset{\alpha-1}{\sim} Y'$. Suppose that

$$f_{\alpha-1}(P * E * Q;\ Y,\ Y') = R * D'D_1 * S\,, \tag{33}$$

where $D \mathrel{\bar{\bar{\circ}}} D_1D_2$ is an image of period A in the occurrence $R * D'D_1 * S$. Set $X_1 \rightleftharpoons D^{13}D'D_1D_2D^{13}$ and $U \rightleftharpoons D^{13} * D'D_1 * D_2D^{13}$. By 4.3 we have

$$X \in \mathrm{Per}(\alpha,\ D) \text{ and } \mathrm{MutNorm}_{\alpha-1}(R * D'D_1 * S,\ U). \tag{34}$$

Then by *IV.3.4* there exists a word Y_1 such that $X_1 \overset{\alpha-1}{\sim} Y_1$ and, for some P_1 and Q_1,

$$f_{\alpha-1}(U;\, X_1,\, Y_1) = P_1 * E * Q_1\,. \tag{35}$$

Suppose that $D \mathrel{\bar{\bar{\circ}}} B^r$, where B is a simple word. Then by 1.14, $X_1 \in \mathrm{Per}(\alpha, B)$. Clearly, the period B of the word $D'D_1$ may be considered as an image of itself in the occurrence U. Then since $Y' \in \mathscr{M}_{\alpha-1}$ and (34) holds, it follows by 4.3 that the period B is minimal in rank α. Since D is an image of the period A in the occurrence $R * D'D_1 * S$ and $r > 0$, we have

$$\begin{aligned} l_{\alpha,B}(P_1 * E * Q_1) = l_{\alpha,B}(U) \geqslant l_{\alpha,D}(U) = \\ = l_{\alpha,A}(R * D'D_1 * S) = l_{\alpha,A}(P * E * Q)\,. \end{aligned}$$

The relation $\mathrm{MutNorm}_{\alpha-1}(P * E * Q, P_1 * E * Q_1)$ follows from (33), (34) and (35).

4.7. *In every semi-integral* 9*-power of rank* α *there occurs an elementary* 9*-power of rank* α.

This follows from 4.6, 3.7 and 2.14.

4.8. *If in the elementary word* E *of rank* α *there occurs no elementary* q*-power of rank* $< \alpha$, *and* $l_\alpha(E) \geqslant 3$, *then* $E \in \mathrm{Per}(A)$, *where* A *is an elementary period of rank* α.

Let $P * E * Q$ be a normal generating occurrence in a word $Y \in \mathrm{Int}(X, \alpha, A)$, where A is an elementary period of rank α. By 2.5 and *IV.1.3*, Y is of type I. If $\alpha = 1$ then $Y \mathrel{\bar{\bar{\circ}}} X$, so that $E \in \mathrm{Per}(A)$. If $\alpha > 1$, then by 2.17 we may assume that $Y \in \mathrm{Int}(X, \alpha - 1, A)$. By induction on α we get $Y \mathrel{\bar{\bar{\circ}}} X$. In the general case, $P * E * Q$ is contained in some normal generating occurrence of rank α.

4.9. *Every minimal period of rank* $\alpha > 1$ *is a minimal period of rank* $\alpha - 1$.

Suppose that $\alpha > 1$ and that $X \in \mathrm{Per}(\alpha, A)$. Then $X \in \mathrm{Per}(\alpha - 1, A)$, by 1.2. By *IV.3.12*, we can find a word $Y \in \mathscr{M}_{\alpha-2}$ such that $X \overset{\alpha-2}{\sim} Y$. Let $P * A'A_1 * Q$ be a normal generating occurrence of rank α in X and suppose that $t \geqslant 3$. By 4.1 we have $f_{\alpha-2}(P * A'A_1 * Q;\, X,\, Y) = P_1 * D'D_1 * Q_1$, where D is an image of the period A in the occurrence $P_1 * D'D_1 * Q_1$. Assume that A is not minimal in rank $\alpha - 1$. Then $D \mathrel{\bar{\bar{\circ}}} B^r$ by 4.3, where $r > 1$. On the other hand, since $f_{\alpha-1}(P * A'A_1 * Q, X, Y)$

$= P_1 * D^t D_1 * Q_1$ by *IV.2.18(b)*, the minimality of A in rank α yields that D is a simple word. We have thus arrived at a contradiction.

§ 5. Elementary Words of Rank α

In this section we shall consider certain properties of elementary words of rank α and their occurrences. Many of these properties are true for bases of arbitrary generating occurrences of rank α. However, we shall formulate them for elementary words of rank α only, since we do not need them for the general case.

5.1. For a given elementary word E of rank α and arbitrary occurrences $R * E * S$ of it, we shall be first of all interested in: the concept of pseudokernel of rank $\alpha - 1$, the relation $\mathrm{Corr}_\alpha(U, V)$ for pseudokernels, the density $\rho_\alpha(R * E * S)$ and the number $l_\alpha(R * E * S)$ of segments, which by definition I.4.11 are connected with some generating occurrence or other of the word E.

Since an elementary period of rank α is a special case of a minimal period of rank α, 3.9 gives that *for elementary 9-powers of rank α, all these concepts are independent of the choice of the original elementary period A and the corresponding generating occurrence $P * E * Q$.* Therefore elementary 9-powers of rank α may be considered without reference to the defining generating occurrence, and in addition the convention agreed in I.4.11 concerning the lowering of the index indicating the original period A causes no misunderstanding. Moreover, in those cases where some elementary word D of rank α that is not a 9-power of rank α is looked upon as a definite segment of an elementary 9-power uDv of rank α, we shall again not indicate a generating occurrence for the word D, meaning by this the generating occurrence of D connected with the word uDv. In particular, if $R * E * S$ is an occurrence of an elementary 9-power E of rank α in a word Z, then for any occurrence W in the word Z the number of segments of the occurrence $R * E * S$ contained in it does not depend on the choice of the original elementary period A for E nor on the generating occurrence $P * E * Q$.

In all remaining cases we shall consider elementary words of rank α in connection with some generating occurrence or other.

5.2. In view of the convention agreed for elementary words, assertion 2.8 may be reformulated as follows:

If E is an elementary 9-power of rank α and $E \overset{\circ}{=} E_1E_2$, then

$$l_\alpha(E_1) + l_\alpha(E_2) - 1 \leqslant l_\alpha(E) \leqslant l_\alpha(E_1) + l_\alpha(E_2) + 2\,,$$

so that, for any occurrence $V \rightleftharpoons P * E_1E_2 * Q$,

$$l_\alpha(V_1) + l_\alpha(V_2) - 1 \leqslant l_\alpha(V) \leqslant l_\alpha(V_1) + l_\alpha(V_2) + 2\,,$$

where $V_1 \rightleftharpoons P * E_1 * E_2Q$, $V_2 \rightleftharpoons PE_1 * E_2 * Q$.

5.3. *Every normal generating occurrence V of rank α in a word $Y \in \mathrm{Int}(X, \alpha, A)$ with elementary period A is a normalised occurrence of its base.*

Indeed $V \in \mathrm{Reg}(\alpha - 1, Y)$ by 2.1, and $\mathrm{MutNorm}_{\alpha-1}(V, V)$ by *IV.3.1*.

5.4. *If $W \in \mathrm{Norm}(\alpha, Z, 9)$, then $W \in \mathrm{Reg}(\alpha - 1, Z)$. In addition all kernels of rank $\alpha - 1$ of Z contained in W are pseudokernels of rank $\alpha - 1$ of W, and these are the only such pseudokernels.*

In fact, by I.4.13 we have $\mathrm{Bas}(V) \eqcirc \mathrm{Bas}(W)$ and $\mathrm{MutNorm}_{\alpha-1}(V, W)$ for some normal generating occurrence V of rank α. Then $W \in \mathrm{Reg}(\alpha - 1, Z)$. The second assertion for V follows from 2.1, and that for W from *IV.3.3*.

If $W \in \mathrm{Norm}(\alpha, Z, 9)$, then by assumption IV.3.6 every occurrence $V \in \mathrm{Reg}(\alpha - 1, Z)$ contained in W may be considered as a normalised occurrence of its base, and in addition W is a continuation of V.

5.5. *Suppose that $P * E * Q \in \mathrm{Norm}(\alpha, Z, r)$ and that $F * E * H$ is an occurrence in some word $Z_1 \in \mathscr{R}_{\alpha-1}$.*

Then for $r \geqslant 2$,

$$\mathrm{MutNorm}_{\alpha-1}(P * E * Q,\ F * E * H) \Rightarrow F * E * H \in \mathrm{Norm}(\alpha,\ Z_1,\ r)\,,$$

and for $r \geqslant 9$,

$$F * E * H \in \mathrm{Norm}(\alpha, Z_1, 9) \Rightarrow \mathrm{MutNorm}_{\alpha-1}(P * E * Q, F * E * H)\,.$$

This follows from 3.9 and the transitivity of the relation $\mathrm{MutNorm}_{\alpha-1}(V, W)$.

In particular, by 5.3 this means that whether an occurrence of an elementary 9-power of rank α is normalised or not does not depend on the choice of generating occurrence for its base (of course, provided that this generating occurrence contains not less than 9 segments).

5.6. *Let V be an occurrence of a normal elementary 9-power of rank α in a word Z, and V_1 a continuation of V. Then there is a generating occurrence W_1 of rank α in some word $Y_1 \in \mathrm{Int}(X_1, \alpha, B)$ with elementary period B such that $\mathrm{Bas}(W_1) \eqcirc \mathrm{Bas}(V_1)$ and the following relations hold for every generating occurrence W in a word $Y \in \mathrm{Int}(X, \alpha, A)$ with arbitrary elementary period A such that $\mathrm{Bas}(W) \eqcirc \mathrm{Bas}(V)$ and $l_{\alpha,A}(W) \geqslant 9$:*

$$\mathrm{MutNorm}_{\alpha-1}(W, \phi(V;\ V_1,\ W_1)),\quad l_{\alpha,A}(W) = l_{\alpha,B}(\phi(V;\ V_1,\ W_1))\,, \tag{1}$$

$$\rho_{\alpha,A}(X) = \rho_{\alpha,B}(X_1)\,. \tag{2}$$

By I.4.12, for some generating occurrence W of rank α of the word $\mathrm{Bas}(V)$ there exists a generating occurrence W_1 of the base of V_1 satisfying relations (1), whence (2) follows from 3.4 since $l_{\alpha,A}(W) \geqslant 9$. Then, for every generating occurrence W of this same base such that $l_{\alpha,B}(W) \geqslant 9$, relations (1) and (2) hold because of 3.9.

Consequently, *the continuability of an arbitrary occurrence of an elementary 9-power of rank α also does not depend on the choice of generating occurrence for its* *base*. Of course, we have in mind here continuations relative to generating occurrences of the given elementary 9-power that contain not less than 9 segments.

5.7. *Suppose that* $V \in \mathrm{Norm}(\alpha, Z, 9)$, $W \in \mathrm{Norm}(\alpha, Z, 2)$. *If* V *is contained in* W, *then* W *is a continuation of* V. *If* $\mathrm{Rel}(V, W)$ *or* $\mathrm{Rel}(V, W^{-1})$, *then the same is true already when* $l_\alpha(V) \geqslant 2$.

Let V_1 and W_1 be generating occurrences in words $Y_1 \in \mathrm{Int}(X_1, \alpha, A)$ and $Y_2 \in \mathrm{Int}(X_2, \alpha, B)$ respectively with elementary periods A and B, where

$$\mathrm{Bas}(V_1) \overset{\circ}{=} \mathrm{Bas}(V),\ \mathrm{MutNorm}_{\alpha-1}(V, V_1),\ l_{\alpha,A}(V_1) \geqslant 9\,,$$
$$\mathrm{Bas}(W_1) \overset{\circ}{=} \mathrm{Bas}(W),\ \ \mathrm{MutNorm}_{\alpha-1}(W,\ W_1).$$

Since V is contained in W, we get from $\mathrm{MutNorm}_{\alpha-1}(W, W_1)$ and *IV.3.6* that $\mathrm{MutNorm}_{\alpha-1}(V, \phi(V; W, W_1))$, and therefore that $\mathrm{MutNorm}_{\alpha-1}(V_1, \phi(V; W, W_1))$. Since $l_{\alpha,A}(V_1) \geqslant 9$, we have $\rho_{\alpha,A}(X_1) = \rho_{\alpha,B}(X_2)$ by I.4.10. Then by 2.10 we have $l_{\alpha,A}(V_1) = l_{\alpha,B}(\phi(V; W, W_1))$. Thus W is a continuation of V.

If $\mathrm{Rel}(V, W)$ or $\mathrm{Rel}(V, W^{-1})$, we may assert that $A \overset{\circ}{=} B$ or $A^{-1} \overset{\circ}{=} B$ respectively. Then the relation $\rho_{\alpha,A}(X_1) = \rho_{\alpha,A}(X_2)$ is satisfied because of 1.18. The rest of the argument is the same.

5.8. *If* $W \rightleftharpoons P * E * Q$ *and* $W_1 \rightleftharpoons R * E * S$, *then for any occurrences* U *and* V *in* W *of elementary words of rank* α,

$$\mathrm{Comp}(U, V) \Longleftrightarrow \mathrm{Comp}(\phi(U; W, W_1), \phi(V; W, W_1))\,.$$

This follows immediately from Definitions I.4.12 and I.4.14.

5.9. *Let* V *be an occurrence of an elementary* 9*-power of rank* α *in a word* Z, *suppose that* V *is contained in* U, U *is contained in* W *and the base of* U *is a normal elementary word of rank* α. *Then*

$$\mathrm{Comp}(V, W) \Rightarrow \mathrm{Comp}(V, U).$$

This follows from Definitions I.4.12 and 3.9 in view of the equation $\phi(V; U, \phi(U; W, W_1)) = \phi(V; W, W_1)$, which holds for any occurrence W_1 such that $\mathrm{Bas}(W_1) \overset{\circ}{=} \mathrm{Bas}(W)$.

The normality condition on the base of the occurrence U in assertion 5.9 is connected with the fact that, in I.4.12, the relation "U is a continuation of V" is defined only for occurrences of normal elementary words of rank α. In 5.14 this relation will be extended to occurrences of arbitrary elementary words of rank α, after which the condition indicated in 5.9 may be omitted.

5.10. *Let* V, V_1, V_2 *be occurrences of elementary words of rank* α *in a word* Z *such that* V *is contained in* V_1 *and* V_1 *contained in* V_2.

If $l_\alpha(V) \geqslant 9$ *and* V_1 *is a continuation of* V, *then*

$$\mathrm{Comp}(V, V_2) \Longleftrightarrow \mathrm{Comp}(V_1, V_2)\,.$$

If $l_\alpha(V_1)$, $l_\alpha(V_2) \geqslant 9$, *$W_1$ is a generating occurrence of rank α in a word* $Y_1 \in \mathrm{Int}(X_1, \alpha, A_1)$, $\mathrm{Bas}(W_1) \overline{\overline{\circ}}\, \mathrm{Bas}(V_1)$, *$V_2$ is related to V_1 or to* V_1^{-1}, $l_{\alpha,A_1}(\phi(V; V_1, W_1)) \geqslant 2$ *and V_2 is a continuation of V relative to the generating occurrence* $\phi(V; V_1, W_1)$, *then V_2 is a continuation of V_1 relative to W_1.*

Suppose that $l_\alpha(V) \geqslant 9$ and that V_1 is a continuation of V. By 5.6, we may assert that a generating occurrence W for the base of V is contained in a generating occurrence W_1 in a word $Y_1 \in \mathrm{Int}(X_1, \alpha, A)$ for the base of V_1, that is, we have $W = \phi(V; V_1, W_1)$. If $\mathrm{Comp}(V, V_2)$, then by 5.6 there exists a generating occurrence W_2 in some word $Y_2 \in \mathrm{Int}(X_2, \alpha, A_2)$ such that $\mathrm{Bas}(W_2) \overline{\overline{\circ}}\, \mathrm{Bas}(V_2)$, $\mathrm{MutNorm}_{\alpha-1}(W, \phi(V; V_2, W_2))$ and $l_{\alpha,A_2}(\phi(V; V_2, W_2)) = l_{\alpha,A_1}(W) \geqslant 9$. Then $\rho_{\alpha,A_1}(X_1) = \rho_{\alpha,A_2}(X_2)$ by 3.4, and from 3.8 we have

$$\mathrm{MutNorm}_{\alpha-1}(W_1, \phi(V_1; V_2, W_2)) \text{ and } l_{\alpha,A_1}(W_1) = l_{\alpha,A_2}(\phi(V_1; V_2, W_2)),$$

that is, $\mathrm{Comp}(V_1, V_2)$. By *IV.3.6* and 5.7, $\mathrm{Comp}(V, V_2)$ follows from $\mathrm{Comp}(V_1, V_2)$.

Suppose that $l_\alpha(V_1)$, $l_\alpha(V_2) \geqslant 9$ and that V_2 is related to V_1 or to V_1^{-1}. Then for the base of V_2 we can find a generating occurrence W_2 in some word $Y_2 \in \mathrm{Int}(X_2, \alpha, A_2)$ such that A_2 is either A_1 or A_1^{-1}. Then by 1.18 and the principle of symmetry, we have $\rho_{\alpha,A_1}(X_1) = \rho_{\alpha,A_2}(X_2)$. If $l_{\alpha,A_1}(\phi(V; V_1, W_1)) \geqslant 2$ and V_2 is a continuation of the occurrence V relative to the generating occurrence $\phi(V; V_1, W_1)$, then by 3.9 and *IV.3.6* we have $\mathrm{MutNorm}_{\alpha-1}(\phi(V; V_1, W_1), \phi(V; V_2, W_2))$, whence by 3.8 we again get that V_2 is a continuation of V_1 relative to the generating occurrence W_1.

5.11. *If V is an occurrence of an elementary 9-power of rank α in a word Z, then any two of its continuations are compatible.*

Let U_1, U_2 be continuations of V. If one is contained in the other, then we may quote 5.10. Thus we may assert that $U_1 < U_2$. By 5.9, the intersection U of the occurrences U_1 and U_2 is a continuation of V. By 5.10 we may assert that $U = V$, that is, $\mathrm{Bas}(V) = E$, $\mathrm{Bas}(U_1) = E_1E$ and $\mathrm{Bas}(U_2) = EE_2$ for some E, E_1, E_2. Then by 5.6 there exist generating occurrences $W \rightleftharpoons P * E * Q$, $W_1 \rightleftharpoons P_1 * E_1E * Q_1$ and $W_2 \rightleftharpoons P_2 * EE_2 * Q_2$ in words $Y \in \mathrm{Int}(X, \alpha, A)$, $Y_1 \in \mathrm{Int}(X_1, \alpha, A_1)$, $Y_2 \in \mathrm{Int}(X_2, \alpha, A_2)$ respectively with elementary periods of rank α, such that $l_{\alpha,A}(W) \geqslant 9$,

$$\rho_{\alpha,A_1}(X_1) = \rho_{\alpha,A}(X) = \rho_{\alpha,A_2}(X_2)\,,$$

$$\mathrm{MutNorm}_{\alpha-1}(W, PE_1 * E * Q_1),\ \mathrm{MutNorm}_{\alpha-1}(W, P_2 * E * E_2Q_2)\,,$$

and hence

$$\mathrm{MutNorm}_{\alpha-1}(P_1E_1 * E * Q_1,\ P_2 * E * E_2Q_2).$$

Then by 4.4, we can find a generating occurrence $R * E_1EE_2 * S$ in some word $Y' \in \mathrm{Int}(X', \alpha, D)$ with elementary period D of rank α, such that

$$\rho_{\alpha,D}(X') = \rho_{\alpha,A_1}(X_1) = \rho_{\alpha,A_2}(X_2)\,, \tag{3}$$

$$\mathrm{MutNorm}_{\alpha-1}(W_1,\ R * E_1E * E_2S),\ \mathrm{MutNorm}_{\alpha-1}(W_2,\ RE_1 * EE_2 * S)\,. \tag{4}$$

From (3) and (4) it follows by 2.10 that

$$l_{\alpha,A_1}(W_1) = l_{\alpha,D}(R * E_1E * E_2S),\ l_{\alpha,A_2}(W_2) = l_{\alpha,D}(RE_1 * EE_2 * S).$$

Hence, $\mathrm{Comp}(U_1, U_2)$.

5.12. *Let V_1, V_2, V_3 be occurrences of elementary 9-powers of rank α in a word Z. Then*

$$\mathrm{Comp}(V_1, V_2)\ \&\ \mathrm{Comp}(V_2, V_3) \Rightarrow \mathrm{Comp}(V_1, V_3).$$

We use U_1 to denote the union of V_1 and V_2 and U_2 to denote that of V_2 and V_3. Suppose that $\mathrm{Comp}(V_1, V_2)$ and $\mathrm{Comp}(V_2, V_3)$. Then by 5.11, $\mathrm{Comp}(U_1, U_2)$. By 5.10 the union U of the occurrences U_1 and U_2 is a continuation of V_1 and of V_3, from which we have $\mathrm{Comp}(V_1, V_3)$ by 5.9.

5.13. From 5.10 and 5.11 it follows that if two occurrences of normal elementary 9-powers of rank α are compatible then they have one and the same maximal continuation. Therefore *for every occurrence of a normal elementary 9-power of rank α, its maximal continuation and also its maximal continuations to the left and to the right are uniquely defined.*

We shall prove in 7.3 that, if two normalised occurrences U, V of elementary 9-powers of rank α are compatible, then their union is also normalised. It follows from this that the occurrences U and V also have one and the same maximal normalised continuation. This means also that, for each occurrence $V \in \mathrm{Norm}(\alpha, Z, 9)$, its maximal normalised continuation and its maximal normalised continuations to the left and to the right are uniquely defined.

5.14. Up until now, by virtue of I.4.12, we have only considered the relations "*U is a continuation of V*" and $\mathrm{Comp}(U, V)$ for occurrences of normal elementary words of rank α. We shall now look at these relations for occurrences of arbitrary elementary words of rank α.

If V is an occurrence of an arbitrary elementary word of rank α, then we denote its maximal normal part by $\hat{V}$, that is, a normal occurrence of an elementary word of rank α contained in V that cannot be continued to the left or to the right without going outside V. It is clear that $\hat{V}$ is contained in $\hat{W}$ if V is contained in W. It follows from 5.13 that, for every occurrence V of an elementary 9-power of rank α, its maximal normal part $\hat{V}$ is uniquely defined, and also that $l_\alpha(\hat{V}) = l_\alpha(V)$.

Suppose that V and W are occurrences of arbitrary elementary words of rank α in some word Z. We shall say that W is a *continuation* of V, if V is contained in W and $\hat{W}$ is a continuation of $\hat{V}$ in the sense of I.4.12. As before, let $\mathrm{Comp}(U, V)$ denote the fact that the union of the occurrences U and V is a continuation of each of them, but in the wider sense this time. It is clear that, for arbitrary occurrences U and V of elementary 9-powers of rank α in a word Z, we have

$$\mathrm{Comp}(U, V) \Rightarrow \mathrm{Comp}(\hat{U}, \hat{V}). \tag{5}$$

It is easy to convince oneself that the converse implication is also true:

$$\mathrm{Comp}(\hat{U},\ \hat{V}) \Rightarrow \mathrm{Comp}(U,\ V)\,. \tag{6}$$

Indeed, suppose that $\mathrm{Comp}(\hat{U}, \hat{V})$, and let W_1 be the union of $\hat{U}$ and $\hat{V}$ and W the union of U and V. We shall prove that the base of W is an elementary 9-power of rank α and that $\hat{W} \overset{\circ}{=} W_1$.

Clearly, we may assume that $U < V$. Then $\hat{U} < \hat{V}$. Suppose that

$$\begin{aligned} &\mathrm{Bas}(\hat{U}) \overset{\circ}{=} E, \qquad &&\mathrm{Bas}(U) \overset{\circ}{=} TET_1\,,\\ &\mathrm{Bas}(\hat{V}) \overset{\circ}{=} F, \qquad &&\mathrm{Bas}(V) \overset{\circ}{=} H_1FH\,,\end{aligned}$$
$$\mathrm{Bas}(W_1) \overset{\circ}{=} EQ \overset{\circ}{=} PF\,. \tag{7}$$

From $\mathrm{Comp}(\hat{U}, \hat{V})$ it follows by 5.8 that

$$\mathrm{Comp}(*\,E*Q,\quad P*F*)\,. \tag{8}$$

By I.4.7, the words TET_1 and H_1EH occur in some normal elementary words $STET_1S_1$ and R_1H_1FHR of rank α. Then STE and FHR are normal elementary 9-powers of rank α so that we have

$$\mathrm{Comp}(*\,STE*,\ ST*E*) \text{ and } \mathrm{Comp}(*\,FHR*,\ *F*HR) \tag{9}$$

in the sense of I.4.12. From (7), (8) and (9) it follows by 5.8 and 5.12 that

$$\mathrm{Comp}(*\,STE*QHR,\ STP*FHR*),$$

that is, $STEQHR$ is a normal elementary word of rank α. Then the base $TEQH$ of the occurrence W is an elementary word of rank α and EQ is the base of the maximal normal part of the occurrence W, that is, $\hat{W} \overset{\circ}{=} W_1$.

By (5) and (6), *all the properties of the relation* $\mathrm{Comp}(U, V)$ *that were proved in* 5.8–5.12 *for occurrences of normal elementary words of rank* α *are also true for occurrences of arbitrary elementary words of rank* α. *Moreover it is now possible to omit the normality condition on the base of the occurrence* U *in* 5.9.

5.15. *Suppose that* $\mathrm{MutNorm}_{\alpha-1}(P*E*Q,\ R*E*S)$. *If* $PF*D*TQ \in \mathrm{Norm}(\alpha, PEQ, 9)$, *then* $RF*D*TS \in \mathrm{Norm}(\alpha, RES, 9)$.

In addition, if the word F *is non-empty and* $PF*D*TQ$ *has no proper normalised continuation to the left, then* $RF*D*TS$ *also has no proper normalised continuation to the left. The analogous statement regarding normalised continuability to the right holds if* T *is non-empty.*

The first assertion follows from 5.5 and *IV.3.6*.

Suppose that F is non-empty and that the occurrence $U \rightleftharpoons RF*D*TS$ has a proper normalised continuation to the left. Then, for some non-empty H, $RF \overset{\circ}{=} R_1H$ and the occurrence $U_1 \rightleftharpoons R_1*HD*TS$ is a normalised continuation of U, that is, $U_1 \in \mathrm{Reg}(\alpha-1, RES)$ and $\partial_{\alpha-1}(U_1) > \partial_{\alpha-1}(U)$. We may clearly assert that $\partial_{\alpha-1}(U_1) = \partial_{\alpha-1}(U) + 1$. Then, since $R*E*S \in \mathrm{Reg}(\alpha-1, RES)$ and F

is a non-empty word, it follows that U_1 is contained in $R * E * S$ and hence, by 5.8 and the assertion already proved, $\phi(U_1;\ R * E * S,\ P * E * Q)$, is a normalised continuation of the occurrence V to the left.

5.16. *Suppose that* $Z \overset{\alpha-1}{\simeq} Y$. *If* $V \in \mathrm{Norm}(\alpha, Z, 2)$, *then*

$$f_{\alpha-1}(V;\ Z,\ Y) \in \mathrm{Norm}(\alpha,\ Y,\ 2),\ l_{\alpha,A}(f_{\alpha-1}(V;\ Z,\ Y)) = l_{\alpha,A}(V)\,, \tag{10}$$

and, for every elementary period A of rank α,

$$\mathrm{Bas}(V) \in \mathrm{El}(\alpha,\ A) \Longleftrightarrow \mathrm{Bas}(f_{\alpha-1}(V;\ Z,\ Y)) \in \mathrm{El}(\alpha,\ A)\,. \tag{11}$$

If $V \in \mathrm{Norm}(\alpha,\ Z,\ 9)$ *and W is a normalised continuation of V, then* $f_{\alpha-1}(W;\ Z,\ Y)$ *is a normalised continuation of the occurrence* $f_{\alpha-1}(V;\ Z,\ Y)$. *For arbitrary occurrences* $U, V \in \mathrm{Norm}(\alpha, Z, 9)$,

$$\mathrm{Comp}(U,\ V) \Longleftrightarrow \mathrm{Comp}(f_{\alpha-1}(U;\ Z,\ Y), f_{\alpha-1}(V;\ Z,\ Y))\,. \tag{12}$$

If $V \in \mathrm{Norm}(\alpha,\ Z,\ 9)$ *and V has no proper normalised continuation to the left (to the right), then* $f_{\alpha-1}(V;\ Z,\ Y)$ *also has no proper normalised continuation to the left (to the right respectively), that is,*

$$V \in \mathrm{MaxNorm}(\alpha,\ Z,\ 9) \Longleftrightarrow f_{\alpha-1}(V;\ Z,\ Y) \in \mathrm{MaxNorm}(\alpha,\ Y,\ 9). \tag{13}$$

The relations (10) and (11) hold with $V \in \mathrm{Norm}(\alpha,\ Z,\ 2)$, by 2.14.

Suppose that W is a normalised continuation of the occurrence $V \in \mathrm{Norm}(\alpha,\ Z,\ 9)$. Then by *IV.2.18* the occurrence $V_1 \rightleftharpoons f_{\alpha-1}(V;\ Z,\ Y)$ is contained in the occurrence $W_1 \rightleftharpoons f_{\alpha-1}(W;\ Z,\ Y)$, where, if V is a start (or end) of W, then V_1 is a start (end respectively) of W_1. Since $V_1,\ W_1 \in \mathrm{Norm}(\alpha,\ Y,\ 9)$, it follows by 5.7 that W_1 is a continuation of V_1. Relations (12) and (13) now follow by *IV.2.18*.

5.17. *If U, V are occurrences of elementary* 9*-powers of rank* α *in a word Z, then*

$$\mathrm{Comp}(U,\ V) \Rightarrow \mathrm{Rel}(U,\ V)\,.$$

Indeed, if U and V are compatible then the base of their union belongs to some set $\mathrm{El}(\alpha,\ A)$, which clearly contains also the bases of U and V.

5.18. *Suppose that* $V \in \mathrm{Norm}(\alpha,\ Z,\ 9)$ *and that* $\mathrm{MutNorm}_{\alpha-1}(V,\ W)$, *where W is an occurrence in a word Y. Then* $W \in \mathrm{Norm}(\alpha,\ Y,\ 9)$, $\mathrm{Rel}(V,\ W)$ *and* $l_\alpha(W) = l_\alpha(V)$.

By I.4.31, it is possible to find a word X such that $Y \overset{\alpha-1}{\simeq} X$ and $\mathrm{Bas}(f_{\alpha-1}(W;\ Y, X)) \overset{\circ}{=} \mathrm{Bas}(V)$. Set $U \rightleftharpoons f_{\alpha-1}(W;\ Y,\ X)$. Since $\mathrm{MutNorm}_{\alpha-1}(V,\ U)$ then, by 5.5, $U \in \mathrm{Norm}(\alpha,\ X,\ 9)$, whence it follows by 5.16 that $W \in \mathrm{Norm}(\alpha,\ Y,\ 9)$, $\mathrm{Rel}(W,\ U)$ and $l_\alpha(W) = l_\alpha(U)$, that is, $\mathrm{Rel}(W,\ V)$ and $l_\alpha(W) = l_\alpha(V)$.

5.19. *Suppose that W and* W_1 *are normal generating occurrences of rank* α *in a word* $Y \in \mathrm{Int}\ (X,\ \alpha,\ A)$ *and that* $Y_1 \in \mathrm{Int}\ (X_1,\ \alpha,\ A)$, $l_{\alpha,A}(W) \geqslant 9$ *and* $l_{\alpha,A}(W_1)$

$\geqslant 9$. *If an occurrence* $V \in \mathrm{Reg}(\alpha - 1, Y)$ *is a start of* W *and* $l_{\alpha,A}(V) < l_{\alpha,A}(W_1)$, *then an occurrence* V_1 *may be found, contained in* W_1, *such that* $\mathrm{MutNorm}_{\alpha-1}(V, V_1)$ *and* $l_{\alpha,A}(V_1) = l_{\alpha,A}(V)$.

Set $U' \rightleftharpoons f_{\alpha-1}(V; Y, X)$ and $U_1 \rightleftharpoons f_{\alpha-1}(W_1; Y_1, X_1)$. Since $l_{\alpha,A}(U') < l_{\alpha,A}(U_1)$, U_1 contains at least one occurrence U which corresponds to U' in phase in the sense of I.2.6. Then, by 1.8, $\mathrm{MutNorm}_{\alpha-1}(U', U)$ and $l_{\alpha,A}(U) = l_{\alpha,A}(U')$. Clearly $f_{\alpha-1}(U; X_1, Y_1)$ is the desired occurrence V_1.

5.20. *The relation* $\mathrm{Rel}(E, F)$ *is transitive for elementary* 9*-powers of rank* α.

Suppose that E, F, H are elementary 9-powers of rank α, $\mathrm{Rel}(E, F)$ and $\mathrm{Rel}(F, H)$. We shall prove that $\mathrm{Rel}(E,H)$.

By I.4.15, we can find generating occurrences W and W_1 in some words $Y \in \mathrm{Int}(X, \alpha, A)$ and $Y_1 \in \mathrm{Int}(X_1, \alpha, A)$ respectively such that

$$\mathrm{Bas}(W) = F,\ l_\alpha(W) \geqslant 9,\ \mathrm{Bas}(W_1) = H,\ l_\alpha(W_1) \geqslant 9\,,$$

and generating occurrences W_2 and W_3 in some words $Y_2 \in \mathrm{Int}(X_2, \alpha, B)$ and $Y_3 \in \mathrm{Int}(X_3, \alpha, B)$ respectively such that

$$\mathrm{Bas}(W_2) = E,\ l_\alpha(W_2) \geqslant 9,\ \mathrm{Bas}(W_3) = F,\ l_\alpha(W_3) \geqslant 9\,.$$

Clearly, we may take all of these 4 generating occurrences to be normal. It follows from 3.9 that

$$\mathrm{MutNorm}_{\alpha-1}(W, W_3),\ \rho_{\alpha,B}(X_3) = \rho_{\alpha,A}(X)\,. \tag{14}$$

Let V be some start of W such that $V \in \mathrm{Reg}(\alpha - 1, Y)$, $l_{\alpha,A}(V) = 8$ and $V_3 \leftrightharpoons \phi(V; W, W_3)$. Then from (14) we have $\mathrm{MutNorm}_{\alpha-1}(V, V_3)$. By 2.10, $l_{\alpha,B}(V_3) = l_{\alpha,A}(V) = 8$. According to 5.19, we can find occurrences V_1 and V_2 contained in W_1 and W_2 respectively, such that

$$\begin{gathered}\mathrm{MutNorm}_{\alpha-1}(V, V_1),\quad \mathrm{MutNorm}_{\alpha-1}(V_2, V_3)\,,\\ l_{\alpha,A}(V_1) = 8,\ l_{\alpha,B}(V_2) = 8\,.\end{gathered} \tag{15}$$

$\mathrm{MutNorm}_{\alpha-1}(V_1, V_2)$ now follows from (15) and $\mathrm{MutNorm}_{\alpha-1}(V, V_3)$.

By *IV.3.12* there is a word $Z_1 \in \mathcal{M}_{\alpha-1}$ such that $Y_1 \overset{\alpha-1}{\approx} Z_1$. Set

$$U_1 \rightleftharpoons f_{\alpha-1}(W_1; Y_1, Z_1),\ U' \rightleftharpoons f_{\alpha-1}(V_1; Y_1, Z_1)\,.$$

Then by *IV.3.17*, there is a word $Z_2 \in \mathcal{M}_{\alpha-1}$ such that $Y_2 \overset{\alpha-1}{\approx} Z_2$ and

$$\mathrm{Bas}\,(f_{\alpha-1}(V_2; Y_2, Z_2)) \overset{\circ}{=} \mathrm{Bas}\,(U')\,.$$

Set $U'' \rightleftharpoons f_{\alpha-1}(V_2; Y_2, Z_2)$ and $U_2 \rightleftharpoons f_{\alpha-1}(W_2; Y_2, Z_2)$. Clearly, U' is contained in U_1, and U'' in U_2.

By 4.3, we can find elementary periods C and D of rank α such that

$$\operatorname{Bas}(U_1) \in \operatorname{El}(\alpha, C), \quad \operatorname{Bas}(U_2) \in \operatorname{El}(\alpha, D), \tag{16}$$

where C is an image of the period A in the occurrence U_1, and D is an image of B in U_2. Let C' and D' be images of A and B in U' and U'' respectively. It is clear that C' is a cyclic shift of C and D' is a cyclic shift of D. Since C' and D' are simple words, $\operatorname{Bas}(U') \overset{\circ}{=} \operatorname{Bas}(U'')$ and $\partial(\operatorname{Bas}(U')) > 2\operatorname{Max}(\partial(C'), \partial(D'))$, and it follows from I.2.9 that $C' \overset{\circ}{=} D'$. Consequently, D is a cyclic shift of C, and by 3.5, $\operatorname{El}(\alpha, D) = \operatorname{El}(\alpha, C)$. Then by 5.16 it follows from (16) that $E \in \operatorname{El}(\alpha, D)$ and $H \in \operatorname{El}(\alpha, D)$, that is, $\operatorname{Rel}(E, H)$.

5.21. *If B is an elementary 9-power of rank α, then the words E and E^{-1} are not related and hence we have* $\neg \operatorname{Rel}(V, V^{-1})$ *for every occurrence V with* $\operatorname{Bas}(V) = E$.

Suppose that some elementary 9-power E of rank α is related to E^{-1}, that is, there exist generating occurrences $P * E^{-1} * Q$ and $R * E * S$ in words $Y \in \operatorname{Int}(X, \alpha, A)$ and $Y_1 \in \operatorname{Int}(X_1, \alpha, A)$ respectively, where the period A is elementary and

$$l_{\alpha,A}(P * E^{-1} * Q) \geqslant 9,\ l_{\alpha,A}(R * E * S) \geqslant 9\,.$$

Then, by the principle of symmetry, $Q^{-1} * E * P^{-1}$ is a generating occurrence in the word $Y^{-1} \in \operatorname{Int}(X^{-1}, \alpha, A^{-1})$ and $l_{\alpha,A^{-1}}(Q^{-1} * E * P^{-1}) = l_{\alpha,A}(P * E^{-1} * Q) \geqslant 9$. Without loss of generality, we may assume that $R * E * S$ is a normal generating occurrence. By 3.9 we have

$$\operatorname{MutNorm}_{\alpha-1}(R * E * S, Q^{-1} * E * P^{-1})\,. \tag{17}$$

By virtue of 2.14 and *IV.2.20*, on replacing Y_1 by some word in $\mathscr{M}_{\alpha-1}$, we can ensure that Y^{-1} is also in $\mathscr{M}_{\alpha-1}$; this because of (17) and *IV.3.14*.

True, instead of E, we now have some other elementary 9-power of rank α. In order to avoid introducing new notation we shall assume by the principle of symmetry that

$$Y, Y_1 \in \mathscr{M}_{\alpha-1}\,. \tag{18}$$

By 3.5, we may assert that

$$f_{\alpha-1}(P * E^{-1} * Q;\ Y, X) = P_1 * A^t A_1 * Q_1\,,$$

where A_1 is a start of A. Suppose in addition that

$$f_{\alpha-1}(R * E * S;\ Y_1, X_1) = R_1 * A' A^r A'' * S_1\,,$$

where A' is an end of A and A'' is a start of A. Then, by 1.8, $R_1 A' * A^r A'' * S_1 \in \operatorname{Reg}(\alpha - 1, X_1)$. Since by *IV.3.3* and *IV.2.18 (j)* it follows from (17) that $\partial_{\alpha-1}(P_1 * A^t A_1 * Q) = \partial_{\alpha-1}(R_1 * A' A^r A'' * S_1)$, we have by 1.8 that $A^r A''$ is a start of $A^t A_1$, that is, for some F we have $A^t A_1 \overset{\circ}{=} A^r A'' F$ and

$$\mathrm{MutNorm}_{\alpha-1}(P_1 * A'A'' * FQ_1, R_1A' * A'A'' * S_1)\,. \tag{19}$$

By *IV.2.18* (*i*), we have

$$f_{\alpha-1}(P_1 * A'A'' * FQ_1; X, Y) = P * D * HQ \tag{20}$$

and

$$f_{\alpha-1}(R_1A' * A'A'' * S_1; X_1, Y_1) = RH_1 * D_1 * S\,, \tag{21}$$

where $DH \overset{\circ}{=} E^{-1}$ and $H_1D_1 \overset{\circ}{=} E$. From (19), (20) and (21) we get from *IV.3.2* that

$$\mathrm{MutNorm}_{\alpha-1}(P * D * HQ, RH_1 * D_1 * S)\,. \tag{22}$$

From (17) and (22) we get, by virtue of *IV.3.6* and *IV.3.4*,

$$\partial_{\alpha-1}(Q^{-1}H_1 * D_1 * P^{-1}) = \partial_{\alpha-1}(RH_1 * D_1 * S) =$$
$$\partial_{\alpha-1}(P * D * HQ) = \partial_{\alpha-1}(Q^{-1}H^{-1} * D^{-1} * P^{-1})\,.$$

Hence $D_1 \overset{\circ}{=} D^{-1}$. Then, from (22) and (17), it follows by *IV.3.6* that

$$\mathrm{MutNorm}_{\alpha-1}(P * D * HQ, Q^{-1}H^{-1} * D^{-1} * P^{-1})\,. \tag{23}$$

By (18) and *IV.3.18* we can find $Y_2 \in \mathcal{M}_{\alpha-1}$ such that $Y \overset{\alpha-1}{\eqsim} Y_2$ and

$$f_{\alpha-1}(P * D * HQ; Y, Y_2) = P_2 * D^{-1} * Q_2\,. \tag{24}$$

Since $l_\alpha(P * D * HQ) > 3$, then by 2.6, $D \overset{\circ}{=} BuC$, where

$$P * B * uCHQ,\ PBu * C * HQ \in \mathrm{Ker}(\alpha - 1, Y)\,. \tag{25}$$

Applying the symmetry principle and proposition *IV.3.3*, it follows from (23) and (24) that

$$P_2 * C^{-1} * u^{-1}B^{-1}Q_2,\quad P_2C^{-1}u^{-1} * B^{-1} * Q_2 \in \mathrm{Ker}\,(\alpha - 1, Y_2)\,. \tag{26}$$

From (24), (25) and (26) it follows by means of *IV.2.18* (*a*) and (*i*) and *IV.1.2* that

$$f_{\alpha-1}(P * B * uCHQ;\ Y, Y_2) = P_2 * C^{-1} * u^{-1}B^{-1}Q_2\,.$$

Then by *IV.2.26* a word $Y_3 \in \mathcal{M}_{\alpha-1}$ can be found such that $Y \overset{\alpha-1}{\eqsim} Y_3$,

$$f_{\alpha-1}(P * B * uCHQ, Y, Y_3) = P_3 * C^{-1} * vCHQ$$

and

$$f_{\alpha-1}(PBu * C * HQ, Y, Y_3) = P_3C^{-1}v * C * HQ .$$

Then

$$f_{\alpha-1}(P * D * HQ; Y, Y_3) = P_3 * C^{-1}vC * HQ , \tag{27}$$

whence, by the symmetry principle,

$$f_{\alpha-1}(Q^{-1}H^{-1} * D^{-1} * P^{-1}; Y^{-1}, Y_3^{-1}) = Q^{-1}H^{-1} * C^{-1}v^{-1}C * P_3^{-1} .$$

Further, by (23) and *IV.3.18*, a word $Y_4 \in \mathcal{M}_{\alpha-1}$ can be found such that $Y \overset{\alpha-1}{\approx} Y_4$ and

$$f_{\alpha-1}(P * D * HQ; Y, Y_4) = P_4 * C^{-1}v^{-1}C * Q_4 . \tag{28}$$

From (28) and (27) there follows $Y_3 \overset{\alpha-1}{\approx} Y_4$ and

$$f_{\alpha-1}(P_3 * C^{-1}vC * HQ; Y_3, Y_4) = P_4 * C^{-1}v^{-1}C * Q_4 .$$

Since in addition the kernels $P_3 * C^{-1} * vCHQ$ and $P_3C^{-1}v * C * HQ$ transform into the kernels $P_4 * C^{-1} * v^{-1}CQ_4$ and $P_4C^{-1}v^{-1} * C * Q_4$ respectively, we have $v \overset{\alpha-1}{\approx} v^{-1}$ by *V.2.2*. But this is impossible by *IV.2.36*, since v is non-empty. Hence our original assumption that E and E^{-1} were related was false.

5.22. *If* $E \in \mathrm{El}(\alpha, A)$, $D \in \mathrm{El}(\alpha, A^{-1})$, $l_\alpha(E) \geqslant 9$ *and* $l_\alpha(D) \geqslant 9$, *then* $\neg$ Rel (E,D).

Indeed, it follows that $D^{-1} \in \mathrm{El}(\alpha, A)$ whenever $D \in \mathrm{El}(\alpha,A^{-1})$. Assume that Rel$(E, D)$. Then from Rel$(E, D^{-1})$, it follows by 5.20 that Rel(D,D^{-1}), which, by 5.21, is impossible.

5.23. *If* $\mathrm{MutNorm}_{\alpha-1}(P * E * Q, F * D * H)$ *and* $P * E * Q \in \mathrm{Norm}(\alpha, Z, r)$, *then* $F * D * H \in \mathrm{Norm}(\alpha, FDH, r)$.

By *IV.3.18*, a word Z_1 can be found such that $Z \overset{\alpha-1}{\approx} Z_1$ and $f_{\alpha-1}(P * E * Q; Z, Z_1) = R * D * S$. Then by *IV.3.2* and *IV.3.1*, we have $\mathrm{MutNorm}_{\alpha-1}(R * D * S, F * D * H)$. Now we need only quote 5.16 and 5.5.

5.24. *If E is an elementary 9-power of rank α, then* $\partial(E) > 5^\alpha$.

This is clear for $\alpha = 1$, while for $\alpha > 1$, there are $\geqslant 5$ pairwise non-intersecting pseudokernels of rank $\alpha - 1$ of the word E, by 2.7. Since a base of each of these pseudokernels is an elementary 9-power of rank $\alpha - 1$, the induction assumption gives that $\partial(E) > 5.5^{\alpha-1}$.

§ 6. Local Properties of Normalisability

As was remarked in 5.5, any two normalised occurrences of one and the same elementary 9-power of rank α are mutually normalised in rank $\alpha - 1$. However, it is not always the case that two occurrences with the same base are mutually normalised in rank $\alpha - 1$, and therefore not every occurrence of a given normal elementary word of rank α is normalised. This depends on the "context" in which the given base is found. In this section we shall adduce sufficient conditions for a given regular occurrence of rank $\alpha - 1$ to transform, under change of context, into an occurrence with which it is mutually normalised. For this it turns out to suffice that identical bases of the two occurrences be separated from the changed context by the same local context which contains an elementary word of rank α with a sufficiently large number of segments. Such a local context is required only on that side away from the base where the (global) context is changed.

6.1. If $X \underset{\circ}{=} FH$, then $F * H$ will be called a *section* of the word X. Suppose that U, V are occurrences in a word X and $U \ll V$. A section $F * H$ of X is said to be *separating for the pair* (U, V) if U is contained in $* F * H$ and V is contained in $F * H *$.

By induction on the rank α, we may extend this concept to pairs of intersecting occurrences of elementary words of rank α. Suppose that U and V are intersecting occurrences of elementary 2-powers of rank α in a word X with $U < V$. The section $F * H$ of X is called *separating for the pair* (U, V) if it is separating for the pair (V_1, V_2), where V_i is the i-th left segment of rank α of V.

If W is an occurrence of some elementary word of rank α in a word X, V_1 and V_2 are pseudokernels of rank $\alpha - 1$ of the occurrence W, $\mathrm{Corr}_\alpha(V_1, V_2)$ *and $V_1 < V_2$, then a section $F * H$ of X can be found that is separating for the pair (V_1, V_2).*

This is clear for $\alpha = 1$, since pseudokernels of rank 0 do not intersect. The result is completed by induction on α.

It is clear that, for a given pair of occurrences (U, V) in a word X, there may exist several separating sections of X.

6.2. *Let $\alpha \geqslant 2$ and let $P * E * Q$ be an occurrence in a word $Z \in \mathscr{R}_{\alpha-2}$ of an elementary word E of rank α generated by an occurrence $P_1 * E * Q_1$ in a word $Y \in$* $\mathrm{Int}\,(X, \alpha, A)$ *such that*

$$\mathrm{MutNorm}_{\alpha-2}(P * E * Q, P_1 * E * Q_1)\,. \tag{1}$$

Let U, V be pseudokernels of rank $\alpha - 1$ of the occurrence $P * E * Q$ such that $U < V$ and $\mathrm{Corr}_{\alpha,A}(U, V)$, *let $\bar{U}$ be the start of the occurrence U which ends with its third right segment of rank $\alpha - 1$ and $\bar{V}$ the end of the occurrence V which starts with its third left segment.*

If the section $F * H$ of the word Z is separating for the pair (U, V), then $\bar{V}$ is contained in $F * H *$ and, for $\alpha > 2$, we have

$$\mathrm{MutNorm}_{\alpha-3}(\bar{V}, \phi(\bar{V}; F * H *, R * H *))\,, \tag{2}$$

for an arbitrary word $RH \in \mathscr{R}_{\alpha-3}$; analogously $\bar{U}$ is contained in $* F * H$, and for $\alpha > 2$ we have

$$\mathrm{MutNorm}_{\alpha-3}(\bar{U}, \phi(\bar{U}; * F * H, * F * R))\,, \tag{3}$$

for an arbitrary word $FR \in \mathcal{R}_{\alpha-3}$.

From relation (1), by virtue of proposition *IV.3.6*, it follows that $\text{MutNorm}_{\alpha-2}(U, \phi(U; P * E * Q, P_1 * E * Q_1))$. Then by *II.5.5*, $U \in \text{Norm}(\alpha - 1, Z, 9)$. Similarly we have $V \in \text{Norm}(\alpha - 1, Z, 9)$.

If U and V do not intersect, then by 6.1, U is contained in $* F * H$ and V is contained in $F * H *$. The required relations (2) and (3) then follow from (1) using *IV.3.9* and *II.2.21*.

Assume, then, that U and V do intersect. Let U_i be the i-th right segment of rank $\alpha - 1$ of the occurrence U and V_i the i-th left segment of rank $\alpha - 1$ of the occurrence V. Then the section $F * H$ is separating for the pair (V_1, V_2). By the induction assumption, the end $\bar{V}_2$ of V_2 that begins with its third left segment of rank $\alpha - 2$ is contained in $F * H *$. By *II.2.6*, $\bar{V}$ is situated to the right of the fourth left segment of the occurrence $\bar{V}_2$, that is, $\bar{V}$ is also contained in $F * H *$. If $\alpha = 3$ then relation (2) is clear. If $\alpha > 3$ the induction assumption gives for any word $RH \in \mathcal{R}_{\alpha-4}$,

$$\text{MutNorm}_{\alpha-4}(\bar{V}_2, \phi(\bar{V}_2; F * H * , R * H *)) ,$$

whence, by *II.6.9*, relation (2) follows for words $RH \in \mathcal{R}_{\alpha-3}$, as required.

We move on to the proof of (3). By 2.6, $U_2 < V_1$. By the induction assumption, the start $\bar{V}_1$ of the occurrence V_1 that ends with its third right segment of rank $\alpha - 2$ is contained in $* F * H$ and, for $\alpha < 3$ and an arbitrary word $FR \in \mathcal{R}_{\alpha-4}$,

$$\text{MutNorm}_{\alpha-4}(\bar{V}_1, \phi(\bar{V}_1; * F * H, * F * R)) . \tag{4}$$

Let U_2' and U_2'' be the first two left segments of the occurrence U_2 and let V_1', V_1'' be the first two left segments of the occurrence V_1. If $\rho_{\alpha-2}(U_1) \leqslant \rho_{\alpha-2}(V_1)$, then it follows by *II.2.9* that $U_2'' < V_1''$, since $U_2' < V_1'$. Then, since $U_3 < U_2''$, it follows that $\bar{U} < V_1''$, that is, $\bar{U}$ is contained in $* F * H$. If $\alpha > 3$, then for $FR \in \mathcal{R}_{\alpha-3}$, relation (3) follows from (4) by *II.6.9*.

Suppose that $\rho_{\alpha-2}(V_1) \leqslant \rho_{\alpha-2}(U_1)$ and that $\bar{U}_2$ is the start of U_2 that ends with its third right segment $U_2^{(3)}$ of rank $\alpha - 2$. Because $U_2 < V_1$, it follows from 2.9 that the third right segment $V_1^{(3)}$ of rank $\alpha - 2$ of the occurrence V_1 is situated to the right of $U_2^{(3)}$, that is, $\bar{U}_2 < V_1^{(3)}$. Then $\bar{U}_2$ is contained in $* F * H$.

From the fact that $V \in \text{Norm}(\alpha - 1, Z, 9)$ it follows by *II.5.4* that $V_1 \in \text{Ker}(\alpha - 2, Z)$, that is, $V_1 \in \text{Norm}(\alpha - 2, Z, 9)$. Then $\bar{V}_1 \in \text{Reg}(\alpha - 3, Z)$ and, in consequence, for $\alpha > 3$ we have by *II.7.1* that $\bar{V}_1 \in \text{ComReg}(\alpha - 4, Z)$. We have similarly for $\alpha > 3$ that $\bar{U}_2 \in \text{ComReg}(\alpha - 4, Z)$. Therefore, for $\alpha > 3$ and an arbitrary word $RH \in \mathcal{R}_{\alpha-4}$, it follows from (4) by *II.7.4* that

$$\text{MutNorm}_{\alpha-4}(\bar{U}_2, \phi(\bar{U}_2; * F * H, * F * R)) ,$$

whence we obtain relation (3) by *II.6.9*.

6.3. *Suppose that* $\alpha \geqslant 2$, *and that* $W \rightleftarrows P * E * Q$ *is an occurrence in a word* $Z \in \mathcal{R}_{\alpha-2}$ *of an elementary r-power E of rank* α, *generated by the occurrence* $P_1 * E * Q_1$ *in a word* $Y \in \text{Int}(X, \alpha, A)$, $r \geqslant 4$, *such that* $\text{MutNorm}_{\alpha-2}(P_1 * E * Q_1, W)$.

For $i = 1, 2, \ldots, r$ *denote by* U_i *the i-th left segment of* W *of rank* α; *for* $1 \leqslant j \leqslant r-1$ *denote by* W_1 *the start of* W *that ends with the sixth right segment of rank* $\alpha - 1$ *of* U_j, *and by* W_2 *the end of* W *which starts with the sixth left segment of rank* $\alpha - 1$ *of* U_{j+1}.

If the section $F * H$ *of the word* Z *is separating for the pair* (U_j, U_{j+1}), *then* W_1 *is contained in* $* F * H$, W_2 *is contained in* $F * H *$, *and moreover, for an arbitrary word* $FR \in \mathscr{R}_{\alpha-2}$ *and an arbitrary occurrence* $V \in \mathrm{Reg}(\alpha - 2, Z)$ *contained in* W_1, *we have*

$$\mathrm{MutNorm}_{\alpha-2}(V, \phi(V; * F * H, * F * R)), \tag{5}$$

and for an arbitrary word $RH \in \mathscr{R}_{\alpha-2}$ *and an arbitrary occurrence* $V \in \mathrm{Reg}(\alpha-2, Z)$ *contained in* W_1 *we have*

$$\mathrm{MutNorm}_{\alpha-2}(V, \phi(V; F * H *, R * H *)). \tag{6}$$

Let $\bar{U}_j$ be the start of U_j ending with its third right segment. By 6.2, $\bar{U}_j$ is contained in $* F * H$ and, for $\alpha > 2$ and any word $FR \in \mathscr{R}_{\alpha-2}$, we have

$$\mathrm{MutNorm}_{\alpha-2}(\bar{U}_j, \phi(\bar{U}_j; * F * H, * F * R)),$$

from which (5) follows on using *II.6.9*, since W_1 ends with the fourth right segment of $\bar{U}_j$. For $\alpha = 2$ relation (5) is clear. Relation (6) is established in a similar fashion.

6.4. *Let* $P * E * Q$ *be an occurrence in a word* $Z \in \mathscr{P}_{\alpha-1}$ *of an elementary 7-power* E *of rank* α *that is generated by an occurrence* $P_1 * E * Q_1$ *in an integral word* Y *of rank* α *such that* $\mathrm{MutNorm}_{\alpha-2}(P * E * Q, P_1 * E * Q_1)$. *Then the occurrence* W *in* Z *which starts with the fourth left segment and ends with the fourth right segment of rank* α *of* $P * E * Q$ *is stable in a* q_1*-reversal of an arbitrary occurrence* $U \in \mathrm{Norm}(\alpha - 1, Z, q_1)$, *where either* $U < P * E * Q$ *or* $P * E * Q < U$.

By the obvious symmetry, it is enough to consider the case $U < P * E * Q$. Suppose that $Z \to Z_1$ is a q_1-reversal of the occurrence $U \in \mathrm{Norm}(\alpha - 1, Z, q_1)$, where $U < P * E * Q$. By definition I.4.19 and propositions *II.5.16* and *III.1.4*, we may assert that this transition is a simple q_1-reversal of U, that is,

$$Z \mathrel{\overset{\circ}{=}} RTA^t A_1 T_1 S \text{ and } Z_1 \mathrel{\overset{\circ}{=}} RGB^h B_1 G_1 S,$$

where $A \mathrel{\overset{\circ}{=}} A_1 A_2$ is an elementary period of rank $\alpha - 1$ and the occurrence U is compatible with $RT * A^t A_1 * T_1 S$. We set

$$D \rightleftharpoons TA^t A_1 T_1 \text{ and } D' \rightleftharpoons GB^h B_1 G_1.$$

By *III.1.6*, D is an elementary word of rank $\alpha - 1$ and U is contained in $R * D * S$. Let $R_1 * D * S_1$ be a generating occurrence of the word D in an integral word Y_1 of rank $\alpha - 1$. By *II.5.5*, we have

$$\mathrm{MutNorm}_{\alpha-2}(U, \phi(U; R * D * S, R_1 * D * S_1)). \tag{7}$$

From the fact that $Y, Z \in \mathscr{R}_{\alpha-2}$, it follows by *II.6.13* that $RDS_1 \in \mathscr{R}_{\alpha-2}$. Since $U \in \text{ComReg}(\alpha - 2, Z)$ by *II.7.1*, it then follows from (7) by *II.7.5* that

$$\text{MutNorm}_{\alpha-2}(U, \phi(U; R * D * S, R * D * S_1)) . \tag{8}$$

We denote by V_i the i-th left segment of rank α of the occurrence $P * E * Q$. By hypothesis the occurrence W starts with a pseudokernel V_4 and ends with some other pseudokernel of rank $\alpha - 1$ of the occurrence $P * E * Q$. Without loss of generality, we may assume that V_1 is a start of $P * E * Q$.

By 6.1, we can find a section $F * H$ of the word Z which is separating for the pair (V_2, V_3). We shall prove in addition that H is an end of S. Assume that this is not so. Suppose that W_1 is the start of the occurrence $P * E * Q$ ending with the sixth right segment of rank $\alpha - 1$ of V_2 and let $\bar{V}_1$ be the maximal normalised continuation of V_1 *to the right*. By 2.6, $\bar{V}_1$ contains less than two segments of V_2, that is, $\bar{V}_1$ is a proper start of W_1. If $\bar{V}_1$ is contained in U, then by *II.5.7* we have Comp $(\bar{V}_1, U)$ and $\bar{V}_1$ is then an end of U. In each case it follows from $U < P * E * Q$ that $U < W_1$. By 6.3, W_1 is contained in $R * D * S$ and so

$$\text{MutNorm}_{\alpha-2}(W_1, \phi(W_1; R * D * S, R * D * S_1)) , \tag{9}$$

whence it follows by *IV.3.6* that

$$\text{MutNorm}_{\alpha-2}(V_1, \phi(V_1; R * D * S, R * D * S_1)) . \tag{10}$$

Let U_1 be the union of the occurrences U and V_1 and U_2 that of U and W_1. It follows from (8) and (10) using *II.7.1* and *II.7.3* that

$$\text{MutNorm}_{\alpha-2}(U_1, \phi(U_1; R * D * S, R * D * S_1)) ,$$

whence it follows from (9) by *IV.3.25* that

$$\text{MutNorm}_{\alpha-2}(U_2, \phi(U_2; R * D * S, R * D * S_1)) .$$

Furthermore, we get from (7) and (8) that

$$\text{MutNorm}_{\alpha-2}(\phi(U; R * D * S, R * D * S_1), \phi(U; R * D * S, R_1 * D * S_1),$$

whence if follows from *II.7.4* that

$$\text{MutNorm}_{\alpha-2}(\phi(U_2; R * D * S, R * D * S_1), \quad \phi(U_2; R * D * S; R_1 * D * S_1)) .$$

Hence we have

$$\text{MutNorm}_{\alpha-2}(U_2, \ \phi(U_2; \ R * D * S, \ R_1 * D * S_1)) ,$$

whence it follows by *II.5.15* that $\phi(\bar{V}_1; R * D * S, R_1 * D * S_1)$ is normalised and has no proper normalised continuation to the right. But this is false since this occurrence is contained in the normal generating occurrence $\phi(U_2; R * D * S, R_1 * D * S_1)$

of rank $\alpha - 1$ in the word Y_1 and is not an end of it. Hence H is an end of the word S, that is, $S \stackrel{\circ}{=} LH$ for some L. Since $W \in \mathrm{Reg}(\alpha - 2, Z)$, by 6.3 W is contained in $F * H *$ and

$$\mathrm{MutNorm}_{\alpha-2}(W, \phi(W; F * H *, RD'L * H *)),$$

that is, according to I.4.17, W is stable under the reversal $Z \to Z_1$.

6.5 *Let $W \rightleftharpoons P * E * Q$ be an occurrence in a word $Z \in \mathscr{P}_{\alpha-1}$ of an elementary 7-power E of rank α generated by an occurrence $W' \rightleftharpoons P' * E * Q'$ in a word $Y \in$* Int (X, α, A), *and suppose that $E \stackrel{\circ}{=} uDv$ and that the occurrence $Pu * D * vQ$ in Z starts with the fourth left and ends with the fourth right segment of W of rank α. If* $\mathrm{MutNorm}_{\alpha-2}(W, W')$, *then for an arbitrary occurrence V contained in $Pu * D * vQ$ we have*

$$V \in \mathrm{Ker}(\alpha - 1, Z) \Longleftrightarrow \phi(V; P * E * Q, P' * E * Q') \in \mathrm{Ker}(\alpha - 1, Y), \quad (11)$$

and if $Z \in \mathscr{R}_{\alpha-1}$ we have in addition that

$$\mathrm{MutNorm}_{\alpha-1}(Pu * D * vQ, P'u * D * vQ'). \quad (12)$$

If Y is a word of type II, then (11) and (12) hold by 2.20.

Suppose that Y is of type I. We set

$$W_1 \rightleftharpoons Pu * D * vQ \text{ and } W_2 \rightleftharpoons P'u * D * vQ'.$$

By *IV.3.6*, it follows from $\mathrm{MutNorm}_{\alpha-2}(W, W')$ that $\mathrm{MutNorm}_{\alpha-2}(W_1, W_2)$. We denote by V_i the i-th left segment and by U_i the i-th right segment of rank α of W. Since $\mathrm{MutNorm}_{\alpha-2}(W, W')$ it follows by *IV.3.6* and *II.5.5* that

$$V_i, U_i \in \mathrm{Norm}(\alpha - 1, Z, 9).$$

Let V be an arbitrary active kernel of rank $\alpha - 1$ of the word Z. Then by *IV.1.3*,

$$V \in \mathrm{Norm}(\alpha - 1, Z, q).$$

We shall prove that the kernel V is not contained in the occurrence W_1.

Assume that $V_2 < V < U_2$. Then by *II.5.15*, $\phi(V; W, W') \in \mathrm{Norm}(\alpha - 1, Y, q)$. By 2.2, 2.22 and 2.19, we can find supporting kernels V' and V'' of rank $\alpha - 1$ contained in W' such that

$$V', V'' \in \mathrm{Norm}(\alpha, Y, q),$$
$$V' < \phi(V; W, W') < V'',$$
$$\neg\, \mathrm{Comp}(V', \phi(V; W, W')), \neg\, \mathrm{Comp}(V'', \phi(V, W, W')).$$

Then $\phi(V'; W', W), \phi(V''; W', W) \in \mathrm{Norm}(\alpha - 1, Z, q)$. From *III.3.30* it follows that $\phi(V; W, W')$ is an active occurrence of rank $\alpha - 1$. By *IV.1.5*, it is compatible with some active kernel $\bar{V}$ of rank $\alpha - 1$. But by *IV.1.3*, the kernel $\bar{V}$ cannot intersect the supporting kernels $\phi(V_2; W, W')$ and $\phi(U_2; W, W')$, and by 2.5 it cannot be contained in W'. Hence V is not situated between V_2 and U_2. This means, in particular, that there are no active kernels of rank $\alpha - 1$ in W_1.

If the kernel V were compatible with V_2, then it would be contained in it, since $V_2 \in \text{MaxNorm}(\alpha - 1, Z, 9)$ by 2.18 and *II.5.15*. Then we would have $l_{\alpha-1}(V_2) \geqslant 9$, which would produce a contradiction in an analogous way to the foregoing. The proof that V is not compatible with U_2 is exactly the same. Hence either $V < V_2$ or $U_2 < V$.

We shall prove in both of these two cases that W_1 is stable in a reversal of V. We may clearly confine ourselves to considering the case $V < V_2$ and $\neg\,\text{Comp}\,(V, V_2)$.

If $V < W$, then by 6.4, W_1 is stable in a reversal of V. Suppose that V is contained in W. Then $\phi(V; W, W') \in \text{Norm}(\alpha - 1, Y, q)$ and by *IV.1.4*, $\phi(V; W, W')$ is compatible with some kernel V' of rank $\alpha - 1$ of Y. By *IV.1.7*, $l_{\alpha-1}(V') \geqslant q - 34$. Set $V_2' \rightleftharpoons \phi(V_2; W, W')$ and $V_3' \rightleftharpoons \phi(V_3; W, W')$. It follows from $\neg\,\text{Comp}(V, V_2)$ that $\neg\,\text{Comp}(V', V_2')$, that is, that $V' < V_2'$. By 2.4, a supporting kernel V'' of rank $\alpha - 1$ can be found such that

$$V' < V'' < V_3' \text{ and } \text{Corr}_{\alpha,A}(V', V'')\,.$$

Then by 2.19, $l_{\alpha-1}(V'') \geqslant q - 34$. Set $U' \rightleftharpoons \phi(V''; W', W)$. Clearly, $U' \in \text{Norm}(\alpha - 1, Z, q - 34)$. From $\neg\text{Comp}(V', V'')$ and $\neg\text{Comp}(V'', V_3')$ there follow $\neg\,\text{Comp}(V, U')$ and $\neg\,\text{Comp}(U', V_3)$. By *III.2.10*, some end U'' of the occurrence U' with $l_{\alpha-1}(U'') \geqslant q - 51$ is stable in a reversal of V. Since $\neg\,\text{Comp}(U'', V_3)$, $U'' < V_3 < W_1$. Thus by *II.7.1* and *III.2.4*, the occurrence W_1 is also stable in a reversal of V.

Thus we have proved that there are no active kernels of rank $\alpha - 1$ in W_1 and that W_1 is stable in a reversal of any active kernel V of rank $\alpha - 1$ of the word Z.

For an arbitrary occurrence U contained in W_2 we have by 2.18 that

$$U \in \text{Ker}(\alpha - 1,\ Y) \Longleftrightarrow U \in \text{MaxNorm}(\alpha - 1,\ Y,\ 9)\,. \tag{13}$$

By *II.6.14* we have

$$U \in \text{MaxNorm}(\alpha - 1,\ Y,\ 9) \Longleftrightarrow$$
$$\Longleftrightarrow \phi(U;\ W_2,\ W_1) \in \text{MaxNorm}(\alpha - 1,\ Z,\ 9)\,. \tag{14}$$

In particular, it follows from (13) and (14) that

$$V_4,\ U_4 \in \text{MaxNorm}(\alpha - 1,\ Z,\ 9)\,. \tag{15}$$

Suppose that U is contained in W_1 and that $U \in \text{Ker}(\alpha - 1,\ Z)$. Then $U \in \text{Norm}(\alpha - 1,\ Z,\ 9)$. By (15) and *II.5.7*, its maximal normalised continuation $\bar{U}$

is also contained in W_1. By *III.2.3*, $\bar{U}$ is stable in a reversal of an arbitrary active kernel of rank $\alpha - 1$ of the word Z. Then by *IV.1.5*, $\bar{U} \in \mathrm{Ker}(\alpha - 1, Z)$, that is, $\bar{U} \mathrel{\underset{\circ}{=}} U$. Hence, for an arbitrary occurrence U contained in W_1, we have

$$U \in \mathrm{Ker}(\alpha - 1, Z) \Longleftrightarrow U \in \mathrm{MaxNorm}(\alpha - 1, Z, 9)\,,$$

from which the desired relation (11) follows in view of (13) and (14). Since W_1 and W_2 do not contain active kernels of rank $\alpha - 1$ and $W_2 \in \mathrm{Reg}(\alpha - 1, Y)$, relation (12) follows for $Z \in \mathscr{R}_{\alpha-1}$ from (11) and $\mathrm{MutNorm}_{\alpha-2}(W_1, W_2)$, by *IV.3.9*.

6.6. *Let $P * E * Q$ be an occurrence of an elementary r-power of rank α in a word $Z \in \mathscr{R}_{\alpha-1}$, where $r \geqslant 9$. If $E \mathrel{\underset{\circ}{=}} uDv$, where the occurrence $Pu * D * vQ$ starts with the fifth left and ends with the fifth right segment of $P * E * Q$ of rank α, then*

$$\mathrm{MutNorm}_{\alpha-1}(Pu * D * vQ,\ Pu * D * vS)\,,$$

where $R * uDv * S$ is an arbitrary generating occurrence for E.

This result follows directly from 2.21 and 6.5.

6.7. *Let $P * E * Q$ and $R * D * S$ be occurrences of elementary* 10*-powers of rank α in the word $X \in \mathscr{R}_{\alpha-1}$ such that*

$$\neg\ \mathrm{Comp}(P * E * Q,\ R * D * S)\,.$$

Then the common part of these two occurrences contains less than 17 *segments of either of them, and if* $\mathrm{Rel}(E, D)$ *or* $\mathrm{Rel}(E, D^{-1})$, *it contains less than* 10 *segments of either of them.*

Let $P_1 * E * Q_1$ and $R_1 * D * S_1$ be the corresponding generating occurrences in words $Y_1 \in \mathrm{Int}(X_1, \alpha, A)$ and $Y_2 \in \mathrm{Int}(X_2, \alpha, B)$. According to 5.14, we may assume without loss of generality that these generating occurrences are normal.

Assume that $R * D * S$ is contained in $P * E * Q$, that is, $E \mathrel{\underset{\circ}{=}} FDH$, $R \mathrel{\underset{\circ}{=}} PF$ and $S \mathrel{\underset{\circ}{=}} HQ$. Then by 6.6, the occurrence $P_1F * D * HQ_1$ is compatible with some occurrence $V \in \mathrm{Norm}(\alpha, Y_1, l_\alpha(D) - 8)$ contained in it. If $l_\alpha(D) \geqslant 17$ (or $l_\alpha(D) \geqslant 10$ if $\mathrm{Rel}(E, D)$ or $\mathrm{Rel}(E, D^{-1})$), then, by 5.7, $P_1 * E * Q_1$ is a generating occurrence of V, that is, we have $\mathrm{Comp}(V, P_1 * E * Q_1)$, whence, by 5.10, it follows that $P_1F * D * HQ$ and $P_1 * E * Q_1$ are compatible. But this, by 5.8, implies that $\mathrm{Comp}\,(R * D * S,\ P * E * Q)$, which is false by hypothesis. Hence we may assert that the occurrences under consideration are not contained in one another, that is, one of them is to the left of the other.

Suppose, for example, that $R * D * S < P * E * Q$ and that $U \rightleftharpoons RF * G * HQ$ is their common part, that is, $D \mathrel{\underset{\circ}{=}} FG$ and $E \mathrel{\underset{\circ}{=}} GH$. If U contains $\geqslant 17$ segments of the occurrence $P * E * Q$, that is, if $l_\alpha(G) \geqslant 17$, then, by what has been proved above, both of the occurrences $P * E * Q$ and $R * D * S$ are continuations of one and the same occurrence U when considered as being generated by the occurrence $R_1F * G * S_1$. Then by 5.11, we again have $\mathrm{Comp}(R * D * S,\ P * E * Q)$, which contradicts the hypothesis.

All the remaining cases are dealt with in a similar fashion.

6.8. *Let $R * E * S$ be a normal generating occurrence of rank α in a word $Y \in \mathrm{Int}(X, \alpha, A)$ with $l_{\alpha,A}(R * E * S) \geqslant 7$, suppose that the period A is elementary in rank α, and that $P_1 * E * Q_1$ and $P_2 * E * Q_2$ are occurrences in words $Z_1 \in \mathscr{P}_{\alpha-1}$ and $Z_2 \in \mathscr{P}_{\alpha-1}$ respectively such that*

$$\mathrm{MutNorm}_{\alpha-2}(P_1 * E * Q_1, R * E * S),$$
$$\mathrm{MutNorm}_{\alpha-2}(P_2 * E * Q_2, R * E * S) \tag{16}$$

and suppose that $V_0 \in \mathrm{Act}(\alpha - 1, Z_1)$.

If $P_2 \overset{\circ}{=} P_1$ and $V_0 < V'$, where V' is the fourth right segment of rank α of the occurrence $P_1 * E * Q_1$, then

$$\phi(V_0; * P_1E * Q_1, * P_1E * Q_2) \in \mathrm{Act}(\alpha - 1, Z_1).$$

If $Q_2 \overset{\circ}{=} Q_1$ and $V' < V_0$, where V' is the fourth left segment of rank α of $P_1 * E * Q_1$, then

$$\phi(V_0; P_1 * EQ_1 *, * P_2 * EQ_1) \in \mathrm{Act}(\alpha - 1, Z_2).$$

These two assertions are symmetric. We therefore confine ourselves to proving one of them.

Suppose that $P_2 \overset{\circ}{=} P_1$, V' is the fourth right segment of $P_1 * E * Q_1$ and that $V_0 < V'$. If $\mathrm{Comp}(V_0, V')$, then by *III.3.26* we have $V' \in \mathrm{Act}(\alpha - 1, Z_1)$, whence by 6.5 and *IV.3.3* it follows that

$$\phi(V'; * P_1E * Q_1, * P_1E * Q_2) \in \mathrm{Act}(\alpha - 1, Z_2).$$

Thus $\phi(V_0; * P_1E * Q_1, * P_1E * Q_2) \in \mathrm{Act}(\alpha - 1, Z_2)$. Suppose that $\neg\, \mathrm{Comp}(V_0, V')$. Then by *IV.1.3*, the maximal normalised continuation of V_0 contains $\geqslant q$ segments and is also situated to the left of V'. Hence by *III.3.26*, we may assert that $V_0 \in \mathrm{Norm}(\alpha - 1, Z_1, q)$.

We introduce the following notation:

$$U_0 \rightleftharpoons \phi(V_0; * P_1E * Q_1, * P_1E * Q_2),\ U' \rightleftharpoons \phi(V'; * P_1E * Q_1, * P_1E * Q_2),$$
$$W_1 \rightleftharpoons P_1 * E * Q_1, \quad W_2 \rightleftharpoons P_1 * E * Q_2.$$

Clearly, U' is the fourth right segment of the occurrence W_2. It follows from (16) that

$$\mathrm{MutNorm}_{\alpha-2}(W_1, W_2), \tag{17}$$

whence $U_0 \in \mathrm{Norm}(\alpha - 1, Z_2, q)$ by *II.7.4* and *II.5.5*.

Suppose that Y is of type II. Then, by 2.3, 2.4 and 2.6, an active kernel V'' of rank $\alpha - 1$ can be found that is contained in the occurrence $R * E * S$, is situated strictly to the right of $\phi(V'; P_1 * E * Q_1, R * E * S)$, and is not compatible with it.

Since in addition $V'' \in \text{Norm}(\alpha - 1, Y, Q)$, it follows from (16) that $\phi(V''; R * E * S, P_1 * E * Q_1) \in \text{Norm}(\alpha - 1, Z_1, q)$ and that $\phi(V''; R * E * S, P_1 * E * Q_2) \in \text{Norm}(\alpha - 1, Z_2, q)$. Then, since $V_0 \in \text{Act}(\alpha - 1, Z_1)$, it follows by *III.3.29* that $U_0 \in \text{Act}(\alpha - 1, Z_2)$.

If Y is of type I and V_0 is contained in $P_1 * E * Q_1$, then it follows from (16) that $\phi(V_0; P_1 * E * Q_1, R * E * S) \in \text{Norm}(\alpha - 1, Y, q)$. Then by 2.22, there is a supporting kernel of rank $\alpha - 1$ in $R * E * S$ containing $\geqslant q$ segments. By 2.4, 2.6 and 2.19 there exists an occurrence $V'' \in \text{Norm}(\alpha - 1, Y, q)$ contained in $R * E * S$, such that $\phi(V'; P_1 * E * Q_1, R * E * S) \ll V''$. We then apply *III.3.29*.

Hence, it remains only for us to consider the case when the integral word Y has type I and $V_0 < P_1 * E * Q_1$.

By *IV.3.12*, we can find a word $Z_1' \in \bar{\mathscr{M}}_{\alpha-2}$ such that $Z_1 \overset{\alpha-2}{\approx} Z_1'$. Set

$$V_0' \rightleftharpoons f_{\alpha-2}(V_0; Z_1, Z_1'), f_{\alpha-2}(P_1 * E * Q_1; Z_1, Z_1') = P' * E' * Q_1'.$$

By *IV.2.18(f)*, $V_0' < P' * E' * Q_1'$. By 2.15, there is a generating occurrence $W' \rightleftharpoons R' * E' * S'$ in some word $Y' \in \text{Int}(X, \alpha, A)$ such that $l_{\alpha,A}(R' * E' * S') = l_{\alpha,A}(R * E * S)$ and $\text{MutNorm}_{\alpha-2}(R' * E' * S', P' * E' * Q_1')$. By (17), (16), *II.7.6* and *II.7.2*, there is a word $Z_2' \in \mathscr{R}_{\alpha-2}$ such that $Z_2 \overset{\alpha-2}{\approx} Z_2'$,

$$f_{\alpha-2}(P_1 * E * Q_2; Z_2, Z_2') = P' * E' * Q_2',$$

and, by *II.7.8*,

$$f_{\alpha-2}(U_0; Z_2, Z_2') = \phi(V_0'; * P'E' * Q_1', * P'E' * Q_2').$$

By *IV.1.24*, $Z_1', Z_2' \in \mathscr{P}_{\alpha-1}$. By *III.3.27*, the occurrence V_0' is active, and we need only prove that $U_0' \rightleftharpoons f_{\alpha-2}(U_0; Z_2, Z_2')$ is active. In other words, we can restrict our attention to the case where $Z_1 \in \bar{\mathscr{M}}_{\alpha-2}$.

Since Y has type I, it follows by 2.17 that there is a word $Y_1 \in \text{Int}(X, \alpha - 1, A)$ such that $Y \overset{\alpha-1}{\approx} Y_1$,

$$f_{\alpha-1}(R * E * S; Y, Y_1) = R_1 * E * S_1,$$

and for an arbitrary regular occurrence V of rank $\alpha - 1$ contained in $f_{\alpha-1}(R * E * S; Y, X)$,

$$f_{\alpha-1}(V; X, Y) = \phi(f_{\alpha-2}(V; X, Y_1); R_1 * E * S_1, R * E * S). \tag{18}$$

Since $\text{MutNorm}_{\alpha-2}(P_1 * E * Q_1, R_1 * E * S_1)$ by (16), it follows by *II.4.1* that

$$E \,\overline{\underline{\circ}}\, C^r C_1, \tag{19}$$

where C is an image of the period A in the generating occurrence $R_1 * E * S_1$ of rank $\alpha - 1$, that is, for arbitrary regular occurrences U and V of rank $\alpha - 2$ contained in $R_1 * E * S_1$ we have

$$\text{Corr}_{\alpha-1,A}(U, V) \Longleftrightarrow \text{Corr}_C(U, V) . \tag{20}$$

Then, from (18), (20) and definition I.4.6, it follows that, for arbitrary regular occurrences U and V of rank $\alpha - 1$ contained in $R * E * S$,

$$\text{Corr}_{\alpha,A}(U, V) \Longleftrightarrow \text{Corr}_C(U, V) . \tag{21}$$

By 6.4 and *III.2.3*, V' is stable in a q-reversal of V_0. By *III.2.2*, these occurrences do not intersect, that is, the word Z_1 has the form $FDTBE'Q_1$, where

$$V_0 = F * D * TBE'Q_1 \text{ and } V' = FDT * B * E'Q_1 .$$

By *III.2.14*, we can find a local q-reversal of V_0. By *III.2.12* this reversal has the form

$$Z_1 \; \overline{\underline{\circ}} \; FDTBE'Q_1 \rightarrow F_1GT_1BE'Q_1 \rightleftharpoons Z_3 , \tag{22}$$

where $V_0' \rightleftharpoons F_1 * G * T_1BE'Q_1$ is a maximal image of V_0 and $f_{Z_1 \rightarrow Z_3}(V') = F_1GT_1 * B * E'Q_1$. Since (22) is a local q-reversal of rank $\alpha - 1$ and $\text{MutNorm}_{\alpha-2}(V',$ $FDT * B * E'Q_2)$, then by *III.2.17* the transition

$$Z_2 \; \overline{\underline{\circ}} \; FDTBE'Q_2 \rightarrow F_1GT_1BE'Q_2 \rightleftharpoons Z_4 \tag{23}$$

is a local q-reversal of $U_0 \rightleftharpoons F * D * TBE'Q_2$, the occurrence $U_0' \rightleftharpoons F_1 * G * T_1BE'Q_2$ is a maximal image of U_0 and $f_{Z_2 \rightarrow Z_4}(FDT * B * E'Q_2) = F_1GT_1 * B * E'Q_2$.

According to *III.3.28*, the transition (22) is a real reversal of rank $\alpha - 1$, that is, V_0 and V_0' are compatible with some completely stable occurrences of rank $\alpha - 1$. To avoid introducing new notation we shall assume by *III.1.3* that V_0 and V_0' are completely stable (we do not need the maximality of the occurrences V_0' and U_0': what is essential for the sequel is the relation $U_0' = \phi(V_0'; * F_1GT_1 * BE'Q_1,$ $* F_1GT_1 * BE'Q_2)$).

Since $Z_1, Z_2 \in \mathcal{P}_{\alpha-1}$, it follows from *III.3.23* that $Z_3, Z_4 \in \mathcal{N}_{\alpha-1}$. It is sufficient for us to prove that U_0 and U_0' are completely stable occurrences of rank $\alpha - 1$. That they are completely stable from the left follows because V_0 and V_0' are completely stable from the left, which is a consequence of *III.3.12* and *III.2.15*. The proof of complete stability from the right is more complicated.

First of all we shall prove that for any right cascade

$$\{U_i | i = 1, 2, \ldots, k\} , \tag{24}$$

of rank $\alpha - 1$ of U_0, all of the elements U_i are contained in the occurrence $* PE * Q_2$ and, what is more, the set

$$\{\phi(U_i; * P_1E * Q_2, * P_1E * Q_1) | i = 1, 2, \ldots, k\} \tag{25}$$

of occurrences in Z_1 forms a right cascade of V_0 of rank $\alpha - 1$.

Suppose that (24) is a right cascade of rank $\alpha - 1$ of U_0. We assume that there is no normalised occurrence of an elementary q_1-power of rank $\alpha - 1$ in $P_1 * E * Q_2$. Then by I.4.22(a) we have, for $i = 1, 2, \ldots, k$, either $U_i < P_1 * E * Q_2$ or $P_1 * E * Q_2 < U_i$. We note that the occurrence $U' \rightleftharpoons FDT * B * E'Q_2$ is the fourth right segment of rank α of $P_1 * E * Q_2$. If $P_1 * E * Q_2 < U_1$, then, by 6.4, U' is stable in q_1-reversals of the occurrences U_0 and U_1. But this is impossible by *III.3.1*, since, according to I.4.22(b), U_0 and U_1 adjoin one another. Hence $U_1 < P_1 * E * Q_2$. It is clear that $U_i < P_1 * E * Q_2$ for $1 < i \leqslant k$, for the same reason. Then by *III.2.2* we have $U_k \ll U'$, whence, by *III.3.13*, it follows that the set (25) is a right cascade of V_0 of rank $\alpha - 1$.

It now remains to consider the case where $P_1 * E * Q_2$ contains some occurrence $\bar{U} \in \mathrm{Norm}(\alpha - 1,\ Z_2,\ q_1)$. In this case $\phi(\bar{U};\ P_1 * E * Q_2,\ R * E * S) \in \mathrm{Norm}(\alpha - 1,\ Y,\ q_1)$. By 2.22, for some supporting kernel V'' of rank $\alpha - 1$ of the occurrence $R * E * S$, we have $l_{\alpha-1}(V'') \geqslant q_1$. Since $l_{\alpha,A}(R * E * S) \geqslant 7$, we can find, by 2.4, supporting kernels

$$V_1',\ V_2',\ V_3',\ V_4',\ V_5',\ V_6',\ V_7'$$

of $R * E * S$ of rank $\alpha - 1$ such that $V_i' < V_{i+1}'$ and

$$\mathrm{Corr}_{\alpha,A}(V_j',\ V'') \text{ for } 1 \leqslant j \leqslant 6\,. \tag{26}$$

Then by 2.18 we have for $2 \leqslant j \leqslant 6$,

$$V_j' \in \mathrm{MaxNorm}(\alpha - 1,\ Y,\ q_1)\,. \tag{27}$$

We denote by V_i the pseudokernels $\phi(V_i';\ R * E * S,\ P_1 * E * Q_2)$ of the ocurrence $P_1 * E * Q_2$, for $1 \leqslant i \leqslant 7$. Then

$$V_1 < V_2 < V_3 < V_4 < V_5 < V_6 < V_7\,,$$

where by (27), (16) and *II.5.15*,

$$V_j \in \mathrm{MaxNorm}(\alpha - 1,\ Z_2,\ q_1) \text{ for } 2 \leqslant j \leqslant 6\,, \tag{28}$$

and by (26) and (21), we have

$$\mathrm{Corr}_C(V_j,\ V_{j+1}) \text{ for } 1 \leqslant j \leqslant 6\,. \tag{29}$$

Clearly, we may assume that, for $1 \leqslant j < 6$, the occurrence V_{j+1} is the result of shifting V_j to the right by one period C. For $j = 2,3,4,5$ we denote the union of V_{j-1} and V_{j+1} by W_j. Then

$$\mathrm{Corr}_C(W_j,\ W_{j+1}) \text{ for } j = 2,3,4\,. \tag{30}$$

From (16) we find by *IV.3.6* and *II.5.5* that $V_i \in \text{Norm}(\alpha - 1, Z_2, 9)$ for $1 \leqslant i \leqslant 7$. Therefore it follows from (29) by *II.5.5* that $\text{MutNorm}_{\alpha-2}(V_j, V_{j+1})$ for $1 \leqslant j \leqslant 5$, whence we have by *II.7.3* that

$$\text{MutNorm}_{\alpha-2}(W_j, W_{j+1}) \quad \text{for } j = 2,3,4\,. \tag{31}$$

Let us prove that $U_k \ll V_7$. Assume that this is not so. Then we have by *II.5.7* that

$$V_6 < U_k \quad \text{or} \quad \text{Comp}(V_6, U_k)\,. \tag{32}$$

For $1 \leqslant i \leqslant k$ we shall denote the maximal normalised continuation and the maximal normalised continuation of U_i to the right by $\bar{U}_i$ and U_i'' respectively. By I.4.22(b), U_{i-1} and U_i adjoin one another for $1 \leqslant i \leqslant k$. Since $U_0 < V_2$ and $\neg\, \text{Comp}(U_0, V_2)$, (32) and *III.3.6* give that each of the occurrences (28) is compatible with some element U_i of the cascade (24) and hence coincides with the corresponding occurrence $\bar{U}_i$. Set

$$V_2 = \bar{U}_r,\ V_3 = \bar{U}_{r+h}\,. \tag{33}$$

Then the occurrences

$$\bar{U}_r,\ \bar{U}_{r+1}, \ldots, \bar{U}_{r+h}$$

are contained in W_2. From (30) and (31) it follows by *II.5.15* that, for $r \leqslant i \leqslant r+h$, $1 \leqslant j \leqslant 3$,

$$\phi(\bar{U}_i;\ W_2,\ W_{2+j}) \in \text{MaxNorm}(\alpha - 1,\ Z_2,\ q_1)\,. \tag{34}$$

Since, in addition, by *III.3.5* the occurrences $\phi(\bar{U}_i;\ W_2,\ W_{2+j})$ and $\phi(\bar{U}_{i+1};\ W_2, W_{2+j})$ adjoin one another, it follows from (33) by *III.3.7* that $\text{Comp}\,(U_{i+jh}, \phi(\bar{U}_i;\ W_2,\ W_{2+j}))$ for the values of i, j indicated. Then by *II.5.13*, we have from (34),

$$\phi(\bar{U}_i;\ W_2,\ W_{2+j}) = \bar{U}_{i+jh}\,. \tag{35}$$

By I.4.22(c) for $r < i \leqslant r+h$ and $j = 0,1,2$, the occurrence U''_{i+jh} is a maximal end of $\bar{U}_{i+jh}$ that is stable in a reversal of U_{i+jh-1} and U_{i+jh} is a maximal start of U''_{i+jh} that is stable in a reversal of U_{i+jh+1}. It therefore follows from (35) by *III.2.15* that, for $r < i \leqslant r+h$ and $j = 1,2$,

$$\phi(U_i'';\ W_2,\ W_{2+j}) = U''_{i+jh}$$

and

$$\phi(U_i;\ W_2,\ W_{2+j}) = U_{i+jh}\,.$$

Hence, for $r < i \leqslant r + h$ and $j = 1,2$

$$l_{\alpha-1}(U''_{i+jh}) = l_{\alpha-1}(U''_i),\ l_{\alpha-1}(U_{i+jh}) = l_{\alpha-1}(U_i)\,.$$

Then, by I.4.22(a), for $r < i \leqslant r + h$ and $j = 1,2$ we have the inequalities

$$q_{t_{i+jh}} \leqslant l_{\alpha-1}(U_{i+jh}) \leqslant l_{\alpha-1}(U''_i) < q_{t_i} + 17$$

and

$$q_{t_i} \leqslant l_{\alpha-1}(U_i) \leqslant l_{\alpha-1}(U''_{i+jh}) < q_{t_{i+jh}} + 17\,,$$

whence, since $q_2 = q_1 + 17$, it follows that

$$t_i = t_{i+h} = t_{i+2h}$$

for $r + 1 \leqslant i \leqslant r + h$, that is, some segment

$$t_{r+1}, t_{r+2}, \ldots, t_{r+h}$$

of the integer sequence (2) mentioned in § 3, Ch. I. is repeated three times in succession. This contradicts the definition of the sequence.

Consequently, $U_k \ll V_7$. Thus by *III.3.13*, the set (25) of occurrences forms a right cascade of rank $\alpha - 1$ for V_0.

Since by assumption the occurrence V_0 is completely stable from the right, it is stable in a reversal of the first element $\phi(U_1;\ * P_1E * Q_2, * P_1E * Q_1)$ of the cascade (25). Thus by *III.2.15*, the occurrence U_0 is stable in a reversal of the first element U_1 of its right cascade (24), that is, U_0 is also completely stable from the right.

If U_0 does not have right cascades of rank $\alpha - 1$, it is completely stable from the right by definition.

We prove next that the occurrence $U'_0 = F_1 * G * T_1BE'Q_2$ is completely stable from the right. Assume that some set

$$\{U'_i \mid i = 1,2, \ldots, k\} \tag{36}$$

of occurrences in a word Z_3 is a right cascade of rank $\alpha - 1$ of U'_0, and that U'_1 is the first element of this cascade. By *III.1.2*, the transition

$$F_1GT_1BE'Q_2 \rightarrow FDTBE'Q_2 \tag{37}$$

is a local q-reversal of the occurrence U'_0, and the occurrence $U_0 \rightleftharpoons F * D * TBE'Q_2$ is an image of U'_0 in this reversal. By I.4.22(c), all the occurrences (36) are stable in the reversal (37). Hence they are all contained in $F_1G * T_1BE'Q_2 *$. If U'_1 is a start of $F_1GN_1 * HBE'Q_2 *$, where $T_1 \mathrel{\underline{\circ}} N_1H$, then both $T \mathrel{\underline{\circ}} NH$ and the occurrences

$$U_i \rightleftarrows \phi(U_i'; \; F_1GN_1 * HBE'Q_2 *, \; FDN * HBE'Q_2 *) \tag{38}$$

are images of the occurrences U_i' in the reversal (37). By *III.3.16*, the occurrences (38) form a right cascade of rank $\alpha - 1$ of U_0. Thus, by what was proved above, all these occurrences are contained in $FDN * HBE' * Q_2$, and the corresponding occurrences (25) form a right cascade of rank $\alpha - 1$ of the occurrence $V_0 \rightleftarrows F * D * NHBE'Q_1$. Since the reversal (22) is local, we have

$$\begin{aligned} &f_{Z_1 \to Z_3}(\phi(U_i'; * P_1E * Q_2, * P_1E * Q_1)) \\ &= \phi(\phi(U_i; * P_1E * Q_2, * P_1E * Q_1); FDB * HBE' * Q_1, F_1GN_1 * HBE' * Q_1) \\ &= \phi(U_i'; \; F_1GN_1 * HBE' * Q_2, \; F_1GN_1 * HBE' * Q_1) \end{aligned}$$

for $i = 1, 2, \ldots, k$. By *III.3.16*, these occurrences form a right cascade of $V_0' \rightleftarrows F_1 * G * N_1HBE'Q_1$. By assumption, V_0' is completely stable from the right, that is, it is stable in a reversal of the occurrence

$$\phi(U_1'; \; F_1GN_1 * HBE' * Q_2, F_1GN_1 * HBE' * Q_1) .$$

Then, by *III.2.15*, U_0' is stable in a reversal of U_1', that is, U_0' is completely stable from the right. Hence the transition (23) is a real reversal of U_0.

6.9. *Suppose that $R * E * S$ is a normal generating occurrence of rank α in a word $Y \in \mathrm{Int}(X, \alpha, A)$ where $l_{\alpha,A}(R * E * S) \geqslant 7$ and the period A is elementary in rank α, that $P * E * Q$ is an occurrence in a word $Z \in \mathscr{R}_{\alpha-1}$ such that* $\mathrm{MutNorm}_{\alpha-2}(P * E * Q, \; R * E * S)$ *and* $E \mathrel{\overline{\underline{\circ}}} FDH$, *where the occurrence $PF * D * HQ$ starts with the fourth left and ends with the fourth right segments of the occurrence $P * E * Q$.*

Then, for every occurrence V in an arbitrary word $PET \in \mathscr{P}_{\alpha-1}$ that is contained in $* PFD * HQ$ such that

$$\mathrm{MutNorm}_{\alpha-2}(P * E * T, \; P * E * Q) , \tag{39}$$

we have

$$V \in \mathrm{Ker}(\alpha - 1, \; Z) \Longleftrightarrow \phi(V; * PE * Q, * PE * T) \in \mathrm{Ker}(\alpha - 1, \; PET) \tag{40}$$

and if in addition $PET \in \mathscr{R}_{\alpha-1}$, then the following relation is also satisfied:

$$V \in \mathrm{Reg}(\alpha - 1, Z) \Longleftrightarrow \mathrm{MutNorm}_{\alpha-1}(V, \phi(V; * PE * Q, * PE * T)) . \tag{41}$$

For every occurrence U in a word $TEQ \in \mathscr{P}_{\alpha-1}$ that is contained in $PF * DHQ *$ and is such that $\mathrm{MutNorm}_{\alpha-2}(P * E * Q, \; T * E * Q)$ *we have*

$$U \in \mathrm{Ker}(\alpha - 1, \; Z) \Longleftrightarrow \phi(U; \; P * EQ * , \; T * EQ *) \in \mathrm{Ker}(\alpha - 1, \; TEQ) ,$$

and if in addition $TEQ \in \mathscr{R}_{\alpha-1}$, then we have

$$U \in \text{Reg}(\alpha - 1, Z) \Rightarrow \text{MutNorm}_{\alpha-1}(U, \phi(U; P * EQ *, T * EQ *)) .$$

Suppose that $PET \in \mathscr{P}_{\alpha-1}$ and that (39) is satisfied. Set

$$U_1 \rightleftharpoons PF * D * HQ, \; V_1 \rightleftharpoons PF * D * HT .$$

By 6.5, it follows from the conditions of the lemma that

$$U_1 \in \text{Reg}(\alpha - 1, Z), \; V_1 \in \text{Reg}(\alpha - 1, PET) ,$$

while if $PET \in \mathscr{R}_{\alpha-1}$ we have moreover that

$$\text{MutNorm}_{\alpha-1}(U_1, V_1) . \tag{42}$$

If V is contained in U_1, then it satisfies (40) and (41), by 6.5, (42) and *IV.3.6*. If $V < U_1$, then (40) follows for it from 6.8 and *IV.3.22*, while (41) follows from 6.8 and *IV.3.23*.

6.10. *Let $P * E * Q$ be an occurrence of an elementary 9-power E of rank α in a word $Z \in \mathscr{R}_{\alpha-1}$ such that $E \mathrel{\overline{\underline{\circ}}} FDH$, where the occurrence $PF * D * HQ$ starts with the fifth left and ends with the fifth right segments of $P * E * Q$.*
Then the relation

$$V \in \text{Ker}(\alpha - 1, \; Z) \Longleftrightarrow \phi(V; * PE * Q, * PE * T) \in \text{Ker}(\alpha - 1, PET) \tag{43}$$

holds for an arbitrary word $PET \in \mathscr{P}_{\alpha-1}$ and an arbitrary occurrence V contained in $* PFD * HQ$, and if in addition, $PET \in \mathscr{R}_{\alpha-1}$ then

$$V \in \text{Reg}(\alpha - 1, Z) \Rightarrow \text{MutNorm}_{\alpha-1}(V, \phi(V; * PE * Q, * PE * T)) . \tag{44}$$

For an arbitrary word $TEQ \in \mathscr{P}_{\alpha-1}$ and an arbitrary occurrence U contained in $PF * DHQ *$,

$$U \in \text{Ker}(\alpha - 1, \; Z) \Longleftrightarrow \phi(U; \; P * EQ *, \; T * EQ *) \in \text{Ker}(\alpha - 1, \; TEQ),$$

and if in addition $TEQ \in \mathscr{R}_{\alpha-1}$ we have that

$$U \in \text{Reg}(\alpha - 1, \; Z) \Longleftrightarrow \text{MutNorm}_{\alpha-1}(U, \phi(U; \; P * EQ *, \; T * EQ *)) .$$

This follows from 2.21 and 6.9.

6.11. *Let $P * E * Q$ be an occurrence of an elementary 9-power E of rank α in a word $Z \in \mathscr{R}_{\alpha-1}$ such that $E \mathrel{\overline{\underline{\circ}}} FDH$, where the occurrence $PF * D * HQ$ starts with the fifth left and ends with the fifth right segments of $P * E * Q$.*
If $PET \in \mathscr{R}_{\alpha-1}$, then the relation

$$V \in \text{Norm}(\alpha, Z, 9) \Longleftrightarrow \phi(V; * PE * Q, * PE * T) \in \text{Norm}(\alpha, PET, 9) \tag{45}$$

holds for an arbitrary occurrence V contained in $* PFD * HQ$, and an occurrence $U \in$ Norm$(\alpha, Z, 9)$ contained in $* PF * DHQ$ has a proper normalised continuation to the left (right) if and only if the occurrence $\phi(U; * PE * Q, * PE * T)$ has a proper normalised continuation to the left (to the right respectively).

If $TEQ \in \mathscr{R}_{\alpha-1}$, then

$$V \in \text{Norm}(\alpha, Z, 9) \Longleftrightarrow \phi(V; P * EQ *, T * EQ *) \in \text{Norm}(\alpha, TEQ, 9)$$

for every occurrence V contained in $TF * DHQ *$, and an occurrence $U \in$ Norm$(\alpha, Z, 9)$ contained in $TFD * HQ *$ has a proper normalised continuation to the left (to the right) if and only if the occurrence $\phi(U; P * EQ *, T * EQ *)$ has a proper normalised continuation to the left (to the right respectively).

Suppose that $PET \in \mathscr{R}_{\alpha-1}$. Then (45) follows from 6.10 and 5.5. Let $U \in$ Norm$(\alpha, Z, 9)$ be an occurrence contained in $* PF * DHQ$. If U has a proper normalised continuation to the left (to the right), there exists a normalised continuation U_1 of U to the left (to the right respectively) different from U and contained in $* PFD * HQ$. Then $\phi(U_1; * PE * Q, * PE * T) \in$ Norm$(\alpha, PET, 9)$ by (45), that is, the occurrence $\phi(U_1; * PE * Q, * PE * T)$ is a normalised continuation of $\phi(U; * PE * Q, * PE * T)$ to the left (to the right respectively).

6.12. *Suppose that $P * E * Q \in$ Norm(α, Z, r), $r \geqslant 13$ and $E \mathrel{\underline{\circ}} FDH$, where $PF * D * HQ$ begins with the fifth left and ends with the fifth right segments of $P * E * Q$. Then for arbitrary words $PER \in \mathscr{R}_{\alpha-1}$ and $TEQ \in \mathscr{R}_{\alpha-1}$,*

$$P * FD * HR \in \text{Norm}(\alpha, PER, r-4)\,,$$
$$TF * DH * Q \in \text{Norm}(\alpha, TEQ, r-4)\,.$$

Moreover, $P * E * Q$ has a proper normalised continuation to the left (to the right) if and only if the occurrence $P * FD * HR$ ($TF * DH * Q$ respectively) has a proper normalised continuation to the left (to the right respectively).

This follows immediately from 6.11 and 5.4.

6.13. *Suppose that the word E contains an occurrence of a semi-integral 9-power of rank α. If the words PEQ and RES both lie in one of the sets $\mathscr{R}_{\alpha-1}$, $\mathscr{K}_{\alpha-1}$, $\mathscr{L}_{\alpha-1}$ or $\mathscr{M}_{\alpha-1}$, then the word REQ lies in the same set. If $PEQ \in \mathscr{M}_{\alpha-1}$, then $E \in \mathscr{L}_{\alpha-1}$.*

If $R * A^t A_1 * Q \in$ Norm$(\alpha, X, 9)$, where $A \mathrel{\underline{\circ}} A_1 A_2$ is an elementary period of rank α and the word X lies in one of the sets $\mathscr{K}_{\alpha-1}$, $\mathscr{L}_{\alpha-1}$ or $\mathscr{M}_{\alpha-1}$, then $RA^r A_1 Q$ lies in the same set for each $r \geqslant 9$.

Everything is clear for $\alpha = 1$. Suppose that $\alpha > 1$. By 4.7 we may assert that E is an elementary 9-power of rank α.

Suppose that $RES \in \mathscr{R}_{\alpha-1}$, $PEQ \in \mathscr{R}_{\alpha-1}$ and $E \mathrel{\underline{\circ}} FDH$, where $V' \rightleftharpoons RF * D * HQ$ is the fifth left segment of the occurrence $R * E * Q$. Since D is an elementary 9-power of rank $\alpha - 1$, the inductive hypothesis gives that $REQ \in \mathscr{R}_{\alpha-2}$.

Assume that $V \in$ Norm$(\alpha - 1, REQ, n - q)$. By *IV.1.18*, the base of V cannot occur in the word RE nor in EQ, and by 2.12 the word E does not occur in the base of V. Consequently, $REQ \in \mathscr{P}_{\alpha-1}$.

By 6.6 and *IV.3.3*, $V' \in$ Ker$(\alpha - 1, REQ)$ and $RF * D * HS \in$ Ker$(\alpha - 1, RES)$.

Thus $l_{\alpha-1}(V') \leqslant n - 176$. Suppose that $V \in \text{Ker}(\alpha - 1, REQ)$. If $V \neq V'$, then either $V < V'$ or $V' < V$. If $V < V'$, by 6.10 we have $\phi(V; *RE*Q, *RE*S) \in \text{Ker}(\alpha - 1, RES)$, whence $l_{\alpha-1}(V) \leqslant n - 176$ because of the condition $RES \in \mathscr{R}_{\alpha-1}$. If $V' < V$, the same thing follows from the condition $PEQ \in \mathscr{R}_{\alpha-1}$. Thus $REQ \in \mathscr{R}_{\alpha-1}$.

The proof of this fact for $\mathscr{K}_{\alpha-1}$, $\mathscr{L}_{\alpha-1}$ and $\mathscr{M}_{\alpha-1}$ is similar.

If $REQ \in \mathscr{M}_{\alpha-1}$, then $RE, EQ \in \mathscr{L}_{\alpha-1}$ by *IV.1.21*, and what has been proved already shows that $E \in \mathscr{L}_{\alpha-1}$.

Suppose that $R * A^t A_1 * Q \in \text{Norm}(\alpha, X, 9)$, where $A \mathrel{\overline{\circ}} A_1 A_2$ is an elementary period of rank α and $X \in \mathscr{K}_{\alpha-1}$. If $l_\alpha(A^t A_1) > 9$, then $l_\alpha(A^{t-1} A_1) \geqslant 9$. Taking $A^{t-1} A_1$ as E, the first assertion that we proved shows that $RA^{t-1} A_1 Q \in \mathscr{K}_{\alpha-1}$ and $RA^{t+1} A_1 Q \in \mathscr{K}_{\alpha-1}$, since $X \in \mathscr{K}_{\alpha-1}$. If $l_\alpha(A^t A_1) = 9$, the relation $RA^{t+1} A_1 Q \in \mathscr{K}_{\alpha-1}$ follows by I.4.26 and 6.9, since $RA^t A_1 Q \in \mathscr{K}_{\alpha-1}$.

6.14. *Set*

$$X \rightleftharpoons PETDQ \text{ and } Y \rightleftharpoons RETDS\,,$$

where E, D are elementary 9-powers of rank α, X, $Y \in \mathscr{R}_{\alpha-1}$ and the occurrence $PE_1 * E_2 TD_2 * D_1 Q$ starts with the fifth left segment of $P * E * TDQ$ and ends with the fifth right segment of $PET * D * Q$.

Then

$$V \in \text{Reg}(\alpha - 1, X) \Rightarrow \text{MutNorm}_{\alpha-1}(V, \phi(V; P * ETD * Q, R * ETD * S))\,, \quad (47)$$

$$V \in \text{Norm}(\alpha, X, 9) \Longleftrightarrow \phi(V; P * ETD * Q, R * ETD * S) \in \text{Norm}(\alpha, Y, 9) \quad (48)$$

for every occurrence V contained in $PE_1 * E_2 TD_2 * D_1 Q$, while

$$U \in \text{MaxNorm}(\alpha, X, 9) \Longleftrightarrow \phi(U; PE * T * DQ, RE * T * DS) \in \text{MaxNorm}(\alpha, Y, 9)$$

for every occurrence U contained in $PE * T * DQ$.

By 6.13 we have $RETDQ \in \mathscr{R}_{\alpha-1}$. Applying 6.10 to the words $PETDQ$ and $RETDQ$, and then to the words $RETDQ$ and $RETDS$, we get relation (47). The two remaining relations follow in an analogous way from 6.11.

6.15. *Suppose that X, $Y \in \mathscr{R}_{\alpha-1}$, $P * E * Q \in$ Norm$(\alpha, X, 9)$ and $R * E * S \in$ Norm$(\alpha, Y, 9)$. Then $REQ \in \mathscr{R}_{\alpha-1}$ and $R * E * Q \in$ Norm$(\alpha, REQ, 9)$.*

In addition, if $P * Eu * Q_1 (R_1 * vE * S)$ is a normalised continuation of the occurrence $P * E * Q$ (of $R * E * S$ respectively), then $R * Eu * Q_1$ $(R_1 * vE * Q$ respectively) is a normalised continuation of $R * E * Q$.

By 6.13, $REQ \in \mathscr{R}_{\alpha-1}$ since $X, Y \in \mathscr{R}_{\alpha-1}$.

Set $E \mathrel{\overline{\circ}} FDH$, where $PF * D * HQ$ starts with the fifth left and ends with the fifth right segments of $P * E * Q$. Then by 6.10 we have

$$\text{MutNorm}_{\alpha-1}(R * FD * HQ, \;\; R * FD * HS) \quad (48)$$

and

$$\mathrm{MutNorm}_{\alpha-1}(RF * DH * Q,\ PF * DH * Q)\,. \tag{49}$$

Moreover, by 5.5 we have $\mathrm{MutNorm}_{\alpha-1}(P * E * Q,\ R * E * S)$, whence by (48) and *IV.3.6* it follows that

$$\mathrm{MutNorm}_{\alpha-1}(R * FD * HQ,\ P * FD * HQ)\,. \tag{50}$$

Because of proposition *IV.3.25*, it follows from (49) and (50) that $\mathrm{MutNorm}_{\alpha-1}(R * E * Q,\ P * E * Q)$, that is, $R * E * Q \in \mathrm{Norm}(\alpha, REQ, 9)$.

Let $P * Eu * Q_1$ be a normalised continuation of $P * E * Q$. Then $PF * DHu * Q_1 \in \mathrm{Reg}(\alpha - 1, X)$, and we have $\mathrm{MutNorm}_{\alpha-1}(PF * DHu * Q_1,\ RF * DHu * Q_1)$ by 6.10, whence by (50) and *IV.3.25* it follows that $\mathrm{MutNorm}_{\alpha-1}(P * Eu * Q_1,\ R * Eu * Q_1)$, that is, $R * Eu * Q_1 \in \mathrm{Norm}(\alpha, REQ, 9)$. It remains only to refer to 5.7.

The case where $R_1 * vE * S$ is a normalised continuation of $R * E * S$ is to be considered in a symmetrical fashion.

6.16. *Let E and D be elementary 9-power of rank α. If*

$$P * E * TDQ \in \mathrm{Norm}(\alpha, X, 9),\ R * E * TDS \in \mathrm{Norm}(\alpha, Y, 9) \tag{51}$$

and the occurrence $P * E * TDQ$ has a proper normalised continuation to the right, then $R * E * TDS$ also has a proper normalised continuation to the right.

If

$$PET * D * Q \in \mathrm{Norm}(\alpha, X, 9) \text{ and } RET * D * S \in \mathrm{Norm}(\alpha, Y, 9)\,,$$

and the occurrence $PET * D * Q$ has a proper normalised continuation to the left, then $RET * D * S$ also has a proper normalised continuation to the left.

Suppose that (51) is satisfied and that the occurrence $V \rightleftharpoons P * E * TDQ$ has a proper normalised continuation to the right. We denote by U the fifth right segment of $PET * D * Q$, and by W the union of V and U. Since $V \in \mathrm{Reg}(\alpha - 1, X)$, it follows from 6.6 that $W \in \mathrm{Reg}(\alpha - 1, X)$. Clearly, some occurrence $P * Eu * Q' \rightleftharpoons V_1$ contained in W is a normalised continuation of V to the right. Since

$$\mathrm{MutNorm}_{\alpha-1}(W,\ \phi(W;\ P * ETD * Q,\ R * ETD * S)$$

by 5.5, 6.14 and *IV.3.25*, it follows from 5.15 and 5.8 that the occurrence $\phi(V_1;\ P * ETD * Q,\ R * ETD * S)$ is a normalised continuation of $R * E * TDS$ to the right.

6.17. *Let $V \rightleftharpoons P * E * TDQ$, $V' \rightleftharpoons PET * D * Q$, $W \rightleftharpoons R * E * TDS$, $W' \rightleftharpoons RET * D * S$ be normalised occurrences of elementary 9-powers of rank α in words $X, Y \in \mathscr{R}_{\alpha-1}$ with $\neg\ \mathrm{Comp}\,(V, V')$. Then every normalised continuation U of V to the right (of V' to the left) is contained in $P * ETD * Q$, and moreover the occurrence*

$$\phi(U; P * ETD * Q, R * ETD * S) \tag{52}$$

is a normalised continuation of W to the right (of W' to the left respectively).

Let $U \rightleftarrows P * Eu * Q_1$ be a normalised continuation of V to the right. Since $\neg$ Comp(U, V'), it follows from 5.7 that V' is not contained in U. Then U is contained in $P * ETD * Q$. Further, repeating the arguments in 6.16, we find that (52) is a normalised continuation of W.

6.18. *If $V \in$ Norm $(\alpha, X, 9)$, V has no proper normalised continuation to the left (right), and W is its maximal continuation to the left (to the right respectively), then $l_\alpha(W) \leqslant l_\alpha(V) + 4$.*

Suppose that V has no proper normalised continuation to the left. Assume that W is a continuation to the left with $l_\alpha(W) > l_\alpha(V) + 4$. Set $V = PD * E * Q$, $W = P * DE * Q$, and suppose that $R * DE * S$ is a generating occurrence of the elementary word DE. Then by 6.15, we have that $PD * E * S \in$ Norm$(\alpha, PDES, 9)$. Comparing the occurrences $PD * E * S$ and $RD * E * S$, we get by virtue of 6.11 and the inequality $l_\alpha(DE) > l_\alpha(E) + 4$ that $PD * E * S$ has a proper normalised continuation to the left. By 6.15 it then follows that V also has a proper normalised continuation to the left, which contradicts our assumption. The case of continuations of V to the right is to be considered in the analogous fashion.

6.19. *Let $P * E * Q$ and $R * D * S$ be occurrences of elementary 9-powers of rank α in a word $X \in \mathcal{R}_{\alpha-1}$ such that $\neg$ Comp $(P * E * Q, R * D * S)$. If $P * E * Q$ is normalised and does not intersect an occurrence $V \in$ MaxNorm $(\alpha, X, 9)$ with Comp$(V, R * D * S)$, then $R * D * S$ contains less than 13 segments of the occurrence $P * E * Q$; if Rel(E, D) or Rel$(E, D)^{-1}$, it contains less than 6 segments of $P * E * Q$.*

Clearly, we may assume that V is contained in $R * D * S$. Suppose that $P * E * Q \ll V$.

Assume that $R * D * S$ contains $\geqslant 13$ segments of the occurrence $P * E * Q$, that is, we have $E \mathrel{\overset{\circ}{=}} FG$ and $D \mathrel{\overset{\circ}{=}} GH$, where $PF * G * HS$ is their common part. We consider a generating occurrence $R_1 * GH * S_1$ of D in the corresponding integral word Y. Let $U \rightleftarrows PFN * G_1 * HS$ be the end of $PF * G * HS$ starting with its fifth left segment. Since $P * E * Q$ is normalised, $U \in$ Norm $(\alpha, X, 9)$. Then by 6.14 we have $R_1N * G_1 * HS_1 \in$ Norm $(\alpha, Y, 9)$, whence by 5.7 and 5.14 we get that $R_1 * NG_1H * S_1$ is a continuation of $R_1N * G_1 * HS_1$. By 5.8, we get from this that Comp$(U, R * D * S)$. Then Comp $(P * E * Q, R * D * S)$ by 5.11, and this contradicts our hypothesis.

All remaining cases are to be considered in an analogous fashion.

§ 7. Completely Regular Occurrences of Rank $\alpha - 1$ and Further Properties of Elementary Words

7.1. We say that an occurrence $P * E * Q$ in a word $X \in \mathcal{R}_{\alpha-1}$ is a *completely regular occurrence of rank $\alpha - 1$*, if either $\alpha = 1$ and E is non-empty, or $\alpha > 1$ and there is a start U and an end V of $P * E * Q$ such that $U, V \in$ Norm $(\alpha, X, 9)$. The set of all completely regular occurrences of rank $\alpha - 1$ in a given word $X \in \mathscr{R}_{\alpha-1}$ will be denoted by

$$\text{ComReg}\,(\alpha - 1, X)\,.$$

The following inclusions are obvious:

$$\text{Norm}(\alpha, X, 9) \subset \text{ComReg}\,(\alpha - 1, X) \subset \text{Reg}\,(\alpha - 1, X)\,.$$

If $\alpha > 1$, we have in addition the inclusion

$$\text{Reg}(\alpha - 1, X) \subset \text{ComReg}(\alpha - 2, X)\,.$$

7.2. *Suppose that* $V \in \text{ComReg}\,(\alpha - 1, X)$. *If* $X \overset{\alpha-1}{\approx} Y$, *then* $f_{\alpha-1}(V; X, Y) \in \text{ComReg}\,(\alpha - 1, Y)$. *If* $\text{MutNorm}_{\alpha-1}(V, W)$, *where* W *is an occurrence in a word* $Z \in \mathscr{R}_{\alpha-1}$, *then* $W \in \text{ComReg}\,(\alpha - 1, Z)$.

The first assertion follows from 5.16 and *IV.2.18(i)*, and the second from 5.23 and *IV.3.7.*

7.3. *Let* $P * E * Q$ *and* $R * E * S$ *be occurrences in words* $X, Y \in \mathscr{R}_{\alpha-1}$ *respectively, let* U *be a start of* $P * E * Q$ *and* W *an end of* $P * E * Q$ *such that* $U, W \in \text{ComReg}\,(\alpha - 1, X)$. *If*

$$\text{MutNorm}_{\alpha-1}(U, \phi(U; P * E * Q, R * E * S))\ \& \\ \&\ \text{MutNorm}_{\alpha-1}(W, \phi(W; P * E * Q, R * E * S))\,, \qquad (1)$$

then

$$\text{MutNorm}_{\alpha-1}(P * E * Q, R * E * S)\,.$$

By *IV.3.21* and *IV.3.3*, it is enough to prove the relation

$$\text{MutNorm}_{\alpha-1}(V, \phi(V; P * E * Q, R * E * S)) \qquad (2)$$

for an arbitrary kernel V of rank $\alpha - 1$ contained in $P * E * Q$, and to prove the relation

$$\text{MutNorm}_{\alpha-1}(V, \phi(V; R * E * S, P * E * Q))$$

for an arbitrary kernel $V \in \text{Ker}\,(\alpha - 1, Y)$ contained in $R * E * S$. By the completeness of the analogy, we may restrict to the proof of (2). If V is contained in U or W, then (2) follows from (1) because of *IV.3.6*; otherwise it is satisfied because of 6.14.

Remark. By 5.5 and 7.1, it follows from the assertion just proved that when $U, V \in \text{Norm}\,(\alpha, Z, 9)$ and $\text{Comp}\,(U, V)$, the union of the occurrences U and V is also normalised. Hence, occurrences U and V like this have one and the same maximal normalised continuation. This means also that, for every occurrence $V \in \text{Norm}\,(\alpha, Z, 9)$, the maximal normalised continuation is uniquely determined, as well as its maximal normalised continuations to the left and to the right.

7.4. *Suppose that* $P * E * Q \in \text{ComReg}\,(\alpha - 1, X)$ *and that*

$$\text{MutNorm}_{\alpha-1}(P * E * Q, R * E * S)\,. \tag{3}$$

If $P \overline{\underline{\circ}} R$, *then for every occurrence* $V \in \text{Reg}\,(\alpha - 1, X)$ *contained in* $* PE * Q$ *we have*

$$\text{MutNorm}_{\alpha-1}(V, \phi(V; * PE * Q, * PE * S))\,. \tag{4}$$

If $Q \overline{\underline{\circ}} S$, *then for every occurrence* $V \in \text{Reg}\,(\alpha - 1, X)$ *contained in* $P * EQ *$ *we have*

$$\text{MutNorm}_{\alpha-1}(V, \phi(V; P * EQ *, R * EQ*))\,.$$

Suppose that $P \overline{\underline{\circ}} R$, that U is an end of $P * E * Q$, $U \in \text{Norm}\,(\alpha, X, 9)$ and that U' is the fifth right segment of U. By *IV.3.21*, it is enough for us to prove (4) for $V \in \text{Ker}\,(\alpha - 1, X)$. If $V < U'$, then (4) is satisfied because of 6.10, otherwise it follows from (3) by *IV.3.6*.

The case $Q \overline{\underline{\circ}} S$ is to be considered in an analogous fashion.

7.5. *Suppose that* $\text{MutNorm}_{\alpha-1}(P * E * Q, R * E * S)$. *If* $P * E * Q \in \text{ComReg}\,(\alpha - 1, X)$, *then* $R * E * Q \in \text{ComReg}\,(\alpha - 1, REQ)$ *and* $\text{MutNorm}_{\alpha-1}\,(P * E * Q, R * E * Q)$.

By 7.1, we have $E \overline{\underline{\circ}} FD$, where

$$P * F * DQ \in \text{Norm}\,(\alpha, X, 9)\,.$$

Then by *IV.3.6* and 5.5 we have $R * F * DS \in \text{Norm}\,(\alpha, RES, 9)$, whence by 6.15 and 5.5 we obtain the relations $RFDQ \in \mathscr{R}_{\alpha-1}$ and $\text{MutNorm}_{\alpha-1}\,(P * F * DQ, R * F * DQ)$. Thus, by 7.1 and 7.4, we have $\text{MutNorm}_{\alpha-1}\,(P * E * Q, R * E * Q)$, so that $R * E * Q \in \text{ComReg}\,(\alpha - 1, REQ)$ by 7.2.

7.6. *Suppose that* $P * E * Q \in \text{ComReg}(\alpha - 1, X)$, $Y \overline{\underline{\circ}} RES$, $X \overset{\alpha-1}{\simeq} X_1$, $\text{MutNorm}_{\alpha-1}\,(P * E * Q, R * E * S)$, *and*

$$f_{\alpha-1}(P * E * Q;\ X,\ X_1) = P_1 * E_1 * Q_1\,.$$

If $P \overline{\underline{\circ}} R$, *then there is a word* $Y_1 \overset{\alpha-1}{\simeq} Y$ *such that*

$$f_{\alpha-1}(P * E * S;\ Y,\ Y_1) = P_1 * E_1 * S_1\,,$$

and, for every occurrence $PEF * D * H \in \text{Reg}\,(\alpha - 1, Y)$,

$$f_{\alpha-1}(PEF * D * H;\ Y,\ Y_1) = P_1E_1F_1 * D * H\,.$$

If $Q \mathrel{\underline{\overline{\circ}}} S$, *then there is a word* $Y_2 \overset{\alpha-1}{\simeq} Y$ *such that*

$$f_{\alpha-1}(R * E * Q;\ Y,\ Y_2) = R_1 * E_1 * Q_1$$

and, for every occurrence $F * D * HEQ \in \mathrm{Reg}\,(\alpha - 1,\ Y)$,

$$f_{\alpha-1}(F * D * HEQ;\ Y,\ Y_2) = F * D * H_1F_1Q_1\,.$$

In addition, if X_1 *lies in* $\mathscr{K}_{\alpha-1}$, $\mathscr{L}_{\alpha-1}$, $\mathscr{M}_{\alpha-1}$ *or* $\bar{\mathscr{M}}_{\alpha-1}$, *then the desired words* Y_1 *and* Y_2 *can be taken to lie in the same set.*

Suppose that $P \mathrel{\underline{\overline{\circ}}} R$. We consider a maximal occurrence $W \in \mathrm{Reg}\,(\alpha - 1, X)$ ending with the occurrence $P * E * Q$. Set

$$U \rightleftharpoons \phi(W;\ * PE * Q,\ * PE * S)\,.$$

By 7.4, we have $\mathrm{MutNorm}_{\alpha-1}\,(W,\ U)$. Since there are no kernels of rank $\alpha - 1$ to the left of W, we may now use Propositions *IV.3.20* and *IV.2.34*.

The case $Q \mathrel{\underline{\overline{\circ}}} S$ is to be considered in an analogous fashion.

7.7. *Suppose that* $P * E * Q \in \mathrm{ComReg}\,(\alpha - 1,\ X)$.
If $X \overset{\alpha-1}{\simeq} PET$, *then* $f_{\alpha-1}(P * E * Q;\ X,\ PET) = P * E * T$.
If $X \overset{\alpha-1}{\simeq} TEQ$, *then* $f_{\alpha-1}(P * E * Q;\ X,\ TEQ) = T * E * Q$.

Suppose that $X \overset{\alpha-1}{\simeq} PET$. By 7.1, there is an end W of the occurrence $P * E * Q$ such that $W \in \mathrm{Norm}\,(\alpha, X, 9)$. Let V be the start of W ending with its fifth right segment, and $V = PF * D * HQ$, say. Then we have $V \in \mathrm{Reg}\,(\alpha - 1,\ X)$ by 5.4, and $PF * D * HT \in \mathrm{Reg}\,(\alpha - 1,\ PET)$ by 6.10, whence $f_{\alpha-1}\,(V;\ X,\ PET) = PF * D * HT$ by *IV.2.29*. Further, by *IV.2.18(i)* and *IV.2.27* we get that $f_{\alpha-1}\,(P * FD * HQ; X,\ PET) = P * FD * HT$ and $f_{\alpha-1}\,(PF * DH * Q;\ X,\ PET) = PF * DH * T$, that is, $f_{\alpha-1}\,(P * E * Q;\ X,\ PET) = P * E * T$.

The case $X \overset{\alpha-1}{\simeq} TEQ$ is to be considered in an analogous fashion.

7.8. *Suppose that* $P * E * Q \in \mathrm{ComReg}\,(\alpha - 1,\ X)$, $Y \mathrel{\underline{\overline{\circ}}} RES$, $\mathrm{MutNorm}_{\alpha-1}(P * E * Q,\ R * E * S)$, $Y \overset{\alpha-1}{\simeq} Y_1$, $X \overset{\alpha-1}{\simeq} X_1$ *and* $f_{\alpha-1}\,(P * E * Q; X, X_1) = P_1 * D * Q_1$.

If $P \mathrel{\underline{\overline{\circ}}} R$ *and* $Y_1 \mathrel{\underline{\overline{\circ}}} P_1DS_1$, *then* $f_{\alpha-1}\,(P * E * S;\ Y,\ Y_1) = P_1 * D * S_1$, *and*

$$\begin{aligned} &f_{\alpha-1}(\phi(V;\ * PE * Q,\ * PE * S);\ Y,\ Y_1)\\ &\quad = \phi(f_{\alpha-1}(V;\ X,\ X_1);\ * P_1D * Q_1,\ * P_1D * S_1)\,,\end{aligned}$$

for every occurrence $V \in \mathrm{Reg}\,(\alpha - 1,\ X)$ *contained in* $* PE * Q$.

If $Q \mathrel{\underline{\overline{\circ}}} S$ *and* $Y_1 \mathrel{\underline{\overline{\circ}}} R_1DQ_1$, *then* $f_{\alpha-1}\,(R * E * Q;\ Y,\ Y_1) = R_1 * D * Q_1$, *and*

$$\begin{aligned} &f_{\alpha-1}(\phi(V;\ P * EQ *,\ R * EQ *);\ Y,\ Y_1)\\ &\quad = \phi(f_{\alpha-1}\,(V;\ X,\ X_1);\ P_1 * DQ_1 *,\ R_1 * DQ_1 *)\end{aligned}$$

for every occurrence $V \in \mathrm{Reg}\,(\alpha - 1,\ X)$ *contained in* $P * EQ *$.

Suppose that $P \stackrel{\circ}{=} R$ and $Y_1 \stackrel{\circ}{=} P_1DS_1$. By 7.2, we have $P * E * S \in$ ComReg$(\alpha - 1, Y)$. By 7.6, there is a word $Y_2 \overset{\alpha-1}{\simeq} Y$ such that $f_{\alpha-1}(P * E * S, Y, Y_2) = P_1 * D * S_2$. Then $Y_2 \overset{\alpha-1}{\simeq} Y_1$, and $f_{\alpha-1}(P_1 * D * S_2; Y_2, Y_1) = P_1 * D * S_1$ by 7. 7. Hence $f_{\alpha-1}(P * E * S; Y, Y_1) = P_1 * D * S_1$.

Consider a maximal regular occurrence W of rank $\alpha - 1$ in a word X ending with the occurrence $P * E * Q$. By 7.4 and *IV.2.18* (*f*), (*d*) there are no kernels of rank $\alpha - 1$ to the left of the occurrences $U \rightleftharpoons \phi(W; * PE * Q, * PE * S)$, $W' \rightleftharpoons f_{\alpha-1}(W; X, X_1)$ and $U' \rightleftharpoons f_{\alpha-1}(U; Y, Y_1)$. Hence $U' = \phi(W'; * P_1D * Q_1, * P_1D * S_1)$. Since an arbitrary occurrence $V \in$ Reg$(\alpha - 1, X)$ contained in $* PE * Q$ is also contained in W, it only remains to use *IV.3.5*.

The case where $Q \stackrel{\circ}{=} S$ and $Y_1 \stackrel{\circ}{=} R_1 DQ_1$ is to be considered in an analogous fashion.

7.9. *If* $A_1A'A_2 \in$ Per(α, A), *then for every word* $X \rightleftharpoons A_1A'A_2$ *the inequality* $\partial(X) \geqslant 27\, \partial(A)$ *implies that* $X \in$ Per (α, A).

Indeed, by 6.13 we get that $X \in \mathscr{R}_{\alpha-1}$ since $A_1A'A_2 \in \mathscr{R}_{\alpha-1}$, whence by 1.8 it follows that there exists a $V \in$ Ker $(\alpha - 1, X)$ such that $V \in$ Inn (X,A) and $\partial($Bas$(V)) < 2\,\partial(A)$. Hence $X \in$ Per (α, A).

7.10. *Suppose that* $X \in \mathscr{M}_{\alpha-1}$ *and* $P * A'A_1 * Q \in$ Norm$(\alpha,X,9)$, *where* $A \stackrel{\circ}{=} A_1A_2$ *is an elementary period of rank* α. *If* $l_{\alpha,A}(A'A_1) \geqslant 9$ *and* $Y \rightleftharpoons PA'A_1Q$, *then* $Y \in \mathscr{M}_{\alpha-1}$ *and* $P * A'A_1 * Q \in$ Norm $(\alpha, Y, 9)$.

Clearly, it is enough for us to consider the case where $|r - t| = 1$.

Suppose that $r = t + 1$, $W_1 \rightleftharpoons P * A'A_1 * A_2A_1Q$ and $W_2 \rightleftharpoons PA * A'A_1 * Q$. From 6.13 it follows that $Y \in \mathscr{M}_{\alpha-1}$, while by 6.10 and *IV.3.25*,

$$\text{MutNorm}_{\alpha-1}(P * A'A_1 * Q, W_1) \text{ and } \text{MutNorm}_{\alpha-1}(P * A'A_1 * Q, W_2).$$

Assume that $Y \notin \mathscr{M}_{\alpha-1}$. Then $Y \in \mathscr{M}_{\beta-1}$ and $Y \notin \mathscr{M}_\beta$ for some β, $0 < \beta \leqslant \alpha - 1$. This means that there exist $U,V \in$ Ker (β, Y) such that MutNorm$_\beta(U,V)$ and $\neg$ Rel(U,V). Since $W_1, W_2 \in$ ComReg(β, Y), U is contained either in $* PA'A_1 * A_2A_1Q$ or in $PA * A'A_1\, Q *$. We denote by U' the occurrence $\phi(U; * PA'A_1 * A_2\,A_1Q, * PA'A_1 * Q)$ in the first of these cases, and $\phi(U; PA * A'A_1Q *, P * A'A_1Q *)$ in the second. In both cases it follows from 7.4 and *IV.3.9.* that $U' \in$ Ker(β, X) and MutNorm$_\beta(U, U')$. Similarly there exists a $V \in$ Ker(β, X) such that Bas$(V') \stackrel{\circ}{=}$ Bas(V) and MutNorm$_\beta(V, V')$. Then MutNorm$_\beta(U', V')$, whence Rel(U', V') because of the assumption that $X \in \mathscr{M}_\beta$. By *II.5.20*, this contradicts the assumption $\neg$ Rel (U, V). Hence $Y \in \mathscr{M}_{\alpha-1}$.

The case $r = t - 1$ is considered in the analogous way to that where $r = t + 1$.

7.11. *Let* $A \stackrel{\circ}{=} A_1A_2$ *be an elementary period of rank* α, *and suppose that the occurrence* $P * A^{10}A_1 * Q$ *in a word* $X \in \mathscr{K}_{\alpha-1}$ *is periodised in rank* $\alpha - 1$, *that is, for some* $RA^{10}A_1S \in \mathscr{M}_{\alpha-1}$ *we have*

$$\text{MutNorm}_{\alpha-1}(P * A^{10}A_1 * Q, R * A^{10}A_1 * S)\,. \tag{5}$$

If $P * A^{10}A_1\, D * Q' \in$ Norm (α, X, q_1) *and* $r \geqslant l_\alpha(A^{10}A_1D) + 11$, *then there exist a word* F *and a number* h *such that* $A^r \overset{\alpha-1}{\simeq} A^{10}A_1DFA^h$ *and* $A^{10}A_1DFA^h \in \mathscr{K}_{\alpha-1}$, *where* $h \geqslant r - 2 - l_\alpha(A^{10}A_1D)$.

If $P' * DA^{10}A_1 * Q \in \text{Norm}(\alpha, X, q_1)$ *and* $r \geqslant l_\alpha(DA^{10}A_1) + 11$, *then there exists a word F such that* $A^r \overset{\alpha-1}{\simeq} A^h FDA^{10}A_1$, *where* $A^h FDA^{10}A_1 \in \mathscr{K}_{\alpha-1}$ *and* $h \geqslant r - 2 - l_\alpha(DA^{10}A_1)$.

Since the second assertion is symmetrical with the first, we can restrict attention to the first.

By (5) and *IV.3.14* there exists a word $Z \in \mathscr{M}_{\alpha-1}$ such that $X \overset{\alpha-1}{\simeq} Z$ and

$$f_{\alpha-1}(P * A^{10}A_1 * DQ'; X, Z) = P_1 * A^{10}A_1 * Q_1 . \tag{6}$$

By 5.16 and 4.1, we have for some t and E,

$$f_{\alpha-1}(P * A^{10}A_1D * Q'; X, Z) = P_1 * E^tE_1 * G ,$$

where E is an image of the period A. Since A and E are simple words by 4.3, we have $E \overset{\circ}{=} A$ by I.2.9, that is,

$$f_{\alpha-1}(P * A^{10}A_1D * Q'; X, Z) = P_1 * A^tA_3 * G , \tag{7}$$

where $A \overset{\circ}{=} A_3A_4$ and $A_2A^{t-11}A_3G \overset{\circ}{=} Q_1$.

It follows from 7.10 and *IV.1.21* that $A^r \in \mathscr{K}_{\alpha-1}$. By 6.6. we have

$$A^4 * A^{t-4}A_3 * A_4A^{r-t-1} \in \text{Norm}(\alpha, A^r, q_1 - 4) .$$

Then by 5.5,

$$\text{MutNorm}_{\alpha-1}(A^4 * A^{t-4}A_3 * A_4A^{r-t-1}, P_1A^4 * A^{t-4}A_3 * G) . \tag{8}$$

By *IV.2.27* and *IV.2.18* (*i*), it follows from (6) and (7) that

$$f_{\alpha-1}(P_1A^4 * A^{t-4}A_3 * G; Z, X) = PA^4 * A^6A_1D * Q . \tag{9}$$

Clearly, $A^4 * A^6A_1 * A_2A^{r-11}$ and $A^{t+1} * A^4A_1 * A_2A^{r-t-6}$ are regular occurrences of rank $\alpha - 1$.

By (8), (9) and *IV.3.18*, there is a word $Y \in \mathscr{K}_{\alpha-1}$ such that $A^r \overset{\alpha-1}{\simeq} Y$ and

$$f_{\alpha-1}(A^4 * A^{t-4}A_3 * A_4A^{r-t-1}; A^r, Y) = T * A^6A_1D * FA^{r-t-1} .$$

Then by *IV.2.18*(*i*) and *IV.2.27*, we have

$$f_{\alpha-1}(A^4 * A^6A_1 * A_2A^{r-11}; A^r, Y) = T * A^6A_1 * DFA^{r-t-1},$$

whence, by *IV.2.33*,

$$A^r \overset{\alpha-1}{\simeq} A^{10}A_1DFA^{r-t-1} \text{ and } A^{10}A_1DFA^{r-t-1} \in \mathscr{K}_{\alpha-1} .$$

By 5.16, it follows from (7) that $l_\alpha(A^{10}A_1D) \geqslant t - 1$, that is, we may take $h = r - t - 1$.

7.12. *Set* $Z = [X, Y]_{\alpha-1}$. *If* $r, t \geqslant 9$, $\mathrm{Norm}(\alpha, X, r) = \varnothing$ *and* $\mathrm{Norm}(\alpha, Y, t) = \varnothing$, *then* $\mathrm{Norm}(\alpha, Z, r + t + 8)$ *is likewise empty.*

Suppose that $X \overset{\alpha-1}{\approx} PT$, $Y \overset{\alpha-1}{\approx} T^{-1}Q$ and $Z \mathrel{\overline{\overline{\circ}}} PQ$. Assume that $W \in \mathrm{Norm}(\alpha, Z, r + t + 8)$. By V_i we denote the i-th left segment of W, by W_i the start of W ending with V_i, and by U_i the end of W starting with V_i. Let $F * H$ be a section of the word PQ that is separating for the pair $(V_{r+4}\ V_{r+5})$. If F is a start of P, then by 6.3 we have $\mathrm{MutNorm}_{\alpha-2}(W_{r+3}, \phi(W_{r+3}; * P * Q, * P * T))$, so that $\mathrm{MutNorm}_{\alpha-1}(W_r, \phi(W_r; * P * Q, * P * T))$ by 6.9, that is, $\phi(W_r; * P * Q, * P * T) \in \mathrm{Norm}(\alpha, PT, r)$ by 5.5. But by 5.16, this contradicts the assumption that $\mathrm{Norm}(\alpha, X, r)$ be empty. If H is an end of Q, we get a similar contradiction to the assumption that $\mathrm{Norm}(\alpha, Y, t)$ is empty.

7.13. *Set*

$$Z = [[P, E]_{\alpha-2}, Q]_{\alpha-2},$$

where $P, Q \in \mathscr{A}_{\alpha-1}$ *and the word* $E \in \mathscr{R}_{\alpha-2}$ *either occurs in some word in* $\mathscr{A}_{\alpha-1}$ *or else it occurs in some elementary word of rank* $\alpha - 1$, *and in addition* $l_{\alpha-1}(E) \leqslant (n + 1)/2$. *Then* $Z \in \mathscr{A}_\alpha$.

By definition I.4.36 we have

$$P \overset{\alpha-2}{\approx} P_1T,\ E \overset{\alpha-2}{\approx} T^{-1}E_1,\ P_1E_1 \in \mathscr{R}_{\alpha-2},$$

$$P_1E_1 \overset{\alpha-2}{\approx} RS,\ Q \overset{\alpha-2}{\approx} S^{-1}Q_1,\ RQ_1 \mathrel{\overline{\overline{\circ}}} Z.$$

By 7.12 and *IV.1.19*, $Z \in \mathscr{R}_{\alpha-1}$.

Assume that $Z \notin \mathscr{A}_\alpha$, that is, that there exists an occurrence

$$W \in \mathrm{Norm}(\alpha, Z, 9).$$

If W' is a generating occurrence for the base of W, we have $\mathrm{MutNorm}_{\alpha-1}(W, W')$. By *IV.3.9*, $\mathrm{MutNorm}_{\alpha-2}(W, W')$. We denote the i-th right segment of the occurrence W by V_i, and the start of W ending with the pseudokernel V_i of rank $\alpha - 1$ by W_i. Let $F * H$ be a section of the word RQ_1 that is separating for the pair (V_3, V_2).

If H is an end of Q_1, then by 6.3 we have

$$\mathrm{MutNorm}_{\alpha-2}(V_1, \phi(V_1; R * Q_1 *, S^{-1} * Q_1 *)),$$

that is,

$$\phi(V_1; R * Q_1 *, S^{-1} * Q_1 *) \in \mathrm{Norm}(\alpha - 1, S^{-1}Q_1, 9).$$

But this is impossible by virtue of the assumption $Q \in \mathscr{A}_{\alpha-1}$ and proposition *II.5.16*.

Hence F is a start of R. Set

$$U \rightleftharpoons \phi(W_4; * R * Q_1, * R * S).$$

By 6.3 we have $\text{MutNorm}_{\alpha-2}(U, W_4)$, and consequently

$$\text{MutNorm}_{\alpha-2}(U, \phi(Q_4; W, W')).$$

Clearly, $l_\alpha(U) = l_\alpha(W_4) \geqslant 6$. By 2.15, we may assert that $P_1E_1 \overset{\circ}{=} RS$. Let $F' * H'$ be a section of the word P_1E_1 that is separating for the pair (U_2, U_3) consisting of the second and third left segments of rank α of the occurrence U. In a way similar to the foregoing, we get from 6.3 and the condition $P \in \mathscr{A}_{\alpha-1}$ that H' is an end of E_1 and

$$\text{MutNorm}_{\alpha-2}(U', \phi(U'; P_1 * E_1 *, T^{-1} * E_1 *)),$$

where U' is the end of U starting with its fourth left segment of rank α. Then by 2.15, the base of the occurrence $f_{\alpha-2}(\phi(U'; P_1 * E_1 *, T^{-1} * E_1 *); T^{-1}E_1, E)$ is a semi-integral 3-power of rank α, which is impossible by 2.11. If E occurs in an elementary word of rank $\alpha - 1$, this is impossible by 2.12. Hence $Z \in \mathscr{A}_\alpha$ in both cases.

7.14. *Suppose that* $\alpha > 1$, $X, Y \in \mathscr{K}_{\alpha-1}$, $P * E * Q \in \text{Reg}(\alpha - 1, X)$, $X \overset{\alpha-1}{\approx} Y$ *and* $f_{\alpha-1}(P * E * Q; X, Y) = R * D * S$.

Then there exist words $T, F \in \mathscr{A}_\alpha$ *such that*

$$E \overset{\alpha-1}{\approx} [T, D, F]_{\alpha-1}, \quad R \overset{\alpha-1}{\approx} [P, T]_{\alpha-1}, \quad S \overset{\alpha-1}{\approx} [F, Q]_{\alpha-1}. \tag{9}$$

In addition, if $P * E * Q$ *does not contain active kernels of rank* $\alpha - 1$, *then* $T, F \in \mathscr{A}_{\alpha-1}$ *and we may replace* $\alpha - 1$ *by* $\alpha - 2$ *in* (9).

By *V.2.5*, there exist words $T, F \in \mathscr{R}_{\alpha-1}$ satisfying all the conditions required, including the condition $T, F \in \mathscr{A}_\alpha$, which is satisfied by 7.13.

7.15. *Let* $W \rightleftharpoons P * A'A_1 * Q$ *and* $W' \rightleftharpoons R * D'D_1 * S$ *be occurrences in words* $Y, Z \in \mathscr{K}_{\alpha-1}$, *and let* $P_1 * A'A_1 * Q_1$ *and* $R_1 * D'D_1 * S_1$ *be generating occurrences in words* $X \in \text{Per}(\alpha, A)$ *and* $X' \in \text{Per}(\alpha, D)$, *where*

$$\text{MutNorm}_{\alpha-1}(W, P_1 * A'A_1 * Q), \quad \text{MutNorm}_{\alpha-1}(W', R_1 * D'D_1 * S_1), \tag{10}$$

$$Y \overset{\alpha-1}{\approx} Z \text{ and } f_{\alpha-1}(W; Y, Z) = W'. \tag{11}$$

If $t \geqslant 4$ *and*

$$\text{Corr}_A(U, V) \Longleftrightarrow \text{Corr}_D(f_{\alpha-1}(U; Y, Z), f_{\alpha-1}(V; Y, Z)) \tag{12}$$

for all occurrences $U, V \in \text{Ker}(\alpha - 1, Y)$ *contained in* W, *then there exists a word* $T \in \mathscr{A}_\alpha$ *such that*

$$A \overset{\alpha-1}{\approx} [T, D, T^{-1}]_{\alpha-1}, \tag{13}$$

$$R \overset{\alpha-1}{\approx} [P, T]_{\alpha-1}, \quad D_1S \overset{\alpha-1}{\approx} [T^{-1}, A_1Q]_{\alpha-1}. \tag{14}$$

For $\alpha = 1$, (11) and (12) give that $R \overset{\circ}{=} P$, $A \overset{\circ}{=} D$ and $A_1Q \overset{\circ}{=} D_1S$. Then T can be taken to be the empty word.

Suppose that $\alpha > 1$. For $i = 1$ we let U_i stand for the initial kernel of rank $\alpha - 1$ of the occurrence $P * A^t A_1 * Q$, and for $i = 2,3,4$ we let it stand for the result of shifting U_1 by $i - 1$ periods A to the right. Set $B \rightleftharpoons \mathrm{Bas}\,(U_1)$. Two cases are possible:

$$\text{I. } \partial(B) \leqslant \partial(A); \quad \text{II. } \partial(B) > \partial(A)\,.$$

Case I. In this case the kernels U_1 and U_2 do not intersect. Set $V_i \rightleftharpoons f_{\alpha-1}\,(U_i; Y, Z)$ for $i = 1,2,3,4$, and $C \rightleftharpoons \mathrm{Bas}\,(V_1)$. By relation (12), for $i = 2,3,4$ the kernel V_i is the result of shifting kernel V_1 by $i - 1$ periods D to the right. By *IV.2.18(g)*, the kernels V_1 and V_2 do not intersect, that is, $\partial(C) \leqslant \partial\,(D)$. Hence we have $A \overset{\circ}{=} Bu$, $D \overset{\circ}{=} Cv$, and

$$U_i = PA^{i-1} * B * uA^{t-i}A_1Q\,,$$
$$V_i = RD^{i-1} * C * vD^{t-i}D_1S$$

for $i = 1,2,3,4$. Since $\mathrm{MutNorm}_{\alpha-1}\,(U_1, U_2)$, by *IV.3.17* there exists a word $Y_1 \in \mathscr{K}_{\alpha-1}$ such that $Y \overset{\alpha-1}{\eqsim} Y_1$ and

$$f_{\alpha-1}\,(U_2; Y, Y_1) = PBu_1 * C * u_2BuA^{t-3}A_1Q\,. \tag{15}$$

Then by 7.14, there exist words T, $F \in \mathscr{A}_\alpha$ such that

$$B \overset{\alpha-1}{\eqsim} [T, C, F]_{\alpha-1}\,, \tag{16}$$
$$PBu_1 \overset{\alpha-1}{\eqsim} [PBu, T]_{\alpha-1}, \; u_2BuA^{t-3}A_1Q \overset{\alpha-1}{\eqsim} [F, uBuA^{t-3}A_1Q]_{\alpha-1}\,.$$

By *V. 1.11* and *IV. 1.21*, it follows from the last two relations that

$$u_1 \overset{\alpha-1}{\eqsim} [u, T]_{\alpha-1}, \; u_2 \overset{\alpha-1}{\eqsim} [F, u]_{\alpha-1}\,. \tag{17}$$

Since $Y_1 \overset{\alpha-1}{\eqsim} Z$ and $f_{\alpha-1}\,(PBu_1 * C * u_2A^{t-2}A_1Q; Y_1, Z) = V_2$, from *IV.2.35* we get

$$PBu_1 \overset{\alpha-1}{\eqsim} RCv \text{ and } u_2A^{t-2}A_1Q \overset{\alpha-1}{\eqsim} vD^{t-2}D_1S\,. \tag{18}$$

By 1.9 we have

$$\mathrm{MutNorm}_{\alpha-1}\,(P * BuBuB * uA^{t-3}A_1Q, PBu * BuBuB * A^{t-4}A_1Q)\,,$$

whence, setting $Y_2 \rightleftharpoons PBuBu_1Cu_2BuA^{t-4}A_1Q$, we get from (15) and *IV.3.26* that $Y_2 \in \mathscr{K}_{\alpha-1}$, $Y \overset{\alpha-1}{\eqsim} Y_2$ and

$$f_{\alpha-1}(U_3; Y, Y_2) = PBuBu_1 * C * u_2BuA^{t-4}A_1Q\,.$$

Then $Z \overset{\alpha-1}{\eqsim} Y_2$ and $f_{\alpha-1}\,(V_3; Z, Y_2) = f_{\alpha-1}\,(U_3; Y, Y_2)$, whence by *IV. 2. 35* it follows that

$$PBuBu_1 \overset{\alpha-1}{\approx} RCvCv\,. \tag{19}$$

Using (19), (16), (17), (18) and the properties of coupling of rank $\alpha-1$ (*V.1.1*, *V.1.7* and *V.1.8*), we get

$$RCvCv \overset{\alpha-1}{\approx} PBuBu_1 \overset{\alpha-1}{\approx} [PBu,\ T,\ C,\ F,\ u_1]_{\alpha-1} \overset{\alpha-1}{\approx}$$
$$\overset{\alpha-1}{\approx} [PB,\ u_1,\ C,\ Fu_1]_{\alpha-1} \overset{\alpha-1}{\approx} [RCv,\ C,\ F,\ u_1]_{\alpha-1}\,,$$

whence, by *V.1.11*, it follows that

$$v \overset{\alpha-1}{\approx} [F,\ u_1]_{\alpha-1} \overset{\alpha-1}{\approx} [F,\ u,\ T]_{\alpha-1}\,. \tag{20}$$

Then

$$A \overset{\circ}{=} Bu \overset{\alpha-1}{\approx} [[T,\ C,\ F]_{\alpha-1},\ u]_{\alpha-1} \overset{\alpha-1}{\approx} [T,\ C,\ v,\ T^{-1}]_{\alpha-1} \overset{\alpha-1}{\approx} [T,\ D,\ T^{-1}]_{\alpha-1}\,,$$

that is, relation (13) holds.

We find from (18), (16) and (20) that

$$RCv \overset{\alpha-1}{\approx} PBu_1 \overset{\alpha-1}{\approx} [P,\ T,\ C,\ F,\ u_1]_{\alpha-1} \overset{\alpha-1}{\approx} [P,\ T,\ Cv]_{\alpha-1}\,,$$

whence the first of the relations (14) follows by virtue of *V.1.11*.

Further, from (18), (13), (17) and (20) we get

$$vD^{t-2}D_1S \overset{\alpha-1}{\approx} u_2A^{t-2}A_1Q \overset{\alpha-1}{\approx} [F,\ u,\ T,\ D^{t-2},\ T^{-1},\ A_1Q]_{\alpha-1} \overset{\alpha-1}{\approx}$$
$$\overset{\alpha-1}{\approx} [vD^{t-2},\ T^{-1},\ A_1Q]_{\alpha-1}\,,$$

that is, the second relation in (14) also holds.

Case II. Since the kernels U_1 and U_2 intersect, there are no active kernels of rank $\alpha-1$ in W. Then by *IV.2.14*, we have Rel $(V, f_{\alpha-1}(V;\ Y,\ Z))$ for every kernel $V \in \mathrm{Ker}\,(\alpha-1,\ Y)$ contained in W. By *IV.2.24*, there is a word $Y' \in \mathscr{K}_{\alpha-1}$ such that $Y \overset{\alpha-2}{\approx} Y'$,

$$f_{\alpha-2}(P * A^tA_1 * Q;\ Y,\ Y') = R' * D^tD_1 * S' \tag{21}$$

and, for every occurrence $U \in \mathrm{Reg}\,(\alpha-1,\ \mathrm{Y})$ contained in $P * A^tA_1 * Q$,

$$f_{\alpha-2}(U;\ Y,\ Y') = \phi(f_{\alpha-1}(U;\ Y,\ Z);\ R * D^tD_1 * S,\ R' * D^tD_1 * S')\,. \tag{22}$$

By *IV. 2.11* we have $Y \overset{\alpha-1}{\approx} Y'$. By *IV. 2.18*, it follows from (11) and (21) that

$$f_{\alpha-1}(R * D^tD_1 * S;\ Z,\ Y') = R' * D^tD_1 * S'\,,$$

whence, by *IV.2.35* we get

$$R \overset{\alpha-1}{\approx} R' \text{ and } S \overset{\alpha-1}{\approx} S' . \tag{23}$$

It follows from (12) and (22) that

$$\mathrm{Corr}_A(U, V) \Longleftrightarrow \mathrm{Corr}_D(f_{\alpha-2}(U; Y, Y'), f_{\alpha-2}(V; Y, Y')) \tag{24}$$

for all occurrences of kernels $U, V \in \mathrm{Ker}\,(\alpha - 1, Y)$ contained in $P * A'A_1 * Q$.

By *IV.3.9*, *IV.3.2* and *IV.3.1* it follows from (10), (11), (21) that

$$\mathrm{MutNorm}_{\alpha-2}(W; P_1 * A'A_1 * Q_1)$$

and

$$\mathrm{MutNorm}_{\alpha-2}(R' * D'D_1 * S', R_1 * D'D_1 * S_1) .$$

Let W_1 be the union of the kernels U_1 and U_2, $V_1' \rightleftharpoons f_{\alpha-2}(U_1; Y, Y')$, $V_2' \rightleftharpoons f_{\alpha-1}(U_2; Y, Y')$ and $W_1' \rightleftharpoons f_{\alpha-2}(W_1; Y, Y')$. We find from I.4.8, *IV.3.3* and *IV.2.18(j)* that

$$\rho_{\alpha-2,A}(X) = \partial_{\alpha-2}(W_1) - \partial_{\alpha-2}(U_2) = \partial_{\alpha-2}(W_1') - \partial_{\alpha-2}(V_2') = \rho_{\alpha-2,D}(X') .$$

By *II.2.9*, this means that (24) is satisfied also for arbitrary kernels U, V of rank $\alpha - 2$ contained in $P * A'A_1 * Q$. Hence we may use the induction hypothesis, according to which there is a word $T \in \mathscr{A}_{\alpha-1}$ such that

$$A \overset{\alpha-2}{\approx} [T, D, T^{-1}]_{\alpha-2} ,$$
$$R' \overset{\alpha-2}{\approx} [P, T]_{\alpha-2}, \quad D_1 S' \overset{\alpha-2}{\approx} [T^{-1}, A_1 Q]_{\alpha-2} .$$

Then by (23), *V.1.1*, *V.2.1* and *IV.2.11*, the desired relations (13) and (14) hold.

7.16. *Suppose that* $X \in \mathrm{Per}\,(\alpha - 1, A)$ *and* $X \in \mathscr{K}_{\alpha-2}$. *Then one can find a word* $T \in \mathscr{A}_{\alpha-1}$ *and a minimal period* B *of rank* $\alpha - 1$ *such that* B^q *occurs in some word in* $\mathscr{M}_{\alpha-2}$ *and, for some* $k > 0$,

$$A \overset{\alpha-2}{\approx} [T, B^k, T^{-1}]_{\alpha-2} .$$

In addition, if no elementary r-power of rank $\alpha - 1$ *(no r-power of rank* α*) occurs in the word* A^2*, then no elementary* $(r + 8)$*-power of rank* $\alpha - 1$ *(no* $(r + 2)$*-power of rank* α *respectively) occurs in* B^2.

Firstly, by *IV.3.12* we find a word $Y \in \mathscr{M}_{\alpha-2}$ such that $X \overset{\alpha-2}{\approx} Y$. Let $W \rightleftharpoons P * A'A_1 * Q$ be a normal generating occurrence of rank $\alpha - 1$ in X, where $l_{\alpha-1,A}(W) \geqslant 10$ and $f_{\alpha-2}(W; X, Y) = R * E * S$. By 4.1 we have $E \overset{\circ}{=} D'D_1$, where D is an image of the period A in the occurrence $R * E * S$. By 4.3, for some generating occurrence $R' * D'D_1 * S'$ in a word $X' \in \mathrm{Per}\,(\alpha - 1, D)$ we have $X' \in \mathscr{K}_{\alpha-2}$ and $\mathrm{MutNorm}_{\alpha-2}(R * D'D_1 * S, R' * D'D_1 * S')$. Then by 7.15, there is a word $T \in \mathscr{A}_{\alpha-1}$ such that

$$A \overset{\alpha-2}{\approx} [T, D, T^{-1}]_{\alpha-2} .$$

Suppose that $D \mathrel{\overline{\circ}} B^k$, where B is a simple word. By 1.14, $X' \in \text{Per}(\alpha - 1, B)$. Clearly, the period B of the word E can be considered as an image of itself in the occurrence $R' * E * S'$. Thus, by 4.3, B is a minimal period of rank $\alpha - 1$.

If $kt \geqslant q$, then B^q occurs in a word $Y \in \mathscr{M}_{\alpha-2}$. In the contrary case, using $\text{MutNorm}_{\alpha-2}(R * D^t D_1 * S,\ D^{2q} * D^t D_1 * D_2 D^{2q-t-1})$ and *IV.3.14* with $X \rightleftharpoons D^{4q}$ we find a word $Z \in \mathscr{M}_{\alpha-2}$ such that $X' \overset{\alpha-2}{\sim} Z$ and

$$f_{\alpha-2}(D^{2q} * D^t D_1 * D_2 D^{2q-t-1};\ X', Z) = F * D^t D_1 * H .$$

Then, by *IV.2.18(i)* we have

$$f_{\alpha-2}(D^q * D^{q+t} D_1 * D_2 D^{2q-t-1};\ X', Z) = F_1 * G D^t D_1 * H ,$$

where $GD^t D_1 \in \text{Per}(B)$ because of *II.4.1* and I.2.9, that is, B^q occurs in the word $Z \in \mathscr{M}_{\alpha-2}$.

Assume that some elementary $(r + 8)$-power of rank $\alpha - 1$ occurs in B^2. Then, by *II.6.6*, $R * B^2 * B^{kt-2} D_1 S$ contains some occurrence $V \in \text{Norm}(\alpha - 1, Y, r)$; by 5.16, $f_{\alpha-2}(V;\ Y, X) \in \text{Norm}(\alpha - 1, X, r)$; and by *IV.2.18(g)* and *(h)*, $f_{\alpha-2}(V;\ Y, X)$ is contained in $P * A^2 * A^{t-2} A_1 Q$.

If some elementary $(r + 2)$-power of rank α occurs in B^2, then, using assertions 2.21 and 2.15 in place of 6.6 and 5.16, we prove similarly that an elementary r-power of rank α occurs in A^2.

7.17. *Let $P * E * Q$ be a generating occurrence of rank α in a word $Y \in \text{Int}(X, \alpha, A)$ such that $l_{\alpha,A}(P * E * Q) = r \geqslant q$, $E \mathrel{\overline{\circ}} E_1 D E_2$ and the occurrence $PE_1 * D * E_2 Q$ starts with the i-th and ends with the $(i + 6)$-th left segments of rank α of the occurrence $P * E * Q$, where $q_1 < i < r - q_1$. Then there exists a section $PE_1 D_1 * D_2 E_2 Q$ of Y such that*

$$PE_1 D_1 \in \mathscr{R}_{\alpha-1} \text{ and } D_2 E_2 Q \in \mathscr{R}_{\alpha-1} . \tag{25}$$

For $\alpha = 1$ everything is clear. Suppose that $\alpha > 1$.

We consider first the case where Y is of type I. By 2.17, in this case there exists a generating occurrence $R * E * S$ of rank $\alpha - 1$ in some word $Y_1 \in \text{Int}(X, \alpha - 1, A)$ such that $Y \overset{\alpha-1}{\sim} Y_1$, $f_{\alpha-1}(P * E * Q;\ Y, Y_1) = R * E * S$, $l_{\alpha-1,A}(R * E * S) \geqslant q$, and in addition the i-th and $(i + 6)$-th left segments of rank $\alpha - 1$ of the occurrence $R * E * S$ are contained in the occurrence $RE_1 * D * E_2 S$. Then by induction there exists a section $F * H$ of Y_1 such that $F, H \in \mathscr{R}_{\alpha-2}$, $F = RE_1 D_1$ and $H = D_2 E_2 S$. Since $Y_1 \in \mathscr{R}_{\alpha-1}$, by *IV.1.18* we have $F, H \in \mathscr{P}_{\alpha-1}$.

Assume that $F \notin \mathscr{R}_{\alpha-1}$. Then for some kernel $U \in \text{Ker}(\alpha - 1, F)$, we will have $l_{\alpha-1}(U) > n - 176$. Since 2.5 gives that there are no active kernels of rank $\alpha - 1$ in $R * E * S$, it follows from *IV.1.20* that U is not contained in the occurrence $R * E_1 D_1 *$. Since $i > q_1$, by 2.12 there exist G and T such that $E_1 D_1 \mathrel{\overline{\circ}} GT$, $l_{\alpha,A}(T) > 9$ and $U \ll RG * T *$. By 4.7, some elementary 9-power of rank α occurs in T. Then by 6.10, $\phi(U;\ * F *,\ * F * H) \in \text{Ker}(\alpha - 1, Y_1)$ since $U \in \text{Ker}(\alpha - 1, F)$, which is impossible as $Y_1 \in \mathscr{R}_{\alpha-1}$. Hence $F \in \mathscr{R}_{\alpha-1}$.

In a similar way one proves that $H \in \mathscr{R}_{\alpha-1}$. Further, by 6.13, relations (25) follow since $Y, F, H \in \mathscr{R}_{\alpha-1}$.

Suppose now that Y is of type II. Then by 2.5 and 2.4 there are supporting kernels

$$V_1, V_2, V_3, V_4, V_5, V_6, V_7,$$

of Y of rank $\alpha - 1$ contained in $PE_1 * D * E_2Q$ such that $V_i < V_{i+1}$ for $1 \leqslant i \leqslant 6$ and $V_j \in \text{Act}(\alpha - 1, Y)$ for $2 \leqslant j \leqslant 6$, whence it follows by *IV.1.3* that $V_j \in \text{Norm}(\alpha - 1, Y, q)$ and the maximal normalised continuation $\bar{V}_j$ of V_j does not intersect V_{j-1} nor V_{j+1}. Set $V_4 = PE_1D' * T * D''E_2Q$. By *II.7.18*, there exists a section $F * H$ of Y such that $F, H \in \mathscr{R}_{\alpha-2}$, $F = PE_1D'T_1$ and $H = T_2D''E_2Q$, where T_1, T_2 are elementary 9-powers of rank $\alpha - 1$. For $i = 2, 3$ we set

$$U_i \rightleftharpoons \phi(V_i; * F * H, * F *).$$

By *II.6.11*, $U_2, U_3 \in \text{Norm}(\alpha - 1, F, q)$.

Assume that $F \notin \mathscr{R}_{\alpha-1}$. Since $F \in \mathscr{P}_{\alpha-1}$, there exists a $U \in \text{Ker}(\alpha - 1, F)$ such that $l_{\alpha-1}(U) > n - 176$.

By *IV.1.23*, we may assert that there is no occurrence of an elementary $4q$-power of rank $\alpha - 1$ in the word D. Then $U < U_2$. Since $V_2 \in \text{Act}(\alpha - 1, Y)$ and $V_2 \ll V_3$, it follows from *III.3.29* that $U_2 \in \text{Act}(\alpha - 1, F)$, whence by *IV.1.3* and *IV.1.14* it follows that $\phi(U; * F *, * F * H) \in \text{Ker}(\alpha - 1, Y)$. But this is impossible since $Y \in \mathscr{R}_{\alpha-1}$. Hence $F \in \mathscr{R}_{\alpha-1}$.

The relation $H \in \mathscr{R}_{\alpha-1}$ is proved in the analogous fashion.

7.18. *Let $Z \in \mathscr{R}_{\alpha-1}$, $Z \mathrel{\bar{\bar{\circ}}} RES$, where E is an elementary q-power of rank α, $E \mathrel{\bar{\bar{\circ}}} E_1DE_2$, and the occurrence $RE_1 * D * E_2S$ begins with the i-th left and ends with the $(i + 6)$-th left segments of the occurrence $R * E * S$, where $q_1 < i < l_\alpha(E) - q_1$. Then there is a section $RE_1D_1 * D_2E_2S$ of Z such that*

$$RE_1D_1 \in \mathscr{R}_{\alpha-1} \text{ and } D_2E_2S \in \mathscr{R}_{\alpha-1}. \tag{26}$$

Let $P * E * Q$ be a generating occurrence in the word $Y \in \text{Int}(X, \alpha, A)$. Then by 7.17, there exists a section $PE_1D_1 * D_2E_2Q$ of the word Y such that $PE_1D_1 \in \mathscr{R}_{\alpha-1}$, $D_2E_2Q \in \mathscr{R}_{\alpha-1}$, $l_{\alpha,A}(E_1) > 9$ and $l_{\alpha,A}(E_2) > 9$. We get relation (26) from this in view of 6.13.

7.19. *Suppose that $X \in \mathscr{K}_{\alpha-1}$, $P * EDF * Q \in$ Norm $(\alpha, X, 31)$, $l_\alpha(E) \geqslant 9$, $PE * D * FQ \in$ Norm $(\alpha, X, 9)$, and $l_\alpha(F) \geqslant 9$. Then one can find words E_1, D_1 and F_1 such that, if the occurrence $R * EDF * S$ is normalised, then the relation*

$$REDFS \overset{\alpha-1}{\simeq} RE_1D_1F_1S$$

holds for every word $REDFS \in \mathscr{K}_{\alpha-1}$, where the occurrence $RE * D * FS$ transforms locally to $RE_1 * D_1 * F_1S$, the occurrence $RE_1 * D_1 * F_1S$ is periodised in rank $\alpha - 1$, and the word $RE_1D_1F_1S$ lies in the same set $\mathscr{K}_{\alpha-1}$, $\mathscr{L}_{\alpha-1}$ or $\mathscr{M}_{\alpha-1}$ that $REDFS$ lies in.

If in addition the word RED is a start (or DFS is an end) of some word $Y \in \mathcal{L}_{\alpha-1}$, *then* $RE_1D_1 \in \mathcal{L}_{\alpha-1}$ *(or* $D_1F_1S \in \mathcal{L}_{\alpha-1}$ *respectively).*

Suppose that $E \mathrel{\overline{\overset{\circ}{=}}} E'u$ and $F \mathrel{\overline{\overset{\circ}{=}}} vF'$, where

$$P * E' * uDFQ \in \text{Norm}(\alpha, X, 9),$$
$$PEDv * F' * Q \in \text{Norm}(\alpha, X, 9),$$

and that each kernel $V \in \text{Ker}(\alpha - 1, X)$ not intersecting $PE * D * FQ$ is contained either in $* PE' * uDFQ$ or in $PEDv * F'Q *$. By *IV.3.12*, we can find a word $X_1 \in \mathcal{M}_{\alpha-1}$ such that $X \overset{\alpha-1}{\rightleftarrows} X_1$. Suppose that $f_{\alpha-1}(PE * D * FQ; X, X_1) = P_1 * D_1 * Q_1$. Then by *IV.2.26*, there is a word $X_2 \in \mathcal{K}_{\alpha-1}$, belonging to the same set $\mathcal{L}_{\alpha-1}$ or $\mathcal{M}_{\alpha-1}$ to which X belongs and such that

$$X \overset{\alpha-1}{\rightleftarrows} X_2,\ f_{\alpha-1}(PE * D * FQ;\ X, X_2) = P_2 * D_1 * Q_2$$

and the occurrence $PE * D * FQ$ transforms into $P_2 * D_1 * Q_2$ locally in rank $\alpha - 1$, that is, we have

$$f_{\alpha-1}(P * E' * uDFQ;\ X, X_2) = P * E' * u_1D_1Q_2$$

and

$$f_{\alpha-1}(PEDv * F' * Q;\ X, X_2) = P_2D_1v_1 * F' * Q.$$

By 5.5, $P_2 * D_1 * Q_2$ is periodised in rank $\alpha - 1$. Hence $E'u_1$, D_1 and v_1F' are the words required for the occurrence $P * EDF * Q$. It remains for us to show that they are suitable even for an arbitrary word $REDFS \in \mathcal{K}_{\alpha-1}$, where $R * EDF * S$ is normalised.

We remark that we may assume by *IV.2.23* that $l_\beta(U) \leqslant (n+1)/2$ for each kernel U of rank β contained in $PE' * u_1D_1v_1 * F'Q$, $0 < \beta \leqslant \alpha - 1$. We set

$$Z \rightleftharpoons RE'uDvF'S \text{ and } Z_1 \rightleftharpoons RE'u_1D_1v_1F'S.$$

Since $Z, X_2 \in \mathcal{K}_{\alpha-1}$, it follows by 6.13 that $Z_1 \in \mathcal{K}_{\alpha-1}$. Furthermore, by virtue of the remark made earlier and with the aid of the arguments used in 6.13, it is easily shown that Z_1 belongs to whichever of the sets $\mathcal{L}_{\alpha-1}$ or $\mathcal{M}_{\alpha-1}$ to which Z belongs, and if *RED* is a start (or *DFS* is an end) of some word $Y \in \mathcal{L}_{\alpha-1}$, then $RE'u_1D_1 \in \mathcal{L}_{\alpha-1}$ ($D_1v_1F'S \in \mathcal{L}_{\alpha-1}$, respectively). Then since $X \overset{\alpha-1}{\rightleftarrows} X_2$, it follows by 6.13, *IV.1.21* and *V.2.1* that $Z \overset{\alpha-1}{\rightleftarrows} Z_1$. Since $R * E' * uDvF'Q \in \text{ComReg}(\alpha - 1, Z)$ we have by 7.7 that

$$f_{\alpha-1}(R * E' * uDvF'Q;\ Z, Z_1) = R * E' * u_1D_1v_1F'S.$$

In an analogous way we obtain

$$f_{\alpha-1}(RE'uDv * F' * Q;\ Z, Z_1) = RE'u_1D_1v_1 * F' * S.$$

Then $f_{\alpha-1}(R * EDF * Q; Z, Z_1) = R * E'u_1D_1v_1F' * S$, whence by *IV.3.5*

$$f_{\alpha-1}(RE * D * FQ; Z, Z_1) = RE'u_1 * D_1 * v_1F'S.$$

Since each kernel $V \in \text{Ker}(\alpha - 1, Z)$ not intersecting $RE * D * FS$ is contained in one of the occurrences $* RE' * uDFS$ or $REDv * F'S *$, it follows that $RE * D * FS$ transforms to $RE'u_1 * D_1 * v_1F'S$ locally in rank $\alpha - 1$, as was to be proved.

7.20. *Suppose that* $X \in \mathscr{K}_{\alpha-1}$, $R * A^tA_1 * Q \in \text{Norm}(\alpha, X, 9)$, *that the occurrence* $V \rightleftharpoons R * A^tA_1 * Q$ *is periodised in rank* $\alpha - 1$, *and that* $A \mathrel{\underset{\circ}{\equiv}} A_1A_2$ *is an elementary period of rank* α.

If $h \geqslant l_\alpha(W) - l_\alpha(V) + 28$, *where W is a maximal normalised continuation of V to the left, then words*

$$R_1FA^{10} \in \mathscr{K}_{\alpha-1} \text{ and } A^{-10}F^{-1}NA^{-r} \in \mathscr{K}_{\alpha-1}$$

can be found satisfying

$$RA^{10} \overset{\alpha-1}{\simeq} R_1FA^{10},\ A^{-h} \overset{\alpha-1}{\simeq} A^{-10}F^{-1}NA^{-r}$$

and

$$R_1NA^{-r} \in \mathscr{K}_{\alpha-1},$$

where $r = h - l_\alpha(W) + l_\alpha(V) - 18$.

If in addition $X \in \mathscr{L}_{\alpha-1}$, *then for every regular occurrence* $P * E * G$ *of rank* $\alpha - 1$ *contained in* $* R_1 * FA^{10}$ (*or in* $A^{-10}F^{-1} * NA^{-r} *$), *the word PE is a start of R* (*EG is an end of* A^{-h}, *respectively*).

If $h \geqslant l_\alpha(U) - l_\alpha(V) + 28$, *where U is a maximal normalised continuation of V to the right, then words*

$$A^{10}A_1FQ_1 \in \mathscr{K}_{\alpha-1} \text{ and } A^{-r}NF^{-1}A_1^{-1}A^{-10} \in \mathscr{K}_{\alpha-1}$$

can be found that satisfy

$$A^{10}A_1Q \overset{\alpha-1}{\simeq} A^{10}A_1FQ_1,\quad A^{-h} \overset{\alpha-1}{\simeq} A^{-r}NF^{-1}A_1^{-1}A^{-10}$$

and

$$A^{-r}NQ_1 \in \mathscr{K}_{\alpha-1},$$

where $r = h - l_\alpha(U) + l_\alpha(V) - 19$.

If, in addition, $X \in \mathscr{L}_{\alpha-1}$, *then for every regular occurrence* $P * E * S$ *of rank* $\alpha - 1$ *contained in* $A^{10}A_1F * Q_1 *$ (*or in* $* A^{-r}N * F^{-1}A_1A^{-10}$), *the word ES is an end of Q* (*the word PE is a start of* A^{-h}, *respectively*).

Generally speaking, the second half of the lemma is not fully symmetric with the first. For full symmetry one has to take instead of the left period $A \overset{\circ}{=} A_1A_2$ of $A'A_1$, its right period A_2A_1. From the symmetric assertions

$$(A_2A_1)^{10}FQ_1,\ (A_2A_1)^{-r}NF^{-1}(A_2A_1)^{-10},\ (A_2A_1)^{-r}NQ_1 \in \mathscr{K}_\alpha,$$
$$(A_2A_1)^{10}Q \overset{\alpha-1}{\sim} (A_2A_1)^{10}FQ_1$$

and

$$(A_2A_1)^{-h} \overset{\alpha-1}{\sim} (A_2A_1)^{-r}NF^{-1}(A_2A_1)^{-r}$$

thus obtained, the required relations follow by 6.13 and *V.2.1*. In particular if, using *V.2.1* we multiply both sides of the last relationship by A_2^{-1} on the left and by A_1^{-1} on the right we get

$$A^{-h-1} \overset{\alpha-1}{\sim} A^{-r}A_2^{-1}NF^{-1}A_1^{-1}A^{-r},$$

after which we can take $A_2^{-1}N$ as a new N. This is why the difference between h and r in the second half changed by 1. Hence we may confine ourselves to proving the first half of the lemma.

Suppose that W is a maximal normalised continuation of the occurrence V to the left and that $h \geqslant l_\alpha(W) - l_\alpha(V) + 28$. Since $A'A_1$ occurs in some word $Y \in \mathscr{M}_{\alpha-1}$, we have by 6.13 that $A'A_1 \in \mathscr{L}_{\alpha-1}$ and $A^k \in \mathscr{L}_{\alpha-1}$ for arbitrary $k \geqslant 10$. Since $X \in \mathscr{K}_{\alpha-1}$ it follows that $RA^{10} \in \mathscr{K}_{\alpha-1}$. By 6.10 we have

$$\text{MutNorm}_{\alpha-1}(R * A^4A_1 * A_2A^5,\ R * A^4A_1 * A_2A^{t-5}A_1Q),$$

by 5.5 and *IV.3.6* we have

$$\text{MutNorm}_{\alpha-1}(R * A^4A_1 * A_2A^{t-5}A_1Q,\ A^{14} * A^4A_1 * A_2A^{t+5})$$

and by 6.6,

$$\text{MutNorm}_{\alpha-1}(A^{14} * A^4A_1 * A_2A^{t+5},\ A^4 * A^4A_1 * A_2A^5).$$

Hence

$$\text{MutNorm}_{\alpha-1}(R * A^4A_1 * A_2A^5,\ A^4 * A^4A_1 * A_2A^5). \tag{27}$$

By *IV.2.22*, there exists an R' such that $RA^{10} \overset{\alpha-1}{\sim} R'A^{10}$, $f_{\alpha-1}(R * A^4A_1 * A_2A^5;\ RA^{10}, R'A^{10}) = R' * A^4A_1 * A_2A^5$ and, for every kernel V of rank β contained in $* R' * A^{10}$, $0 < \beta \leqslant \alpha - 1$, we have $l_\beta(V) \leqslant (n+1)/2$. Then the fact that $R'A^{10} \in \mathscr{L}_{\alpha-1}$ follows from (27) and the fact that $A^{14} \in \mathscr{L}_{\alpha-1}$.

If $X \in \mathscr{L}_{\alpha-1}$, then by 6.13 we have $RA^{10} \in \mathscr{L}_{\alpha-1}$. It is enough to consider this case since in the contrary case we can consider the word $R'A^{10}$ instead of RA^{10}.

By the principle of symmetry, $A^{-h} \in \mathscr{L}_{\alpha-1}$. By *V.1.3*, there are words $R_1 T \in \mathscr{K}_{\alpha-1}$ and $T^{-1} Y_1 \in \mathscr{K}_{\alpha-1}$ such that

$$RA^{10} \overset{\alpha-1}{\simeq} R_1 T, \; A^{-h} \overset{\alpha-1}{\simeq} T^{-1} Y_1 , \tag{28}$$

$R_1 Y_1 \in \mathscr{K}_{\alpha-1}$ and conditions (a), (b) and (c) of *V.1.2* are satisfied by the relations (28) for rank $\alpha - 1$. It then follows from (c) that $T \overset{\circ}{=} FA^{10}$ for some F. By (a) and (c), the word PE is a start of the word R for every occurrence $P * E * G \in \text{Reg}(\alpha - 1, R_1 T)$ contained in $* R_1 * T$, and EG is an end of A^{-h} for every occurrence $P * E * G \in \text{Reg}(\alpha - 1, T^{-1} Y)$ contained in $T^{-1} * Y_1 *$. Suppose that

$$r = h - l_\alpha(W) + l_\alpha(V) - 18 . \tag{29}$$

We consider the occurrences

$$U_1 \rightleftharpoons A^{-4} A_2^{-1} * A_1^{-1} A^{-h+r+6} * A^{-r-1}$$

and

$$U_2 \rightleftharpoons A^{-h+r+1} A_2^{-1} * A_1^{-1} A^{-r+4} * A^{-4} .$$

Since $U_2 \in \text{Reg}(\alpha - 1, A^{-h})$, by (a) it remains only to prove that $f_{\alpha-1}(U_2; A^{-h}, T^{-1} Y_1)$ is contained in $T^{-1} * Y_1 *$.

Assume that this is not so. Then, by *IV.2.18(g)*, the occurrence $f_{\alpha-1}(U_1; A^{-h}, T^{-1} Y_1)$ is contained in $* T^{-1} * Y_1$. Since

$$f_{\alpha-1}(A^{-4} A_2^{-1} * A_1^{-1} A^{-5} * A^{-h+10}; A^{-h}, T^{-1} Y_1) = A^{-4} A_2^{-1} * A_1^{-1} A^{-5} * F^{-1} Y_1$$

by (c), then according to *IV.2.18(i)* we have

$$f_{\alpha-1}(U_1; A^{-h}, T^{-1} Y_1) = A^{-4} A_2^{-1} * A_1^{-1} A^{-5} H^{-1} * L^{-1} Y_1 ,$$

where $H^{-1} L^{-1} \overset{\circ}{=} F^{-1}$. Therefore, by *V.2.1* and *IV.2.29*, $A^{-10} H^{-1}$ is an elementary word of rank α with initial period A^{-1}, where $l_\alpha(A^{-10} H^{-1}) \geqslant h - r - 2$. We therefore have $T \overset{\circ}{=} LHA^{10}$, where HA^{10} is an elementary word of rank α with initial period A and

$$l_\alpha(HA^{10}) \geqslant h - r - 2 . \tag{30}$$

From the facts that $R_1 LHA^{10} \in \mathscr{K}_{\alpha-1}$ and $X \in \mathscr{K}_{\alpha-1}$, it follows by 6.13 that $R_1 LHA^t A_1 Q \in \mathscr{K}_{\alpha-1}$, whence we get, by (28), *IV.1.21* and *V.2.1*,

$$R_1 LHA^t A_1 Q \overset{\alpha-1}{\simeq} X .$$

Then by 7.7, we have

$$f_{\alpha-1}(V; X, R_1 LHA^t A_1 Q) = R_1 LH * A^t A_1 * Q .$$

Let W_1 be the maximal normalised continuation to the left of the word $R_1LH * A^tA_1 * Q$. By 5.16, we have $l_\alpha(W_1) = l_\alpha(W)$. Since $R_1L * HA^tA_1 * Q$ is a continuation of $R_1LH * A^tA_1 * Q$ to the left, then by 6.18 we have

$$l_\alpha(W_1) \geqslant l_\alpha(HA^tA_1) - 4 \geqslant l_\alpha(HA^{10}) + l_\alpha(V) - 15,$$

whence

$$l_\alpha(W) \geqslant h - r + l_\alpha(V) - 17$$

follows by (30). But this contradicts (29). Hence A^{-r} is an end of Y_1, that is, $Y_1 \overset{\circ}{\equiv} NA^{-r}$ for some N.

7.21. *Suppose that* $P * E * Q \in$ Norm $(\alpha, X, 9)$, $R * D * S \in$ Norm $(\alpha, Y, 9)$, $E \in$ Per (A), *and that* $D \in$ Per (A) *and* A *is an elementary period of rank* α. *If the occurrence* $P * E * Q$ *is periodised in rank* $\alpha - 1$ *and* $\text{MutNorm}_{\alpha-1}(P * E * Q, R * D * S)$, *then* $E \overset{\circ}{\equiv} D$.

Suppose that $P * E * Q$ is periodised in rank $\alpha - 1$, that is, $\text{MutNorm}_{\alpha-1}(P * E * Q, P_1 * E * Q_1)$ for some $P_1EQ_1 \in \mathscr{M}_{\alpha-1}$. We may clearly assume that A is a left period of E, that is, $E \overset{\circ}{\equiv} A^tA_1$, where $A \overset{\circ}{\equiv} A_1A_2$. Set $Z \rightleftharpoons P_1A^rA_1Q_1$, where $r = t + \partial(D)$. Then, by 7.10, we have $Z \in \mathscr{M}_{\alpha-1}$ and $P_1 * A^rA_1 * Q_1 \in$ Norm $(\alpha, Z, 9)$. It is clear that D occurs in A^rA_1. Suppose that $A^rA_1 \overset{\circ}{\equiv} FDH$. By 5.4 and 5.5 we have

$$\text{MutNorm}_{\alpha-1}(P * E * Q, P_1 * E * A_2A^{r-t-1}A_1Q_1)$$

and

$$\text{MutNorm}_{\alpha-1}(R * D * S, P_1F * D * HQ_1),$$

whence, since $\text{MutNorm}_{\alpha-1}(P * E * Q, R * D * S)$, there follows

$$\text{MutNorm}_{\alpha-1}(P_1 * E * A_2A^{r-t-1}A_1Q_1, P_1F * D * HQ_1).$$

By *IV.3.19* we then have $E \overset{\circ}{\equiv} D$.

7.22. *Suppose that* A *and* B *are elementary periods of rank* α *and that the words* A^q *and* B^q *occur in some word in* $\mathscr{M}_{\alpha-1}$. *If* Rel (A^q, B^q), *then a word* $T \in \mathscr{R}_{\alpha-1}$ *can be found such that*

$$A \overset{\alpha-1}{\simeq} [T, B, T^{-1}]_{\alpha-1}.$$

Suppose that $PA^qQ \in \mathscr{M}_{\alpha-1}$ and $RB^qS \in \mathscr{M}_{\alpha-1}$. From Rel (A^q, B^q) it follows that we can find generating occurrences $P_1 * A^q * Q_1$ and $R_1 * B^q * S_1$ in integral words $Y_1 \in$ Int (X_1, α, D) and $Y_2 \in$ Int (X_2, α, D) respectively, where D is an elementary period of rank α and $X_1, X_2 \in$ Per (α, D). Without loss of generality we may assume that $V \rightleftharpoons P_1A^4 * A^{10}A_1 * A_2A^{q-15}Q_1$ is a normal generating occurrence. Suppose that $f_{\alpha-1}(V; Y_1, X_1) = P_2 * E * Q_2$. Since $l_{\alpha,D}(E) \leqslant 11$, there is an occurrence $R_2 * E * S_2$ in X_2 which corresponds to $P_2 * E * Q_2$ in phase in the sense of I.2.6 and such that the occurrence $f_{\alpha-1}(R_2 * E * S_2; X_2, Y_2)$ is contained in $R_1 B^4 * B^{q-8} * B^4 S_1$. Suppose that

$$f_{\alpha-1}(R_2 * E * S_2; X_2, Y_2) = R_3 * F * S_3 .$$

By 1.8 we have

$$\mathrm{MutNorm}_{\alpha-1}(P_2 * E * Q_2, R_2 * E * S_2),$$

whence by *IV.3.2* and *IV.3.1* there follows $\mathrm{MutNorm}_{\alpha-1}(V, R_3 * F * S_3)$. Then by 6.6 and *IV.3.1* we have

$$\mathrm{MutNorm}_{\alpha-1}(PA^4 * A^{10}A_1 * A_2A^{q-15}Q,$$
$$\phi(R_3 * F * S_3; R_1 * B^q * S_1, R * B^q * S)) .$$

Consequently, by *IV.3.18*, there is a word $Z \in \mathscr{M}_{\alpha-1}$ such that $PA^qQ \overset{\alpha-1}{\approx} Z$ and

$$f_{\alpha-1}(PA^4 * A^{10}A_1 * A_2A^{q-15}Q; PA^qQ, Z) = P_3 * F * Q_3 . \tag{31}$$

By 4.1, $F \overset{\circ}{=} C^{10}C_1$, where C is an image of the period A in the occurrence $P_3 * F * Q_3$. Moreover, since $F \in \mathrm{Per}(B)$ by hypothesis, C is a cyclic shift of the period B, that is $B \overset{\circ}{=} B_1B_2$ and $C \overset{\circ}{=} B_2B_1$ for some B_1 and B_2. Further, from (31) it follows by 4.3 and 7.15 that there exists a word $H \in \mathscr{A}_\alpha$ such that

$$A \overset{\alpha-1}{\approx} [H, C, H^{-1}]_{\alpha-1} .$$

By *V.1.4*, it is possible to implement the coupling $[H, B_1^{-1}]_{\alpha-1}$. Let $T = [H, B_1^{-1}]_{\alpha-1}$. Then by *V.1.8*, *V.1.9* and *V.1.1* we have

$$[T, B, T^{-1}]_{\alpha-1} \overset{\alpha-1}{\approx} [H, B_1^{-1}, B_1B_2, B_1, H^{-1}]_{\alpha-1} \overset{\alpha-1}{\approx}$$
$$\overset{\alpha-1}{\approx} [H, B_2B_1, H^{-1}]_{\alpha-1} \overset{\alpha-1}{\approx} A ,$$

as required.

Chapter III. Reversals of Rank α

§ 1. Elementary Properties of Reversals

1.1. *Suppose that $r \geqslant 9$, the transition $X \to Y$ is a 9-reversal of an occurrence $W \in$ Norm $(\alpha, X, 9)$ and U is an image of W in this reversal. The transition $X \to Y$ is an r-reversal of W if and only if the maximal normalised continuations of W and U contain not less than r segments.*

If $r \geqslant k \geqslant 9$ and the transition $X \to Y$ is an r-reversal of rank α of W, then it is a k-reversal of rank α of this same occurrence.

For simple reversals this follows at once from definitions I.4.16 and I.4.17, and in the general case from I.4.19 and II.5.16.

1.2. *If $X \to Y$ is an r-reversal of rank α of the occurrence W and if W_1 is an image of W in this reversal, then the inverse transition $Y \to X$ is an r-reversal of W_1 and W is an image of W_1 in it. If in addition the occurrence $V \in$ Reg $(\alpha - 1, X)$ is stable in the reversal $X \to Y$ and $f_{X \to Y}(V) = U$, then U is stable in $Y \to X$ and $f_{Y \to X}(U) = V$.*

Let $X \to Y$ be a simple r-reversal of rank α of W and let W_1 be an image of W in this reversal, that is, we have occurrences

$$RT * A^t A_1 * T_1 Q \in \text{MaxNorm}\,(\alpha, X, r)\,,$$

$$RH * D^h D_1 * H_1 Q \in \text{MaxNorm}\,(\alpha, Y, r)$$

and

$$P * A^t A_1 * S \in \text{Norm}\,(\alpha, Z, 9)\,, \tag{1}$$

where $Z \in \mathscr{M}_{\alpha-1}$,

$$\text{Comp}\,(W, RT * A^t A_1 * T_1 Q),\ \text{Comp}\,(W_1, RH * D^h D_1 * H_1 Q)\,,$$

and all the conditions listed in I.4.16 hold. Since D is a cyclic shift of A^{-1}, and A is an elementary period, it follows from the principle of symmetry and II.3.5 that D is also an elementary period of rank α. Furthermore, $D^h D_1$ occurs in the word $A_1^{-1} A^{-h-2}$, that is, for some F and G we have $A_1^{-1} A^{h-2} \mathrel{\overset{\circ}{=}} F D^h D_1 G$. From $Z \in \mathscr{M}_{\alpha-1}$ and (1) it follows by II.7.10 and the principle of symmetry that $S^{-1} A_1^{-1} A^{-h-2} P^{-1} \in \mathscr{M}_{\alpha-1}$ and $S^{-1} * A_1^{-1} A^{-h-2} * P^{-1} \in$ Norm $(\alpha,\ S^{-1} A_1^{-1} A^{-h-2} P^{-1},\ 9)$. Then by II.5.4, we have $S^{-1} F * D^h D_1 * G P^{-1} \in$ Norm $(\alpha,\ S^{-1} A_1^{-1} A^{-h-2} P^{-1},\ 9)$. Hence all the conditions are fulfilled that, according to I.4.16, are necessary to ensure that the transition $Y \to X$ can be regarded as a simple reversal of W_1.

The second assertion about the simple r-reversal $X \to Y$ follows trivially from I.4.17.

For an arbitrary r-reversal of rank α our assertions follow from what we have proved by I.4.19 and the fact that the mapping $V_1 = f_{\alpha-1}(V; X, X_1)$ (see *IV.2.18(d)*) is symmetric.

1.3. *Suppose that $U, W \in$ Norm $(\alpha, X, 9)$ and Comp (U, W). Then an arbitrary r-reversal $X \to Y$ of the occurrence U can be regarded as an r-reversal of W. Hence the occurrences $V \in$ Reg $(\alpha - 1, X)$ which are stable in a reversal of the occurrence U are precisely those occurrences which are stable in a reversal of W.*

If U' is an image of U in this reversal then all occurrences $W' \in$ Norm $(\alpha, Y, 9)$ which are compatible with U', and only they, are the images of W in this reversal.

There exists a unique maximal image of W in its reversal $X \to Y$.

For simple reversals all these assertions, except the last, follow from II.5.12 and the corresponding definitions. In the general case we still need to take advantage of relation (12) from II.5.16. The last assertion holds by virtue of the remark on II.7.3 and definition I.4.20.

1.4. *Suppose that $X \overset{\alpha-1}{\simeq} X_1$, $Y \overset{\alpha-1}{\simeq} Y_1$, $X \to Y$ is an r-reversal of an occurrence $W \in$ Norm $(\alpha, X, 9)$ and U is an image of W in this reversal. Then the transition $X_1 \to Y_1$ is an r-reversal of the occurrence $f_{\alpha-1}(W; X, X_1)$ and $f_{\alpha-1}(U; Y, Y_1)$ is an image of it in this reversal. If, in addition, the occurrence $V \in$ Reg $(\alpha - 1, X)$ is stable in the reversal $X \to Y$, then $f_{\alpha-1}(V; X, X_1)$ is stable in the reversal $X_1 \to Y_1$ and*

$$f_{X_1 \to Y_1}(f_{\alpha-1}(V; X, X_1)) = f_{\alpha-1}(f_{X \to Y}(V); Y, Y_1)$$

(see *IV.2.18(c)*).

This follows directly from definition I.4.19 and the fact that the relation $X \overset{\alpha-1}{\simeq} X_1$ and the mapping $f_{\alpha-1}(V; X, X_1)$ are transitive (see *IV.2.18(e)*).

1.5. *Suppose that $X \to Y$ is a 9-reversal of rank α of the occurrence W. If U is an image of W in this reversal, then* Rel (W, U^{-1}) *and* $\neg$ Rel (W, U).

Suppose that $X \overset{\alpha-1}{\simeq} RTA^t A_1 T_1 Q$, $Y \overset{\alpha-1}{\simeq} RHD^h D_1 H_1 Q$ and that the transition

$$RTA^t A_1 T_1 Q \to RHD^h D_1 H_1 Q$$

is a simple 9-reversal of the occurrence $f_{\alpha-1}(W; X, RTA^t A_1 T_1 Q)$. Then we have Rel $(A^t A_1, D_1^{-1} D^{-h})$, whence it follows by II.5.20, II.5.17 and (11) of II.5.16 that Rel (W, W^{-1}). Then by II.5.21 we have $\neg$ Rel (W, U).

1.6. *Suppose that the transition*

$$RTA^t A_1 T_1 Q \to RHD^h D_1 H_1 Q \tag{2}$$

is a simple 9-reversal of rank α, that is, all the conditions in I.4.16 *hold. Then $TA^t A_1 T_1$ is an elementary word of rank α,* Comp $(R * TA^t A_1 T_1 * Q, RT * A^t A_1 * TQ)$

and every occurrence $W \in \text{Norm}(\alpha, RTA^tA_1T_1Q, 9)$ *that is compatible with* $R * TA^tA_1T_1 * Q$ *is contained in* $RT * A^tA_1 * T_1Q$ *and is periodised in rank* $\alpha - 1$. *The same is true of* $R * HD^hD_1H_1 * Q$.

This follows from II.5.14, the remark on II.7.3, II.4.5 and 1.2.

1.7. *Suppose that the occurrence* $V \in \text{Norm}(\alpha, RTA^tA_1T_1Q, 9)$ *is stable in a simple 9-reversal* (2) *of rank* α *and* V_1 *is an image of* V *in this reversal. If the maximal normalised continuation of* V *intersects* $W \rightleftharpoons R * TA^tA_1T_1 * Q$, *then the maximal normalised continuation of* V_1 *does not intersect* $W_1 \rightleftharpoons R * HD^hD_1H_1 * Q$.

Suppose that $V < W$. Then for some R_1, E and R_2 we have

$$V = R_1 * E * R_2TA^tA_1T_1Q \quad \text{and} \quad V_1 = R_1 * E * R_2HD^hD_1H_1Q\,.$$

Suppose that the maximal normalised continuations of these occurrences intersect W and W_1 respectively. Then there are normalised continuations

$$U \rightleftharpoons R_1 * ER_2F * GT_1Q \quad \text{and} \quad U_1 \rightleftharpoons R_1 * ER_2F_1 * G_1H_1Q$$

of them to the right such that $FG \overset{\circ}{=} TA^tA_1$, $F_1G_1 \overset{\circ}{=} HD^hD_1$ and the words F and F_1 are non-empty. In this case, by virtue of II.3.10, the word $F^{-1}F_1$ must fail to occur in any word in $\mathscr{K}_{\alpha-1}$. However it does occur in the word $D^{h+10}D_1H_1T_1^{-1}A_1^{-1}A^{-t}T^{-1}HD^{h+10}$, which belongs to $\mathscr{K}_{\alpha-1}$ by I.4.16 and II.6.13. We have obtained a contradiction.

The case $W < V$ is treated in an analogous way.

1.8. *Suppose that the transition* $X \to Y$ *is an r-reversal of the occurrence* $W \in \text{MaxNorm}(\alpha, X, 9)$ *and that the occurrence* $U \in \text{MaxNorm}(\alpha, Y, 9)$ *is a maximal image of it in this reversal. Then the following relations hold:*

$$\begin{gathered} r \leqslant l_\alpha(W) \leqslant n - r + 1\,, \\ r \leqslant l_\alpha(U) \leqslant n - r + 1\,, \\ n - 18 \leqslant l_\alpha(W) + l_\alpha(U) \leqslant n + 1\,. \end{gathered} \tag{3}$$

In the light of (10) and (13), by II.5.16 we may confine attention to the case where $X \to Y$ is a simple reversal of W, that is,

$$\begin{aligned} &X \overset{\circ}{=} RTA^tA_1T_1Q\,, && Y \overset{\circ}{=} RHD^hD_1H_1Q\,, \\ &W = RT * A^tA_1 * T_1Q\,, && U = RH * D^hD_1 * H_1Q_1\,, \end{aligned}$$

where t, $h > 0$, $l_\alpha(W) \geqslant r$, $l_\alpha(U) \geqslant r$, $A \overset{\circ}{=} A_1A_2$ is an elementary period of rank α, the word $D \overset{\circ}{=} D_1D_2$ is a cyclic shift of A^{-1},

$$A^{n+20} \overset{\alpha-1}{\sim} A^{t+10}A_1T_1H_1^{-1}D_1^{-1}D^{-h}H^{-1}TA^{10} \tag{4}$$

and the words A^{n+20} and $G \rightleftharpoons A^{t+10}A_1T_1H_1^{-1}D_1^{-1}D^{-h}H^{-1}TA^{10}$ belong to $\mathscr{K}_{\alpha-1}$. It only remains for us to prove relation (3), since the first two follow from it.

Suppose that B is the base of an initial kernel of rank $\alpha - 1$ of W. Then by II.1.11, we have $A^2 \mathrel{\underline{\circ}} BC$ for some C. Since the occurrences

$$V \rightleftharpoons A^{10} * B * CA^{n+8} \text{ and } V_1 \rightleftharpoons A^{10} * B * CA^{t-2}A_1T_1H_1^{-1}D_1^{-1}D^{-h}H^{-1}TA^{10}$$

are supporting kernels of rank $\alpha - 1$ of the words $A^{n+20} \in \mathrm{Per}(\alpha, A)$ and $G \in \mathrm{Int}(A^{n+20}, \alpha, A)$, then by *IV. 2.29* it follows from (4) that

$$f_{\alpha-1}(V; A^{n+10}, G) = V_1 \,. \tag{5}$$

In view of *V.2.1* and II.7.7, it follows from (4) that

$$f_{\alpha-1}(A^{n+10} * BCA^6B * CA^8; A^{n+28}, GA^8)$$
$$= A^{t+10}A_1T_1H_1^{-1}D_1^{-1}D^{-h}H^{-1}T * BCA^6B * CA^8 \,.$$

Furthermore by *IV. 2.27*,

$$A^{t+10}A_1T_1H_1^{-1}D_1^{-1}D^{-h}H^{-1}T * B * CA^{16} \in \mathrm{Ker}(\alpha - 1, GA^8) \,.$$

Then from (4) by II.6.10 and *IV.2.29* we get

$$f_{\alpha-1}(A^{n+10} * B * CA^8;\ A^{n+20},\ G) = A^{t+10}A_1T_1H_1^{-1}D_1^{-1}D^{-h}H^{-1}T * B * CA^8. \tag{6}$$

From (5) and (6), it follows by I.4.30 that

$$f_{\alpha-1}(A^{10} * A^nB * CA^8;\ A^{n+20},\ G) = A^{10} * A^tA_1T_1H_1^{-1}D_1^{-1}D^{-h}H^{-1}TB * CA^8,$$

that is, the generating occurrence

$$W_1 \rightleftharpoons A^{10} * A^tA_1T_1H_1^{-1}D_1^{-1}D^{-h}H^{-1}TB * CA^8,$$

like the occurrence $A^{10} * A^nB * CA^8$, contains precisely $n + 1$ supporting kernels of rank $\alpha - 1$ which correspond in phase relative to the initial supporting kernel V_1 of the original period A. We denote the i-th left of these supporting kernels by V_i.

Clearly, $l_\alpha(W) = l_\alpha(A^tA_1)$ is the number of kernels V_i contained in the occurrence $A^{10} * A^tA_1 * T_1H_1^{-1}D_1^{-1}D^{-h}H^{-1}TBCA^8$. Since

$$U_1 \rightleftharpoons A^{t+10}A_1T_1H_1^{-1} * D_1^{-1}D^{-h} * H^{-1}TBCA^8$$

is a generating occurrence of the word $D_1^{-1}D^{-h}$ and

$$l_\alpha(U) = l_\alpha(D^hD_1) = l_\alpha(D_1^{-1}D^{-h}) \leqslant l_\alpha(D_1^{-1}D^{-h}H^{-1}TB) \,,$$

it follows that $l_\alpha(U)$ is not greater than the number of kernels V_i which are contained in $A^{t+10}A_1T_1H_1^{-1} * D_1^{-1}D^{-h}H^{-1}TB * CA^8$. Hence we have

$$l_\alpha(U) + l_\alpha(W) \leqslant n + 1 .$$

Let $l_\alpha(W) = k$. By II.6.1, we can find a section $F * L$ of $A^{t+10}A_1T_1H_1^{-1}D_1^{-1}D^{-h}H^{-1}TA^{10}$ which is separating for the pair of kernels (V_{k+5}, V_{k+6}). If F were a start of the word $A^{t+10}A_1T_1$, then by II.6.3. and II.6.5 the occurrence W would have a proper normalised continuation to the right. Hence L is an end of $H_1^{-1}D_1^{-1}D^{-h}H^{-1}TA^{10}$. Furthermore, since by the principle of symmetry the occurrence $Q^{-1}H_1^{-1} * D_1^{-1}D^{-h} * H^{-1}R^{-1}$ has no proper normalised continuation to the left then, by II.6.3 and II.6.5, the kernel V_{k+10} must be contained in U_1.

From (4) it follows by *V.2.1* that

$$A^{n+20+t} \overset{\alpha-1}{\sim} A^{t+10}A_1T_1H_1^{-1}D_1^{-1}D^{-h}H^{-1}TA^{t+10} , \tag{7}$$

that is, $A^{t+10}A_1T_1H_1^{-1}D_1^{-1}D^{-h}H^{-1}T * A^tA_1 * A_2A^9$ is also a generating occurrence of the elementary word A^tA_1. Starting at the same point, we could clearly consider relation (7) instead of relation (4). Then, on considering a section $F_1 * L_1$ of the word $A^{t+10}A_1T_1H_1^{-1}D_1^{-1}D^{-h}H^{-1}TA^{t+10}$ which is separating for the pair of kernels (V_{n-5}, V_{n-4}), we deduce from the fact that W has no proper normalised continuation to the left and using II.6.3 and II.6.5 that F_1 is a start of the word $A^{t+10}A_1T_1H_1^{-1}D_1^{-1}D^{-h}H^{-1}$, whence, by the fact that $Q^{-1}H_1^{-1} * D_1^{-1}D^{-h} * H^{-1}R^{-1}$ has no proper normalised continuation to the right, it follows in an analogous way that the kernel V_{n-9} is contained in U_1.

Thus we have proved that U_1 does not contain less than $n - 9 - (k + 10) + 1 = n - k - 18$ kernels V_i, that is, $l_\alpha(U_1) \geqslant n - k - 18$. Hence

$$l_\alpha(U) + l_\alpha(W) \geqslant n - 18 .$$

1.9. *Suppose that* $W \in$ Norm $(\alpha, X, 9)$ *and* $\bar{W}$ *is the maximal normalised continuation of* W. *If*

$$18 \leqslant r \leqslant l_\alpha(\bar{W}) \leqslant n - r - 18 , \tag{8}$$

then there exists an r-reversal of W.

In addition, if the occurrence W is periodised in rank $\alpha - 1$, *then a corresponding simple reversal* $X_1 \to Y_1$ *of the occurrence* $f_{\alpha-1}(W; X, X_1)$ *can be chosen in such a way that the bases of W and* $f_{\alpha-1}(W; X, X_1)$ *coincide.*

By 1.3, we may assume that $W = \bar{W}$. By 1.4, *IV.3.12* and *IV.3.14* we may also assume that $X \in \mathscr{M}_{\alpha-1}$. In this case we have by II.4.1 and II.4.3 that $W = P * A^tA_1 * S$, where $A \overset{\circ}{=} A_1A_2$ is an elementary period of rank α and $A^{n+20} \in$ Per (α, A). Since $X \in \mathscr{M}_{\alpha-1}$ it follows from II.6.13 that

$$PA^{10} \in \mathscr{L}_{\alpha-1},\ A^{10}A_1S \in \mathscr{L}_{\alpha-1},\ A^{n+20-t} \in \mathscr{L}_{\alpha-1} . \tag{9}$$

By (9), the symmetry principle and lemma II.7.20, words

$$RTA^{10} \in \mathscr{K}_{\alpha-1},\ A^{-10}T^{-1}NA^{-1} \in \mathscr{K}_{\alpha-1},\ RNA^{-k} \in \mathscr{K}_{\alpha-1} , \tag{10}$$

can be found which satisfy the relations

$$PA^{10} \overset{\alpha-1}{\sim} RTA^{10}, \tag{11}$$

$$A^{-n-20+t} \overset{\alpha-1}{\sim} A^{-10}T^{-1}NA^{-k}, \tag{12}$$

where $k = n + 2 - t$, and for every occurrence $F * E * G \in \text{Reg}(\alpha - 1, A^{-10}T^{-1}NA^{-k})$ contained in $A^{-10}T^{-1} * NA^{-k} *$, the word EG is an end of $A^{-n-20+t}$. Using the second half of lemma II.7.20, we can similarly find words

$$A^{10}A_1T_1Q,\ A^{-s}N_1T_1^{-1}A_1^{-1}A^{-10},\ A^{-s}N_1Q \in \mathscr{K}_{\alpha-1}, \tag{13}$$

which satisfy the relations

$$A^{10}A_1S \overset{\alpha-1}{\sim} A^{10}A_1T_1Q, \tag{14}$$

$$A^{-k} \overset{\alpha-1}{\sim} A^{-s}N_1T_1^{-1}A_1^{-1}A^{-10}, \tag{15}$$

where $s = k - 19 = n - 17 - t$ and, for every occurrence $F * E * G \in \text{Reg}(\alpha - 1, A^{-s}N_1T_1^{-1}A^{-10})$ contained in $* A^{-s}N_1 * T_1^{-1}A^{-10}$, the word FE is a start of A^{-k}.

Since $l_\alpha(\bar{W}) \geqslant t - 1$, it then follows from (8) that

$$s = n - t - 17 \geqslant n - l_\alpha(\bar{W}) - 18 \geqslant r \geqslant 18. \tag{16}$$

It follows from (10) and (13) by II.6.13 that

$$RTA^tA_1T_1Q \in \mathscr{K}_{\alpha-1} \text{ and } RNA^{-s}N_1Q \in \mathscr{K}_{\alpha-1}. \tag{17}$$

From (11) and (14), using II.6.13, *IV.1.21* and *V.2.1* we find that

$$X \overset{\alpha-1}{\sim} RTA^tA_1S \overset{\alpha-1}{\sim} RTA^tA_1T_1Q. \tag{18}$$

Similarly from (12) and (15) we obtain

$$A^{-n-20+t} \overset{\alpha-1}{\sim} A^{-10}T^{-1}NA^{-k} \overset{\alpha-1}{\sim} A^{-10}T^{-1}NA^{-s}N_1T_1^{-1}A_1^{-1}A^{-10},$$

and from this by the principle of symmetry there follows

$$A^{n+20-t} \overset{\alpha-1}{\sim} A^{10}A_1T_1N_1^{-1}A^sN^{-1}TA^{10}.$$

Applying *V.2.1* a further time, we deduce that

$$A^{n+20} \overset{\alpha-1}{\sim} A^{t+10}A_1T_1N_1^{-1}A^sN^{-1}TA^{10}. \tag{19}$$

Suppose that $A^2 \mathrel{\overline{\underline{\circ}}} BC$, where B is the base of an initial kernel of rank $\alpha - 1$ of the occurrence W. We look at the occurrence

$$V \rightleftharpoons RNA^{-4}C^{-1} * B^{-1}A^{-s+10} * A^{-4}N_1Q\,.$$

By (16) and II.6.6, we have $V \in \text{Norm}\,(\alpha, RNA^{-s}N_1Q, 9)$.

We shall assume that the maximal normalised continuation V_1 of the occurrence V to the left is not contained in $R * NA^{-s}N_1 * Q$. Then we have $V_1 = R_1 * LNA^{-s+4} * A^{-4}N_1Q$, where $R_1L \mathrel{\overline{\underline{\circ}}} R$. Since

$$A^{-8}C^{-1} * B^{-1}T^{-1}NA^{-s} * N_1T_1^{-1}A_1^{-1}A^{-10}$$

is a normal generating occurrence of rank α, by II.3.10 the word LTB cannot occur in any word in $\mathscr{K}_{\alpha-1}$. However, it does occur in the word $RTA^tA_1T_1Q \in \mathscr{K}_{\alpha-1}$. Hence V_1 is contained in $R * NA^{-s}N_1 * Q$, that is, $V_1 = RH * E_1A^{-s+4} * A^{-4}N_1\,Q$, where $HE_1 \mathrel{\overline{\underline{\circ}}} N$. We then have

$$A^{-10}T^{-1}H * E_1A^{-s} * A^{-k+s} \in \text{Reg}\,(\alpha - 1,\ A^{-10}T^{-1}NA^{-k})\,,$$

whence, according to the above hypothesis for $F * E * G \in \text{Reg}\,(\alpha - 1, A^{-10}T^{-1}NA^{-k})$, it follows that E_1A^{-k} is an end of $A^{-n-20+t}$, that is, $E_1A^{-s} \in \text{Per}(A^{-1})$. It is proved similarly that the maximal normalised continuation V_2 of V_1 to the right has the form $RH * E_1A^{-s}E_2 * H_1Q$, where $E_2H_1 \mathrel{\overline{\underline{\circ}}} N_1$ and $A^{-s}E_2 \in \text{Per}\,(A^{-1})$. Hence, $E_1A^{-s}E_2 \in \text{Per}\,(A^{-1})$, that is, $E_1A^{-s}E_2 \mathrel{\overline{\underline{\circ}}} D^hD_1$, where $D \mathrel{\overline{\underline{\circ}}} D_1D_2$ is a cyclic shift of A^{-1}. Then the maximal normalised continuation of the occurrence V has the form $RH * D^hD_1 * H_1Q$, that is,

$$RH * D^hD_1 * H_1Q \in \text{MaxNorm}\,(\alpha,\ RNA^{-s}N_1Q, 9)\,. \tag{20}$$

Moreover, we have by hypothesis that

$$P * A^tA_1 * S \in \text{MaxNorm}\,(\alpha, X, r), \tag{21}$$

where $X \in \mathscr{M}_{\alpha-1}$. From this and (18), on applying II.7.7 twice we get

$$\mathcal{f}_{\alpha-1}\,(P * A^tA_1 * S;\ X,\ RTA^tA_1T_1Q) = RT * A^tA_1 * T_1Q\,. \tag{22}$$

By II.5.16, from (22) and (21) there follows

$$RT * A^tA_1 * T_1Q \in \text{MaxNorm}\,(\alpha,\ RTA^tA_1T_1Q, r)\,. \tag{23}$$

Since $E_1A^{-s}E_2 \mathrel{\overline{\underline{\circ}}} D^hD_1$, it follows that $E_1A^{-s} \mathrel{\overline{\underline{\circ}}} D^sE'$ for some E'. Then for every $i \geqslant 10$ we have

$$E_1A^{-i} \mathrel{\overline{\underline{\circ}}} D^iE'\,. \tag{24}$$

Since $k = n - t + 2$, from (12) we get by *V.2.1* that

$$A^{-18} \overset{\alpha-1}{\approx} T^{-1}HE_1A^{-10} . \tag{25}$$

If follows analogously from (15) that

$$E_1A^{-n-20} \overset{\alpha-1}{\approx} E_1A^{-s-10}E_2H_1T_1^{-1}A_1^{-1}A^{-t-18} ,$$

whence, using (24) and (25), we obtain

$$D^{n+20}E^{\iota} \overset{\alpha-1}{\approx} D^{h+10}D_1H_1T_1^{-1}A_1^{-1}A^{-t}T^{-1}HD^{10}E^{\iota} .$$

Finally, it follows from this by *V.2.1* that

$$D^{n+20} \overset{\alpha-1}{\approx} D^{h+10}D_1H_1T_1^{-1}A_1^{-1}A^{-t}T^{-1}HD^{10} . \tag{26}$$

Thus we have the words (17) and the occurrences (20), (23) and (21), where $X \in \mathscr{M}_{\alpha-1}$ and relations (19) and (26) hold, the left and right hand sides of which belong, by II.6.13, to $\mathscr{K}_{\alpha-1}$. Hence the transition

$$RTA^tA_1T_1Q \rightarrow RHD^hD_1H_1Q \tag{27}$$

is a simple 9-reversal of W. Then by 1.8, we have the relation

$$l_\alpha(A^tA_1) + l_\alpha(D^hD_1) \geqslant n - 18 ,$$

whence $l_\alpha(D^hD_1) \geqslant r$ by (8), that is, (27) is an r-reversal of W of rank α.

1.10. *Suppose that $X \rightarrow Y$ is a 9-reversal of the occurrence $P * B^rB_1 * S \in$ Norm(α, X, 9), where $X \in \mathscr{K}_{\alpha-1}$ and $B \overset{\circ}{=} B_1B_2$ is an elementary period of rank α. Then*

$$Y \overset{\alpha-1}{\approx} [PB^{10}, B^{-n+r-20}, B^{10}B_1S]_{\alpha-1} \overset{\alpha-1}{\approx} [P, B^{-n+r}, B_1S]_{\alpha-1} . \tag{28}$$

Since $X \in \mathscr{K}_{\alpha-1}$, it follows by II.6.13 that $PB^{n+20-r}B_1S \in \mathscr{K}_{\alpha-1}$. Then by *V.1.8* and *V.1.10*, we have

$$[PB^{10}, B^{-n-20+r}, B^{10}B_1S]_{\alpha-1} \overset{\alpha-1}{\approx} [P, B^{-n+r-10}, B^{10}B_1S]_{\alpha-1}$$
$$\overset{\alpha-1}{\approx} [P, B^{-n-r}, B_1S]_{\alpha-1} .$$

We shall establish the first of the relations (28) first of all in the case where $X \rightarrow Y$ is a simple reversal of rank α, that is,

$$X \overset{\circ}{=} RTA^tA_1T_1Q,\ Y \overset{\circ}{=} RHD^hD_1H_1Q ,$$
$$\text{Comp}(P * B^rB_1 * S,\ RT * A^tA_1 * T_1Q) , \tag{29}$$

and all the conditions in I.4.16 are fulfilled, including the relation

$$A^{n+20} \stackrel{\alpha-1}{\sim} A^{t+10}A_1T_1H_1^{-1}D_1^{-1}D^{-h}H^{-1}TA^{10},$$

from which it follows by *V.2.1* and the principle of symmetry that

$$A^{-n-20+t} \stackrel{\alpha-1}{\sim} A^{-10}T^{-1}HD^hD_1H_1T_1^{-1}A_1^{-1}A^{-10}. \tag{30}$$

Since $P * B'B_1 * S$ is contained in $RT * A'A_1 * T_1Q$ by 1.6, it follows from (29) that A is a cyclic shift of B, and for some E, E_1 we have

$$A'A_1 \overset{\circ}{=} EB'E_1,\ P \overset{\circ}{=} RTE,\ B_1S \overset{\circ}{=} E_1T_1Q, \tag{31}$$

whence there easily follows

$$EB \overset{\circ}{=} AE \text{ and } EB^{10}E_1 \overset{\circ}{=} A^{t+10-r}A_1. \tag{32}$$

As a result we have

$$B^{-n+r} = [E^{-1}, A^{-n+r}, E]_0. \tag{33}$$

Now, using (31), (32), (33), (30) and the properties *V.1.8*, *V.1.1*, *V.1.7* and *V.1.10* of coupling of rank $\alpha - 1$, we obtain

$$\begin{aligned}
&[PB^{10}, B^{-n-20+r}, B^{10}B_1S]_{\alpha-1}\\
&\quad\stackrel{\alpha-1}{\sim} [RTEB^{10}, [E^{-1}, A^{-n-20+r}, E]_{\alpha-1}, B^{10}E_1T_1Q]_{\alpha-1}\\
&\quad\stackrel{\alpha-1}{\sim} [RTA^{10}E, [E^{-1}, A^{-n-20+r}]_{\alpha-1}, EB^{10}E_1T_1Q]_{\alpha-1}\\
&\quad\stackrel{\alpha-1}{\sim} [RTA^{10}, [A^{-n-20+t}, A^{10}A_1T_1Q]_{\alpha-1}]_{\alpha-1}\\
&\quad\stackrel{\alpha-1}{\sim} [RTA^{10}, A^{-10}T^{-1}HD^hD_1H_1Q]_{\alpha-1} \stackrel{\alpha-1}{\sim} RHD^hD_1H_1Q \overset{\circ}{=} Y.
\end{aligned}$$

In general we have $X \stackrel{\alpha-1}{\sim} RTA'A_1T_1Q$ and

$$Y \stackrel{\alpha-1}{\sim} RHD^hD_1H_1Q, \tag{34}$$

where the transition

$$RTA'A_1T_1Q \to RHD^hD_1H_1Q \tag{35}$$

is a simple 9-reversal of the occurrence

$$W \rightleftharpoons f_{\alpha-1}(P * B'B_1 * S;\ \ X, RTA'A_1T_1Q).$$

Then we have by 1.6 that $W = P_1 * C'C_1 * S$, where $C \overset{\circ}{=} C_1C_2$ is a cyclic shift of the period A. Moreover, by what we have proved for the reversal (35), the relation

$$RHD^hD_1H_1Q \overset{\alpha-1}{\sim} [P_1, C^{-n+r}, C_1S_1]_{\alpha-1} \tag{36}$$

holds. By II.7.15, there exists a word $F \in \mathscr{A}_\alpha$ such that

$$B \overset{\alpha-1}{\sim} [F, C, F^{-1}]_{\alpha-1}, \tag{37}$$

$$P_1 \overset{\alpha-1}{\sim} [P, F]_{\alpha-1}, \quad C_1S_1 \overset{\alpha-1}{\sim} [F^{-1}, B_1S]_{\alpha-1}. \tag{38}$$

Using (34), (36), (38), (37) and the properties of coupling of rank $\alpha - 1$ we obtain

$$Y \overset{\alpha-1}{\sim} [P_1, C^{-n+r}, C_1S_1]_{\alpha-1} \overset{\alpha-1}{\sim} [P, F, C^{-n+r}, F^{-1}, B_1S]_{\alpha-1}$$
$$\overset{\alpha-1}{\sim} [P, [F, C^{-n+r}, F^{-1}]_{\alpha-1}, B_1S]_{\alpha-1} \overset{\alpha-1}{\sim} [P, B^{-n+r}, B_1S]_{\alpha-1}.$$

1.11. *Suppose that the transition $X \to Y$ is a q_1-reversal of rank α of an occurrence $W \in$ Norm$(\alpha, X, 9)$, the occurrence $U \in$ Norm$(\alpha, Y, 9)$ is an image of W in this reversal and $X \to Z$ is an arbitrary 9-reversal of W of rank α.*

Then $Y \overset{\alpha-1}{\sim} Z$, the transition $X \to Z$ is a q_1-reversal of the occurrence W and, independent of the choice of the intermediate simple reversal for $X \to Z$, the occurrence $f_{\alpha-1}(U; Y, Z)$ is an image of W in the reversal $X \to Z$, an arbitrary occurrence $V \in$ ComReg$(\alpha - 1, X)$ which is stable in the reversal $X \to Y$ is stable in $X \to Z$ and it satisfies

$$f_{X\to Z}(V) = f_{\alpha-1}(f_{X\to Y}(V);\ Y, Z). \tag{39}$$

By 1.4, we may assume that $X \to Y$ is a simple q_1-reversal of W and $X \overset{\alpha-1}{\sim} X_1$, where $X_1 \to Z$ is a simple 9-reversal of the occurrence $f_{\alpha-1}(W; X, X_1)$. Suppose that these simple reversals have the form

$$X \overset{\circ}{=} PTA^tA_1T_1Q \to PHD^hD_1H_1Q \overset{\circ}{=} Y \tag{40}$$

and

$$X_1 \overset{\circ}{=} RT_2C^sC_1T_3S \to RH_2B^rB_1H_3S \overset{\circ}{=} Z, \tag{41}$$

where A, C are elementary periods of rank α, D and B are cyclic shifts of the words A^{-1} and C^{-1},

$$PT * A^tA_1 * T_1Q \in \text{MaxNorm}(\alpha, X, q_1), \quad \text{Comp}(W, PT * A^tA_1 * T_1Q),$$
$$PH * D^hD_1 * H_1Q \in \text{MaxNorm}(\alpha, Y, q_1), \quad \text{Comp}(U, PH * D^hD_1 * H_1Q),$$
$$RT_2 * C^sC_1 * T_3S \in \text{MaxNorm}(\alpha, X_1, 9),$$
$$\text{Comp}(f_{\alpha-1}(W; X, X_1), RT_2 * C^sC_1 * T_3S)$$

and

$$RH_2 * B^rB_1 * H_3S \in \text{MaxNorm}(\alpha, Z, 9).$$

Then we may assert by 1.3 that

$$W = PT * A^t A_1 * T_1 Q \text{ and } U = PH * D^h D_1 * H_1 Q,$$

whence by II.5.16 and II.5.13 there follows

$$\begin{aligned} &f_{\alpha-1}(W; X, X_1) = RT_2 * C^s C_1 * T_3 S, \\ &l_\alpha(C^s C_1) = l_\alpha(A^t A_1) \geqslant q_1 \text{ and } s = t, \end{aligned} \tag{42}$$

if, in addition, we assume that A_1 and C_1 are distinct from A and C. By 1.10, we have the relations

$$Y \overset{\alpha-1}{\approx} [PT, A^{-n+t}, A_1 T_1 Q]_{\alpha-1}$$

and

$$Z \overset{\alpha-1}{\approx} [RT_2, C^{-n+t}, C_1 T_3 S]_{\alpha-1}.$$

By II.7.15, there exists a word $N \in \mathscr{A}_\alpha$ such that

$$\begin{gathered} A \overset{\alpha-1}{\approx} [N, C, N^{-1}]_{\alpha-1}, \\ RT_2 \overset{\alpha-1}{\approx} [PT, N]_{\alpha-1}, \; C_1 T_3 S \overset{\alpha-1}{\approx} [N^{-1}, A_1 T_1 Q]_{\alpha-1}. \end{gathered}$$

Hence we have

$$\begin{aligned} Y &\overset{\alpha-1}{\approx} [PT, [N, C^{-n+t}, N^{-1}]_{\alpha-1}, A_1 T_1 Q]_{\alpha-1} \\ &\overset{\alpha-1}{\approx} [[PT, N]_{\alpha-1}, C^{-n+t}, [N^{-1}, A_1 T_1 Q]_{\alpha-1}]_{\alpha-1} \overset{\alpha-1}{\approx} Z. \end{aligned}$$

We shall first of all prove that the occurrence

$$U_1 \rightleftharpoons f_{\alpha-1}(U; Y, Z)$$

coincides with the occurrence

$$W_1 \rightleftharpoons RH_2 * B^r B_1 * H_3 S.$$

By II.5.16 we have

$$l_\alpha(U_1) = l_\alpha(U) \geqslant q_1 = 37. \tag{43}$$

From (42), we have $\operatorname{Rel}(A^t A_1, C^s C_1)$ by II.5.16, whence we obtain from the principle of symmetry that $\operatorname{Rel}(D^h D_1, B^r B_1)$ and, in consequence, $\operatorname{Rel}(U_1, W_1)$.

Assume that $\neg \operatorname{Comp}(U_1, W_1)$. Then one of these occurrences is, by II.5.7, situated to the left of the other. We shall consider the case where $U_1 < W_1$. The other case is treated in an analogous way.

Let U_i, for $1 \leqslant i < l_\alpha(U_1)$, be the start of U_1 which ends with its i-th right segment. Then by 1.6, II.6.7 and II.2.7 we have

$$U_{12} = R_1 * F * R_2H_2B^rB_1H_3S ,$$

where $R_1FR_2 \mathrel{\underline{\overline{\circ}}} R$. Let $F \mathrel{\underline{\overline{\circ}}} F_1F_2F_3$, where

$$U_{16} = R_1 * F_1F_2 * F_3R_2H_2B^rB_1H_3S$$

and

$$U_{24} = R_1 * F_1 * F_2\, F_3R_2H_2B^rB_1H_3S .$$

We consider the occurrences

$$V_{16} \rightleftharpoons R_1 * F_1F_2 * F_3R_2T_2C^sC_1T_3S$$

and

$$V_{24} \rightleftharpoons R_1 * F_1 * F_2F_3R_2T_2C^sC_1T_3S .$$

Clearly all U_i are normalised. By II.6.10 we have

$$\text{MutNorm}_{\alpha-1}(V_{16},\, U_{16}) . \tag{44}$$

By 1.6 we have $\neg$ Comp(V_{16}, $RT_2 * C^sC_1 * T_3S$). Set

$$W_{16} \rightleftharpoons f_{\alpha-1}(V_{16};\, X_1,\, X) \text{ and } W_{24} \rightleftharpoons f_{\alpha-1}(V_{24};\, X_1,\, X) .$$

Since $V_{16} \in \text{Norm}(\alpha, X_1, 9)$, we have by II.5.16 that $W_{16} \in \text{Norm}(\alpha, X, 9)$ and $\neg$ Comp(W_{16}, W). Since $l_\alpha(W_{16}) = l_\alpha(V_{16}) \geqslant q_1 - 15 = 22$ by (43), by II.6.19 and II.2.7 the start W_{24} of W_{16} does not intersect $P * TA^tA_1T_1 * Q$, that is, we have

$$f_{\alpha-1}(V_{24};\, X_1,\, X) = P_1 * G * P_2TA^tA_1T_1Q , \tag{45}$$

where $P_1GP_2 \mathrel{\underline{\overline{\circ}}} P$. In addition $l_\alpha(V_{24}) = l_\alpha(U_1) - 23 > 9$, that is, $V_{24} \in$ ComReg($\alpha-1, X_1$). Since MutNorm$_{\alpha-1}(V_{24}, U_{24})$ follows from (44) by *IV.3.6*, we have

$$f_{\alpha-1}(U_{24};\, Z,\, Y) = P_1 * G * P_2HD^hD_1H_1Q ,$$

by (45) and II.7.8. On the other hand, by *IV.2.18(i)*, $f_{\alpha-1}(U_{24}; Z, Y)$ must be a start of the occurrence $f_{\alpha-1}(U_1;\, Z,\, Y) = U = PH * D^hD_1 * H_1Q$.

We have reached a contradiction. Hence U_1 and W_1 are compatible, that is, U_1 is an image of W in the reversal $X \to Z$. Then $U_1 = W_1$.

Since $l_\alpha(W_1) = l_\alpha(U) \geqslant q_1$ and $l_\alpha(C^sC_1) = l_\alpha(W) \geqslant q_1$ by II.5.16, the transition $X \to Z$ is a q_1-reversal of the occurrence W.

Suppose that some occurrence $V \in \mathrm{ComReg}(\alpha - 1, X)$ is stable in the reversal $X \to Y$. Without loss of generality we may assume that $V \ll W$, that is, $V = P_3 * L * P_4 T A^t A_1 T_1 Q$, where $P_3 L P_4 \stackrel{\circ}{=} P$, and for $V' \rightleftarrows P_3 * L * P_4 H D^h D_1 H_1 Q$ we have

$$\mathrm{MutNorm}_{\alpha-1}(V, V') . \tag{46}$$

Suppose that

$$f_{\alpha-1}(V; X, X_1) = R' * L' * S' \tag{47}$$

and

$$f_{\alpha-1}(V'; Y, Z) = R'' * L'' * S'' . \tag{48}$$

By II.7.6, (46) and (47) there exists a word $Z_0 \in \mathscr{K}_{\alpha-1}$ such that $Y \overset{\alpha-1}{\sim} Z_0$ and $f_{\alpha-1}(V'; Y, Z_0) = R' * L' * S_0$. Then $Z \overset{\alpha-1}{\sim} Z_0$ and

$$f_{\alpha-1}(R'' * L'' * S''; Z, Z_0) = R' * L' * S_0 . \tag{49}$$

By *IV.2.18(g)*, we have

$$R' * L' * S' \ll RT_2 * C^s C_1 * T_3 S \text{ and } R'' * L'' * S'' \ll RH_2 * B^r B_1 * H_3 S ,$$

that is, $R'L'$ is a start of RT_2 and $R''L''$ is a start of RH_2.

We shall show that neither of them is a start of R, that is, $R'L' \stackrel{\circ}{=} RM_1$ and $R''L'' \stackrel{\circ}{=} RM_2$, where $T_2 \stackrel{\circ}{=} M_1 T_2'$, $H_2 \stackrel{\circ}{=} M_2 H_2'$, the words M_1, M_2 being non-empty. Then it follows from *V.2.4* and (49) that the word

$$[(R''L'')^{-1}, R'L']_0 = [M_2^{-1} R^{-1}, RM_1]_0 = [M_2^{-1}, M_1]_0$$

does not occur in any word in $\mathscr{K}_{\alpha-1}$. On the other hand, since (41) is a simple reversal, by I.4.16 the word $B^{r+10} B_1 H_3 T_3^{-1} C_1^{-1} C^{-s} T_2^{-1} H_2 B^{10}$ belongs to $\mathscr{K}_{\alpha-1}$ and the word $M_1^{-1} M_2$ occurs in it. Hence, one of the words $R'L'$ and $R''L''$ is a start of R. Then one of them is a start of the other, and from (49) it follows by II.7.7 that $R'' \stackrel{\circ}{=} R'$ and $L'' \stackrel{\circ}{=} L'$. From (46), (47) and (48) it follows that $\mathrm{MutNorm}_{\alpha-1}(R' * L' * S', R'' * L'' * S'')$, that is, the occurrence V is stable in the reversal $X \to Z$ and

$$f_{X \to Z}(V) = R'' * L'' * S'' . \tag{50}$$

Relation (39) follows from (50) and (48).

1.12. *If $X \to Y$ is a q_1-reversal of rank α, then X is not equivalent in rank $\alpha - 1$ either to Y or to Y^{-1}.*

It is clearly sufficient to prove these assertions for simple reversals. Suppose that $X \to Y$ is a simple reversal of the occurrence $RT * A^t A_1 * T_1 Q \in$ MaxNorm(α, X, q_1), where $A \mathrel{\overline{\underline{\circ}}} A_1 A_2$ is an elementary period of rank α. Then by 1.10, we have the relation

$$Y \overset{\alpha-1}{\approx} [RT, A^{-n+t}, A_1 T_1 Q]_{\alpha-1} . \tag{51}$$

We assume that $X \overset{\alpha-1}{\approx} Y$. Then, using the properties *V.1.7*, *V.1.8* and *V.1.9* of coupling of rank $\alpha - 1$, we obtain

$$A^{-n+t} \overset{\alpha-1}{\approx} [T^{-1}R^{-1}, X, Q^{-1}T_1^{-1}A_1^{-1}]_{\alpha-1} \overset{\alpha-1}{\approx} A^t ,$$

that is, $A \overset{\alpha-1}{\approx} A^{-n}$. But this is impossible, by *IV.2.16*. Hence $\neg\, (X \overset{\alpha-1}{\approx} Y)$.

Assume that $X \overset{\alpha-1}{\approx} Y^{-1}$. We shall prove that this assumption also leads to a contradiction.

Since $X \to Y$ is by hypothesis a simple reversal of

$$W \rightleftharpoons RT * A^t A_1 * T_1 Q ,$$

we have $Y \mathrel{\overline{\underline{\circ}}} RHD^h D_1 H_1 Q$, where $U \rightleftharpoons RH * D^h D_1 * H_1 Q$ is a maximal image of the occurrence W in this reversal and $D \mathrel{\overline{\underline{\circ}}} D_1 D_2$ is a cyclic shift of A^{-1}. Then we have

$$W \in \text{MaxNorm}(\alpha, X, q_1) \quad \text{and} \quad U \in \text{MaxNorm}(\alpha, Y, q_1) ,$$

from which it follows by the principle of symmetry that

$$U^{-1} \in \text{MaxNorm}\,(\alpha, Y^{-1}, q_1) .$$

Assume that U^{-1} is compatible with $f_{\alpha-1}(W; X, Y^{-1})$. In this case we have by II.5.16 and II.5.13 that

$$f_{\alpha-1}(W; X, Y^{-1}) = U^{-1} , \tag{52}$$

and therefore $\text{MutNorm}_{\alpha-1}(W, U^{-1})$. Since W is periodised in rank $\alpha - 1$, by II.7.21 we have $A^t A_1 \mathrel{\overline{\underline{\circ}}} D_1^{-1} D^{-h}$. Since A and D are simple words,

$$A \mathrel{\overline{\underline{\circ}}} D_1^{-1} D_2^{-1},\; A_1 \mathrel{\overline{\underline{\circ}}} D_1^{-1} \text{ and } t = h . \tag{53}$$

It follows from (52) by *IV.2.35* that

$$RT \overset{\alpha-1}{\approx} Q^{-1} H_1^{-1} \text{ and } T_1 Q \overset{\alpha-1}{\approx} H^{-1} R^{-1} ,$$

whence, using *IV.1.21* and properties of coupling of rank $\alpha - 1$, we obtain

$$\begin{aligned} H^{-1}T &\overset{\alpha-1}{\approx} [H^{-1}, T]_{\alpha-1} \overset{\alpha-1}{\approx} [[T_1 Q, R]_{\alpha-1}, [R^{-1}, Q^{-1}H_1^{-1}]_{\alpha-1}]_{\alpha-1} \\ &\overset{\alpha-1}{\approx} [T_1 Q, Q^{-1} H_1^{-1}]_{\alpha-1} \overset{\alpha-1}{\approx} T_1 H_1^{-1} . \end{aligned} \tag{54}$$

By definition I.4.16 and equations (53), we have the relation

$$A^{n+20} \overset{\alpha-1}{\approx} A^{t+10}A_1T_1H_1^{-1}A^tA_1H^{-1}TA^{10}\,, \tag{55}$$

both sides of which lie in $\mathscr{K}_{\alpha-1}$. We set

$$Z_1 \rightleftharpoons A^{t+10}A_1T_1H_1^{-1}A^tA_1H^{-1}TA^{10}\,.$$

Clearly, $Z_1 \in \mathrm{Int}(A^{n+20}, \alpha, A)$ and the occurrence

$$V \rightleftharpoons A^{t+10}A_1T_1H_1^{-1} * A^tA_1 * H^{-1}TA^{10}$$

is a normal generating occurrence of rank α. Set

$$f_{\alpha-1}(V;\, Z_1,\, A^{n+20}) = P_1 * E * S_1\,. \tag{56}$$

Then, by II.7.21, we have $E \underline{\overline{\circ}}\, \mathrm{Bas}(V) \underline{\overline{\circ}}\, A^tA_1$, that is, $P_1 \underline{\overline{\circ}}\, A^i$, $S_1 \underline{\overline{\circ}}\, A_2A^j$, where

$$i + j + t + 1 = n + 20\,. \tag{57}$$

By *IV.2.35*, it follows from (56) that

$$A^{t+10}A_1T_1H_1^{-1} \overset{\alpha-1}{\approx} A^i \quad \text{and} \quad H^{-1}TA^{10} \overset{\alpha-1}{\approx} A_2A^j\,. \tag{58}$$

Using (48), (54) and *V.2.1*, we find that

$$\begin{aligned} A^{i+10} &\overset{\alpha-1}{\approx} A^{t+10}A_1T_1H_1^{-1}A^{10} \overset{\alpha-1}{\approx} A^{t+10}A_1H^{-1}TA^{10} \\ &\overset{\alpha-1}{\approx} A^{t+10}A_1A_2A^j \,\underline{\overline{\circ}}\, A^{t+j+11}\,, \end{aligned}$$

whence it follows that $A^{t+j-i+1} \overset{\alpha-1}{\approx} A$. Then by *IV.2.16*, we have $A^{t+j-i+1} \underline{\overline{\circ}}\, A$, that is, $t + j + 1 = i$, whence by (57) it follows that $n + 20 = 2i$. But n is odd by assumption. Thus the occurrence U^{-1} is not compatible with $f_{\alpha-1}(W;\, X,\, Y^{-1})$.

Suppose that

$$W_1 \rightleftharpoons f_{\alpha-1}(W;\, X,\, Y^{-1}) \quad \text{and} \quad U_1 \rightleftharpoons f_{\alpha-1}(U^{-1};\, Y^{-1},\, X)\,.$$

In view of II.5.16, we have Rel (W, W_1) and Rel (U^{-1}, U_1). Thus Rel (W, U_1) and Rel (W_1, U_1) follow from Rel (W, U^{-1}).

Since $\neg$ Comp (W_1, U^{-1}), it follows from II.5.16 that $\neg$ Comp (W, U_1). By II.5.7, one of the occurrences W and U_1 is situated to the left to the other. Because of the completeness of the analogy, we may restrict attention to the case where $W < U_1$. In such a case, II.6.7 and II.2.7 give that the end of the occurrence U_1 starting with its 12-th left segment has the form

$$U_2 \rightleftharpoons RTA^tA_1T_1Q_1 * E * Q_2\,,$$

where $Q_1EQ_2 \;\overline{\circ}\; Q$ and $l_\alpha(U_2) = l_\alpha(U_1) - 11 \geqslant q_1 - 11 = 26$.

Since in addition $W_1 < U^{-1}$, in view of II.6.7 and II.2.7 the start of the occurrence W_1 ending with its 12-th right segment has the form

$$W_2 \rightleftharpoons Q_3^{-1} * F^{-1} * Q_4^{-1}H_1^{-1}D_1^{-1}D^{-h}H^{-1}R^{-1},$$

where $Q_3^{-1}F^{-1}Q_4^{-1} \;\overline{\circ}\; Q^{-1}$ and $l_\alpha(W_2) = l_\alpha(W) - 11 \geqslant q_1 - 11 = 26$.

Since Rel (W_1, U_1), it follows that Rel (E, F^{-1}). Then, by II.5.21, we have $\neg$ Rel (E, F), whence it follows from II.5.17 that the occurrence

$$W_2' \rightleftharpoons RTA'A_1T_1Q_4 * F * Q_3$$

is not compatible with U_2. By II.6.7, either $U_2 < W_2'$ or $W_2' < U_2$. Since in addition II.5.8 gives that the occurrence

$$U_2' \rightleftharpoons Q_2^{-1} * E^{-1} * Q_1^{-1}H_1^{-1}D_1^{-1}D^{-h}H^{-1}R^{-1}$$

is not compatible with W_2, and $W_2' < U_2$ implies that $U_2' < W_2$, it is enough to prove that the case $U_2 < W_2'$ is impossible, in view of the obvious analogy.

Suppose that $U_2 < W_2'$. Then $W_2 < U_2' \ll U^{-1}$. Since $l_\alpha(U_2') > 17$, we have $W_1 < U_2'$ by II.6.7, and consequently $W_2 \ll U_2'$, Then $U_2 \ll W_2'$, and by II.6.11, $W_2' \in$ Norm $(\alpha, X, 26)$ since $W_2 \in$ Norm $(\alpha, Y^{-1}, 26)$. We consider the occurrence

$$W_3 \rightleftharpoons f_{\alpha-1}(W_2'; X, Y^{-1}).$$

By II.5.16, we have $W_3 \in$ Norm$(\alpha, Y^{-1}, 26)$ and $\neg$Comp(U^{-1}, W_3). By *IV.2.18(g)*, since $U_1 \ll W_2'$ it follows that $U^{-1} \ll W_3$. Since $\neg$Comp$(Q^{-1}*H_1^{-1}D_1^{-1}D^{-h}H_1^{-1}*R^{-1}$, $W_3)$, by II.6.7 and I.2.7 the end of the occurrence W_3 starting with its 12-th left segment does not intersect $Q^{-1} * H_1^{-1}D_1^{-1}D^{-h}H^{-1} * R^{-1}$, so that it has the form

$$W_4 \rightleftharpoons Q^{-1}H_1^{-1}D_1^{-1}D^{-h}H^{-1}R_2^{-1} * G^{-1} * R_1^{-1},$$

where $R_2^{-1}G^{-1}R_1^{-1} \;\overline{\circ}\; R^{-1}$ and $l_\alpha(W_4) = l_\alpha(W_3) - 11 \geqslant 15$.

Suppose that the occurrence $Q^{-1}H_1^{-1}D_1^{-1}D^{-h}H^{-1}R_2^{-1}N^{-1} * L^{-1} * R_1^{-1}$ starts with the fifth left segment of the occurrence W_4. Then by II.6.12 we have

$$Q^{-1}T_1^{-1}A_1^{-1}A^{-\prime}T^{-1}R_2^{-1}N^{-1} * L^{-1} * R_1^{-1} \in \text{Norm}\,(\alpha, X^{-1}, 9),$$

whence, by the principle of symmetry, it follows that the occurrence

$$V_1 \rightleftharpoons R_1 * L * NR_2TA'A_1T_1Q$$

is also normalised. Thus we have found an occurrence $V_1 \in$ Reg $(\alpha - 1, X)$ such that $V_1 \ll W$. Then $f_{\alpha-1}(V_1; X, Y^{-1}) \ll W$, that is,

$$f_{\alpha-1}(V_1;\, X,\, Y^{-1}) = Q_5^{-1} * L_1^{-1} * Q_6^{-1}F^{-1}Q_4^{-1}H_1^{-1}D_1^{-1}D^{-h}H^{-1}R^{-1},$$

where $Q_5^{-1}L_1^{-1}Q_6^{-1} \overset{\alpha}{\simeq} Q_3^{-1}$. Since $f_{\alpha-1}(V_1;\, X,\, Y^{-1}) \in \mathrm{Reg}(\alpha-1,\, Y^{-1})$, it follows from II.6.10 and the principle of symmetry that we have $V_2 \in \mathrm{Reg}\,(\alpha-1,\, X)$ and $W_2' \ll V_2$ for

$$V_2 \rightleftarrows RTA'A_1T_1Q_4FQ_6 * L_1 * Q_5\,.$$

Further, we have $W_3 \ll f_{\alpha-1}(V_2;\, X,\, Y^{-1})$ by *IV.2.18(g)*, that is,

$$f_{\alpha-1}(V_2;\, X,\, Y^{-1}) = Q^{-1}H_1^{-1}D_1^{-1}D^{-h}H^{-1}R_2^{-1}G^{-1}R_3^{-1} * L_2^{-1} * R_4^{-1},$$

where $R_3^{-1}L_2^{-1}R_4^{-1} \overset{\alpha}{\simeq} R_1^{-1}$. Since $l_\alpha(G^{-1}) > 9$, by II.6.10 and the principle of symmetry we have $V_3 \in \mathrm{Reg}\,(\alpha-1,\, X)$ and $V_3 \ll V_1$ for

$$V_3 \rightleftarrows R_4 * L_2 * R_3GR_2TA'A_1T_1Q\,.$$

Clearly, by repeating our arguments sufficiently often, we can find an arbitrarily large number of pairwise non-intersecting occurrences in the word X, and this contradicts the finiteness of the length of X. Thus the relation $X \overset{\alpha-1}{\approx} Y^{-1}$ is false.

§ 2. Stable Occurrences and Local Reversals of Rank α

2.1. *If the occurrence V is stable in the reversal $X \to Y$ of rank α, then* $\mathrm{MutNorm}_{\alpha-1}(V, f_{X\to Y}(V))$.

For a simple reversal $X \to Y$, this holds by I.4.17, and the general case follows in view of I.4.19, *IV.3.2* and *IV.3.1*.

2.2. *Let $X \to Y$ be a reversal of the occurrence $W \in$* Norm $(\alpha, X, 9)$, *and suppose that $U \in$* Norm $(\alpha, Y, 9)$ *is an image of W in this reversal and that the occurrence V is stable in it. Then V does not intersect the maximal normalised continuation of the occurrence W. If in addition $V \ll W$, then $f_{X\to Y}(V) \ll U$. If $W \ll V$, then $U \ll f_{X\to Y}(V)$.*

For a simple reversal this follows from I.4.17 and 1.6, while for the general case we need to use proposition *IV.2.18(g)* as well.

2.3. *Suppose that the occurrence U is stable in the reversal $X \to Y$ of rank α. If $V \in$* Reg $(\alpha-1, X)$ *is an occurrence contained in U, then V is stable in the reversal $X \to Y$ and $f_{X\to Y}(V)$ is contained in $f_{X\to Y}(U)$.*

For a simple reversal, this follows from I.4.17 and *IV.3.6*, while for the general case we need to use propositions *IV.2.18(h)*, *IV.3.2* and *IV.3.1*.

2.4. *Suppose that the occurrence $P * E * Q \in$* ComReg $(\alpha-1, X)$ *is stable in a reversal of rank α of the occurrence W.*

If $P * E * Q < W$ (or $W < P * E * Q$), then every regular occurrence of rank α contained in $* PE * Q$ (in $P * EQ *$ respectively) is likewise stable in that reversal.

For a simple reversal, this follows from I.4.17 and II.7.4, and for the general case it remains only to refer to II.7.2 and *IV.2.18(f)*.

2.5. *For any occurrences U, V stable in a reversal $X \to Y$ of rank α, the following relations are satisfied:*

$$U < V \Longleftrightarrow f_{X \to Y}(U) < f_{X \to Y}(V), \quad U \ll V \Longleftrightarrow f_{X \to Y}(U) \ll f_{X \to Y}(V).$$

If U is a start (or an end) of V, then $f_{X \to Y}(U)$ is a start (an end respectively) of $f_{X \to Y}(V)$, and conversely.

This follows from *IV.2.18*.

2.6. *If the occurrence $V \in$ Norm $(\alpha, X, 9)$ is stable in a reversal $X \to Y$ of rank α, then $f_{X \to Y}(V) \in$ Norm $(\alpha, Y, 9)$, Rel $(V, f_{X \to Y}(V))$ and $l_\alpha(f_{X \to Y}(V)) = l_\alpha(V)$. If the occurrences U, $V \in$ Norm $(\alpha, X, 9)$ are stable in a reversal $X \to Y$ of rank α, then*

$$\text{Comp}\,(U, V) \Longleftrightarrow \text{Comp}\,(f_{X \to Y}(U), f_{X \to Y}(V)).$$

The first assertion follows from 2.1 and II.5.18, and the second from II.5.16 and II.5.8.

2.7. *Suppose that the occurrence $V \in$ Norm $(\alpha, X, 9)$ is stable in a reversal $X \to Y$ of rank α of the occurrence W.*

If $V < W$ (or $W < V$), then V has no proper normalised continuation to the left (to the right) if and only if $f_{X \to Y}(V)$ has no proper normalised continuation to the left (to the right respectively).

Suppose that $V = PF * E * TQ$ and that the occurrence $U \rightleftarrows P * FET * Q$ is stable in the reversal $X \to Y$. If $V < W$ and the word T is non-empty (if $W < V$ and the word F is non-empty) then V has no proper normalised continuation to the right (to the left) if and only if the occurrence $f_{X \to Y}(V)$ has no proper normalised continuation to the right (left respectively).

The first assertion follows from II.5.16 and II.6.15, and the second from II.5.16 and II.5.15.

2.8. *Suppose that the occurrence $V \in$ Norm $(\alpha, X, 9)$ is stable in the reversal $X \to Y$ of rank α of the occurrence W.*

If no normalised continuation of V to the left (to the right) different from V itself is stable in the reversal $X \to Y$, then at least one of the occurrences V and $f_{X \to Y}(V)$ fails to have a proper normalised continuation to the left (to the right respectively).

By symmetry we can restrict to the case where $V < W$. Then the first part of our assertion, that relating to left continuability, follows from 2.4 and 2.7.

To prove the second part, we note first that II.5.16 allows us to restrict to the case where $X \to Y$ is a simple reversal, that is, $X \mathrel{\underline{\underline{\circ}}} PTA^t A_1 T_1 Q$, $Y \mathrel{\underline{\underline{\circ}}} PHD^h D_1 H_1 Q$ and Comp $(W, PT * A^t A_1 * T_1 Q)$, where $A \mathrel{\underline{\underline{\circ}}} A_1 A_2$ is an elementary period of rank α.

Let us assume that the occurrences V and $f_{X \to Y}(V)$ both have proper normalised continuations to the right. Suppose that V_1 and V_2 are normalised continuations of them to the right, where

$$\partial_{\alpha-1}(V_1) = \partial_{\alpha-1}(V) + 1 \text{ and } \partial_{\alpha-1}(V_2) = \partial_{\alpha-1}(f_{X\to Y}(V)) + 1 .$$

By 1.7, either V_1 does not intersect $P * TA^tA_1T_1 * Q$, or else V_2 does not intersect $P * HD^hD_1H_1 * Q$. These two cases are analogous. Suppose that V_1 does not intersect $P * TA^tA_1T_1 * Q$. Consider the ends U_1 and U_2 of V_1 and V_2 respectively that start with the final kernels of rank $\alpha - 1$ of the occurrences V and $f_{X\to Y}(V)$. Clearly, $\partial_{\alpha-1}(U_1) = \partial_{\alpha-1}(U_2) = 2$. By II.2.2, II.2.9, *IV.3.6* and *IV.3.1*, $\text{MutNorm}_{\alpha-1}(V, f_{X\to Y}(V))$ gives $\text{MutNorm}_{\alpha-1}(U_1, U_2)$. Then we have Bas $(U_2) \overset{\circ}{=}$ Bas (U_1) by *IV.3.10*, so that Bas $(V_2) \overset{\circ}{=}$ Bas (V_1), whence $\text{MutNorm}_{\alpha-1}(V_1, V_2)$ follows because of *IV.3.25*. But this means that V_1 is stable in the simple reversal $X \to Y$, that is, we have arrived at a contradiction.

2.9. *Suppose that* $U, W \in$ Norm $(\alpha, X, 21)$, *that* $\neg$ Comp (U, W), *and that* $X \to Y$ *is a 9-reversal of the occurrence* U.

If $U < W$, *then the end of* W *starting with its* 18-*th left segment is stable in the reversal* $X \to Y$.

If $W < U$, *then the start of* W *ending with its* 18-*th right segment is stable in the reversal* $X \to Y$.

Clearly, it is enough to prove the first assertion. Suppose that $U < W$. By 1.4, we may assume that $X \to Y$ is a simple reversal of U, that is, we have

$$PT * A^tA_1 * T_1Q \in \text{MaxNorm}(\alpha, X, 21) ,$$
$$PH * D^hD_1 * H_1Q \in \text{MaxNorm}(\alpha, Y, 9) ,$$

and

$$\text{Comp}(U, PT * A^tA_1 * T_1Q) ,$$

where $A \overset{\circ}{=} A_1A_2$ is an elementary period of rank α and $TA^tA_1T_1$ is an elementary word generated by some occurrence $R * TA^tA_1T_1 * S$ in an integral word Z of rank α(see 1.6). By 1.3, we may assert that $U = PT * A^tA_1 * T_1Q$. Then by II.3.9, we have

$$\text{MutNorm}_{\alpha-1}(U, RT * A^tA_1 * T_1S),$$

whence by II.7.5 it follows that

$$\text{MutNorm}_{\alpha-1}(RT * A^tA_1 * T_1S, PT * A^tA_1 * T_1S). \tag{1}$$

For $i \leqslant 21$, we denote by V_i the i-th left segment of the occurrence W, by W_i the end of W starting with the pseudokernel V_i, and by W_i' the start of W ending with the pseudokernel V_i.

By II.6.1, we can find a section $F * N$ of the word X that is separating for the pair (V_{12}, V_{14}). We show that N is an end of the word Q. Assume that this is not so.

Then F is a start of $PTA'A_1T_1$. Since $U < W$, it follows from II.6.3 for $\alpha > 1$ that W'_{12} is contained in the occurrence $U' \rightleftarrows P * TA'A_1T_1 * Q$ and the following relation is satisfied:

$$\text{MutNorm}_{\alpha-2}(W'_{12}, \phi(W'_{12}; U', P * TA'A_1T_1 * S)).$$

Then by II.6.9 we have

$$\text{MutNorm}_{\alpha-1}(W'_9, \phi(W'_9; U', P * TA'A_1T_1 * S)). \tag{2}$$

Since $\neg$ Comp (U, V), it follows that $U < W'_9$, that is, W'_9 is contained in $PT * A'A_1T_1 * Q$. Then, by II.7.4, we get from (2) and (1) that

$$\text{MutNorm}_{\alpha-1}(W'_9, \phi(W'_9; U', R * TA'A_1T_1 * S)),$$

that is,

$$\phi(W'_9; U', R * TA'A_1T_1 * S) \in \text{Norm}(\alpha, Z, 9). \tag{3}$$

By II.5.7, it follows from (3) that

$$\text{Comp}(R * TA'A_1T_1 * S, \phi(W'_9; U', R * TA'A_1T_1 * S)).$$

Then, in view of II.5.8, we will have Comp(U, W'_9), which contradicts the condition $\neg$ Comp (U, W). Thus N is an end of Q, that is, $Q \,\overline{\underline{\circ}}\, GN$ for some G. Then for $\alpha > 1$, we have by II.6.3 that the occurrence W_{15} is contained in $PTA'A_1T_1G * N *$ and also that

$$\text{MutNorm}_{\alpha-2}(W_{15}, \phi(W_{15}; PTA'A_1T_1 * Q *, PHD^hD_1H_1 * Q *)),$$

whence, in view of II.6.9, we get

$$\text{MutNorm}_{\alpha-1}(W_{18}, \phi(W_{18}; PTA'A_1T_1 * Q *, PHD^hD_1H_1 * Q *)),$$

that is, W_{18} is stable in the simple reversal under consideration.

2.10. *Suppose that $V, W \in$ Norm $(\alpha, X, 21)$, $\neg$ Comp (V, W), $l_\alpha(V) = r$, and that there is a 9-reversal $X \to Y$ of rank α of the occurrence W.*

If $V < W$ and U is a maximal start of V that is stable in the reversal $X \to Y$, then $U \in$ Norm $(\alpha, X, r - 17)$.

If $W < V$ and U is a maximal end of V that is stable in the reversal $X \to Y$, then $U \in$ Norm $(\alpha, X, r - 17)$.

This follows immediately from 2.9.

2.11. *Suppose that $U, W \in$ Norm (α, X, q_1) and that the occurrence V is stable in an arbitrary 9-reversal of the occurrence W. If $U < W < V$ or $V < W < U$, then V is also stable in an arbitrary 9-reversal of U.*

If Comp (U, W), this follows from 1.3, otherwise it follows from 2.9, 2.2 and 2.4.

2.12. Let $X \to Y$ be a 9-reversal of rank α of the occurrence W, and suppose that the occurrence $P * E * Q \in \text{ComReg}(\alpha - 1, X)$ is stable in it. We shall say that the reversal $X \to Y$ is *local with respect to the occurrence* $P * E * Q$, if $f_{X \to Y}(P * E * Q) = P * E * R$ when $P * E * Q < W$ and $f_{X \to Y}(P * E * Q) = T * E * Q$ when $W < P * E * Q$.

We say that the reversal $X \to Y$ is *local from the left* (*from the right*) if it is local with respect to every occurrence $V \in \text{ComReg}(\alpha - 1, X)$ that is stable in it such that $V < W$ (such that $W < V$ respectively).

The reversal $X \to Y$ is said to be *a local reversal* if it is local from the left and from the right.

2.13. *Suppose that the reversal $X \to Y$ of rank α of the occurrence W is local with respect to the occurrence $P * E * Q \in$ ComReg $(\alpha - 1, X)$.*

If $P * E * Q < W$ (if $W < P * E * Q$), then the reversal $X \to Y$ is local with respect to every occurrence $V \in$ ComReg $(\alpha - 1, X)$ contained in $* PE * Q$ (in $P * EQ *$ respectively).

Indeed, suppose that $P * E * Q < W$, $f_{X \to Y}(P * E * Q) = P * E * R$, $V \in \text{ComReg}(\alpha - 1, X)$, and that V is contained in $* PE * Q$. Then, by 2.4, V is stable in the reversal $X \to Y$, and by II.7.8 we have

$$f_{X \to Y}(V) = \phi(V; * PE * Q, * PE * R).$$

The case $W < P * E * Q$ is similar.

It follows from what has been proved that the reversal $X \to Y$ of W is local from the left (from the right) if it is local with respect to the completely regular occurrence of rank $\alpha - 1$ that is nearest to W on the left (right) and stable in it.

2.14. *Suppose that $W \in$ Norm $(\alpha, X, 9)$. If one can find a 9-reversal of the occurrence W, then one can find a local reversal $X \to Z$ such that a maximal image of W in this reversal is periodised in rank $\alpha - 1$.*

Suppose that $X \overset{\alpha-1}{\simeq} PTA^t A_1 T_1 Q$ and that the transition

$$PTA^t A_1 T_1 Q \to PHD^h D_1 H_1 Q \rightleftharpoons Y \tag{4}$$

is a simple 9-reversal of the occurrence $f_{\alpha-1}(W; X, PTA^t A_1 T_1 Q)$, where the occurrence

$$W' \rightleftharpoons PH * D^h D_1 * H_1 Q$$

is a maximal image of W in the reversal $X \to Y$.

First of all, there exists a word Y_1 such that $X \to Y_1$ is a reversal of W that is local from to the left. Let $R * E * S$ be the completely regular occurrence of rank $\alpha - 1$ in the word X that is nearest to W on the left and stable in the reversal $X \to Y$, that is, $R * E * S < W$ and every occurrence $V \in \text{ComReg}(\alpha - 1, X)$ that is stable in the reversal $X \to Y$, and such that $V < W$, is contained in $* RE * S$. We set

$$V_1 \rightleftharpoons f_{\alpha-1}(R * E * S;\ X,\ PTA^t A_1 T_1 Q)\,. \tag{5}$$

Since $V_1 < PT * A^t A_1 * T_1 Q$ and V_1 is stable in the simple reversal (4), it follows that $V_1 = R_1 * E_1 * P_1 TA^t A_1 T_1 Q$, where $R_1 E_1 P_1 \mathrel{\underset{\circ}{\equiv}} P$ and

$$\mathrm{MutNorm}_{\alpha-1}(V_1,\ R_1 * E_1 * P_1 HD^h D_1 H_1 Q)\,. \tag{6}$$

By II.7.6, it follows from (5) and (6) that $Y \overset{\alpha-1}{\approx} Y_1$ for some Y_1,

$$f_{\alpha-1}(R_1 * E_1 * P_1 HD^h D_1 H_1 Q;\ Y,\ Y_1) = R * E * P_2 D^h D_1 H_1 Q \tag{7}$$

and

$$f_{\alpha-1}(W';\ Y,\ Y_1) = REP_2 * D^h D_1 * H_1 Q\,. \tag{8}$$

Then, according to I.4.19, the transition $X \to Y_1$ is a 9-reversal of the occurrence W, and $f_{X \to Y_1}(R * E * S) = R * E * P_2 D^h D_1 H_1 Q$. By 2.13, this reversal is local from the left.

Now we consider the occurrence $F * L * G \in \mathrm{ComReg}\,(\alpha - 1, X)$ that is nearest to W on the right and stable in the reversal $X \to Y$. By I.4.19, we have

$$f_{\alpha-1}(F * L * G;\ X,\ PTA^t A_1 T_1 Q) = PTA^t A_1 Q_1 * L_1 * G_1\,,$$

where $Q_1 L_1 G_1 \mathrel{\underset{\circ}{\equiv}} Q$ and

$$\mathrm{MutNorm}_{\alpha-1}(PTA^t A_1 T_1 Q_1 * L_1 * G_1,\ PHD^h D_1 H_1 Q_1 * L_1 * G_1)\,. \tag{9}$$

Since $\mathrm{MutNorm}_{\alpha-1}(W',\ REP_2 * D^h D_1 * H_1 Q)$ follows from (8), by II.7.4 we have

$$\mathrm{MutNorm}_{\alpha-1}(PHD^h D_1 H_1 Q_1 * L_1 * G_1,\ REP_2 D^h D_1 H_1 Q_1 * L_1 * G_1)\,,$$

whence by (9) it follows that

$$\mathrm{MutNorm}_{\alpha-1}(PTA^t A_1 T_1 Q_1 * L_1 * G_1,\ REP_2 D^h H_1 Q_1 * L_1 * G_1)\,.$$

Further, by II.7.6 there is a word Z such that $Y_1 \overset{\alpha-1}{\approx} Z$,

$$f_{\alpha-1}\,(REP_2 D^h H_1 Q_1 * L_1 * G_1;\ Y_1,\ Z) = REP_2 D^h D_1 Q_2 * L * G \tag{10}$$

and

$$f_{\alpha-1}(REP_2 * D^h D_1 * H_1 Q_1 L_1 G_1;\ Y_1,\ Z) = REP_2 * D^h D_1 * Q_2 LG\,. \tag{11}$$

Then $Y \overset{\alpha-1}{\approx} Z$ and, by II.7.7, it follows from (7) that

$$f_{\alpha-1}(R_1 * E_1 * P_1 HD^h D_1 H_1 Q;\ Y,\ Z) = R * E * P_2 D^h D_1 Q_2 LG\,.$$

Moreover, it follows from (10) in an analogous way that

$$f_{\alpha-1}(R_1E_1P_1HD^hD_1H_1Q_1 * L_1 * G_1;\ Y, Z) = REP_2D^hD_1Q_2 * L * G\,,$$

and from (8) and (11) that

$$f_{\alpha-1}(W';\ Y, Z) = REP_2 * D^hD_1 * Q_2LG\,. \tag{12}$$

Thus, $X \to Z$ is a reversal of the occurrence W,

$$f_{X\to Z}(R * E * S) = R * E * P_2D^hD_1Q_2LG\,,$$

and

$$f_{X\to Z}(F * L * G) = REP_2D^hD_1Q_2 * L * G\,.$$

By 2.13, the reversal $X \to Z$ is local from the left and from the right. By II.5.16, it follows from (12) that the occurrence $W'' \rightleftharpoons REP_2 * D^hD_1 * Q_2LG$ is a maximal image of W in this reversal. By 1.6, W' is periodised in rank $\alpha - 1$. Since $\text{MutNorm}_{\alpha-1}(W', W'')$ follows from (12), W'' is also periodised in rank $\alpha - 1$.

2.15. *Suppose that*

$$P * E * uDQ \in \text{Norm}\,(\alpha, X, q_1)\,, \tag{13}$$

$$R * E * uDS \in \text{Norm}\,(\alpha, Y, q_1)\,, \tag{14}$$

$$PEu * D * Q \in \text{ComReg}\,(\alpha - 1, X)\,, \tag{15}$$

$$\text{MutNorm}_{\alpha-1}\,(PEu * D * Q,\ REu * D * S), \tag{16}$$

that q_1-reversals may be carried out on (13) *and* (14), *and that the occurrence* (15) *is stable in some reversal of the occurrence* (13). *Then the occurrence $REu * D * S$ is stable in every reversal of* (14).

If the transition

$$PEuDQ \to P'E'vDQ \tag{17}$$

is a local reversal of occurrence (13), *where $P' * E' * vDQ$ is a maximal image of occurrence* (13), *then there is a local reversal of occurrence* (14),

$$REuDS \to R'E''vDS\,,$$

where $R' * E'' * vDS$ is a maximal image of occurrence (14) *and one of the words E' and E'' is an end of the other. If in addition the words $P'E'vDQ$ and $REuDS$ both lie in one of the sets $\mathcal{K}_{\alpha-1}$, $\mathcal{L}_{\alpha-1}$ or $\mathcal{M}_{\alpha-1}$, then the word $R'E''vDS$ may be taken to be in the same set.*

The assertion symmetric to this one, that is, the assertion obtained by interchanging the rôles of the words E and D, is proved in an analogous fashion.

We note first of all that, by 2.2, II.7.1 and II.6.17, we may assume that the occurrences (13) and (14) both fail to have proper normalised continuations to the right. By II.7.3 and *IV.3.6*, we may assume also that (15) is the occurrence in ComReg $(\alpha - 1, X)$ that is nearest to (13) on the right and stable in the reversal (17).

By 2.14 and 1.11, there is a local q_1-reversal of (13). Suppose that (17) is such a reversal, that is, there are words $X_1, Z_1 \in \mathscr{K}_{\alpha-1}$ such that

$$X \overset{\alpha-1}{\simeq} X_1, \; P'E'vDQ \overset{\alpha-1}{\simeq} Z_1 ,$$

the transition

$$X_1 \equiv P_1 T A^t A_1 T_1 u D_1 Q_1 \to P_1 H L^h H_1 u_1 D_1 Q_1 \equiv Z_1 \tag{18}$$

is a simple q_1-reversal of an occurrence

$$P_1 T * A^t A_1 * T_1 u_1 D_1 Q_1 \in \text{MaxNorm}\,(\alpha, X_1, q_1), \tag{19}$$

the occurrence

$$P_1 H * L^h L_1 * H_1 u_1 D_1 Q_1 \in \text{MaxNorm}\,(\alpha, Z_1, q_1) \tag{20}$$

is a maximal image of it in (18), $A \equiv A_1 A_2$ is an elementary period of rank α, $L \equiv L_1 L_2$ is a cyclic shift of the word A^{-1} and the following relations are satisfied:

$$A^{n+20} \overset{\alpha-1}{\simeq} A^{t+10} A_1 T_1 H_1^{-1} L_1^{-1} L^{-h} H^{-1} T A^{10} , \tag{21}$$

$$\text{Comp}\,(f_{\alpha-1}(P * E * u\, DQ; X, X_1), P_1 T * A^t A_1 * T_1 u_1 D_1 Q_1), \tag{22}$$

$$f_{\alpha-1}(P' * E' * vDQ; Z, Z_1) = P_1 H * L^h L_1 * H_1 u_1 D_1 Q_1 , \tag{23}$$

$$f_{\alpha-1}(PEu * D * Q; X, X_1) = P_1 T A^t A_1 T_1 u_1 * D_1 * Q_1 , \tag{24}$$

$$f_{\alpha-1}(P'E'v * D * Q; Z, Z_1) = P_1 H L^h L_1 H_1 u_1 * D_1 * Q_1 , \tag{25}$$

and

$$\text{MutNorm}_{\alpha-1}(P_1 T A^t A_1 T_1 u_1 * D_1 * Q_1, P_1 H L^h L_1 H_1 u_1 * D_1 * Q_1) . \tag{26}$$

Since, by II.5.16, the occurrence $f_{\alpha-1}(P * E * uDQ;\; X,\; X_1)$ also has no proper normalised continuation to the right, it follows from (22) and (19) that

$$f_{\alpha-1}(P * E * uDQ; X, X_1) = P_1 T F * G * T_1 u_1 D_1 Q_1 , \tag{27}$$

where

$$A'A_1 \overset{\circ}{=} FG. \tag{28}$$

By II.5.5 and II.7.3, it follows from (13), (14) and (16) that

$$\text{MutNorm}_{\alpha-1}(P * EuD * Q, R * EuD * S). \tag{29}$$

Since $X_1 \in \mathscr{K}_{\alpha-1}$ and $f_{\alpha-1}(P * EuD * Q;\ X,\ X_1) = P_1TF * GT_1u_1D_1 * Q_1$, by *IV.3.17* there is a word $Y_1 \in \mathscr{K}_{\alpha-1}$ such that $Y \overset{\alpha-1}{\sim} Y_1$ and

$$f_{\alpha-1}(R * EuD * S;\ Y,\ Y_1) = R_1 * GT_1u_1D_1 * S_1.$$

In addition, by (29), (27), (24) and *IV.3.5* we have

$$f_{\alpha-1}(R * E * uDS;\ Y,\ Y_1) = R_1 * G * T_1u_1D_1S_1 \tag{30}$$

and

$$f_{\alpha-1}(REu * D * S;\ Y,\ Y_1) = R_1GT_1u_1 * D_1 * S_1. \tag{31}$$

Since by 1.6 the occurrence (27) is periodised in rank $\alpha - 1$, in view of II.5.5 the occurrence (30) is also periodised in rank $\alpha - 1$. Then by 1.4 and 1.9, there is a simple reversal

$$Y_2 \rightleftharpoons R_2NB^rB_1N_1S_2 \to R_2MC^kC_1M_1S_2 \rightleftharpoons Z_2 \tag{32}$$

of rank α of the occurrence

$$R_2N * B^rB_1 * N_1S_2 \in \text{MaxNorm}\,(\alpha, Y_2, q_1), \tag{33}$$

such that $B \overset{\circ}{=} B_1B_2$ is an elementary period of rank α, $C \overset{\circ}{=} C_1C_2$ is a cyclic shift of the word B^{-1}, and the following relations hold:

$$Y_1 \overset{\alpha-1}{\sim} R_2NB^rB_1N_1S_2,$$

$$\text{Comp}\,(f_{\alpha-1}(R_1 * G * T_1u_1D_1S_1;\ Y_1,\ Y_2),\ R_2N * B^rB_1 * N_1S_2), \tag{34}$$

$$R_2M * C^kC_1 * M_1S_2 \in \text{MaxNorm}\,(\alpha, Z_2, q_1) \tag{35}$$

and

$$\text{Bas}\,(f_{\alpha-1}(R_1 * G * T_1u_1D_1S_1;\ Y_1,\ Y_2)) = G.$$

Since by assumption $R * E * uDS$ has no proper normalised continuation to the right, it follows from (34) in view of II.5.16 that $f_{\alpha-1}\,(R * E * uDS;\ Y,\ Y_2)$ is an end of the occurrence (33), that is, we have

$$f_{\alpha-1}(R_1 * G * T_1u_1D_1S_1;\ Y_1,\ Y_2) = R_2NF_1 * G * N_1S_2, \tag{36}$$

where

$$B'B_1 \overset{\circ}{=} F_1G\,. \tag{37}$$

From (20) it follows that $h \geqslant 36$. Since L^{-1} is a cyclic shift of A, we have

$$A \overset{\circ}{=} A_3A_4,\ L^{-1} \overset{\circ}{=} A_4A_3\,, \tag{38}$$

and thus

$$L_1^{-1}L^{-h} \overset{\circ}{=} L_1^{-1}L^{-10}A_4A^{h-21}A_3L^{-10}\,.$$

We set

$$Z' \rightleftharpoons A^{t+10}A_1T_1H_1^{-1}L_1^{-1}L^{-h}H^{-1}TA^{10}\,.$$

It is clear from (19) and (20) that $A^{h-21}A_3$ is a normal elementary 9-power of rank α. Then by (21) and II.3.9, the occurrence

$$V \rightleftharpoons A^{t+10}A_1T_1H_1^{-1}L_1^{-1}L^{-10}A_4 * A^{h-21}A_3 * L^{-10}H^{-1}TA^{10}$$

is a normal generating occurrence in a word $Z' \in \operatorname{Int}(A^{n+20}, \alpha, A)$, that is, we have $V \in \operatorname{Norm}(\alpha, Z', 9)$. Since the occurrence (19) is periodised in rank $\alpha - 1$, by II.7.10 the occurrence V is also periodised in rank $\alpha - 1$. Then by II.7.21, we have $\operatorname{Bas}(f_{\alpha-1}(V; Z', A^{n+20})) = A^{h-21}A_3$, that is,

$$f_{\alpha-1}(V; Z', A^{n+20}) = A^i * A^{h-21}A_3 * A_4A^j\,, \tag{39}$$

where $i + j + h - 20 = n + 20$. By *IV.2.35*, it follows from (39) that

$$A^{t+10}A_1T_1H_1^{-1}L_1^{-1}L^{-10}A_4 \overset{\alpha-1}{\eqsim} A^i \tag{40}$$

and

$$L^{-10}H^{-1}TA^{10} \overset{\alpha-1}{\eqsim} A_4A^j\,.$$

For the simple reversal (32), we find in an analogous way relations

$$B^{r+10}B_1N_1M_1^{-1}C_1^{-1}C^{-10}B_4 \overset{\alpha-1}{\eqsim} B^{i_1}, \tag{41}$$

and

$$C^{-10}M^{-1}NB^{10} \overset{\alpha-1}{\eqsim} B_4B^{j_1}, \tag{42}$$

where

$$B \overset{\circ}{=} B_3B_4,\ C^{-1} \overset{\circ}{=} B_4B_3\,, \tag{43}$$

and

$$i_1 + j_1 + k - 20 = n + 20\,. \tag{44}$$

It follows from (28) and (37) that one of the words $A'A_1$ and $B'B_1$ is an end of the other. Without loss of generality we may assume that $A_1 \overset{\circ}{=} A_5B_1$. Then we will have

$$A_5B \overset{\circ}{=} AA_5 \text{ and } A^{10}A_1 \overset{\circ}{=} A_5B^{10}B_1 . \tag{45}$$

Since the occurrences (19) and (20) do not have proper normalised continuations to the right, by II.6.18 and II.5.2 it follows from (40) that

$$\begin{aligned} i &\leqslant l_\alpha(A^{t+10}A_1T_1) + l_\alpha(H_1^{-1}L_1^{-1}L^{-10}A_4) + 2 \\ &\leqslant t + 15 + 16 + 2 = t + 33. \end{aligned} \tag{46}$$

It follows from relation (41) that $i_1 \geqslant r + 20$.

Using *V.2.1*, we get from (40) and (45) that

$$B^{r+10}B_1T_1H_1L_1^{-1}L^{-10}A_4A_5 \overset{\alpha-1}{\simeq} B^{i+r-t} . \tag{47}$$

In an analogous way, it follows from (42) and (43) that

$$B^{n+29-j_1}B_3M^{-1}NB^{10} \overset{\alpha-1}{\simeq} B^{n+20} . \tag{48}$$

Since $k \geqslant 36$, it follows from (44), (46) and the fact that $i_1 \geqslant r + 20$ that

$$\begin{aligned} n + 29 - j_1 = i_1 + k - 11 &\geqslant r + k + 9 \\ &\geqslant r + 45 \geqslant r + i - t + 12 . \end{aligned} \tag{49}$$

Therefore, using *V.2.1*, we get from (48) and (47) that

$$B^{n+20} \overset{\alpha-1}{\simeq} B^{r+10}B_1T_1H_1L_1^{-1}L^{-10}A_4A_5B^{n+29-j_1-i-r+t}B_3M^{-1}NB^{10} . \tag{50}$$

It follows from (38) and (45) that A is a right period of the word $L_1^{-1}L^{-10}A_4$ and a left period of $A_5B^{n+29-j_1-i-r+t}B_3$. Thus

$$L_1^{-1}L^{-10}A_4A_5B^{n+29-j_1-i-r+t}B_3 \in \mathrm{Per}\,(A) .$$

Since, by (43), C^{-1} is a right period of this word, we have

$$L_1^{-1}L^{-10}A_4A_5B^{n+29-j_1-i-r+t}B_3 \overset{\circ}{=} C_3^{-1}C^{-l} , \tag{51}$$

where $C^{-1} \overset{\circ}{=} C_4^{-1}C_3^{-1}$, and $l \geqslant 22$ by (49). At this point, relation (50) can be rewritten in the form

$$B^{n+20} \overset{\alpha-1}{\simeq} B^{r+10}B_1T_1H_1C_3^{-1}C^{-l}M^{-1}NB^{10} . \tag{52}$$

Using (43), (42), (47), (49), (51), the principle of symmetry and *V.2.1*, we get in a fashion similar to the foregoing that

$$
\begin{aligned}
C^{n+20} &\mathrel{\underset{\circ}{=}} B_3^{-1}B^{-n-19}B_4^{-1} \overset{\alpha-1}{\simeq} B_3^{-1}B^{-n-29+j_1}N^{-1}MC^{10} \\
&\overset{\alpha-1}{\simeq} B_3^{-1}B^{-n-29+j_1+l+r-t-10}A_5^{-1}A_4^{-1}L^{10}L_1H_1^{-1}T_1^{-1}B_1^{-1}B^{-r}N^{-1}MC^{10} \\
&\mathrel{\underset{\circ}{=}} C^{l+10}C_3H_1^{-1}T_1B_1^{-1}B^{-r}N^{-1}MC^{10}\,. \qquad (53)
\end{aligned}
$$

We consider the words

$$Y_3 \rightleftharpoons R_2NB^rB_1T_1u_1D_1S_1 \text{ and } Z_3 \rightleftharpoons R_2MC^lC_3H_1u_1D_1S_1\,.$$

By II.6.13, Y_1, $Y_2 \in \mathscr{K}_{\alpha-1}$ implies that $Y_3 \in \mathscr{K}_{\alpha-1}$.

It follows from (51) that one of the words C^lC_3 and L^hL_1 is an end of the other. Thus Z_1, $Z_2 \in \mathscr{K}_{\alpha-1}$ implies $Z_3 \in \mathscr{K}_{\alpha-1}$. In addition, by II.6.12 and II.7.3 it follows from (35) and (20) that

$$R_2M * C^lC_3 * H_1u_1D_1S_1 \in \text{MaxNorm}\,(\alpha, Z_3, 9). \qquad (54)$$

By *IV.2.35*, it follows from (36) that $N_1S_2 \overset{\alpha-1}{\simeq} T_1u_1D_1S_1$, from which we get $Y_2 \overset{\alpha-1}{\simeq} Y_3$ by *V.2.1*. Then by *II.7.7* we have

$$f_{\alpha-1}(R_2N * B^rB_1 * N_1S_2;\, Y_2, Y_3) = R_2\,N * B^rB_1 * T_1u_1D_1S_1\,, \qquad (55)$$

whence by (33) and II.5.16 it follows that

$$R_2N * B^rB_1 * T_1u_1D_1S_1 \in \text{MaxNorm}(\alpha, Y_3, q_1)\,. \qquad (56)$$

Since $Y_1 \overset{\alpha-1}{\simeq} Y_3$, by (37) and II.7.7 we also have the relations

$$f_{\alpha-1}(R_1 * G * T_1u_1D_1S_1;\, Y_1, Y_3) = R_2NF_1 * G * T_1u_1D_1S_1 \qquad (57)$$

and

$$f_{\alpha-1}(R_1GT_1u_1 * D_1 * S_1;\, Y_1, Y_3) = R_2NF_1GT_1u_1 * D_1 * S_1\,. \qquad (58)$$

According to definition I.4.16, relations (56), (54), (52) and (53) mean that the transition

$$R_2NB^rB_1T_1u_1D_1S_1 \to R_2MC^lC_3H_1u_1D_1S_1 \qquad (59)$$

is a simple 9-reversal of the occurrence (56) and that the occurrence (54) is a maximal image of occurrence (56) in this reversal. Since (32) is a q_1-reversal of occurrence (33) by assumption, by 1.4 and 1.11 it follows from (55) that (59) is also a q_1-reversal. Since occurrences (56) and (57) are compatible, it follows from (30) and (57) that the transition

$$Y \to R_2MC^lC_3H_1u_1D_1S_1 \qquad (60)$$

is a q_1-reversal of the occurrence $R * E * uDS$, and that occurrence (54) is a maximal image of it in this reversal.

Let us show that the occcurrence $REu * D * S$ is stable in the reversal (60). In view of relations (31) and (58), for this it is enough to establish the relation

$$\mathrm{MutNorm}_{\alpha-1}(R_2NB^rB_1T_1u_1 * D_1 * S_1,\ R_2MC^lC_3H_1u_1 * D_1 * S_1)\,. \tag{61}$$

By II.6.10, we have the relation

$$\mathrm{MutNorm}_{\alpha-1}(P_1HL^hL_1H_1u_1 * D_1 * S_1,\ R_2MC^lC_3H_1u_1 * D_1 * S_1)\,. \tag{62}$$

It follows easily from relations (16), (24), (31), (26) and (58) that

$$\mathrm{MutNorm}_{\alpha-1}(P_1HL^hL_1H_1u_1 * D_1 * Q_1,\ R_2NB^rB_1T_1u_1 * D_1 * S_1)\,,$$

whence, by II.7.5, we get

$$\mathrm{MutNorm}_{\alpha-1}(P_1HL^hL_1H_1u_1 * D_1 * S_1,\ R_2NB^rB_1T_1u_1 * D_1 * S_1)\,. \tag{63}$$

The desired relation (61) follows from (62) and (63). Thus we have proved that

$$f_{Y\to Z_3}(REu * D * S) = R_2MC^lC_3H_1u_1 * D_1 * S_1\,.$$

It is clear from (51) that one of the words L^hL_1 and C^lC_3 is an end of the other. Because of the analogy, we shall restrict attention to the case where L^hL_1 is an end of C^lC_3, that is,

$$C^lC_3 \overset{\circ}{=} F_2L^hL_1\,.$$

In such a case it follows from (20) and (54) in view of II.5.5 that

$$\mathrm{MutNorm}_{\alpha-1}(P_1H * L^hL_1 * H_1u_1D_1Q_1,\ R_2MF_2 * L^hL_1 * H_1u_1D_1S_1),$$

whence by (62) and II.7.3 we get

$$\mathrm{MutNorm}_{\alpha-1}(P_1H * L^hL_1H_1u_1D_1 * Q_1,\ R_2MF_2 * L^hL_1H_1u_1D_1 * S_1)\,. \tag{64}$$

Suppose that $R \overset{\circ}{=} R_3Kw$, where $R_3 * K * wEuDS$ is the completely regular occurrence of rank $\alpha - 1$ that is nearest to $R * E * uDS$ on the left and stable in the reversal (60). If there are no such occurrences, then every reversal of occurrence (14) is local from the left. Suppose that

$$f_{\alpha-1}(R_3 * K * wEuDS;\ Y,\ Y_3) = R_4 * K_1 * w_1NB^rB_1T_1u_1D_1S_1\,,$$

where $R_4K_1w_1 \overset{\circ}{=} R_2$. Then we have

$$\mathrm{MutNorm}_{\alpha-1}(R_4 * K_1 * w_1NB^rB_1T_1u_1D_1S_1,\ R_4 * K_1 * w_1MF_2L^hL_1H_1u_1D_1S_1)\,,$$

whence it follows by II.7.6 that, for some Z_4,

$$Z_3 \overset{\alpha-1}{\approx} Z_4$$
$$f_{\alpha-1}(R_4 * K_1 * w_1 M F_2 L^h L_1 H_1 u_1 D_1 S_1;\, Z_3, Z_4) \\ = R_3 * K * w_2 L^h L_1 H_1 u_1 D_1 S_1 \tag{65}$$

and

$$f_{\alpha-1}(R_4 K_1 w_1 M F_2 * L^h L_1 H_1 u_1 D_1 * S_1;\, Z_3, Z_4) = R_3 K w_2 * L^h L_1 H_1 u_1 D_1 * S_1 \,.$$

By (64), it follows from this last relation that

$$\mathrm{MutNorm}_{\alpha-1}(P_1 H * L^h L_1 H_1 u_1 D_1 * Q_1,\, R_3 K w_2 * L^h L_1 H_1 u_1 D_1 * S_1)\,. \tag{66}$$

Since it follows from (23) and (25) that

$$f_{\alpha-1}(P_1 H * L^h L_1 H_1 u_1 D_1 * Q_1;\, Z_1, Z) = P' * E' v D * Q\,,$$

by (66) and *IV.3.17* there is a word Z_5 such that $Z_4 \overset{\alpha-1}{\approx} Z_5$ and

$$f_{\alpha-1}(R_3 K w_2 * L^h L_1 H_1 u_1 D_1 * S_1;\, Z_4, Z_5) = R_3 K w_3 * E' v D * S_3\,.$$

In addition, it follows from (23) and (25) in view of *IV.3.5* that

$$f_{\alpha-1}(R_3 K w_2 * L^h L_1 * H_1 u_1 D_1 S_1;\, Z_4, Z_5) = R_3 K w_3 * E' * v D S_3 \tag{67}$$

and

$$f_{\alpha-1}(R_3 K w_2 L^h L_1 H_1 u_1 * D_1 * S_1;\, Z_4, Z_5) = R_3 K w_3 E' v * D * S_3\,. \tag{68}$$

On the other hand, since by II.5.5 we have

$$\mathrm{MutNorm}_{\alpha-1}(R_3 K w_2 L^h L_1 H_1 u_1 * D_1 * S_1,\, R_1 G T_1 u_1 * D_1 * S_1)\,,$$

by II.7.6 and (31) there is a word Z_6 such that $Z_4 \overset{\alpha-1}{\approx} Z_6$ and

$$f_{\alpha-1}(R_3 K w_2 L^h L_1 H_1 u_1 * D_1 * S_1;\, Z_4, Z_6) = Z_6' * D * S\,. \tag{69}$$

Then, setting

$$Z_7 \rightleftharpoons R_3 K w_3 E' v D S\,,$$

we get from (68) and (69) in view of *IV.2.33* that $Z_5 \overset{\alpha-1}{\approx} Z_7$ and

$$f_{\alpha-1}(R_3 K w_3 E' v * D * S_3;\, Z_5, Z_7) = R_3 K w_3 E' v * D * S\,. \tag{70}$$

Thus $Z_3 \overset{\alpha-1}{\approx} Z_7$. Using (68), (69) and II.7.7 as applied to $Z_3 \overset{\alpha-1}{\approx} Z_4$, we get

$$f_{\alpha-1}(R_4K_1w_1MF_2L^hL_1H_1u_1 * D_1 * S_1; Z_3, Z_7) = R_3Kw_3E'v * D * S.$$

It follows from (67) in an analogous way that

$$f_{\alpha-1}(R_4K_1w_1MF_2 * L^hL_1 * H_1u_1D_1S_1; Z_3, Z_7) = R_3Kw_3 * E' * vDS \tag{71}$$

and from (65),

$$f_{\alpha-1}(R_4 * K_1 * w_1MF_2L^hL_1H_1u_1D_1S_1; Z_3, Z_7) = R_3 * K * w_3E'vDS.$$

Thus the transition

$$REuDS \to R_3Kw_3E'vDS \tag{72}$$

is a q_1-reversal of occurrence (14), and it is local with respect to the occurrences $R_3 * K * wEuDS$ and $REu * D * S$. It follows from 2.13, and the fact that (17) is local, that (72) is a local reversal. By II.6.12, II.5.16 and *IV.2.18(i)*, it follows from (54) and (71) that a maximal image of the occurrence $R * E * uDS$ in the reversal (72) has the form $R_3Kw' * E'' * vDS$, where $E'' \overset{\circ}{=} w''E'$ and $w_3 \overset{\circ}{=} w'w''$.

Finally, we note that if both of the words $P'E'vDQ$ and $REuDS$ lie in $\mathscr{K}_{\alpha-1}$, $\mathscr{L}_{\alpha-1}$ or $\mathscr{M}_{\alpha-1}$, by *IV.3.15* there exists a word Z_8 from the same set such that $Z_7 \overset{\alpha-1}{\simeq} Z_8$,

$$f_{\alpha-1}(R_3 * K * w_3E'vDS; Z_7, Z_8) = R_3 * K * w_4E'vDS$$

and

$$f_{\alpha-1}(R_3Kw_3 * E' * vDS; Z_7, Z_8) = R_3Kw_4 * E' * vDS.$$

Clearly, the transition $Y \to Z_8$ is the desired local reversal of occurrence (14).

2.16. *Suppose that q_1-reversals may be carried out on the occurrences $P * E * Q \in$* Norm (α, X, q_1) *and $R * E * S \in$* Norm (α, Y, q_1), *that the transition*

$$PEQ \to P_1E_1Q_1$$

is a local reversal of $P * E * Q$, and that $P_1 * E_1 * Q_1$ is a maximal image of it in this reversal.

If $Q \overset{\circ}{=} S$ (or $P \overset{\circ}{=} R$), then there is a local reversal

$$RES \to R_1E_2S_1$$

of the occurrence $R * E * S$, where $R_1 * E_2 * S_1$ is a maximal image of $R * E * S$, $S_1 \overset{\circ}{=} Q_1 (R_1 \overset{\circ}{=} P_1)$ and one of the words E_1, E_2 is an end (a start respectively) of the other.

In addition, if the words RES and $P_1E_1Q_1$ both lie in the same one of $\mathscr{K}_{\alpha-1}$, $\mathscr{L}_{\alpha-1}$ or $\mathscr{M}_{\alpha-1}$, then the desired word $R_1E_2Q_1$ ($P_1E_2S_1$) can be taken to lie in that same set.

The proof is analogous to that of 2.15, but essentially simpler. The simplification is connected with the fact that here, instead of the availability in occurrences (13) and (14) of the same piece on the same side (for example, uD on the right), we require that the whole of the contexts on this side are the same ($Q \overset{\circ}{\equiv} S$). On the other hand, if $Q \overset{\circ}{\equiv} uDQ'$, where the occurrence $PEu * D * Q' \in \mathrm{ComReg}\,(\alpha - 1, PEQ)$ is stable in the reversal $PEQ \to P_1E_1Q_1$, then our assertion follows immediately from 2.15. However, this is not true in general, and it is necessary to repeat the corresponding arguments.

2.17. *Suppose that* $r \geqslant q_1$,

$$R * E * uDQ \in \mathrm{Norm}\,(\alpha, X, 9)\,, \tag{73}$$

$$REu * D * Q \in \mathrm{ComReg}\,(\alpha - 1, X) \tag{74}$$

and

$$\mathrm{MutNorm}_{\alpha-1}(REu * D * Q,\ REu * D * S)\,. \tag{75}$$

If the transition

$$X = REuDQ \to R'E'vDQ \rightleftharpoons Z \tag{76}$$

is a local r-reversal of the occurrence (73), *where occurrence* (74) *is stable and* $R' * E' * vDS$ *is a maximal image of* (73), *then the transition*

$$REuDS \to R'E'vDS \tag{77}$$

is a local r-reversal of the occurrence $R * E * uDS$, *where* $R' * E' * vDS$ *is a maximal image of it and the occurrence* $REu * D * S$ *is stable.*

If the words $R'E'vDQ$ *and* $REuDS$ *both lie in* $\mathscr{K}_{\alpha-1}$, $\mathscr{L}_{\alpha-1}$ *or* $\mathscr{M}_{\alpha-1}$, *then by* II.6.13 *the word* $R'E'vDS$ *lies in that same set.*

The symmetric assertion obtained from the preceding by interchanging the rôles of E and D is proved in an analogous fashion.

We set $Y \rightleftharpoons REuDS$. By 2.2 and 1.3, we may assume that $l_\alpha(E) \geqslant r$. Then by II.6.11 we have

$$R * E * uDS \in \mathrm{Norm}\,(\alpha, Y, r)\,, \tag{78}$$

where we may assume that the occurrences (73) and (78) are both maximal normalised occurrences, by 1.3. By assumption we have a simple r-reversal of rank α,

$$X_1 \rightleftharpoons R_1TA^tA_1T_1u_1D_1Q_1 \to R_1HL^hL_1H_1u_1D_1Q \rightleftharpoons Z_1\,, \tag{79}$$

where $X \overset{\alpha-1}{\eqsim} X_1$, $Z_1 \overset{\alpha-1}{\eqsim} R'E'vDQ$ and the relations

$$f_{\alpha-1}(R * E * uDQ; X, X_1) = R_1T * A^tA_1 * T_1u_1D_1Q_1\,, \tag{80}$$

$$f_{\alpha-1}(R' * E' * vDQ;\ Z, Z_1) = R_1\ H * L^h L_1 * H_1 u_1 D_1 Q_1\,, \tag{81}$$

$$f_{\alpha-1}(REu * D * Q;\ X, X_1) = R_1 TA^t A_1 T_1 u_1 * D_1 * Q_1 \tag{82}$$

$$\mathrm{MutNorm}_{\alpha-1}(R_1 TA^t A_1 T_1 u_1 * D_1 * Q_1,\ R_1 HL^h L_1 H_1 u_1 * D_1 * Q_1) \tag{83}$$

and

$$f_{\alpha-1}(R_1 HL^h L_1 H_1 u_1 * D_1 * Q_1;\ Z_1, Z) = R'E'v * D * Q \tag{84}$$

hold. By II.7.6, it follows from (75) and (82) that $Y \overset{\alpha-1}{\sim} Y_1$ for some word $Y_1 \in \mathscr{K}_{\alpha-1}$ and

$$f_{\alpha-1}(REu * D * S;\ Y, Y_1) = R_1 TA^t A_1 T_1 u_1 * D_1 * S_1\,, \tag{85}$$

whence by (80) and II.7.8 it follows that

$$f_{\alpha-1}(R * E * uDS;\ Y, Y_1) = R_1 T * A^t A_1 * T_1 u_1 D_1 S_1\,. \tag{86}$$

We consider the word

$$Z_2 \rightleftharpoons R_1 HL^h L_1 H_1 u_1 D_1 S_1\,.$$

Since $Z_1, Y_1 \in \mathscr{K}_{\alpha-1}$, we have $Z_2 \in \mathscr{K}_{\alpha-1}$. Since (79) is a simple reversal of occurrence (80), and (81) is a maximal image of it, by II.6.11 we have

$$R_1 H * L^h L_1 * H_1 u_1 D_1 S_1 \in \mathrm{MaxNorm}\,(\alpha, Z_2, r)\,. \tag{87}$$

Thus the transition $Y_1 \to Z_2$ is a simple r-reversal of the occurrence (86), with (87) as maximal image of it. By 2.15, occurrence (85) is stable in this reversal, that is we have

$$\mathrm{MutNorm}_{\alpha-1}(R_1 HL^h L_1 H_1 u_1 * D_1 * S_1,\ R_1 TA^t A_1 T_1 u_1 * D_1 * S_1)\,,$$

whence by (85), (75), (82) and (83) it follows that

$$\mathrm{MutNorm}_{\alpha-1}(R_1 HL^h L_1 H_1 u_1 * D_1 * S_1,\ R_1 HL^h L_1 H_1 u_1 * D_1 * Q_1)\,. \tag{88}$$

Then, by II.7.6, it follows from (84) and (85) that there exist words $Z', Z'' \in \in \mathscr{K}_{\alpha-1}$ such that $Z_2 \overset{\alpha-1}{\sim} Z'$, $Z_2 \overset{\alpha-1}{\sim} Z''$,

$$f_{\alpha-1}(R_1 HL^h L_1 H_1 u_1 * D_1 * S_1;\ Z_2, Z') = R'E'v * D * S'$$

and

$$f_{\alpha-1}(R_1 HL^h L_1 H_1 u_1 * D_1 * S_1;\ Z_2, Z'') = R'' * D * S\,.$$

Further, in view of *IV.2.33* we get $Z_2 \overset{\alpha-1}{\approx} R'E'vDS$ and

$$f_{\alpha-1}(R_1HL^hL_1H_1u_1 * D_1 * S_1; Z_2, R'E'vDS) = R'E'v * D * S.$$

Moreover, by II.7.8 it follows from (88), (84) and (81) that

$$f_{\alpha-1}(R_1H * L^hL_1 * H_1u_1D_1S_1; Z_2, R'E'vDS) = R' * E' * vDS.$$

Thus $R' * E' * vDS$ is a maximal image of $R * E * uDS$ in the r-reversal (77). This reversal is local since (76) is local by II.7.8.

2.18. *Suppose that $r \geqslant q_1$ and that the occurrences*

$$P * F * uEvDQ,\ PFu * E * vDQ,\ PFuEv * D * Q, \tag{89}$$

$$R * F * uEvDS \quad \text{and} \quad RFuEv * D * S \tag{90}$$

of elementary 9*-powers of rank α are normalised.*

If the transition

$$X \rightleftharpoons PFuEvDQ \to PFu'E'v'DQ \rightleftharpoons Z \tag{91}$$

is a local r-reversal of the occurrence $PFu * E * vDQ$, where $PFu' * E' * v'DQ$ is a maximal image of it and the occurrences of the words F and D indicated in (89) *are stable, then the transition*

$$Y \rightleftharpoons RFuEvDS \to RFu'E'v'DS \rightleftharpoons Z'$$

is a local r-reversal of the occurrence $RFu * E * vDS$, where $RFu' * E' * v'DS$ is a maximal image of it, and the occurrences (90) *are stable.*

If the words Y and Z both lie in one of the sets $\mathscr{K}_{\alpha-1}$, $\mathscr{L}_{\alpha-1}$ or $\mathscr{M}_{\alpha-1}$, then by II.6.13 *the word Z' lies in that same set.*

The proof is analogous to that of 2.17. Firstly, by 2.2 and 1.3 we may assume that

$$PFu * E * vDQ \in \mathrm{MaxNorm}(\alpha, X, r).$$

Then, by II.6.14, we will have

$$RFu * E * vDS \in \mathrm{MaxNorm}(\alpha, Y, r).$$

By assumption, we have a simple r-reversal

$$X_1 \rightleftharpoons P_1F_1u_1TA^tA_1T_1v_1D_1Q_1 \to P_1F_1u_1HL^hL_1H_1v_1D_1Q_1 \rightleftharpoons Z_1, \tag{92}$$

of rank α, where $X \overset{\alpha-1}{\approx} X_1$, $Z \overset{\alpha-1}{\approx} Z_1$,

$$f_{\alpha-1}(PFu * E * vDQ;\ X, X_1) = P_1F_1u_1T * A'A_1 * T_1v_1D_1Q_1\,, \tag{93}$$

$$f_{\alpha-1}(PFu' * E' * v'DQ;\ Z, Z_1) = P_1F_1u_1H * L^hL_1 * H_1v_1D_1Q_1 \tag{94}$$

and the occurrences of the words F and D indicated in (89) are transformed to the occurrences of the words F_1 and D_1 in X introduced in (92), which are mutually normalised in rank $\alpha - 1$ with the occurrences of the same words in Z_1 that we have introduced, and these last are transformed to the occurrences of F and D in Z introduced in (91). Thus, all the occurrences indicated of elementary words are normalised and the relations

$$f_{\alpha-1}(P * FuEvD * Q;\ X, X_1) = P_1 * F_1u_1TA'A_1T_1v_1D_1 * Q_1 \tag{95}$$

and

$$f_{\alpha-1}(P_1 * F_1u_1HL^hL_1H_1v_1D_1 * Q_1;\ Z_1, Z) = P * Fu'E'v'D * Q \tag{96}$$

hold. By II.5.5 and II.7.3, we have

$$\mathrm{MutNorm}_{\alpha-1}(P * FuEvD * Q,\ R * FuEvD * S)\,.$$

Then, by *IV.3.17* it follows from (95) that $Y \overset{\alpha-1}{\simeq} Y_1$ for some word $Y_1 \in \mathscr{K}_{\alpha-1}$, and

$$f_{\alpha-1}(R * FuEvD * S; Y, Y_1) = R_1 * F_1u_1TA'A_1T_1v_1D_1 * S_1\,. \tag{97}$$

We set $Z_2 \rightleftharpoons R_1F_1uHL^hL_1H_1v_1D_1S_1$ By II.6.13, $Z_2 \in \mathscr{K}_{\alpha-1}$ since $Y_1, Z_1 \in \mathscr{K}_{\alpha-1}$. By *IV.3.5*, it follows from (95), (97) and (93) that

$$f_{\alpha-1}(RFu * E * vDS;\ Y, Y_1) = R_1F_1u_1T * A'A_1 * T_1v_1D_1S_1\,. \tag{98}$$

By II.5.5 and II.7.5, we have

$$\mathrm{MutNorm}_{\alpha-1}(P_1 * F_1u_1HL^hL_1H_1v_1D_1 * Q_1,\ R_1 * F_1u_1HL^hL_1H_1v_1D_1 * S_1)\,,$$

whence, by (96) and *IV.3.17*, it follows that $Z_2 \overset{\alpha-1}{\simeq} Z_3$ for some word Z_3 and

$$f_{\alpha-1}(R_1 * F_1u_1HL^hL_1H_1v_1D_1 * S_1;\ Z_2, Z_3) = R_3 * Fu'E'v'D * S_3\,. \tag{99}$$

Since (92) is a simple r-reversal of the occurrence (93), and (94) is a maximal image of it in this reversal, by II.6.14 the transition $Y_1 \to Z_2$ is a simple r-reversal of occurrence (98) and the occurrence $R_1F_1u_1H * L^hL_1 * H_1v_1D_1S_1$ is a maximal image of it in this reversal.

Since, by 2.15, the occurrences (90) are stable in the r-reversal $Y \to Z_2$, and by (97) and proposition *IV.3.5* they transform to the occurrences

$$R_1 * F_1 * u_1TA'A_1T_1v_1D_1S_1 \quad \text{and} \quad R_1F_1u_1TA'A_1T_1v_1 * D_1 * S_1\,,$$

it follows that the latter are mutually normalised in rank $\alpha - 1$ with the occurrences

$$R_1 * F_1 * u_1HL^hL_1H_1v_1D_1S_1 \quad \text{and} \quad R_1F_1u_1HL^hL_1H_1v_1 * D_1 * S_1$$

respectively. Moreover it follows from (96) and (99) by *IV.3.5* that

$$f_{\alpha-1}(R_1 * F_1 * u_1HL^hL_1H_1v_1D_1S_1; Z_2, Z_3) = R_3 * F * u'E'v'DS_3$$

and

$$f_{\alpha-1}(R_1F_1u_1HL^hL_1H_1v_1 * D_1 * S_1; Z_2, Z_3) = R_3Fu'E'v' * D * S_3 .$$

From this, on applying II.7.6 and *IV.2.33* twice, in an analogous way to that used at the end of section 2.17, we obtain the relations

$$R_1Fu_1HL^hL_1H_1v_1D_1S_1 \overset{\alpha-1}{\approx} RFu'E'v'DS_3 \overset{\alpha-1}{\approx} RFu'E'v'DS ,$$

where the occurrences of the words F_1 and D_1 in the first word transform into the isolated occurrences of the words F and D in the last word. Hence the transition $Y \to RFu'E'v'DS$ is an r-reversal of the occurrence $RFu * E * vDS$ and by (94) and *IV.3.5*, $RFu' * E' * v'DS$ is a maximal image of it in this reversal.

The fact that the reversal $Y \to RFu'E'v'DS$ is local follows from the localness of the reversal (91), by 2.13 and *IV.3.5*.

2.19. *Suppose that $X \to Y$ is a q_1-reversal of the occurrence $W \in$ Norm$(\alpha, X, 9)$ and that the occurrences $U_1 \in$ ComReg $(\alpha - 1, Y)$ and $V_1 \in$ Norm $(\alpha, Y, 9)$ are images of occurrences $U \in$ ComReg $(\alpha - 1, X)$ and $V \in$ Norm $(\alpha, X, 9)$ in this reversal. If $\neg$ Comp (U, W) and it is possible to carry out q-reversals of V and V_1, then U is stable in a reversal of V if and only if U_1 is stable in a reversal of V_1.*

For Comp (V, W) this follows from 1.2. Suppose that $\neg$ Comp (V, W). Then by definitions I.4.17 and I.4.19, the occurrences U and V must be stable in the reversal $X \to Y$. In this case the required result follows for simple reversals from 2.15 and 2.11 and in the general case we need to refer to 1.4.

2.20. *Suppose that $r \geqslant q_1$, each of the occurrences*

$$P * E * uDQ,\ PEu * D * Q \in \text{Norm}\,(\alpha, X, r)$$

is stable in a reversal of the other, the transition

$$PEuDQ \to Y_1 \tag{100}$$

is an r-reversal of the occurrence $P * E * uDQ$, the occurrence $V_1 \in$ Norm (α, Y_1, r) is an image of $P * E * uDQ$ in the reversal (100), *and is stable in a q_1-reversal of the occurrence $W_1 \rightleftharpoons f_{X \to Y_1}$ $(PEu * D * Q)$; suppose further that the transition*

$$PEuDQ \to Y_2 \tag{101}$$

is an r-reversal of the occurrence $PEu * D * Q$, *the occurrence* $W_2 \in \text{Norm}(\alpha, Y_2, r)$ *is an image of* $PEu * D * Q$ *in the reversal* (101) *and is stable in a reversal of the occurrence* $V_2 \rightleftharpoons f_{X \to Y_2}(P * E * uDQ)$.

Then a word $Z \in \mathscr{R}_{\alpha-1}$ *can be found such that the transitions* $Y_1 \to Z$ *and* $Y_2 \to Z$ *are r-reversals of the occurrences* W_1 *and* V_2 *and the occurrence* $f_{Y_1 \to Z}(V_1)$ *is an image of* V_2 *in the reversal* $Y_2 \to Z$.

We set

$$V \rightleftharpoons P * E * uDQ \text{ and } W \rightleftharpoons PEu * D * Q .$$

By *IV.3.12* and 1.4, we may assume that $X \in \mathscr{M}_{\alpha-1}$. Then the occurrences V and W are periodised in rank $\alpha - 1$, that is,

$$E \overset{\circ}{=} A^t A_1 \text{ and } D \overset{\circ}{=} B^k B_1 ,$$

where $A \overset{\circ}{=} A_1 A_2$ and $B \overset{\circ}{=} B_1 B_2$ are elementary periods of rank α. By 2.14, 1.12 and 1.4, we may assert that (100) and (101) are local reversals of the corresponding occurrences.

Suppose that

$$Y_1 \overset{\circ}{=} P'E'vDQ \text{ and } Y_2 \overset{\circ}{=} PEwD'Q'$$

where

$$V_1 = P' * E' * vDQ,\ W_1 = P'E'v * D * Q ,$$
$$V_2 = P * E * wD'Q',\ W_2 = PEw * D' * Q' .$$

By 2.4, we may assume that W_2 is an end of a maximal image of W in the reversal (101) and that V_1 is a start of a maximal image of V in the reversal (100). Since by hypothesis we can carry out q_1-reversals on W_1 and V_1 in which they are stable, by 2.16 it is possible to find a local reversal

$$Y_1 \to P'E'v_1 D''Q' \rightleftharpoons Z_1 ,$$

of W_1, where $P'E'v_1 * D'' * Q'$ is a maximal image of W_1 and one of the words D' and D'' is an end of the other.

Since V and W_1 are periodised in rank $\alpha - 1$, we have by 1.10 that

$$Y_1 \overset{\alpha-1}{\simeq} [P, A^{-n+t}, A_1 uDQ]_{\alpha-1} \tag{102}$$

and

$$Z_1 \overset{\alpha-1}{\simeq} [P'E'v, B^{-n+k}, B_1 Q]_{\alpha-1} .$$

It follows from (102) by *IV.1.21* and *V.2.1* that

$$P'E'v \overset{\alpha-1}{\simeq} [P, A^{-n+t}, A_1 u]_{\alpha-1} ,$$

that is,

$$Z_1 \overset{\alpha-1}{\sim} [P, A^{-n+t}, A_1u, B^{-n+k}, B_1Q]_{\alpha-1}\,. \tag{103}$$

In a similar way we can find a local reversal

$$Y_2 \to P'E''w_1D'Q' \rightleftharpoons Z\,, \tag{104}$$

of V_2 where the occurrence

$$P' * E'' * w_1D'Q' \in \text{MaxNorm}(\alpha, Z, q_1) \tag{105}$$

is an image of W_1, one of the words E' and E'' is a start of the other and the relation

$$Z \overset{\alpha-1}{\sim} [P, A^{-n+t}, A_1u, B^{-n+k}, B_1Q]_{\alpha-1} \tag{106}$$

holds. It follows from (103) and (106) that $Z \overset{\alpha-1}{\sim} Z_1$.

We assume that E'' is a proper start of E', that is $E' \overset{\circ}{=} E''T$, where T is non-empty. Then it follows from (105) by II.7.7 that

$$f_{\alpha-1}(P' * E'' * w_1D'Q'; Z, Z_1) = P' * E'' * Tv_1D''Q'\,,$$

and this, by II.5.16, contradicts (105), since $P' * E' * v_1D''Q'$ is normalised. Hence, E' is a start of E'', that is, $E'' \overset{\circ}{=} E'H$ for some H. Then $l_\alpha(E'') \geqslant l_\alpha(E') = l_\alpha(V_1) \geqslant r$, whence we deduce by 1.1 that the transition (104) is an r-reversal of the occurrence V_2.

The fact that the transition $Y_1 \to Z_1$ is an r-reversal of W_1 is proved analogously, and from this it follows by 1.4 that $Y_1 \to Z$ is also an r-reversal of W_1. Since we have in addition

$$f_{Y_1 \to Z}(V_1) = P' * E' * Hw_1D'Q'$$

by II.7.7, it follows from 1.3 that this occurrence is an image of V_2 in the reversal (104).

§ 3. Cascades and Real Reversals

We shall first of all note some properties of the symmetric relation of two words "adjoining one another", which was defined in I.4.21.

3.1. *If $X \in \mathcal{N}_\alpha$ and the occurrence $V \in$ Norm$(\alpha, X, 9)$ is stable in reversals of the occurrences U, $W \in$ Norm(α, X, q_1) where $U < V < W$, then U and W do not adjoin one another.*

This property follows easily from 2.2, 2.4, 2.6, 2.19 and 1.2. In [4] it was used as the definition of "adjunction" (definition 10 on page 506). In addition, the proofs of all the necessary properties of this relation were carried out based on that definition, (lemmas 41–47 on pages 506–508). These proofs are somewhat easier than those adduced below in 3.2–3.9. However, we introduce this new version here since it allows us to lower the value of the parameter q_1 to 37. In the old version it was necessary to take $q_1 \geqslant 43$ in the proof of 3.6. Thus the lower bound for the exponent n is reduced by 18, since q_1 appears in the expression for this bound with coefficient 3.

3.2. *Suppose that $U, V, W \in \mathrm{Norm}(\alpha, X, 9)$, where $X \in \mathscr{N}_\alpha$ and $\mathrm{Comp}(V, W)$. If U and V adjoin one another, then so also do U and W.*

This follows from 1.3 and II.5.13.

3.3. *Suppose that $U, V \in \mathrm{Norm}(\alpha, X, q_1)$, where $X \in \mathscr{N}_\alpha$ and $X \overset{\alpha-1}{\simeq} Y$. Then $Y \in \mathscr{N}_\alpha$. If U and V adjoin one another then so also do $f_{\alpha-1}(U; X, Y)$ and $f_{\alpha-1}(V; X, Y)$.*

This follows from II.5.16 and 1.4.

3.4. *Suppose that $V_0, W_0 \in \mathrm{Norm}(\alpha, X_0, q_1)$, where $X_0 \in \mathscr{N}_\alpha$ and V_0 and W_0 do not adjoin one another. If there exists a sequence*

$$X_0 \to X_1 \to X_2 \to \cdots \to X_\lambda, \tag{1}$$

where for $i > 0$ the transition $X_{i-1} \to X_i$ is a q_1-reversal of one of the occurrences V_{i-1} and W_{i-1}, and V_i and W_i are maximal normalised continuations of images of these occurrences in the reversal $X_{i-1} \to X_i$, then it is possible to find a q_1-reversal

$$X_\lambda \to X_{\lambda+1}$$

of each of the two occurrences V_λ and W_λ in which the other is stable.

By 3.2 and 1.3 we may assume that

$$V_0, W_0 \in \mathrm{MaxNorm}(\alpha, X, q_1).$$

The proof is by induction on λ. If $\lambda \leqslant 1$, the required result holds by I.4.21. For by I.4.21 we have two sequences

$$X_0 \to Y_1 \to Y_2 \text{ and } X_0 \to Z_1 \to Z_2$$

of type (1), where in the first one we first of all carry out a q_1-reversal of the occurrence V_0 and then a q_1-reversal of $f_{X_0 \to Y_1}(W_0)$, while in the second we first of all carry out a reversal of W_0 and then one of $f_{X_0 \to Z_1}(V_0)$.

By I.4.21, 1.2 and 2.6, the occurrences $f_{X_0 \to Y_1}(W_0)$ and $f_{X_0 \to Z_1}(V_0)$ coincide with their maximal normalised continuations.

By 2.20 and 1.11, we have $Y_2 \overset{\alpha-1}{\simeq} Z_2$. From 1.11, 1.4 and 1.2 it easily follows that for an arbitrary sequence (1) with the above properties the word X_1 is equivalent in rank $\alpha - 1$ to one of the words X_0 and Y_2. Then by 1.4 and 1.2, the transition $X_2 \to Y_1$ is a q_1-reversal of one of the occurrences V_2 and W_2, $X_2 \to Z_1$ is a

q_1-reversal of the other and, in each of these reversals, whichever of the occurrences V_2 or W_2 is not reversed is stable. Hence we may take Y_1 or Z_1 to be the desired X_3. Moreover, V_3 and W_3 will also coincide with the occurrences of V_1 and W_1 in the corresponding word Y_1 or Z_1.

Clearly, the same reasoning can be used in the general case. As a result we get that in an arbitrary sequence of type (1), each word X_i is equivalent in rank $\alpha - 1$ to one of the words X_0, Y_1, Z_1 and Y_2.

3.5. *Suppose that* $X_0,\ Y \in \mathcal{K}_\alpha,\ V_0 \rightleftharpoons P * E * uDQ,\ W_0 \rightleftharpoons PEu * D * Q,$ $V_0,\ W_0 \in \mathrm{Norm}(\alpha,\ X_0,\ q_1),\ V \rightleftharpoons R * D * uDS,\ W \in REu * D * S,\ V,\ W \in$ $\mathrm{Norm}(\alpha,\ Y,\ q_1)$.

If the occurrences V and K adjoin one another, then so also do V_0 *and* W_0.

By II.5.8, $\neg\mathrm{Comp}(V_0, W_0)$ follows from $\neg\mathrm{Comp}(V, W)$. We assume that V_0 and W_0 do not adjoin one another. Then by 3.4, it is possible to find for them a sequence of words of type (1) with $\lambda = 3$. We shall show that for any such sequence we can find a similar sequence which begins with the word Y and corresponds to the reversals of U and V and their images.

By 2.14, we may assume that all the reversals $X_{i-1} \to X_i$ in (1) are local.

Let, for example, $X_1 \overset{\circ}{=} P_1E_1u_1DQ$ and $X_2 \overset{\circ}{=} P_1EvD_1Q_1$, where $P_1 * E_1 * u_1DQ$ is a maximal image of V_0 in its reversal $X_0 \to X_1$ and $P_1E_1v * D_1 * Q_1$ is a maximal image of the occurrence $W_1 \rightleftharpoons P_1E_1u_1 * D * Q$ in its reversal $X_1 \to X_2$. Then by 2.15, we can find a local reversal $Y \to R_1E_2u_1DS$ of V, where $V' \rightleftharpoons R_1 * E_2 * u_1DS$ is a maximal image of V and one of the words E_1 and E_2 is an end of the other. If, for example, $E_2 \overset{\circ}{=} TE_1$, then the occurrence $R_1T * E_1 * u_1DS$ is normalised, whence it follows by 2.15, 2,4 and the properties of the sequence (1) that $R_1 * E_2 * u_1DS$ is stable in a reversal of the occurrence $R_1E_2u_1 * D * S$. The latter ocurrence is stable in a reversal of $R_1 * E_2 * u_1DS$, by 1.2. Taking the local reversal $X_1 \to X_2$ into account, we find in an analogous way a reversal

$$R_1TE_1u_1DS \to R_1TE_1vD_2S_1$$

of $R_1TE_1u_1 * D * S$, where $R_1TE_1v * D_2 * S_1$ is a maximal image of it and one of the words D_1, D_2 is a start of the other. Further, one can convince oneself as in the foregoing that each of the occurrences $R_1 * TE_1 * vD_2S_1$ and $R_1TE_1v * D_2 * S_1$ is stable in a reversal of the other.

It follows from what was proved in 1.11 that V and W do not adjoin one another, which contradicts the hypothesis.

3.6. *Suppose that* $U,\ V,\ W \in \mathrm{Norm}(\alpha,\ X,\ q_1)$, *where* $X \in \mathcal{K}_\alpha$. *If U and W adjoin one another and V is not compatible with them, then V cannot be situated between them.*

It follows from $\neg\mathrm{Comp}(U, W)$ by II.5.7 that either $U < W$ or $W < U$. These cases are similar.

We suppose that $U < V < W$. We shall prove that U and W do not then adjoin one another. By 3.2, we may assume that

$$U,\ W \in \mathrm{MaxNorm}(\alpha,\ X,\ q_1)\,.$$

According to I.4.21, we have to prove that each of the occurrences U and W is stable in a reversal of the other and that the same holds for the maximal normalised continuations of their images in a reversal $X \to Y$ of either of them and for the images of their images in a succeeding reversal.

Let $X \to Y$ be a q_1-reversal of the occurrence U. We may assume by II.5.16 and 3.3 that it is a simple reversal.

For $1 \leqslant i \leqslant l_\alpha(V)$, we denote by V_i the i-th left segment of V. By II.2.7, the kernels V_{18} and V_{20} do not intersect.

Suppose that

$$V \rightleftharpoons R * E_1FGuLHE_2 * Q ,$$

where

$$V_{18} = RE_1F * G * uLHE_2Q, \; V_{20} = RE_1FGu * L * HE_2Q,$$

V_{15} is a start and V_{23} is an end of the occurrence

$$RE_1 * FGuLH * E_2Q .$$

As was proved in 2.9, in the simple reversal $X \to Y$ the transformation takes place to the left of the occurrence V_{15}, that is, this reversal has the form

$$X = RE_1FGuLHE_2Q \to R'FGuLHE_2Q = Y , \tag{2}$$

and in addition the relation

$$\text{MutNorm}_{\alpha-1}(R'F * GuLHE_2 * Q, RE_1F * GuLHE_2 * Q) \tag{3}$$

holds.

It is clear that W is also stable in the reversal (2).

Suppose that $X \overset{\alpha-1}{\rightleftharpoons} X_1$ and the transition $X_1 \to Y_1$ is a simple q_1-reversal of the occurrence $f_{\alpha-1}(W; X, X_1)$. Set

$$f_{\alpha-1}(R * E_1FGuLH * E_2Q;\; X, X_1) = R_1 * DL_1H_1 * Q_1 \tag{4}$$

and

$$f_{\alpha-1}(V_{20};\; X, X_1) = R_1D * L_1 * H_1Q_1 . \tag{5}$$

Since $q_1 = 37$, the occurrence $f_{\alpha-1}(V_{23}; X, X_1)$ either coincides with the 15-th right segment of $f_{\alpha-1}(V; X, X_1)$ or is situated to the left of it. Therefore, by 2.9, in the simple reversal $X \to Y$ the transformation takes place to the right of the occurrence (4), that is, this reversal has the form

$$X_1 = R_1DL_1H_1Q_1 \to R_1DL_1H_1Q_2 = Y_1 \tag{6}$$

and the relation

$$\text{MutNorm}_{\alpha-1}(R_1 * DL_1 * H_1Q_1, R_1 * DL_1 * H_1Q_2) \tag{7}$$

is satisfied.

By (5) and *IV.2.34*, a word $X_2 \in \mathscr{K}_{\alpha-1}$ can be found such that $X \overset{\alpha-1}{\rightleftarrows} X_2$,

$$f_{\alpha-1}(V_{20};\ X,\ X_2) = RE_1FGv * L_1 * H_1Q_1$$

and

$$f_{\alpha-1}(V_{18};X,X_2) = RE_1F * G * vL_1H_1Q_1\,.$$

Then by *IV.2.18(i)* and *IV.2.27*, we have

$$f_{\alpha-1}(RE_1 * FG * uLHE_2Q;\ X,\ X_2) = RE_1 * FG * vL_1H_1Q$$

and

$$f_{\alpha-1}(RE_1 * FGuLH * E_2Q;\ X,\ X_2) = RE_1 * FGvL_1H_1 * Q_1\,,$$

that is, $RE_1 * FGvL_1H_1 * Q_1 \in \text{Norm}(\alpha, X_2, 9)$.

Suppose that

$$f_{\alpha-1}(RE_1F * GuLHE_2 * Q;\ X,\ X_2) = RE_1F * GvL_1H_1D_1 * T\,,$$

where $D_1T \mathrel{\underline{\circ}} Q_1$. Then by II.7.6 and (3), a word $Y_2 \in \mathscr{K}_{\alpha-1}$ can be found such that $Y \overset{\alpha-1}{\rightleftarrows} Y_2$ and

$$f_{\alpha-1}(R'F * GuLHE_2 * Q;\ Y,\ Y_2) = R_2 * GvL_1H_1D_1 * T\,,$$

whence there follows by II.7.8:

$$f_{\alpha-1}(R'F * G * uLHE_2Q; Y, Y_2) = R_2 * G * vL_1H_1D_1T\,. \tag{8}$$

We set $Z \rightleftharpoons R'FGvL_1H_1Q_1$, $Z_1 \rightleftharpoons RE_1FGvL_1H_1Q_2$ and $Z_2 \rightleftharpoons R'FGvL_1H_1Q_2$. From (8) it follows by *IV.2.33* that $Z \in \mathscr{K}_{\alpha-1}$, $Y \overset{\alpha-1}{\rightleftarrows} Z$ and

$$f_{\alpha-1}(R'F * G * u\,LHE_2\,Q;\ Y,\ Z) = R'F * G * vL_1H_1Q_1.$$

Using relations (4) and (7), we get analogously $Z_1 \in \mathscr{K}_{\alpha-1}$, $Y_1 \overset{\alpha-1}{\rightleftarrows} Z_1$ and

$$f_{\alpha-1}(R_1D * L_1 * H_1Q_2;\ Y_1,\ Z_1) = RE_1FGv * L_1 * H_1Q_2\,.$$

Since $Z, Z_1 \in \mathscr{K}_{\alpha-1}$, by II.6.13 it follows that $Z_2 \in \mathscr{K}_{\alpha-1}$.

Since (2) is a simple reversal of U and since $f_{\alpha-1}(U;\ X,\ X_2)$ is contained in

$* RE_1 * FGvL_1H_1Q_1$ by II.7.7, $X_2 \to Z$ is a simple reversal of $f_{\alpha-1}(U; X, X_2)$. Hence the transition $X \to Z$ is a reversal of U, a maximal image of it in this reversal is contained in $* R' * FGvL_1H_1Q_1$, and the occurrence $f_{X\to Z}(W)$ is equal to

$$\phi(f_{\alpha-1}(W; X, X_2); RE_1FGvL_1H_1 * Q_1 *, R'FGvL_1H_1 * Q_1 *)$$

and coincides with its maximal normalised continuation.

On considering the simple reversal (6) of the occurrence $f_{\alpha-1}(W; X, X_1)$, we find similarly that the transition $X_2 \to Z_1$ is a simple reversal of $f_{\alpha-1}(W; X, X_2)$, from which it follows in an analogous way that the transition

$$Z = R'FGvL_1H_1Q_1 \to R'FGvL_1H_1Q_2 = Z_2 \tag{9}$$

is a simple reversal of $f_{X\to Z}(W)$. If, in addition, U_1 is a maximal image of U in the reversal $X \to Z$, then we have by II.6.10 that

$$\text{MutNorm}_{\alpha-1}(U_1, \phi(U_1; * R' * FGvL_1H_1Q_1, * R' * FGvL_1H_1Q_2)),$$

that is, U_1 is stable in the reversal (9).

In exactly the same way, it can be established that the transition

$$Z_1 = RE_1FGvL_1H_1Q_2 \to R'FGvL_1H_1Q_2 = Z_2 \tag{10}$$

is a simple reversal of $f_{X\to Z_1}(U)$. If, in addition, W_1 is a maximal image of W in the reversal $X \to Z_1$, then by II.6.10, W_1 is stable in the reversal (10). Clearly, the occurrence

$$\phi(U_1; * R' * FGvL_1H_1Q_1, * R' * FGvL_1H_1Q_2),$$

is a maximal image of $f_{X\to Z_1}(U)$ in its reversal (10), and this occurrence is stable in the reversal $Z_2 \to Z$ of $f_{Z_1\to Z_2}(W_1)$, by 1.2.

Thus we have proved the stability of the corresponding occurrences under some choice of a sequence of reversals of U and W:

$$X \to Z \to Z_2 \text{ and } X \to Z_1 \to Z_2 .$$

By 1.11 and 1.4, this is true for any choice of reversals of these occurrences. Hence U and W do not adjoin one another, that is, we have reached a contradiction.

3.7. *Suppose that $U, V, W \in \text{Norm}(\alpha, X, q_1)$, where $X \in \mathcal{N}_\alpha$. If V and W both adjoin U and are situated on one side of it, then V and W are compatible.*

We assume that with the above hypothesis V and W are not compatible. Thus by 3.6, $\neg\text{Comp}(W, U)$ implies that W does not lie between U and V, and $\neg\text{Comp}(V, U)$ implies that V does not lie between U and W. But this contradicts the assumption that V and W are situated on the same side of U.

3.8. *Suppose that $X, Y \in \mathcal{N}_\alpha$, the transition $X \to Y$ is a q_1-reversal of rank α,* *the occurrence $V \in$ Norm(α, X, q_1) is stable in this reversal and the occurrence $W_1 \in$ Norm(α, Y, q_1) is an image of some occurrence $W_1 \in$ Norm(α, X, q_1) in this reversal. If V and W adjoin one another, then so also do $f_{X\to Y}(V)$ and W_1.*

According to definitions I.4.17 and I.4.19, either $X \to Y$ is a reversal of W or W is stable in this reversal and $W_1 = f_{X\to Y}(W)$.

Suppose that W is stable in the reversal $X \to Y$. Then this is a reversal of some occurrence $U \in$ Norm(α, X, q_1), which is compatible neither with W nor with V. By 3.6, W and V are situated on the same side of U. The required result now follows from 3.3 and 3.5.

Let $X \to Y$ be a reversal of W. Then, assuming that $f_{X\to Y}(V)$ and W do not adjoin one another, we get also from 3.5 and definition I.4.21 that V and W do not adjoin one another.

3.9. *Suppose that the set of occurrences*

$$V_i \; (i = 1, 2, \ldots, r) \tag{11}$$

in a word $X \in \mathcal{N}_\alpha$ form a right (left) cascade of rank α of the occurrence V_0. If $0 < j < r$ and W_j is the maximal normalised continuation of V_j, then $l_\alpha(W_j) < q$.

Set $l_\alpha(W_j) = h$. Since we have that $\neg$Comp(V_{j-1}, V_j) by I.4.22(b), by 2.10 if U is the maximal end of W_j which is stable in a reversal of V_{j-1}, $l_\alpha(U) \geqslant h - 17$. Since by I.4.22(c) V_j is stable in reversals of V_{j-1} and V_{j+1}, by 2.4 a start of U which ends with V_j is also stable in reversals of V_{j-1} and V_{j+1}. By I.4.22(c), this means that V_j is a start of U, that is, U is a normalised continuation of V_j to the right. Since $q_1 < q_2$, we have by I.4.22(a) that $h - 17 < q_2 + 17$, that is, $h < q_2 + 34 < q$.

3.10. *Suppose that $V_0, W_0 \in$ Norm(α, X, q), Comp(V_0, W_0), $X \in \mathcal{N}_\alpha$, and that the set of occurrences* (11) *is a right (left) cascade of rank α with head V_0. Then W_0 is also the head of the cascade* (11), *and if further the set of occurrences*

$$W_i \; (i = 1, 2, \ldots, k) \tag{12}$$

is also a right (left, respectively) cascade of rank α with head W_0, then the cascades (11) *and* (12) *are the same, that is, $k = r$ and $W_i = V_i$ for $1 \leqslant i \leqslant r$.*

Since the cases are fully analogous we can restrict ourselves to that of a right cascade. Without loss of generality we may assume that $r \leqslant k$.

By 3.2, V_1 adjoins W_0 and by 1.3 it is stable in a reversal of W_0. Hence W_0 can be regarded as a head of the right cascade (11).

Suppose now that (12) is an arbitrary cascade of rank α with head W_0. Since Comp(U, W), it follows from I.4.22(b), 3.2 and 3.7 that Comp(V_1, W_1) and, furthermore, we obtain by induction on i that Comp(V_i, W_i) for $1 \leqslant i \leqslant r$. By II.5.13, the occurrences V_i and W_i have one and the same maximal normalised continuation U_i.

Since V_r is the last element in the cascade (11), by I.4.22(a) it has no proper normalised continuation to the right, and $l_\alpha(V_r) \geqslant q_{t_r} + 17$. By I.4.22(c) and 2.4, V_r is the maximal end of U which is stable in a reversal of V_{r-1}. Then by 1.3 and 2.4, W_r is a start of V_r, that is, V_r is a normalised continuation of W_r to the right. Since

$l_\alpha(V_r) \geqslant q_{t_r} + 17$, it follows by condition I.4.22(a) that W_r is the last element of the cascade (12), that is, $k = r$ and $W_r = V_r$.

Finally, $\mathrm{Comp}(V_i, W_i)$ for $0 \leqslant i \leqslant r$ implies, by I.4.22(c), 1.3 and 2.4, that for $0 < j < r$ the occurrence W_j, as well as V_j, is a maximal start stable in a reversal of V_{j+1}, of a maximal end of U_j which is stable in a reversal of V_{j-1}. Hence $W_j = V_j$ for $0 < j < r$.

3.11. *Suppose that $X \in \mathcal{N}_\alpha$ and $X \overset{\alpha-1}{\simeq} Y$. If the set of occurrences* (11) *in X is a right (left) cascade of rank α with head V_0, then the set of occurrences*

$$f_{\alpha-1}(V_i;\, X,\, Y) \quad (i = 1, 2, \ldots, r) \tag{13}$$

forms a right (left, respectively) cascade of rank α with head $f_{\alpha-1}(V_0;\, X,\, Y)$.

It is clearly enough to consider a right cascade. By 3.3, $Y \in \mathcal{N}_\alpha$.

Since $V_0 \in \mathrm{Norm}(\alpha, X, q)$, it follows by II.5.16 that $f_{\alpha-1}(V_0;\, X,\, Y) \in \mathrm{Norm}(\alpha, Y, q)$. Since $V_i \ll V_{i+1}$ for $0 \leqslant i < r$, it follows by *IV.2.18(g)* that $f_{\alpha-1}(V_i;\, X,\, Y) \ll f_{\alpha-1}(V_{i+1};\, X,\, Y)$. Condition (a) of definition I.4.22 follows for the set (13) from II.5.16 and the corresponding conditions for the cascade (11). Condition (b) follows from 3.3, and (c) from 1.4 and II.5.16.

3.12. *Suppose that $P * A * Q \in \mathrm{Norm}(\alpha, X, q)$ and $R * A * S \in \mathrm{Norm}(\alpha, Y, q)$, where $X, Y \in \mathcal{N}_\alpha$ and the set* (11) *is a right (left) cascade of rank α of $P * A * Q$.*

If $S \overset{\circ}{=} Q$ (or $R \overset{\circ}{=} P$), then the set of occurrences

$$\phi(V_i;\, W_1,\, W_2) \quad (i = 1,\, 2,\, \ldots,\, r)\,, \tag{14}$$

where $W_1 \rightleftharpoons PA * Q *$ and $W_2 \rightleftharpoons RA * Q *$ ($W_1 \rightleftharpoons * P * AQ$ and $W_2 \rightleftharpoons * P * AS$, respectively) forms a right (left) cascade of rank α of the occurrence $R * A * S$.

We deduce conditions (a), (b) and (c) of definition I.4.22 for the set of occurrences (14) from the corresponding conditions for cascade (11). Suppose that $U_i \rightleftharpoons \phi(V_i;\, W_1,\, W_2)$. By II.6.11, we have $U_i \in \mathrm{Norm}(\alpha, Y, q_{t_i})$ for $1 \leqslant i < r$, $U_r \in \mathrm{Norm}(\alpha, Y, q_{t_r} + 17)$, and U_r has no proper normalised continuation to the right (left). By II.6.17, the maximal normalised continuation of U_i to the right (to the left) contains less than $q_{t_i} + 17$ segments, $1 \leqslant i \leqslant r$. Condition (b) for (14) follows from 3.5, and (c) from 2.15, 2.2 and II.6.17.

3.13. *Suppose that A is an elementary 9-power of rank α. If the occurrences* (11) *in the word $PAQ \in \mathcal{N}_\alpha$ form a right cascade of rank α of the occurrence V_0 and $V_r \ll P * A * Q$, then, for an arbitrary word $PAS \in \mathcal{N}_\alpha$, the occurrences* (14), *where $W_1 \rightleftharpoons * P * AQ$ and $W_2 \rightleftharpoons * P * AS$, form a right cascade of rank α for the occurrence $\phi(V_0;\, * P * AQ,\, * P * AS)$.*

If the occurrences (11) *in the word $PAQ \in \mathcal{N}_\alpha$ form a left cascade of rank α of the occurrence V_0 and $P * A * Q \ll V_r$, then for an arbitrary word $PAQ \in \mathcal{N}_\alpha$ the occurrences* (14), *where $W_1 \rightleftharpoons PA * Q *$ and $W_2 \rightleftharpoons RA * Q *$, form a left cascade of rank α of the occurrence $\phi(V_0;\, PA * Q *\,,\, RA * Q *)$.*

The proof is analogous to that of 3.12.

3.14. *Suppose that $V, W \in \mathrm{Norm}(\alpha, X, q)$, where $X \in \mathcal{N}_\alpha$, $V = P * A * uBQ$, $W = PAu * B * Q$ and $\neg\mathrm{Comp}(U, W)$. If the occurrences* (11) *form a right (left)* *cascade of rank α of V (of W, respectively), then one and only one of the conditions $V_r < W$ and* $\mathrm{Comp}(V_r, W)$ *(one of the conditions $V < V_r$ and* $\mathrm{Comp}(V, V_r)$, *respectively) holds.*

Suppose that (11) is a right cascade of V. Since V_1 adjoins V, by 3.6 either $V_1 < W$ or $\mathrm{Comp}(V_1, W)$. If $\mathrm{Comp}(V_1, W)$, then by 3.9 we have $r = 1$. Then I.4.22 (a) negates the inequality $V_1 < W$. If $\neg\mathrm{Comp}(V_1, W)$, they by 3.6 we have either $V_2 < W$ or $\mathrm{Comp}(V_2, W)$, *etc.*

3.15. *Suppose that $V \rightleftharpoons P * A * uBQ$, $V' \rightleftharpoons PAu * B * Q$, $W \rightleftharpoons R * A * uBS$ and $W' \rightleftharpoons RAu * B * S$ are normalised occurrences of elementary q-powers of rank α in words X, Y in $\mathcal{N}_\alpha$, where* $\neg\mathrm{Comp}(V, V')$.

If V is the head of some right cascade of rank α with first element V_1, then W is the head of some right cascade of rank α with first element W_1, where

$$\mathrm{Comp}(V_1, V') \Rightarrow \mathrm{Comp}(W_1, W') \tag{15}$$

and

$$\neg\mathrm{Comp}(V_1, V') \\ \Rightarrow (V_1 < V' \;\&\; W_1 = \phi(V_1;\, PA * uB * Q,\, RA * uB * S))\,. \tag{16}$$

Similarly, if V' has a left cascade of rank α with first element V_1, then W' has a left cascade of rank α with first element W_1, where

$$\mathrm{Comp}(V, V_1) \Rightarrow \mathrm{Comp}(W, W_1)$$

and

$$\neg\mathrm{Comp}(V, V_1) \Rightarrow (V < V_1 \;\&\; W_1 = \phi(V_1;\, P * Au * BQ,\, R * Au * BS))\,.$$

Suppose that (11) is a right cascade of rank α of V. We set

$$U \rightleftharpoons PA * uB * Q \quad \text{and} \quad U_1 \rightleftharpoons RA * uB * S\,.$$

Assume that $\neg\mathrm{Comp}(V_r, V')$. Then we have $V_r < V'$ by 3.14, that is, all elements of (11) are contained in U.

We put

$$W_i = \phi(V_i;\, U,\, U_1) \quad (i = 1, 2, \ldots, r)\,. \tag{17}$$

It is easy to see that the occurrences (17) form the required right cascade of rank α of W. Condition (a) of definition I.4.22 follows for it from the corresponding condition on the cascade (11), together with II.6.14 and II.6.17. Condition (b) follows from 3.5, and (c) from 2.15, 2.2 and II.6.17. Moreover, relations (15) and (16) are clear.

It remains to consider the case where $\mathrm{Comp}(V_r, V')$. By II.5.8, we have $\neg\mathrm{Comp}(W, W')$. We use $\bar{W}'$ to denote the maximal normalised continuation of W'.

If $r = 1$, then we denote by W_1 the maximal end of $\bar{W}'$ which is stable in a reversal of W. By 2.10 we have $l_\alpha(W_1) \geqslant q - 17 > q_{t_1} + 17$. From 3.2 and 3.5 it follows that W_1 adjoins W. Hence for $r = 1$, W_1 forms a single-element cascade of rank α of W satisfying (15).

Suppose that $r > 1$. Since V_{r-1} is stable in a reversal of V_r, then we have by 2.2 that $V_{r-1} \ll V'$. Set

$$W_i \rightleftharpoons \phi(V_i;\, U,\, U_1) \quad (i = 1, 2, \ldots, r-1)\,. \tag{18}$$

Since $\neg\mathrm{Comp}(V_{r-1}, V_r)$, we have $\neg\mathrm{Comp}(V_{r-1}, V')$, and therefore $\neg\mathrm{Comp}(W_{r-1}, W')$. We denote by W_r the maximal end of $\bar{W}'$ which is stable in a reversal of W_{r-1}. We shall prove that this occurrence W_r together with the occurrences (18) form the required right cascade of W.

Since conditions (a), (b) and (c) of I.4.22 are verified in precisely the same way as for (17), we shall restrict ourselves to verifying that part of these conditions which depends on the last element W_r. By 2.10, we have $l_\alpha(W_r) \geqslant q - 17 > q_2 + 17 \geqslant q_{t_r} + 17$. By 3.2 and 3.5, W_r adjoins W_{r-1}. It follows from 1.3 and 2.15 that W_{r-1} is stable in a reversal of W_r. Finally by 2.2, II.6.17 and 2.15, no normalised continuation to the right of W_{r-1} is stable in a reversal of W_r.

3.16. *Suppose that $X \to Y$ is a q-reversal of rank α of U, where $X, Y \in \mathcal{N}_\alpha$ and $W_0 \in \mathrm{Norm}(\alpha, Y, q)$ is an image of an occurrence $V_0 \in \mathrm{Norm}(\alpha, X, q)$. If V_0 has a right (left) cascade of rank α with first element V_1, then W_0 has a right (left, respectively) cascade of rank α, the first element of which is compatible with the image in the reversal $X \to Y$ of some occurrence $V' \in \mathrm{Norm}(\alpha, X, q_1)$ which is compatible with V_1.*

If $X \to Y$ is a reversal of the occurrence V_0 then the desired cascade of rank α of W_0 consists of the images of the elements of the given cascade of V_0.

Since the case of a left cascade is analogous to that of a right cascade, we only consider the latter. Suppose that (11) is a right cascade of rank α of V_0. If $\neg\mathrm{Comp}(V_0, U)$, then V_0 is stable in $X \to Y$ and $W_0 = f_{X\to Y}(V_0)$. By 2.2, one of the following three cases occurs:

I. $\mathrm{Comp}(V_0, U)$; II. $U \ll V_0$; III. $V_0 \ll U$.

Case I. By 1.3 and 2.4, all elements of the cascade (11) are stable in the reversal $X \to Y$. We shall prove that the occurrences

$$W_i \rightleftharpoons f_{X\to Y}(V_i) \quad (i = 1, 2, \ldots, r) \tag{19}$$

form the desired right cascade of W_0. By 2.6, all the W_i are normalised and $l_\alpha(W_i) = l_\alpha(V_i)$, by 2.5 and 2.2 we have $W_{i-1} \ll W_i$ and by 2.7, W_r has no proper normalised continuation to the right. Let $\bar{V}_i$ be the maximal normalised continuation of V_i to the right. Then by 2.6 and 2.7, $f_{X\to Y}(\bar{V}_i)$ is the maximal normalised continuation of W_i to the right. We have thus verified condition (a) of definition I.4.22. Condition (b) follows from 3.8. By 2.19, each W_i is stable in reversals of the occurrences W_{i-1}

and W_{i+1}. If some normalised continuation W of W_i is stable in reversals of W_{i-1} and W_{i+1}, then by 2.6 and 1.2, $f_{Y\to X}(W)$ is a normalised continuation of V_i, and by 2.19 it is stable in reversals of V_{i-1} and V_{i+1}. Hence (19) is a right cascade of rank α of W_0.

Case II. By what we have proved, we may assert that $\neg\mathrm{Comp}(V_0, U)$. Then $W_0 = f_{X\to Y}(V_0)$ and we can prove, exactly as in Case I, that the occurrences (19) form a right cascade of rank α of W_0.

Case III. We can again assert that $\neg\mathrm{Comp}(V_0, U)$, and hence $W_0 = f_{X\to Y}(V_0)$. Since $X \to Y$ is a q-reversal, by 3.9 we have $\neg\mathrm{Comp}(V_j, U)$ for $1 \leqslant j < r$.

We consider 2 further cases separately:

III'. $\neg\mathrm{Comp}(V_r, U)$, III''. $\mathrm{Comp}(V_r, U)$.

Case III'. By 3.6 we have $V_r < U$. We denote by V_r' the maximal start of V_r which is stable in the reversal $X \to Y$. Since $l_\alpha(V_r) \geqslant q_{t_r} + 17$, it follows by 2.10 that $l_\alpha(V_r') \geqslant q_{t_r}$.

If $l_\alpha(V_r') \geqslant q_{t_r} + 17$, then by analogy with case I it can be established that the occurrences

$$W_j \rightleftharpoons f_{X\to Y}(V_j) \quad (j = 1, 2, \ldots, r-1)\,, \tag{20}$$

together with the maximal normalised continuation of $f_{X\to Y}(V_r')$ to the right, form a right cascade of rank α of W_0. Its first element W_1 for $r = 1$ is compatible with $f_{X\to Y}(V_1')$, and for $r > 1$ it coincides with $f_{X\to Y}(V_1)$.

Suppose that $l_\alpha(V_r') < q_{t_r} + 17$. Then by 2.8, the occurrence $W_r \rightleftharpoons f_{X\to Y}(V_r')$ has no proper normalised continuation to the right. Let U' be a maximal image of U in the reversal $X \to Y$, and let W_{r+1} be a maximal end of U' which is stable in a reversal of W_r. Since $l_\alpha(U') \geqslant q$, by 2.10 we have $l_\alpha(W_{r+1}) \geqslant q - 17 > q_{t_{r+1}} + 17$. By 1.2 and 1.3, W_r is stable in a reversal of W_{r+1}. By I.4.21, U adjoins V_r, whence it follows by 3.2 and 3.8 that W_{r+1} adjoins W_r. It is now easy to convince oneself by repeating the argument of Case I that the occurrences (20) together with W_r and W_{r+1} form the desired right cascade of rank α of W_0.

Case III''. Let U' be a maximal image of U in the reversal $X \to Y$. If $r = 1$, then by 3.8, U' adjoins W_0. In this case the maximal end U_1 of U' which is stable in a reversal of W_0 forms a single-element cascade of rank α of W_0, since by 2.10 we have $l_\alpha(U_1) \geqslant q - 17 \geqslant q_{t_1} + 17$.

Suppose that $r > 1$. Then V_{r-1} is stable in the reversal $X \to Y$. If the maximal normalised continuation of $f_{X\to Y}(V_{r-1})$ to the right contains not less than $q_{t_{r-1}} + 17$ segments, then we show, by analogy with Case III', that together with the occurrences $W_i \rightleftharpoons f_{X\to Y}(V_i)$ for $i = 1, 2, \ldots, r-2$, it forms the desired right cascade of rank α of W_0. In the contrary case it is proved in exactly the same way that the maximal end W_r of U' which is stable in a reversal of $f_{X\to Y}(V_{r-1})$ together with the occurrences (20) forms the desired cascade of rank α of W_0.

3.17. *Suppose that $X \in \mathcal{N}_\alpha$, $V \in \mathrm{Norm}(\alpha, X, 9)$ and W is the maximal normalised continuation of V. If $l_\alpha(W) \geqslant q + 34$, then V is compatible with some completely stable occurrence of rank α.*

We clearly have $\mathrm{Comp}(V, W)$. If W has neither a right nor a left cascade of rank

α, then according to I.4.23, it is completely stable. Suppose that W has a right cascade (2) of rank α with first element W_1. Then $\neg\text{Comp}(W, W_1)$. Suppose that W' is a maximal start of W stable in a reversal of W_1. Then we have by 2.10 that $l_\alpha(W') \geqslant q + 17$. By 3.10, W_1 is the first element of a right cascade of W', and hence W' is completely stable from the right. If W' has a left cascade of rank α with first element V_1, we then consider its maximal end W'' which is stable in a reversal of V_1. By 2.10, $l_\alpha(V') \geqslant q$. Clearly V' is both completely stable from the right and from the left.

3.18. *If V is a completely stable occurrence of rank α in a word $X \in \mathcal{N}_\alpha$ and $X \overset{\alpha-1}{\sim} Y$, then $f_{\alpha-1}(V; X, Y)$ is a completely stable occurrence of rank α in Y.*

This follows from II.5.16, 3.11 and 1.4.

3.19. *Suppose that $V \rightleftharpoons P * A * uBQ$, $W \rightleftharpoons PAu * B * Q$, $V' \rightleftharpoons R * A * uBS$ and $W' \rightleftharpoons RAu * B * S$ are normalised occurrences of elementary $(q-17)$-powers of rank α in words $X, Y \in \mathcal{N}_\alpha$, where $\neg\text{Comp}(V, W)$.*

If $P \overset{\circ}{=} R$, V is a completely stable occurrence of rank α and either $l_\alpha(W) \geqslant q$ or W adjoins V and is stable in a reversal of it, then V' is a completely stable occurrence of rank α.

If $Q \overset{\circ}{=} S$, W is a completely stable occurrence of rank α and either $l_\alpha(V) \geqslant q$, or V adjoins W and is stable in a reversal of it, then W' is a completely stable ocurrence of rank α.

By analogy, we confine our attention to the case $P \overset{\circ}{=} R$. Since V is completely stable, it follows that $l_\alpha(V) \geqslant q$. The complete stability of V' from the left follows from that of V, together with 3.12 and 2.15.

Suppose that W adjoins V and is stable in a reversal of it. Then the inequality $l_\alpha(W) \geqslant q_{t_1} + 17$ implies that the maximal end stable in a reversal of V of the maximal normalised continuation of W forms a single-element cascade of rank α of V. It then follows by 1.3 from the complete stability of V from the right that V is stable in a reversal of W. By 2.15, each of the occurrences V' and W' is stable in a reversal of the other. By 3.5, W' adjoins V'. Since $l_\alpha(W') = l_\alpha(W) \geqslant q_{t_1} + 17$, W' is compatible with the first and only element of some right cascade of V'.

Hence V' is completely stable from the right.

It remains for us to prove the complete stability of V' from the right when $l_\alpha(W) \geqslant q$. Assume that (12) is a right cascade of rank α of V'. Since $l_\alpha(W') = l_\alpha(W)$, by 3.15 a right cascade (11) of rank α of V can be found such that

$$\text{Comp}(W_1, W') \Rightarrow \text{Comp}(V_1, W)$$

and

$$\neg\text{Comp}(W_1, W') \Rightarrow (W_1 < W' \;\&\; V_1 = \phi(W_1;\ PA * uB * Q,\ RA * uB * S)).$$

The complete stability of V' from the right then follows from 1.3 and 2.15.

3.20. *Suppose that $U \rightleftharpoons P * A * uBvCQ$, $V \rightleftharpoons PAu * B * vCQ$, $W \rightleftharpoons PAuBv * C * Q$, $U_1 \rightleftharpoons R * A * uBvCS$, $V_1 \rightleftharpoons RAu * B * vCS$ and $W_1 \rightleftharpoons RAuBv * C * S$ are normalised occurrences of elementary $(q-17)$-powers of rank α in words $X, Y \in \mathcal{N}_\alpha$, where $\neg\text{Comp}(U, V)$, $\neg\text{Comp}(V, W)$, V is a completely* *stable occurrence of rank α, and each of the occurrences U and W either contains not less than q segments or adjoins V and is stable in a reversal of it. Then V_1 is also a completely stable occurrence of rank α.*

From the complete stability of V and the fact that $l_\alpha(V_1) = l_\alpha(V)$, it follows that $V_1 \in \text{Norm}(\alpha, Y, q)$. The complete stability of V_1 from the left and from the right is proved analogously to 3.19.

3.21. *If a completely stable occurrence V of rank α in a word $X \in \mathscr{N}_\alpha$ is not compatible with the occurrence $W \in \text{Norm}(\alpha, X, q)$, then it is stable in a reversal of W.*

By II.5.7, we have either $V < W$ or $W < V$. These cases are similar. Suppose that $V < W$. We assume that V is not stable in a reversal of W. Then W adjoins V and, as we convinced ourselves earlier, it is compatible with the first element of a right cascade of V. Then V must be stable in a reversal of W.

3.22. *Suppose that $X, Y \in \mathscr{N}_\alpha$, V is a completely stable occurrence of rank α in X, and the transition $X \to Y$ is a q-reversal of the occurrence $W \in \text{Norm}(\alpha, X, q)$, where $\neg\,\text{Comp}(V, W)$. Then V is stable in the reversal $X \to Y$ and $f_{X \to Y}(V)$ is a completely stable occurrence of rank α.*

By 3.21, V is stable in the reversal $X \to Y$. Let $U \rightleftharpoons f_{X\to Y}(V)$. By 2.6, $U \in \text{Norm}(\alpha, Y, q)$. By 1.2, $Y \to X$ is a q-reversal of some occurrence $W' \in \text{Norm}(\alpha, Y, q)$ and $V = f_{Y\to X}(U)$. The complete stability of U from the right (left) follows easily from the complete stability of V from the right (left respectively), by 3.16, 2.19 and 1.3.

3.23. *If $X \in \mathscr{P}_\alpha$, then $X \in \mathscr{N}_\alpha$, and for an arbitrary q-reversal $X \to Y$ of rank α we have $Y \in \mathscr{N}_\alpha$.*

Suppose that $X \in \mathscr{P}_\alpha$ and $X \to Y$ is a q-reversal of rank α. Then by I.4.26, the set $\text{Norm}(\alpha, X, n-88)$ is empty, whence by 1.2, 1.8 and 2.10 it follows that the set $\text{Norm}(\alpha, Y, n-71)$ is also empty. It is therefore enough to prove that, if $Y \in \mathscr{R}_{\alpha-1}$ and $\text{Norm}(\alpha, Y, n-71)$ is empty, then $Y \in \mathscr{N}_\alpha$.

Suppose that $U \in \text{MaxNorm}(\alpha, Y, q_1)$. Then $l_\alpha(U) \leqslant n-72$ and by 1.9, we can carry out a q_1-reversal of U. Suppose that $Y \to Z$ is such a reversal and $V \in \text{Norm}(\alpha, Z, q_1)$. If V is an image of U in the reversal $Y \to Z$, then by 1.2, $Z \to Y$ is a q_1-reversal of V. In the contrary case we have by 1.2 and 2.10 that, if W is the maximal normalised continuation of V, then $l_\alpha(W) \leqslant n-55$. By 1.9 it is then possible to perform a q_1-reversal of V. Therefore, $Y \in \mathscr{N}_\alpha$.

3.24. *Suppose that $X \in \mathscr{P}_\alpha$, $W \in \text{Norm}(\alpha, X, 9)$ and $\bar{W}$ is the maximal normalised continuation of W. If $q + 34 \leqslant l_\alpha(\bar{W}) \leqslant n-q-52$, then the occurrence W is active, that is, there exists a real reversal of it.*

By 1.9, we can find some $(q+34)$-reversal of W. Suppose that $X \to Y$ is such a reversal and U is a maximal image of W in this reversal. By 1.8, $l_\alpha(U) \geqslant q + 34$. By 3.23, $Y \in \mathscr{N}_\alpha$. By 3.17, the occurrences U and W are compatible with some completely stable occurrences of rank α, that is, $X \to Y$ is a real reversal of rank α.

3.25. *Suppose that the transition $X \to Y$ is a real reversal of rank α of the occurrence V and that W is an image of V in this reversal. Then $Y \to X$ is a real reversal of rank α of W.*

This follows from 1.2.

3.26. *Suppose that* $V, W \in \mathrm{Norm}(\alpha, X, q)$, *where* $X \in \mathscr{N}_\alpha$ *and* $\mathrm{Comp}(V, W)$. *Then every real reversal of rank* α *of* V *can be regarded as a real reversal of* W *and, in consequence, if* V *is active, so also is* W.

This follows from 1.3.

3.27. *Suppose that* $X \to Y$ *is a real reversal of rank* α *of the occurrence* $V \in \mathrm{Norm}(\alpha, X, 9)$, $X \overset{\alpha-1}{\sim} X_1$ *and* $Y \overset{\alpha-1}{\sim} Y_1$. *Then* $X_1 \to Y_1$ *is a real reversal of rank* α *of the occurrence* $f_{\alpha-1}(V; X, X_1)$ *. *Hence, if* $V \in \mathrm{Act}(\alpha, X)$, *then* $f_{\alpha-1}(V; X, X_1) \in \mathrm{Act}(\alpha, X_1)$.

This follows from 1.4, II.5.16 and 3.18.

3.28. *If an occurrence* V *in a word* $X \in \mathscr{N}_\alpha$ *is active, then each of its 9-reversals is a real reversal of rank* α. *In particular, a word* $Z \in \mathscr{M}_{\alpha-1}$ *can be found such that* $Y \to Z$ *is a real reversal of* V.

By hypothesis, there exists a real reversal $X \to Y$ of rank α of V. Suppose that $X \to Z$ is an arbitrary 9-reversal of V. Then we have by 1.11 that $Y \overset{\alpha-1}{\sim} Z$, from which it follows by 3.27 that $X \to Z$ is a real reversal of rank α. The second assertion follows from 3.27 and *IV.3.12*.

3.29. *Suppose that* $X, Y \in \mathscr{P}_\alpha$, $R * B * S \in \mathrm{Norm}(\alpha, Y, q-17)$, $P * B * Q \in \mathrm{Norm}(\alpha, X, q-17)$, $V \in \mathrm{Norm}(\alpha, X, 9)$, *the occurrences* V *and* $P * B * Q$ *are not compatible, and either* $l_\alpha(P * B * Q) \geqslant q$ *or* $P * B * Q$ *adjoins* V *and is stable in a reversal of it.*

If $V < P * B * Q$ *and* $R \mathrel{\overset{\circ}{=}} P$, *then*

$$V \in \mathrm{Act}(\alpha, X) \Longleftrightarrow \phi(V; *PB*Q, *PB*S) \in \mathrm{Act}(\alpha, Y).$$

If $P * B * Q < V$ *and* $S \mathrel{\overset{\circ}{=}} Q$, *then*

$$V \in \mathrm{Act}(\alpha, X) \Longleftrightarrow \phi(V; P*BQ*, R*BQ*) \in \mathrm{Act}(\alpha, Y).$$

By symmetry, we can confine ourselves to the case $V < P * B * Q$. Suppose that $W \rightleftharpoons P * B * Q$, $W' \rightleftharpoons P * B * S$ and $V' \rightleftharpoons \phi(V; *PB*Q, *PB*S)$. By 5.8, we have $\neg\,\mathrm{Comp}(V', W')$.

We consider first of all the case where W adjoins V and is stable in a reversal of it. We then have by 2.2 that $V = P_1 * A * uBQ$, where $P_1Au \mathrel{\overset{\circ}{=}} P$.

Suppose that $V \in \mathrm{Act}(\alpha, X)$. By 2.14, it is possible to find a local reversal of V,

$$X = P_1AuBQ \to P_2A_1u_1BQ \rightleftharpoons X_1, \tag{21}$$

where $P_2 * A_1 * u_1BQ$ is a maximal image of V and $f_{X \to X_1}(W) = P_2A_1u * B * Q$. By 3.28, the transition (21) is a real reversal, that is, $X_1 \in \mathscr{N}_\alpha$ and completely stable occurrences U and U_1 of rank α can be found such that

$$\mathrm{Comp}(U, V) \quad \text{and} \quad \mathrm{Comp}(U_1, P_2 * A_1 * u_1BQ). \tag{22}$$

By 2.17, the transition

*We note that if $X \mathrel{\overset{\circ}{=}} X_1$, we have by *IV.2.18(c)* that $f_{\alpha-1}(V; X, X_1) = V$.

$$Y = P_1AuBS \to P_2A_1u_1BS \rightleftharpoons Y_1 \tag{23}$$

is a local q-reversal of the occurrence V', where $P_2 * A_1 * u_1\,BS$ is a maximal image of V' and

$$f_{Y\to Y_1}(W') = P_2A_1u_1 * B * S\,.$$

From the fact that $Y \in \mathscr{P}_\alpha$, it follows by 3.23 that $Y_1 \in \mathscr{N}_\alpha$. By 2.2, U is contained in $* P_1Au * BQ$. By 3.2, W adjoins U. By II.6.11, the occurrence $\phi(U; * P_1Au * BQ; * P_1Au * BS)$ is normalised, by 3.19 it is completely stable in rank α, and it is compatible with V' by II.5.8. Since, by 2.6, the occurrences $P_2A_1u_1 * B * Q \rightleftharpoons W_1$. and $P_2A_1u_1 * B * S$ are normalised and since by 3.8 U_1 adjoins W_1, it follows as before from the complete stability of U_1 that $\phi(U_1 * P_2A_1u_1 * BQ, * P_2A_1u_1 * BS)$ is also a completely stable occurrence of rank α, and moreover it is compatible with $P_2 * A_1 * u_1BS$, by II.5.8. Hence the transition (23) is a real reversal of rank α, that is, $V' \in \text{Act}(\alpha, Y)$.

Since by 3.5 W' adjoins V' in the case under consideration and it is stable in a reversal of V' by 2.15, then by what has been proved already, $V' \in \text{Act}(\alpha, Y)$ implies $V \in \text{Act}(\alpha, X)$.

It now remains for us to consider the case where $l_\alpha(W) \geqslant q$. If W is not stable in a reversal of V, then replacing W by its maximal end W_0 which is stable in a reversal of V, we get by 2.10 that $l_\alpha(W_0) \geqslant q - 17$. Since in addition W_0 adjoins V, we are in the case already considered. If W is stable in a reversal of V, we can repeat the above arguments and take advantage in addition of the corresponding part of 3.19.

3.30. *Suppose that* $U \rightleftharpoons P * A * ECQ$, $W \rightleftharpoons PAE * C * Q$, $U' \rightleftharpoons R * A * ECS$ *and* $W' \rightleftharpoons RAE * C * S$ *are normalised occurrences of elementary* $(q-17)$*-powers of rank* α *in words* $X, Y \in \mathscr{P}_\alpha$, $V \in \text{Norm}(\alpha, X, 9)$, $U < V < W$, $\neg\text{Comp}(U, V)$, $\neg\text{Comp}(V, W)$ *and each of the occurrences* U *and* W *either contains not less than* q *segments or adjoins* V *and is stable in a reversal of it. Then*

$$V \in \text{Act}(\alpha, X) \Longleftrightarrow \phi(V;\, P * AEC * Q,\, R * AEC * S) \in \text{Act}(\alpha, Y)\,.$$

The proof is analogous to that of 3.29. However, instead of 2.17, II.6.11 and 3.19, one must use 2.18, II.6.14 and 3.20, respectively.

3.31. *If* $X \to Y$ *is a real reversal of rank* α *and* $X, Y \in \mathscr{P}_\alpha$, *then every occurrence* $V \in \text{Norm}(\alpha, X, q)$ *which is stable in it satisfies the condition*

$$V \in \text{Act}(\alpha, X) \Longleftrightarrow f_{X\to Y}(U) \in \text{Act}(\alpha, Y)\,. \tag{24}$$

By hypothesis, we have completely stable occurrences W and W_1 in words X and Y such that $X \to Y$ is a reversal of W and W_1 is an image of it in this reversal. Suppose that $V \in \text{Act}(\alpha, X)$, that is, we can find a q-reversal $X \to Y_1$ of V and completely stable occurrences U and U_1 in words X and Y_1 such that $\text{Comp}(U, V)$

and U_1 is an image of V in this reversal. Then, by 3.21, W is stable in a reversal of the occurrence $f_{X\to Y}(U)$ and U_1 is stable in a reversal of $f_{X\to Y_1}(W)$.

By 2.20, a word $Z \in \mathscr{R}_{\alpha-1}$ can be found such that the transitions $Y \to Z$ and $Y_1 \to Z$ are q-reversals of occurrences $f_{X\to Y}(U)$ and $f_{X\to Y_1}(W)$, and the occurrence $f_{Y_1\to Z}(U_1)$ is an image of $f_{X\to Y}(U)$ in the reversal $Y \to Z$. By 3.22, $f_{X\to Y}(U)$ and $f_{Y_1\to Z}(U_1)$ are completely stable occurrences of rank α, that is, the transition $Y \to Z$ is a real reversal of $f_{X\to Y}(U)$. We then have that $f_{X\to Y}(V) \in \mathrm{Act}(\alpha, Y)$, by 2.6 and 3.26.

The other part of implication (24) is proved in a similar way.

Chapter IV. Reduced Words and Equivalence in Rank α

§1. Kernels and Reduced Words of Rank α

1.1. By I.4.24, I.4.29 and II.7.1, for every word $X \in \mathcal{N}_\alpha$ we have the inclusions

$$\mathrm{Ker}(\alpha, X) \subset \mathrm{Norm}(\alpha, X, 9),$$
$$\mathrm{Ker}(\alpha, X) \subset \mathrm{Reg}(\alpha, X) \subset \mathrm{ComReg}(\alpha - 1, X) \subset \mathrm{Reg}(\alpha - 1, X).$$

Moreover, it follows directly from I.4.29 that if $U, V \in \mathrm{Reg}(\alpha, X)$, then the union of the occurrences U and V is also a regular occurrence of rank α.

1.2. *If the kernels U and V of rank α of a word $X \in \mathcal{N}_\alpha$ do not coincide, then they are not compatible and neither of them is contained in the other, that is, either $U < V$ or $V < U$.*

Suppose that the kernels U and V are distinct. Assume they are compatible. Then by II.7.3, their union V_1 is a normalised continuation of V and is distinct from V. We consider an arbitrary occurrence $W \in \mathrm{Act}(\alpha, X)$, where $\neg\mathrm{Comp}(V, W)$. Then $\neg\mathrm{Comp}(U, W)$; by I.4.24, U and V are both stable in a reversal of W and by II.5.7 they lie on the same side of W. Then by III.2.4, V_1 is also stable in a reversal of W. But this contradicts I.4.24. We therefore have $\neg\mathrm{Comp}(U, V)$. It only remains to quote II.5.7 once more.

1.3. *If $V \in \mathrm{Ker}(\alpha, X)$ and $V \in \mathrm{Act}(\alpha, X)$, then $V \in \mathrm{Norm}(\alpha, X, q)$ and the maximal normalised continuation of V does not intersect any kernel of rank α of the word X, other than itself. Moreover, each occurrence $W \in \mathrm{Reg}(\alpha, X)$ either contains V or does not intersect it.*

Suppose that $V \in \mathrm{Ker}(\alpha, X)$ and $V \in \mathrm{Act}(\alpha, X)$. Then by I.4.23, we can find a completely stable occurrence U of rank α such that $\mathrm{Comp}(U, V)$. Since, by III.3.21, U is stable in a reversal of an arbitrary $W \in \mathrm{Act}(\alpha, X)$, where $\neg\mathrm{Comp}(W, U)$, it follows by I.4.24 that U is contained in V. Hence $l_\alpha(V) \geqslant l_\alpha(U) \geqslant q$. The second assertion follows by III.2.2, and the last from the second.

1.4. *Suppose that $X \in \mathcal{N}_\alpha$. If V is an occurrence of an elementary* 51*-power of rank α in a word X or $V \in \mathrm{Norm}(\alpha, X, 43)$, then there is a kernel $U \in \mathrm{Ker}(\alpha, X)$ such that* $\mathrm{Comp}(V, U)$.

By II.6.6, we may restrict ourselves to the case where $V \in \mathrm{Norm}(\alpha, X, 43)$. Suppose that V_1 begins with the 18-th left and ends with the 18-th right segments of V. Clearly, $V_1 \in \mathrm{Norm}(\alpha, X, 9)$ and $\mathrm{Comp}(V, V_1)$. In view of III.2.9 and III.2.4, V_1 is stable in a reversal of any occurrence $W \in \mathrm{Act}(\alpha, X)$ that is not compatible with V, and so not compatible with V_1. The desired kernel U is a normalised continuation of the occurrence V_1 that is maximal with respect to this property.

1.5. *Suppose that $X \in \mathcal{N}_\alpha$. If $V \in \mathrm{Act}(\alpha, X)$, then there is a kernel U of rank α such that $\mathrm{Comp}(V, U)$ and $U \in \mathrm{Act}(\alpha, X)$.*

This follows from III.3.26 and 1.4.

By 1.2, this means that *every real reversal $X \to Y$ of rank α is a reversal of a unique kernel $V \in \mathrm{Ker}(\alpha, X)$*. By I.4.23, it follows from this that: *if $V \in \mathrm{Norm}(\alpha, X, 9)$ then $V \in \mathrm{Ker}(\alpha, X)$ if and only if V is stable in a reversal of any active kernel of rank α that is not compatible with it, and no proper normalised continuation of V has this property.*

1.6. *Suppose that $V \in \mathrm{Ker}(\alpha, X)$ and that W is its maximal normalised continuation. If U is the maximal end of the occurrence W stable in a reversal of the active kernel of rank α that is nearest to V on the left, then V is the maximal start of the occurrence U stable in a reversal of the active kernel of rank α that is nearest to V on the right. In addition, if the kernel of rank α of X that is nearest to V on the left (or the right) is not active, then $U = W$ ($V = U$ respectively).*

This follows from 1.5, 1.2, III.2.4 and III.2.7.

1.7. *Suppose that $V \in \mathrm{Ker}(\alpha, X)$. If W is a normalised continuation of V to the left (or to the right), then*

$$l_\alpha(W) \leqslant l_\alpha(V) + 17 \tag{1}$$

and thus, for every normalised continuation U of the kernel V, we have

$$l_\alpha(U) \leqslant l_\alpha(V) + 34 .$$

If W' is a continuation of V to the left (or to the right), then

$$l_\alpha(W') \leqslant l_\alpha(V) + 21 , \tag{2}$$

and thus, for every continuation U' of V we have

$$l_\alpha(U') \leqslant l_\alpha(V) + 42 .$$

Inequality (1) is satisfied because of 1.2, III.2.10 and III.2.4, and (2) follows from (1) and II.6.18.

1.8. *Suppose that $X \in \mathcal{N}_\alpha$ and $X \overset{\alpha-1}{\sim} Y$. Then*

$$V \in \mathrm{Ker}(\alpha, X) \Longleftrightarrow f_{\alpha-1}(V; X, Y) \in \mathrm{Ker}(\alpha, Y) \tag{3}$$

and

$$V \in \mathrm{Act}(\alpha, X) \Longleftrightarrow f_{\alpha-1}(V; X, Y) \in \mathrm{Act}(\alpha, Y) . \tag{4}$$

By III.3.3, $Y \in \mathcal{N}_\alpha$. Relation (4) holds in view of III.3.27, and (3) follows from (4), II.5.16 and III.1.4.

1.9. *Suppose that $X, Y \in \mathcal{P}_\alpha$ and that the transition $X \to Y$ is a real reversal of rank α. Then for every occurrence V stable in this reversal, we have*

$$V \in \mathrm{Ker}(\alpha, X) \Longleftrightarrow f_{X \to Y}(V) \in \mathrm{Ker}(\alpha, Y).$$

Let $X \to Y$ be a reversal of the occurrence W, and suppose that $V \in \mathrm{Ker}(\alpha, X)$ is stable in it. Then $\neg\mathrm{Comp}(W, V)$, and $f_{X \to Y}(V) \in \mathrm{Norm}\,(\alpha, Y, 9)$, by III.2.6. Consider any occurrence $U \in \mathrm{Act}(\alpha, Y)$ that is not compatible with $f_{X \to Y}(V)$. If U is an image of W in the reversal $X \to Y$, then, by III.1.2, $f_{X \to Y}(V)$ is stable in a reversal of it. In the contrary case, by III.3.21 a completely stable occurrence U_1 of rank α that is compatible with U is stable in the reversal $Y \to X$, while by III.3.31, $f_{Y \to X}(U_1) \in \mathrm{Act}(\alpha, X)$. Then, since V is stable in a reversal of $f_{Y \to X}(U_1)$, it follows by III.2.19 that $f_{X \to Y}(V)$ is stable in a reversal of U_1, and this means that it is stable in a reversal of U. Clearly, it follows from what has been proved that $f_{X \to Y}(V) \in \mathrm{Ker}(\alpha, Y)$.

1.10. *Suppose that $X, Y \in \mathcal{P}_\alpha$ and that the transition $X \to Y$ is a real reversal of rank α. Then, for every kernel $V \in \mathrm{Ker}(\alpha, X)$, there exists one and only one kernel $W \in \mathrm{Ker}(\alpha, Y)$ that is an image of V in the reversal $X \to Y$. We shall denote this kernel W by $f_\alpha(V; X, Y)$.*

The kernels $V \in \mathrm{Ker}(\alpha, X)$ and $f_\alpha(V; X, Y)$ are connected by the relations

$$f_\alpha(f_\alpha(V; X, Y); Y, X) = V$$

and

$$V \in \mathrm{Act}(\alpha, X) \Longleftrightarrow f_\alpha(V; X, Y) \in \mathrm{Act}(\alpha, Y). \tag{5}$$

If $X \to Y$ is a real reversal of a kernel $U \in \mathrm{Ker}(\alpha, X)$ different from $V \in \mathrm{Ker}(\alpha, X)$, then

$$f_\alpha(V; X, Y) = f_{X \to Y}(V). \tag{6}$$

According to 1.5, we may assume that $X \to Y$ is a reversal of some kernel U of rank α.

If the kernel V is different from U, then by 1.2 it is stable in the reversal $X \to Y$. In such a case, the desired kernel $f_\alpha(V; X, Y)$ is the occurrence $f_{X \to Y}(V)$. Indeed, by 1.9 we have $f_{X \to Y}(V) \in \mathrm{Ker}(\alpha, Y)$, and relation (5) is satisfied in view of III.3.31.

Suppose that $V = U$ and that V_1 is an image of the occurrence V in the reversal $X \to Y$. Since $V_1 \in \mathrm{Act}(\alpha, Y)$ by III.3.25, in view of 1.5 there exists a word $W \in \mathrm{Ker}(\alpha, Y)$ such that $\mathrm{Comp}(W, V_1)$. Clearly, W is an image of V in the reversal $X \to Y$, and by 1.2 such a kernel W is uniquely defined. Now we need only refer to III.1.2.

1.11. *Suppose that $V_1, V_2 \in \mathrm{Ker}(\alpha, X)$, that $X, Y \in \mathscr{P}_\alpha$, and that $X \to Y$ is a real reversal of rank α. Then*

$$V_1 < V_2 \Longleftrightarrow f_\alpha(V_1; X, Y) < f_\alpha(V_2; X, Y),$$
$$V_1 \ll V_2 \Longleftrightarrow f_\alpha(V_1; X, Y) \ll f_\alpha(V_2; X, Y). \tag{7}$$

If the transition $X \to Y$ is a reversal of one of the kernels V_1 and V_2, then relations (7) hold in view of III.2.2, while in the contrary case they hold by III.2.5.

1.12. *Suppose that $X, Y \in \mathscr{P}_\alpha$, that $X \to Y$ is a real reversal of rank α, and that the kernel $W \in \mathrm{Ker}(\alpha, Y)$ is an image of $V \in \mathrm{Ker}(\alpha, X)$ in this reversal. If $X \overset{\alpha-1}{\eqsim} X_1$ and $Y \overset{\alpha-1}{\eqsim} Y_1$, then the kernel $f_{\alpha-1}(W; Y, Y_1)$ is an image of the kernel $f_{\alpha-1}(V; X, X_1)$ in the real reversal $X_1 \to Y_1$.*

This follows from 1.8 and III.1.4.

1.13. *Suppose that $X, Y \in \mathscr{P}_\alpha$, that $X \to Y$ is a real reversal of rank α, that $W \in \mathrm{Ker}(\alpha, Y)$ is an image of the kernel $V \in \mathrm{Ker}(\alpha, X)$, and that $\bar{V}$ is the maximal normalised continuation of the kernel V.*

If $X \to Y$ is a reversal of the kernel V, then

$$n - 52 \leqslant l_\alpha(\bar{V}) + l_\alpha(W) \leqslant n + 1 \tag{8}$$

and

$$n - 86 \leqslant l_\alpha(V) + l_\alpha(W), \tag{9}$$

and in the contrary case $l_\alpha(W) = l_\alpha(V)$.

Relations (8) and (9) hold in view of III.1.8 and 1.7, and the last relation by 1.9 and III.2.6.

1.14. *Suppose that $X, Y \in \mathscr{P}_\alpha$,*

$$P * E * Q \in \mathrm{Norm}(\alpha, X, q) \cap \mathrm{Act}(\alpha, X)$$

and

$$R * E * S \in \mathrm{Norm}(\alpha, Y, q) \cap \mathrm{Act}(\alpha, Y).$$

If $P \mathrel{\underline{\overline{\circ}}} R$, then for every occurrence $W \in \mathrm{Norm}(\alpha, X, q)$ contained in $* PE * Q$ we have

$$W \in \mathrm{Act}(\alpha, X) \Longleftrightarrow \phi(W; * PE * Q, * PE * S) \in \mathrm{Act}(\alpha, Y) \tag{10}$$

and for every occurrence V contained in $* P * EQ$ we have

$$V \in \mathrm{Ker}(\alpha, X) \Longleftrightarrow \phi(V; * P * EQ, * P * ES) \in \mathrm{Ker}(\alpha, Y). \tag{11}$$

If $Q \mathrel{\underline{\overline{\circ}}} S$, then for every occurrence $W \in \mathrm{Norm}(\alpha, X, q)$ contained in $P * EQ *$ we have

$$W \in \mathrm{Act}(\alpha, X) \Longleftrightarrow \phi(W; P * EQ *, R * EQ *) \in \mathrm{Act}(\alpha, Y),$$

*and for every occurrence V contained in PE * Q * we have*

$$V \in \mathrm{Ker}(\alpha, X) \Longleftrightarrow \phi(V; PE * Q *, RE * Q *) \in \mathrm{Ker}(\alpha, Y).$$

In view of the obvious analogy, we restrict to the case $P \underset{\circ}{=} R$. Suppose that W is contained in $* PE * Q$ and that $W \in \mathrm{Norm}(\alpha, X, q)$. If $\mathrm{Comp}(W, P * E * Q)$, then relation (10) holds by III.3.26 and II.5.8. If $\neg\mathrm{Comp}(W; P * E * Q)$, then it follows from II.5.7 that $W < P * E * Q$. Let U be the maximal end of the occurrence $P * E * Q$ stable in a reversal of W. Clearly, either $U = P * E * Q$, or U adjoins W and also $l_\alpha(U) \geqslant q - 17$, by III.2.10. Since $W \ll U$ by III.2.2, relation (10) holds by II.6.11 and III.3.29.

Suppose that V is contained in $* P * EQ$ and that $V \in \mathrm{Ker}(\alpha, X)$. We set $V' \rightleftharpoons \phi(V; * P * EQ, * P * ES)$. By II.6.11, $V' \in \mathrm{Norm}(\alpha, X, 9)$ since $V \in \mathrm{Norm}(\alpha, X, 9)$. Since V is stable in a reversal of $P * E * Q$, V' is stable in a reversal of $P * E * S$, by III.2.15. Let us prove that V' is stable in a reversal of any occurrence $U \in \mathrm{Act}(\alpha, Y)$ that is not compatible with it. If $\mathrm{Comp}(U, P * E * S)$, this follows from III.1.3. In the contrary case, by II.5.7 we have either $U < P * E * S$ or $P * E * S < U$. The fact that V' is stable in a reversal of U follows from III.2.15 and relation (10), or from III.2.11 respectively.

Let us assume that $\neg(V' \in \mathrm{Ker}(\alpha, Y))$, that is, some normalised continuation V'' of V' and different from V' is also stable in a reversal of any occurrence $U \in \mathrm{Act}(\alpha, Y)$ that is not compatible with V'. Then V'' is contained in $* P * ES$ by III.2.2, whence it follows from what has been proved that $\phi(V''; * P * ES, * P * EQ)$ is stable in a reversal of any occurrence $U \in \mathrm{Act}(\alpha, X)$ that is not compatible with it. Since, by II.5.8, $\phi(V''; * P * ES, * P * EQ)$ is a continuation of V and is different from V, we have obtained a contradiction to the assumption that $V \in \mathrm{Ker}(\alpha, X)$. Thus $V' \in \mathrm{Ker}(\alpha, Y)$. In an analogous way, from $V' \in \mathrm{Ker}(\alpha, Y)$ it may be deduced that $V \in \mathrm{Ker}(\alpha, X)$.

1.15. *Suppose that* $X, Y \in \mathscr{P}_\alpha$,

$$P * A * ECQ, PAE * C * Q \in \mathrm{Norm}(\alpha, X, q) \cap \mathrm{Act}(\alpha, X)$$

and

$$R * A * ECS,\ RAE * C * S \in \mathrm{Norm}(\alpha, Y, q) \cap \mathrm{Act}(\alpha, Y).$$

Then for every occurrence $W \in \mathrm{Norm}(\alpha, X, q)$ *contained in* $P * AEC * Q$, *we have*

$$W \in \mathrm{Act}(\alpha, X) \Longleftrightarrow \phi(W; P * AEC * Q, R * AEC * S) \in \mathrm{Act}(\alpha, Y),$$

*and, for every occurrence V contained in PA * E * CQ, the condition*

$$V \in \mathrm{Ker}(\alpha, X) \Longleftrightarrow \phi(V; PA * E * CQ, RA * E * CS) \in \mathrm{Ker}(\alpha, Y)$$

holds.

This is proved in a way analogous to 1.14, with only one difference; that instead of III.3.29 and II.6.11, we must apply III.3.30 and II.6.14.

1.16. *Suppose that* $X \in \mathscr{P}_\alpha$ *and* $P * ED * Q \in \operatorname{Ker}(\alpha, X)$. *If* $l_\alpha(E) \geqslant q + 21$ *and* $PET \in \mathscr{P}_\alpha$ *then there exists a kernel* $V \in \operatorname{Ker}(\alpha, PET)$ *such that* $l_\alpha(V) \geqslant l_\alpha(E) - 21$, $V \in \operatorname{Act}(\alpha, PET)$ *and* V *is a start of the occurrence* $P * ET *$ *compatible with* $P * E * T$.

If $l_\alpha(D) \geqslant q + 21$ *and* $TDQ \in \mathscr{P}_\alpha$ *then there exists a kernel* $V \in \operatorname{Ker}(\alpha, TDQ)$ *such that* $l_\alpha(V) \geqslant l_\alpha(D) - 21$, $V \in \operatorname{Act}(\alpha, TDQ)$, *and* V *is an end of the occurrence* $* TD * Q$ *that is compatible with* $T * D * Q$.

Suppose that $l_\alpha(E) \geqslant q + 21$ and $PET \in \mathscr{P}_\alpha$. Clearly, we may assume that E is a normal elementary word of rank α. Then by II.5.4, the fact that the kernel $P * ED * Q$ is normalised yields that $P * E * DQ \in \operatorname{Norm}(\alpha, X, q + 21)$.

Suppose that $U \rightleftharpoons P * E * T$ and that U_5 and U_{22} are the starts of U ending with the fifth and twenty-second right segments. By II.6.12, $U_5 \in \operatorname{Norm}(\alpha, PET, q + 17)$, so that $U_{22} \in \operatorname{Norm}(\alpha, PET, q)$. By 1.4, U_{22} is compatible with some kernel $V \in \operatorname{Ker}(\alpha, PET)$. By III.3.29, we have $U_{22} \in \operatorname{Act}(\alpha, PET)$, and, by III.3.27, $V \in \operatorname{Act}(\alpha, PET)$.

By III.2.9, the occurrence U_{22} is stable in a reversal of any occurrence $W \in \operatorname{Act}(\alpha, PET)$ such that $\neg\operatorname{Comp}(W, U_{22})$ and $U_{22} < W$. Suppose that $W \in \operatorname{Act}(\alpha, PET)$, $\neg\operatorname{Comp}(W, U_{22})$ and $W < U_{22}$. Then it follows from III.2.2 and III.3.29 that $\phi(W; * PE * T, * PE * DQ) \in \operatorname{Act}(\alpha, X)$. Since by II.5.8 this last occurrence is not compatible with the kernel $P * ED * Q$, it follows that $P * ED * Q$ is stable in a reversal of it, and thus by III.2.3 and III.2.15, the occurrence U_{22} is stable in a reversal of W. Thus U_{22} is stable in a reversal of any occurrence $W \in \operatorname{Act}(\alpha, PET)$ such that $\neg\operatorname{Comp}(W, U_{22})$.

Analogously, we can convince ourselves that no normalised continuation of U_{22} to the left can be stable in a reversal of any occurrence $W \in \operatorname{Act}(\alpha, PET)$ situated to the left and not compatible with it. Thus a kernel V compatible with U is a start of $P * ET *$ and

$$l_\alpha(V) \geqslant l_\alpha(U_{22}) = l_\alpha(E) - 21\,.$$

1.17. The following relations follow from definitions I.4.26 and I.4.35:

$$\mathscr{R}_\alpha \subset \mathscr{P}_\alpha \subset \mathscr{R}_{\alpha-1},\ \bar{\mathscr{M}}_\alpha \subset \mathscr{M}_\alpha \subset \mathscr{L}_\alpha \subset \mathscr{K}_\alpha \subset \mathscr{R}_\alpha\,.$$

The elements of the set $\mathscr{R}_\alpha$ will be called *reduced words of rank* α.

1.18. *For every occurrence* $V \in \operatorname{Norm}(\alpha, X, 9)$ *and every elementary word* E *of rank* α *that occurs in a word* X, *the following relations are satisfied*:

$$X \in \mathscr{R}_\alpha \Rightarrow l_\alpha(V) \leqslant n - 142 \ \&\ l_\alpha(E) \leqslant n - 134\,,$$
$$X \in \mathscr{K}_\alpha \Rightarrow l_\alpha(V) \leqslant n - 184 \ \&\ l_\alpha(E) \leqslant n - 176\,,$$
$$X \in \mathscr{L}_\alpha \Rightarrow l_\alpha(V) \leqslant \frac{n+1}{2} + 55 \ \&\ l_\alpha(E) \leqslant \frac{n+1}{2} + 63\,,$$

$$X \in \mathcal{M}_\alpha \Rightarrow l_\alpha(V) \leqslant \frac{n+1}{2} + 34 \;\&\; l_\alpha(E) \leqslant \frac{n+1}{2} + 42 .$$

This follows from I.4.26, 1.4 and 1.7.

1.19. *If $X \in \mathcal{R}_{\alpha-1} \backslash \mathcal{R}_\alpha$, then there is an occurrence $V \in \mathrm{Norm}(\alpha, X, n-175)$. Similarly,*

$$X \in \mathcal{K}_{\alpha-1} \backslash \mathcal{K}_\alpha \Rightarrow \exists\, V(V \in \mathrm{Norm}(\alpha, X, n-217)) ,$$

$$X \in \mathcal{L}_{\alpha-1} \backslash \mathcal{L}_\alpha \Rightarrow \exists\, V(V \in \mathrm{Norm}(\alpha, X, \frac{n+3}{2} + 21)) ,$$

$$X \in \mathcal{M}_{\alpha-1} \backslash \mathcal{M}_\alpha \Rightarrow \exists\, V(V \in \mathrm{Norm}(\alpha, X, \frac{n+3}{2})) .$$

This follows from I.4.26.

1.20. *If $X \in \mathcal{R}_\alpha$, then*

$$V \in \mathrm{Norm}(\alpha, X, q+34)$$
$$\Rightarrow \exists\, U(U \in \mathrm{Ker}(\alpha, X) \;\&\; U \in \mathrm{Act}(\alpha, X) \;\&\; \mathrm{Comp}(U, V)).$$

The conclusion of this implication is also true if V is any occurrence of an elementary $(q+42)$-power of rank α in the word X.

Suppose that $V \in \mathrm{Norm}(\alpha, X, q+34)$. By 1.4, there is a $U \in \mathrm{Ker}(\alpha, X)$ such that $\mathrm{Comp}(U, V)$. Let W be its maximal normalised continuation. Then $l_\alpha(W) \leqslant n - 142 = n - q - 52$ by 1.18, whence it follows by III.3.24 that $U \in \mathrm{Act}(\alpha, X)$. The second assertion follows from what has been proved, in view of II.6.6.

1.21.

$$PQ \in \mathcal{M}_\alpha \Rightarrow P \in \mathcal{L}_\alpha \;\&\; Q \in \mathcal{L}_\alpha ;$$

$$PEQ \in \mathcal{K}_\alpha \Rightarrow E \in \mathcal{R}_\alpha ;$$

$$PEQ \in \mathcal{L}_\alpha \Rightarrow E \in \mathcal{K}_\alpha .$$

For $\alpha = 0$ all these implications follow immediately from definition I.4.1, since E is uncancellable whenever PEQ is uncancellable. Suppose that $\alpha > 0$ and the assertions are true for smaller ranks.

Suppose that $PQ \in \mathcal{M}_\alpha$. Then we have $P, Q \in \mathcal{L}_{\alpha-1}$ by the induction hypothesis. By 1.18, $P \in \mathcal{R}_\alpha$. If $P \notin \mathcal{L}_\alpha$, then for some word $V \in \mathrm{Ker}(\alpha, P)$, we have $l_\alpha(V) \geqslant (n+3)/2 + 21$. Let us consider the occurrence $U \rightleftarrows \phi(V; * P *, * P * Q)$. By 1.16, there exists a kernel $W \in \mathrm{Ker}(\alpha, PQ)$ such that $\mathrm{Comp}(W, U)$ and $l_\alpha(W) \geqslant l_\alpha(V) - 21 \geqslant (n+3)/2$, and this contradicts the condition that $PQ \in \mathcal{M}_\alpha$. Consequently $P \in \mathcal{L}_\alpha$. The fact that $Q \in \mathcal{L}_\alpha$ is proved in the analogous fashion.

Suppose that $PEQ \in \mathcal{K}_\alpha$. By the induction hypothesis, $E \in \mathcal{R}_{\alpha-1}$. By 1.18, $E \in \mathcal{P}_\alpha$. If $E \notin \mathcal{R}_\alpha$, then we have $l_\alpha(V) > n - 176$ for some kernel $V \in \mathrm{Ker}(\alpha, E)$. But by 1.18, this contradicts the assumption that $PEQ \in \mathcal{K}_\alpha$. Thus $E \in \mathcal{R}_\alpha$.

Similarly, since $n \geqslant 665$ implies that $(n+1)/2 + 63 < n - 218$, we have $E \in \mathscr{K}_\alpha$ since $PEQ \in \mathscr{L}_\alpha$.

1.22. *If a is a letter in the alphabet under consideration, then*

$$X \in \mathscr{M}_\alpha \Rightarrow [a, X]_0 \in \mathscr{L}_\alpha \;\&\; [X, a]_0 \in \mathscr{L}_\alpha ,$$
$$X \in \mathscr{K}_\alpha \Rightarrow [a, X]_0 \in \mathscr{R}_\alpha \;\&\; [X, a]_0 \in \mathscr{R}_\alpha .$$

For $\alpha = 0$, this is clear. Suppose that $\alpha > 0$ and $X \in \mathscr{M}_\alpha$. We set $Y \rightleftharpoons [a, X]_0$. By definition I.4.36, either $X \overset{\circ}{=} a^{-1}Y$ or $Y \overset{\circ}{=} aX$. In the first case, we have $Y \in \mathscr{L}_\alpha$ by 1.21. Suppose that $Y \overset{\circ}{=} aX$. By induction we have $Y \in \mathscr{L}_{\alpha-1}$. From 1.18 it follows that $Y \in \mathscr{P}_\alpha$. If $Y \notin \mathscr{L}_\alpha$, then we have $l_\alpha(V) > (n+1)/2 + 21$ for some kernel $V \in \mathrm{Ker}(\alpha, Y)$. Since $X \in \mathscr{M}_\alpha$, in view of 1.16, V is not contained in $a * X *$, that is, $V = * aE * Q$, where $l_\alpha(E) \geqslant (n+1)/2 + 21$. Then, by 1.16, some end U of the occurrence $* E * Q$ is a kernel of rank α of the word X. Since in addition there are no active kernels of rank α to the left of U, by 1.6 and II.6.18 we have $l_\alpha(U) \geqslant (n+1)/2 + 17$, which contradicts the condition that $X \in \mathscr{M}_\alpha$. Thus $Y \in \mathscr{L}_\alpha$.
The remaining assertions are proved in an analogous fashion.

1.23. *Suppose that $X \in \mathscr{R}_\alpha$. If $\partial(X) > 1$, then there are non-empty words $Y \in \mathscr{R}_\alpha$ and $Z \in \mathscr{R}_\alpha$ such that $X \overset{\circ}{=} YZ$.*
If $X \overset{\circ}{=} PEQ$, where E is an elementary $(2q+92)$-power of rank α, then there are words F and G such that $E \overset{\circ}{=} FG$, $PF \in \mathscr{R}_\alpha$ and $GQ \in \mathscr{R}_\alpha$.
For $\alpha = 0$, the first assertion is clear. Suppose that $\alpha > 0$. Since $X \in \mathscr{R}_{\alpha-1}$, it follows from the induction hypothesis that $X \overset{\circ}{=} YZ$ for some $Y \in \mathscr{R}_{\alpha-1}$ and $Z \in \mathscr{R}_{\alpha-1}$. If there are no occurrences of an elementary $(n-175)$-power of rank α in X, then by 1.19 we have that $Y \in \mathscr{R}_\alpha$ and $Z \in \mathscr{R}_\alpha$. Thus it only remains for us to prove the second assertion.
Suppose that $X \overset{\circ}{=} PEQ$, where E is an elementary $(2q+92)$-power of rank α and the occurrence $P * E * Q$ has no proper continuation to the left or the right.
Suppose that $i = q + 44$ and $E \overset{\circ}{=} E_1 D E_2$, where the occurrence $PE_1 * D * E_2Q$ begins with the i-th and ends with the $(i+6)$-th left segments of the occurrence $P * E * Q$. By II.7.18, there exist D_1 and D_2 such that $D \overset{\circ}{=} D_1 D_2$, $PE_1D_1 \in \mathscr{R}_{\alpha-1}$ and $D_2E_2Q \in \mathscr{R}_{\alpha-1}$. We shall show that E_1D_1 and D_2E_2 are the desired words F and G.
By II.2.7, we have that $l_\alpha(E_1) \geqslant q + 42$ and $l_\alpha(E_2) \geqslant q + 42$. In addition it is clear that

$$l_\alpha(E_1D_1) \leqslant i + 6 = q + 50 , \tag{12}$$

and it follows from 1.18 and II.5.2 that

$$l_\alpha(D_2E_2) \leqslant l_\alpha(E) - q - 41 \leqslant n - 134 - q - 41 < n - 176 .$$

By 1.18, $PE_1D_1 \in \mathscr{P}_\alpha$. Suppose that the occurrence $Pu * E' * vD_1$ starts with the fifth left segment and ends with the fifth right segment of the occurrence $P * E_1 * D_1$. By II.6.6, we have

$$Pu * E' * vD_1 \in \mathrm{Norm}(\alpha,\ PE_1D_1,\ q + 34)$$

and

$$Pu * E' * vDE_2Q \in \mathrm{Norm}(\alpha,\ X,\ q + 34)\,.$$

Let V and W be the maximal normalised continuations of these occurrences. Since by assumption $P * E * Q$ has no proper continuation to the left, it follows from (12) that $l_\alpha(V) \leqslant q + 50$. By 1.18, we have $l_\alpha(W) \leqslant n - q - 52$. Then, in view of III.3.24, we have

$$Pu * E' * vD_1 \in \mathrm{Act}(\alpha,\ PE_1D_1)$$

and

$$Pu * E' * vDE_2Q \in \mathrm{Act}(\alpha,\ X)\,.$$

We consider an arbitrary kernel $U \in \mathrm{Ker}(\alpha, PE_1D_1)$. If $\mathrm{Comp}(U, Pu * E' * vD_1)$, then $l_\alpha(U) \leqslant l_\alpha(V) \leqslant q + 50$. If $\neg\mathrm{Comp}(U,\ Pu * E' * vD_1)$, then in view of 1.5 and 1.3, we have $U \ll Pu * E' * vD_1$, whence by 1.14 it follows that

$$\phi(U;\, * Pu * E'vD_1,\, * Pu * E'vDE_2Q) \in \mathrm{Ker}(\alpha,\ X)\,.$$

Then

$$l_\alpha(U) = l_\alpha(\phi(U;\, * Pu * E'vD_1,\, * Pu * E'vDE_2Q)) \leqslant n - 176.$$

Thus $PE_1D_1 \in \mathscr{R}_\alpha$.

The relation $D_2E_2Q \in \mathscr{R}_\alpha$ is proved in an analogous fashion.

1.24. *If* $X \overset{\alpha-1}{\simeq} Y$, *then*

$$\begin{aligned} X \in \mathscr{P}_\alpha &\Longleftrightarrow Y \in \mathscr{P}_\alpha\,, \\ X \in \mathscr{R}_\alpha &\Longleftrightarrow Y \in \mathscr{R}_\alpha\,, \\ X \in \mathscr{K}_\alpha \;\&\; Y \in \mathscr{K}_{\alpha-1} &\Longleftrightarrow Y \in \mathscr{K}_\alpha\,, \\ X \in \mathscr{L}_\alpha \;\&\; Y \in \mathscr{L}_{\alpha-1} &\Longleftrightarrow Y \in \mathscr{L}_\alpha\,, \\ X \in \mathscr{M}_\alpha \;\&\; Y \in \mathscr{M}_{\alpha-1} &\Longleftrightarrow Y \in \mathscr{M}_\alpha\,. \end{aligned}$$

This follows from I.4.26, 1.8 and II.5.16.

1.25. *If* $X \in \mathscr{R}_\alpha$ *and* $X \to Y$ *is a real reversal of rank* α, *then* $Y \in \mathscr{P}_\alpha$.

Suppose that $X \to Y$ is a real reversal of a kernel $V \in \mathrm{Ker}(\alpha,\ X)$. If $U \in \mathrm{Norm}(\alpha,\ Y, n - 88)$, then by III.1.8, U is not an image of the occurrence V in the reversal under discussion, whence it follows by III.2.10 that some start (or end) $U_1 \in \mathrm{Norm}(\alpha,\ Y,\ n - 105)$ of the occurrence U is stable in the reversal $Y \to X$. Then $f_{Y\to X}(U_1) \in \mathrm{Norm}(\alpha, X, n - 105)$ by III.2.6, which by 1.18 contradicts the assumption that $X \in \mathscr{R}_\alpha$.

1.26. *Suppose that the transition $X \to Y$ is a real reversal of the kernel $V \in$* Ker (α, X), *and that U is the maximal normalised continuation of V. Then*

$$(X \in \mathcal{R}_\alpha \;\&\; l_\alpha(U) > 176) \Rightarrow Y \in \mathcal{R}_\alpha,$$
$$(X \in \mathcal{K}_\alpha \;\&\; l_\alpha(U) > 218 \;\&\; Y \in \mathcal{K}_{\alpha-1}) \Rightarrow Y \in \mathcal{K}_\alpha,$$
$$\left(X \in \mathcal{L}_\alpha \;\&\; l_\alpha(U) > \left(\frac{n+1}{2} + 21\right) \;\&\; Y \in \mathcal{L}_{\alpha-1}\right) \Rightarrow Y \in \mathcal{L}_\alpha,$$
$$\left(X \in \mathcal{M}_\alpha \;\&\; l_\alpha(U) > \frac{n+1}{2} \;\&\; Y \in \mathcal{M}_{\alpha-1}\right) \Rightarrow Y \in \mathcal{M}_\alpha.$$

Suppose that $X \in \mathcal{R}_\alpha$ and $l_\alpha(U) > 176$. By 1.25, $Y \in \mathcal{P}_\alpha$. Suppose that $W \in \mathrm{Ker}(\alpha, Y)$. If W is an image of the kernel V in the reversal $X \to Y$, then we have $l_\alpha(W) \leqslant n - 176$ by 1.13. Otherwise, by 1.10, W is an image of some kernel $V_1 \in \mathrm{Ker}(\alpha, X)$ different from X. Then $l_\alpha(W) = l_\alpha(V_1) \leqslant n - 176$ by 1.13. Thus $Y \in \mathcal{R}_\alpha$. The remaining three implications are proved in an analogous fashion.

1.27. *If $X \to Y$ is a q_1-reversal of rank α of the occurrence V, where $X, Y \in \mathcal{R}_\alpha$, then $X \to Y$ is a real reversal of rank α.*

By III.1.3, $X \to Y$ may be considered as a q_1-reversal of some occurrence $W \in \mathrm{MaxNorm}(\alpha, X, q_1)$. Let $U \in \mathrm{MaxNorm}(\alpha, Y, q_1)$ be a maximal image of W in this reversal. By 1.18, we have $l_\alpha(W) \leqslant n - 142$ and $l_\alpha(U) \leqslant n - 142$, whence by III.1.8 it follows that

$$l_\alpha(W) \geqslant n - 18 - n + 142 = q + 34.$$

Then W is active by III.3.24, that is, $X \to Y$ is a real reversal of rank α.

§ 2. Equivalence in Rank α

2.1. *Suppose that $X, Y \in \mathcal{P}_\alpha$, that $U, V \in$* Ker(α, X) *and that, in the sequence*

$$X \to Y \to Z$$

of real reversals of rank α, there first comes a reversal of the kernel V, and then a reversal of the kernel $f_\alpha(U; X, Y)$.

If $U = V$, then $Z \overset{\alpha-1}{\simeq} X$.

If $U \neq V$ and $X \to Y_1$ is a real reversal of the kernel U, where $Y_1 \in \mathcal{P}_\alpha$, then the transition $Y_1 \to Z$ is a reversal of the kernel $f_\alpha(V; X, Y_1)$, and the occurrence $f_{Y \to Z}(f_\alpha(V; X, Y))$ is an image of the occurrence $f_\alpha(V; X, Y_1)$ in the reversal $Y_1 \to Z$.

For $U = V$, III.3.25 gives that the transition $Y \to X$ is also a reversal of the kernel $f_\alpha(U; X, Y)$, whence by III.1.11 it follows that $Z \overset{\alpha-1}{\simeq} X$.

Suppose that $U \neq V$, $Y_1 \in \mathcal{P}_\alpha$, $X \to Y_1$ is a real reversal of the kernel U, and set $V_1 \rightleftharpoons f_\alpha(V; X, Y_1)$. By 1.10 we have $f_\alpha(U; X, Y) = f_{X \to Y}(U)$ and $f_\alpha(V; X, Y_1) =$

$f_{X\to Y_1}(V)$, and also V_1 is active. Since the kernel $f_\alpha(V;\, X,\, Y)$ is stable in a reversal of the kernel $f_{X\to Y}(U)$, and $f_\alpha(U;\, X,\, Y_1)$ is stable in a reversal of $V_1 = f_{X\to Y_1}(V)$, by 1.3 and III.2.20 there is a word $Z_1 \in \mathscr{R}_{\alpha-1}$ such that the transitions $Y \to Z_1$ and $Y_1 \to Z_1$ are q-reversals of the kernels $f_{X\to Y_1}(U)$ and V_1 respectively, while the occurrence $f_{Y\to Z_1}(f_\alpha(V;\, X,\, Y))$ is an image of V_1 in the reversal $Y_1 \to Z_1$ of it. By III.3.28, $Y \to Z_1$ and $Y_1 \to Z_1$ are real reversals of rank α. By III.1.11, we have $Z \overset{\alpha-1}{\approx} Z_1$ and

$$f_{Y\to Z}(f_\alpha(V;\, X,\, Y)) = f_{\alpha-1}(f_{Y\to Z_1}(f_\alpha(V;\, X,\, Y));\, Z_1,\, Z). \tag{1}$$

Then in view of III.3.27, the reversal $Y_1 \to Z$ is a real reversal of the kernel V_1, and by 1.12 the occurrence $f_{\alpha-1}(f_{Y\to Z_1}(f_\alpha(V;\, X,\, Y));\, Z_1,\, Z)$ is an image of V_1 in its reversal $Y_1 \to Z$. By (1), this means that $f_{Y\to Z}(f_\alpha(V;\, X,\, Y))$ is an image of V_1 in the reversal $Y_1 \to Z_1$.

2.2. The mapping $f_\alpha(V;\, X,\, Y) = W$ of $\operatorname{Ker}(\alpha, X)$ onto $\operatorname{Ker}(\alpha, Y)$ indicated in 1.10 for a single reversal $X \to Y$ can be extended in a natural way to a sequence of real reversals of rank α. Indeed, for a sequence

$$X_1 \to X_2 \to \ldots \to X_i \to X_{i+1} \to \ldots \to X_r \to X_{r+1} \tag{2}$$

of real reversals of rank α, where $X_i \in \mathscr{P}_\alpha$ for $1 \leqslant i \leqslant r+1$, $V \in \operatorname{Ker}(\alpha, X_1)$ and $U = f_\alpha(V;\, X_1,\, X_r)$, we set

$$f_\alpha(V;\, X_1,\, X_{r+1}) \rightleftharpoons f_\alpha(U;\, X_r,\, X_{r+1})\,.$$

By 1.11, this mapping $f_\alpha(V;\, X_1,\, X_r)$ preserves the relations $<$ and $\ll$. Thus it does not depend on the choice of the sequence (2). By 1.10, we have $f_\alpha(W;\, X_r,\, X_1) = V$ since $f_\alpha(V;\, X_1,\, X_r) = W$, and the following relation is satisfied:

$$V \in \operatorname{Act}(\alpha, X_1) \Longleftrightarrow f_\alpha(V;\, X_1,\, X_r) \in \operatorname{Act}(\alpha, X_r)\,.$$

We shall call the number of reversals occurring in a sequence the *length* of the sequence of reversals. In particular, the length of the sequence (2) is r. It will be convenient to consider the degenerate case where the sequence (2) consists of a single word $X_1 \in \mathscr{P}_\alpha$, and so has length 0. In this case we agree that $f_\alpha(V;\, X_1,\, X_1) \rightleftharpoons V$ for every kernel $V \in \operatorname{Ker}(\alpha, X_1)$.

2.3. Suppose that (2) is a sequence of real reversals of rank α, with $X_i \in \mathscr{P}_\alpha$ for $1 \leqslant i \leqslant r+1$. We shall say that a kernel $V \in \operatorname{Ker}(\alpha, X_j)$, $1 \leqslant j \leqslant r+1$, is *active in the sequence* (2) if there is an i such that $X_i \to X_{i+1}$ is a reversal of the kernel $f_\alpha(V;\, X_j,\, X_i)$. A kernel $V \in \operatorname{Ker}(\alpha, X_j)$ will be said to be *doubly active in the sequence* (2), if there are at least two different values of i such that $X_i \to X_{i+1}$ is a reversal of the kernel $f_\alpha(V;\, X_j,\, X_i)$. A sequence (2) of real reversals of rank α is said to be *simple* if there is no kernel $V \in \operatorname{Ker}(\alpha, X_1)$ that is doubly active in it.

2.4. *If the kernel* $V \in \operatorname{Ker}(\alpha, X_1)$ *is not active in the sequence*

$$X_1 \to X_2 \to \ldots \to X_i \to X_{i+1} \to \ldots \to X_r \tag{3}$$

of real reversals of rank α, where $X_i \in \mathscr{P}_\alpha$ for all i, then

$$\mathrm{MutNorm}_{\alpha-1}(V, f_\alpha(V; X_1, X_r)), \quad \mathrm{Rel}(V, f_\alpha(V; X_1, X_r))$$

and

$$l_\alpha(f_\alpha(V; X_1, X_r)) = l_\alpha(V).$$

For $r = 2$ this follows from 1.10 and III.2.6. Proceed by induction on r.

2.5. *If $X_1 \in \mathscr{R}_\alpha$, then we have $X_{r+1} \in \mathscr{P}_\alpha$ for every sequence (2) of real reversals of rank α.*

If the kernel $V_1 \in$ Ker(α, X_1) is active in the sequence (2), then there is a sequence

$$X_1 \to Y_2 \to \ldots \to Y_i \to Y_{i+1} \to \ldots \to Y_r \to X_{r+1} \tag{4}$$

of real reversals of rank α, in which the transition $X_1 \to Y_2$ (or $Y_r \to X_{r+1}$) is a reversal of V_1 (of $f_\alpha(V_1; X_1, Y_r)$ respectively).

For $r = 1$, this is true by 1.25. Assume that our assertion holds for any sequence of real reversals of rank α of length $< r$.

Let us assume that $X_{r+1} \notin \mathscr{P}_\alpha$. Then there is an occurrence $U \in$ Norm$(\alpha, X_{r+1}, n - 88)$. Let $X_{r+1} \to X_r$ be a reversal of an occurrence $W \in$ Norm(α, X_{r+1}, q). Since $\neg$Comp(U, W) by III.1.8, we have either $U < W$ or $W < U$. These two cases are analogous. If $U < W$, by III.2.10 there exists a start U' of the occurrence U that is stable in the reversal $X_{r+1} \to X_r$, and such that $l_\alpha(U') \geqslant n - 105$. Let $V' \rightleftharpoons f_{X_{r+1}-X_r}(U')$. By III.2.6,

$$V' \in \mathrm{Norm}(\alpha, X_r, n - 105)\,.$$

By induction, we have $X_i \in \mathscr{P}_\alpha$ for $1 \leqslant i \leqslant r$. By 1.4 there exists a $V \in$ Ker (α, X_r) such that Comp(V, V'), and moreover $l_\alpha(V) \geqslant l_\alpha(V') - 34 \geqslant n - 139$ by 1.7. Since $X_1 \in \mathscr{R}_\alpha$, by 2.4 the kernel V is active in the sequence (3) consisting of the first $r - 1$ reversals in the sequence (2) under discussion. Let $X_i \to X_{i+1}$ be the last reversal of the kernel $f_\alpha(V; X_r, X_i)$ in the sequence (3). Since $i < r$, and by induction $Y \in \mathscr{P}_\alpha$ for any real reversal $X_i \to Y$ of rank α, it follows from 2.1 that we may permute the first two reversals in the sequence

$$X_i \to X_{i+1} \to X_{i+2} \to \ldots \to X_r \to X_{r+1}$$

to obtain a sequence

$$X_i \to Y_{i+1} \to X_{i+2} \to \ldots \to X_r \to X_{r+1}\,, \tag{5}$$

of real reversals, where $Y_{i+1} \to X_{i+2}$ is a reversal of the kernel $f_\alpha(f_\alpha(V; X_r, X_i); X_i, Y_{i+1})$, and where the occurrence $f_{X_{i+1} \to X_{i+2}}(f_\alpha(V;\ X_r,\ X_{i+1}))$ is an image of it in this reversal. Then, in exactly the same way, we may permute the second and third reversals in sequence (5), *etc.* Having done the permuting $r - i$ times, we will obtain a sequence

$$X_i \to Y_{i+1} \to Y_{i+2} \to \dots \to Y_r \to X_{r+1}, \tag{6}$$

of real reversals of rank α, where $Y_r \to X_{r+1}$ is a reversal of the kernel $f_\alpha(f_\alpha(V;\ X_r,\ X_i);\ X_i,\ Y_r)$ and the occurrence $f_{X_r \to X_{r+1}}(V)$ is an image of it in this reversal. Since we have $\mathrm{Comp}(f_{X_r \to X_{r+1}}(V),\ U')$ by III.2.6, U' is an image of the occurrence $f_\alpha(f_\alpha(V;\ X_r,\ X_i);\ X_i,\ Y_r)$ in the real reversal $Y_r \to X_{r+1}$. But this contradicts III.1.8, since $l_\alpha(U) \geqslant n - 88 > n - q + 1$ by assumption. Thus $X_{r+1} \in \mathscr{P}_\alpha$.

If the kernel $V_1 \in \mathrm{Ker}(\alpha,\ X_1)$ is active in the sequence (3) and $X_i \to X_{i+1}$ is the last reversal of the kernel $f_\alpha(V_1;\ X_1,\ X_{i+1})$ in (3), then on adding the first $i - 1$ reversals in the sequence (3) on the left of the sequence (6) obtained above, we get the desired sequence (4), in which the transition $Y_r \to X_{r+1}$ is a reversal of the kernel $f_\alpha(V_1;\ X_1,\ Y_r)$. A sequence (4) in which $X_1 \to Y_2$ is a reversal of the kernel V_1 is obtained in an analogous fashion. To this end, we need to consider the first reversal $X_j \to X_{j+1}$ in (3) of the kernel $f_\alpha(V_1;\ X_1,\ X_j)$, and, for $j > 1$, to permute it with the preceding reversals on the basis of 2.1.

2.6. *If $X \in \mathscr{R}_\alpha$ and X is equivalent to Y in rank α, then either $X \overset{\alpha-1}{\sim} Y$ or there is a simple sequence*

$$X \to X_1 \to X_2 \to \dots \to X_r \to Y \tag{7}$$

of real reversals of rank α.

Suppose that $X \in \mathscr{R}_\alpha$ and that there is a sequence (7) of real reversals of rank α. Then $X_i \in \mathscr{P}_\alpha$ for all $i \leqslant r$, by 2.5. If some kernel $V \in \mathrm{Ker}(\alpha,\ X)$ is doubly active in this sequence (7), then by 2.5 there is a sequence

$$X \to Y_1 \to Y_2 \to \dots \to Y_r \to Y$$

of real reversals of rank α, where $Y_{r-1} \to Y_r$ and $Y_r \to Y$ are reversals of the kernels $f_\alpha(V;\ X,\ Y_{r-1})$ and $f_\alpha(V;\ X,\ Y_r)$ respectively (for $r = 1$, we understand X for Y_{r-1}). Then, by 2.1, we have $Y_{r-1} \overset{\alpha-1}{\sim} Y$, that is, $X \overset{\alpha-1}{\sim} Y$ for $r = 1$, while, for $r > 1$, we have by III.3.27 a shorter sequence $X \to Y_1 \to Y_2 \to \dots Y_{r-2} \to Y$ of real reversals of rank α carrying X to Y.

2.7. *The relation $\overset{\alpha}{\sim}$ is reflexive, symmetric and transitive.*
This can be proved by induction on α, using III.3.27 for transitivity.

2.8. Suppose that $X \in \mathscr{R}_\alpha$ and that X is equivalent to Y in rank α, where $\alpha > 0$. If there exists a sequence (7) of real reversals of rank α, then by 2.5 and 2.2 we have a function $f_\alpha(V;\ X,\ Y) = W$ that effects a one-to-one mapping of $\mathrm{Ker}(\alpha,\ X)$ onto $\mathrm{Ker}(\alpha,\ Y)$. On the other hand, when $X \overset{\alpha-1}{\sim} Y$, 1.8 says that the function

$f_{\alpha-1}(V; X, Y) = W$ also effects a one-to-one mapping between these sets. More than that, if both these functions are defined for given words X and Y, then for every kernel $V \in \mathrm{Ker}(\alpha, X)$ we have

$$f_\alpha(V; X, Y) = f_{\alpha-1}(V; X, Y), \tag{8}$$

since, by 2.2 and *IV.2.18(f)*, both these mappings preserve the relation $<$. Finally, if $X \overset{\alpha-1}{\sim} Y$, the set Ker (α, X) is non-empty and there does not exist a sequence of real reversals of rank α transforming X to Y, then for $V \in \mathrm{Ker}\,(\alpha, X)$ we define $f_\alpha(V; X, Y)$ using relation (8). Thus, we get a *function* $W = f_\alpha(V; X, Y)$, *which* for $X \in \mathcal{R}_\alpha$ such that Ker (α, X) is non-empty, and for any Y equivalent to X in rank α, *effects a one-to-one mapping of* Ker (α, X) *onto* Ker (α, Y) *and satisfies the condition*:

$$X \overset{\alpha-1}{\sim} Y \Rightarrow f_\alpha(V; X, Y) = f_{\alpha-1}(V; X, Y). \tag{9}$$

2.9. *When* $X \overset{\alpha}{\sim} Y$, *the function* $f_\alpha(V; X, Y)$ *indicated in* 2.8 *satisfies the following conditions, in which* U *and* V *are arbitrary kernels of rank* α *of* X:

$$\begin{aligned} &f_\alpha(V; X, X) = V, \\ &f_\alpha(V; X, Y) = W \Longleftrightarrow f_\alpha(W; Y, X) = V, \\ &U < V \Longleftrightarrow f_\alpha(U; X, Y) < f_\alpha(V; X, Y), \\ &U \ll V \Longleftrightarrow f_\alpha(U; X, Y) \ll f_\alpha(V; X, Y). \end{aligned}$$

For $X \overset{\alpha-1}{\sim} Y$, these conditions are satisfied in view of (9) and *IV.2.18*; in the contrary case they follow from 2.2.

2.10. *If* $X \overset{\alpha}{\sim} Y$, $Y \overset{\alpha}{\sim} Z$ *and* $V \in \mathrm{Ker}\,(\alpha, X)$, *then*

$$(f_\alpha(V; X, Y) = U \,\&\, f_\alpha(U; Y, Z) = W) \Rightarrow f_\alpha(V; X, Z) = W.$$

This follows from 2.2, III.3.27, (9) and *IV.2.18(e)*.

2.11. *Suppose that* $X \in \mathcal{R}_\alpha$. *If* $X \overset{\alpha-1}{\sim} Y$, *then* $X \overset{\alpha}{\sim} Y$ *and, for any kernel* $V \in \mathrm{Ker}(\alpha, X)$, *the relations*

$$\mathrm{MutNorm}_{\alpha-1}(V, f_\alpha(V; X, Y)), \quad \mathrm{Rel}\,(V, f_\alpha(V; X, Y))$$

and

$$l_\alpha(f_\alpha(V; X, Y)) = l_\alpha(V)$$

hold.

This follows from 1.24, (9), II.5.18 and *IV.3.2*.

2.12. *Suppose that* $X \in \mathcal{R}_\alpha$, *that the word* Y *is equivalent to* X *in rank* α *and that* $V \in \mathrm{Ker}\,(\alpha, X)$. *If* Rel $(V, f_\alpha(V; X, Y))$, *then* $l_\alpha(f_\alpha(V; X, Y)) = l_\alpha(V)$ *and* $\mathrm{MutNorm}_{\alpha-1}(V, f_\alpha(V; X, Y))$.

If $\neg$ Rel $(V, f_\alpha(V; X, Y))$, *then* Rel $(V^{-1}, f_\alpha(V; X, Y))$ *and the inequalities*

$$l_\alpha(f_\alpha(V; X, Y)) \geqslant n - l_\alpha(V) - 86\,, \tag{10}$$

$$n - 52 \leqslant l_\alpha(\bar{V}) + l_\alpha(f_\alpha(V;X,Y)) \leqslant n + 1\,, \tag{11}$$

are satisfied, where $\bar{V}$ *is the maximal normalised continuation of* V.

By 2.6 and 2.11, it remains only to consider the case where there is a simple sequence (7) of real reversals of rank α. If the kernel V is not active in this sequence (7), then everything follows from 2.4. Suppose that V is active in (7). By 2.5, we may assume that the transition $X \to X_1$ in (7) is a reversal of the kernel V. Then the kernel $f_\alpha(V; X, X_1)$ is not active in the sequence $X_1 \to X_2 \to \ldots \to X_r \to Y$, and by 2.4 we have

$$\text{Rel}\,(f_\alpha(V; X, X_1), f_\alpha(V; X, Y))$$

and

$$l_\alpha(f_\alpha(V; X, Y)) = l_\alpha(f_\alpha(V; X, X_1))\,.$$

On the other hand, by III.1.5 we have Rel $(V^{-1}, f_\alpha(V; X, X_1))$ and $\neg$ Rel $(V, f_\alpha(V; X, X_1))$, and by 1.13 that

$$l_\alpha(f_\alpha(V; X, X_1)) \geqslant n - l_\alpha(V) - 86$$

and

$$n - 52 \leqslant l_\alpha(\bar{V}) + l_\alpha(f_\alpha(V; X, X_1)) \leqslant n + 1\,.$$

Thus (10) and (11) are satisfied, and we have Rel $(V^{-1}, f_\alpha(V; X, Y))$ and $\neg$ Rel $(V, f_\alpha(V; X, Y))$ because of the transitivity of Rel (U, V).

2.13. *For* $\alpha > 0$ *and* $X \overset{\alpha}{\sim} Y$, *we denote by* $I_\alpha(X, Y)$ *the number of kernels* $V \in$ Ker (α, X) *such that* $\neg$ Rel $(V, f_\alpha(V; X, Y))$. *This parameter* $I_\alpha(X, Y)$ *satisfies the condition*

$$X \in \mathscr{R}_\alpha \;\&\; X \overset{\alpha-1}{\simeq} Y \Longleftrightarrow X \overset{\alpha}{\sim} Y \;\&\; \alpha > 0 \;\&\; I_\alpha(X, Y) = 0\,.$$

The implication from left to right holds by 2.11.

Suppose that $X \overset{\alpha}{\sim} Y$. If $\neg\,(X \overset{\alpha-1}{\simeq} Y)$, then by 2.6 there is a simple sequence (7) of real reversals of rank α. Then some kernel $V \in$ Ker (α, X) is active in this sequence (7), and as we convinced ourselves in 2.12, we have $\neg$ Rel $(V, f_\alpha(V; X, Y))$ for this kernel V. But this contradicts the condition $I_\alpha(X, Y) = 0$. Thus $X \overset{\alpha-1}{\simeq} Y$.

2.14. *Suppose that* $X \overset{\alpha}{\sim} Y$ *and* $V \in$ Ker (α, X). *If* $\neg$ Rel $(V, f_\alpha(V; X, Y))$, *then* $V \in$ Act (α, X), *and for any reversal* $X \to Z$ *of the kernel* V *we have* $Z \in \mathscr{R}_\alpha$. *If* V *is not active, then* MutNorm$_{\alpha-1}(V, f_\alpha(V; X, Y))$ *and* $l_\alpha(f_\alpha(V; X, Y)) = l_\alpha(V)$.

Suppose that $\neg\text{Rel}(V, f_\alpha(V; X, Y))$. Then, as we convinced courselves in 2.12, there is a simple sequence (7) of real reversals of rank α in which the transition $X \to X_1$ is a reversal of V. In addition $l_\alpha(f_\alpha(V; X, X_1)) = l_\alpha(f_\alpha(V; X, Y))$ and, for any kernel $U \in \text{Ker}(\alpha, X_1)$ different from $f_\alpha(V; X, X_1)$, we have $l_\alpha(U) = l_\alpha(f_\alpha(U; X_1, X))$. Therefore $X_1 \in \mathscr{R}_\alpha$ since $X, Y \in \mathscr{R}_\alpha$. By III.1.11 we have $X_1 \overset{\alpha-1}{\sim} Z$, whence $Z \in \mathscr{R}_\alpha$ by 1.24.

The second assertion follows from 2.4 and 2.11.

2.15. *If $X \overset{\alpha}{\sim} Y$ and X has no active kernels of rank α, then $I_\alpha(X, Y) = 0$ and $X \overset{\alpha-1}{\sim} Y$.*

This follows from 2.14 and 2.13.

2.16. *If there is no occurrence in X of a periodic q-power of rank* 1, *then*

$$X \overset{\alpha}{\sim} Y \Rightarrow X \overset{0}{=} Y.$$

Proceed by induction on α. Suppose that $\alpha > 0$, $X \overset{\alpha}{\sim} Y$ and no q-power of rank 1 occurs in X. Then, by II.4.8, there are no occurrences in X of elementary q-powers of rank $\leqslant \alpha$, whence by 1.3 and 2.15 it follows that $X \overset{\alpha-1}{\sim} Y$. Further, we get $X \overset{0}{=} Y$ by the inductive assumption.

2.17. *If $X \overset{\alpha}{\sim} Y$, then for every kernel $V \in$* Ker (α, X),

$$V \in \text{Act}(\alpha, X) \Longleftrightarrow f_\alpha(V; X, Y) \in \text{Act}(\alpha, Y).$$

This follows from 2.2 and 1.8.

2.18. *For $X \overset{\alpha}{\sim} Y$, the mapping $f_\alpha(V; X, Y)$ indicated in* 2.8 *extends in a natural way (according to* I.4.30) *to the set* Reg (α, X). *As a consequence we get, for $X \overset{\alpha}{\sim} Y$, a one-to-one mapping $f_\alpha(V; X, Y)$ of* Reg (α, X) *onto* Reg (α, Y).

This mapping satisfies the following conditions (a)-(j), *where U and V are arbitrary regular occurrences of rank α in the word X:*

(a) $V \in \text{Ker}(\alpha, X) \Rightarrow f_\alpha(V; X, Y) \in \text{Ker}(\alpha, Y)$,

(b) $X \overset{\alpha-1}{\sim} Y \Rightarrow f_\alpha(V; X, Y) = f_{\alpha-1}(V; X, Y)$,

(c) $f_\alpha(V; X, X) = V$,

(d) $f_\alpha(V; X, Y) = W \Rightarrow f_\alpha(W; Y, X) = V$,

(e) $f_\alpha(V; X, Y) = U \;\&\; f_\alpha(U; Y, Z) = W \Rightarrow f_\alpha(V; X, Z) = W$,

(f) $U < V \Longleftrightarrow f_\alpha(U; X, Y) < f_\alpha(V; X, Y)$,

(g) $U \ll V \Longleftrightarrow f_\alpha(U; X, Y) \ll f_\alpha(V; X, Y)$,

(h) *if U is contained in V, then $f_\alpha(U; X, Y)$ is contained in $f_\alpha(V; X, Y)$,*

(i) *if U is a start (or end) of V, then $f_\alpha(U; X, Y)$ is a start (end respectively) of $f_\alpha(V; X, Y)$,*

(j) $\partial_\alpha(f_\alpha(V; X, Y)) = \partial_\alpha(V)$.

These conditions follow easily from definition I.4.30 and the properties of this mapping for kernels of rank α as indicated in (9), 2.9 and 2.10.

2.19. *If $X \to Y$ is a real reversal of the kernel $U \in$* Ker (α, X), *where $X, Y \in \mathscr{P}_\alpha$ and U is not contained in the occurrence $V \in$* Reg (α, X), *then, by* 1.3 *and relation* (6) *introduced in* 1.10, *we have*

$$f_\alpha(V; X, Y) = f_{X \to Y}(V).$$

2.20. *If $X \in \mathscr{R}_\alpha$, then there is a word $Y \in \mathscr{M}_\alpha$ such that $X \overset{\alpha}{\sim} Y$.*

For $\alpha = 0$, we have $\mathscr{M}_\alpha = \mathscr{R}_\alpha$. For $\alpha > 0$ and under the assumption that our assertion is true for rank $\alpha - 1$, we shall prove it in rank α by induction on the number $m(X)$ of kernels $V \in \operatorname{Ker}(\alpha, X)$ such that $l_\alpha(V) > (n+1)/2$. Suppose that $m(X) = 0$. Since $X \in \mathscr{R}_{\alpha-1}$, by the inductive hypothesis there exists a word $Y \in \mathscr{M}_{\alpha-1}$ such that $X \overset{\alpha-1}{\sim} Y$. Then $X \overset{\alpha}{\sim} Y$ by 2.11, and $m(Y) = m(X) = 0$. Thus $Y \in \mathscr{M}_\alpha$.

If $m(X) > 0$, then we have $l_\alpha(V) > (n+1)/2$ for some kernel $V \in \operatorname{Ker}(\alpha, X)$. Then by 1.20 there exists a real reversal $X \to Z$ of V, where $Z \in \mathscr{P}_\alpha$ in view of 1.26. By 1.13 we have

$$l_\alpha(f_\alpha(V; X, Z)) \leqslant n + 1 - l_\alpha(V) < \frac{n+1}{2}.$$

Since in addition we have, by 2.4, that $l_\alpha(U) = l_\alpha(f_\alpha(U; Z, X))$ for kernels $U \in \operatorname{Ker}(\alpha, X)$ different from $f_\alpha(V; X, Z)$, it follows from $X \in \mathscr{R}_\alpha$ that $Z \in \mathscr{R}_\alpha$ and $m(Z) = m(X) - 1$. Then, by the second inductive hypothesis, there exists a $Y \in \mathscr{M}_\alpha$ such that $Z \overset{\alpha}{\sim} Y$ and so $X \overset{\alpha}{\sim} Y$.

2.21. *Suppose that $X \in \mathscr{R}_\alpha$, $V \in \operatorname{Ker}(\alpha, X)$ and $V \in \operatorname{Act}(\alpha, X)$. Then there is a word $Z \in \mathscr{R}_{\alpha-1}$ such that the transition $X \to Z$ is a local reversal of V and the kernel $f_\alpha(V; X, Z)$ is periodised in rank $\alpha - 1$.*

If in addition X lies in one of the sets $\mathscr{K}_{\alpha-1}$, $\mathscr{L}_{\alpha-1}$, $\mathscr{M}_{\alpha-1}$ or $\bar{\mathscr{M}}_{\alpha-1}$, then Z may be taken to lie in the same set.

If $X \overset{\alpha}{\sim} Y$ and $\neg \operatorname{Rel}(V, f_\alpha(V; X, Y))$, then $Z \overset{\alpha}{\sim} Y$, $I_\alpha(Z, Y) = I_\alpha(X, Y) - 1$ and the desired word Z may be taken in the same set $\mathscr{K}_\alpha$, $\mathscr{L}_\alpha$ or $\mathscr{M}_\alpha$ that contains X and Y.

By 2.14, $V \in \operatorname{Act}(\alpha, X)$ since $\neg \operatorname{Rel}(V, f_\alpha(V; X, Y))$, that is, every 9-reversal of the kernel V is a real reversal of rank α. By III.2.14, there is a local reversal $X \to Z$ of V such that a maximal image W of V is periodised. Since the kernel $f_\alpha(V; X, Z)$ is contained in W, by *IV.3.6* it is also periodised.

By III.1.5 and 2.12, the kernels $f_\alpha(V; X, Z)$ and $f_\alpha(V; X, Y)$ are related to the kernel V^{-1}. Thus $\operatorname{Rel}(f_\alpha(V; X, Y), f_\alpha(V; X, Z))$. Since X is equivalent to Y in rank α and

$$f_\alpha(f_\alpha(V; X, Y); Y, Z) = f_\alpha(V; X, Z),$$

by 2.12 we have

$$l_\alpha(f_\alpha(V; X, Z)) = l_\alpha(f_\alpha(V; X, Y)). \tag{12}$$

In addition, by 2.4 we have for kernels $U \in \operatorname{Ker}(\alpha, X)$ different from V,

$$\operatorname{Rel}(U, f_\alpha(U; X, Z)) \quad \text{and} \quad l_\alpha(f_\alpha(U; X, Z)) = l_\alpha(U). \tag{13}$$

Thus $Z \in \mathscr{R}_\alpha$ and $I_\alpha(Z, Y) = I_\alpha(X, Y) - 1$.

By III.2.1, *IV.3.15* and III.1.4, the word Z may be replaced by a new word equivalent to it in rank $\alpha - 1$ and lying in the same set $\mathscr{K}_{\alpha-1}$, $\mathscr{L}_{\alpha-1}$, $\mathscr{M}_{\alpha-1}$ or $\bar{\mathscr{M}}_{\alpha-1}$ as contains X. (If to the left or right of V there are no completely regular occurrences of rank $\alpha - 1$ stable in a reversal of V, then instead of *IV.3.15* it is necessary to use *IV.3.14* or *IV.3.12*.) For such a choice of Z, if X and Y both lie in one of the sets $\mathscr{K}_\alpha$, $\mathscr{L}_\alpha$ or $\mathscr{M}_\alpha$, then, by equations (12) and (13), Z will lie in the same set.

2.22. *If $X \in \mathscr{R}_\alpha$ and $P * E * Q \in$ Reg (α, X), then there are words R and S such that $X \overset{\alpha}{\sim} REQ$, $X \overset{\alpha}{\sim} PES$, $f_\alpha(P * E * Q;\ X,\ REQ) = R * E * Q$, $f_\alpha(P * E * Q;\ X,\ PES) = P * E * S$ and, for every kernel V of rank β, $0 < \beta \leqslant \alpha$, contained in $* R * EQ$ or $PE * S *$,*

$$l_\beta(V) \leqslant \frac{n+1}{2}. \tag{14}$$

The desired words R and S are found in similar ways. Let us carry out the proof for R. For $\alpha = 0$ we may take $R \rightleftharpoons P$. For $\alpha > 0$ we proceed by induction on the number $\gamma_\alpha(* P * EQ)$ of kernels $V \in$ Ker (α, X) contained in $* P * EQ$ and such that $l_\alpha(V) > (n + 1)/2$. Suppose that $\gamma_\alpha(* P * EQ) = 0$. By the first inductive hypothesis we find an R such that $X \overset{\alpha-1}{\sim} REQ, f_{\alpha-1}(P * E * Q; X, REQ) = R * E * Q$ and relation (14) holds for kernels V of rank β, $0 < \beta < \alpha$, contained in $* R * EQ$. Then, by 2.11 and 2.18 (b), $X \overset{\alpha}{\sim} REQ$ and $f_\alpha\ (P * E * Q;\ X,\ REQ) = R * E * Q$, whence by 2.18 (g) and 2.11 it follows that $\gamma_\alpha\ (* R * EQ) = 0$, that is, (14) holds also for $\beta = \alpha$.

Suppose that $\gamma_\alpha\ (* P * EQ) > 0$, that is, for some kernel $U \in$ Ker (α, X) contained in $* P * EQ$, we have $l_\alpha\ (U) < (n + 1)/2$. By 1.20, $U \in$ Act (α, X). By 2.21, there is a word $Z \in \mathscr{R}_\alpha$ such that $X \to Z$ is a local reversal of the kernel U, that is, $X \overset{\alpha}{\sim} Z$ and $f_\alpha\ (P * E * Q;\ X,\ Z) = P_1 * E * Q$ for some P_1. Then $l_\alpha(f_\alpha(U; X, Z)) < (n + 1)/2$ by 1.13, so that $\gamma_\alpha\ (* P_1 * EQ) = \gamma_\alpha\ (* P * EQ) - 1$ by 2.4. By the second inductive hypothesis, there is a word R such that $Z \overset{\alpha}{\sim} REQ$, $f_\alpha(P_1 * E * Q; Z, REQ) = R * E * Q$, and (14) holds. Then $X \overset{\alpha}{\sim} REQ$ and, by 2.18 (e) we have

$$f_\alpha(P * E * Q;\ X,\ REQ) = R * E * Q\,.$$

2.23. *If $X \in \mathscr{R}_\alpha$, $P * E * FDQ \in$ Reg (α, X) and $PEF * D * Q \in$ Reg (α, X), then there is a word R such that $X \overset{\alpha}{\sim} PERDQ, f_\alpha(P * E * FDQ;\ X,\ PERDQ) = P * E * RDQ$, $f_\alpha(PEF * D * Q;\ X,\ PERDQ) = PER * D * Q$, and condition (14) holds for every kernel V of rank β, $0 < \beta \leqslant \alpha$, contained in $PE * R * DQ$.*

The proof is similar to that of 2.22.

2.24. *Suppose that $\alpha > 0$, $X \overset{\alpha}{\sim} Y$, $P * E * Q \in$ Reg (α, X) and*

$$f_\alpha(P * E * Q;\ X,\ Y) = R * D * S\,.$$

If every kernel $V \in$ Ker (α, X) contained in $P * E * Q$ is related to the kernel $f_\alpha(V;\ X,\ Y)$, then there is a word $Z \in \mathscr{R}_\alpha$ such that $X \overset{\alpha-1}{\sim} Z$,

$$f_{\alpha-1}(P*E*Q; X, Z) = R_1 * D * S_1$$

and, for every occurrence $U \in \operatorname{Reg}(\alpha, X)$ *contained in* $P*E*Q$, *the following condition is satisfied:*

$$f_{\alpha-1}(U; X, Z) = \phi(f_\alpha(U; X, Y); R*D*S, R_1*D*S_1). \tag{15}$$

If in addition the words X and Y both lie in one of the sets $\mathscr{K}_\alpha$, $\mathscr{L}_\alpha$ *or* $\mathscr{M}_\alpha$, *then Z may be taken to lie in the same set.*

Proceed by induction on the parameter $I_\alpha(X, Y)$. When $I_\alpha(X, Y) = 0$, by 2.13 we have $X \overset{\alpha-1}{\approx} Y$, and in view of 2.18 (b) we have $f_{\alpha-1}(P*E*Q; X, Y) = R*D*S$, and for $Z \rightleftharpoons Y$ the desired relation (15) is a tautology.

Suppose that $I_\alpha(X, Y) > 0$, that is, for some kernel $V \in \operatorname{Ker}(\alpha, Y)$ we have $\neg \operatorname{Rel}(V, f_\alpha(V; Y, X))$. Then $V \in \operatorname{Act}(\alpha, Y)$ and Y is not contained in $R*D*S$. By 2.21, there is a word $Y_1 \in \mathscr{R}_\alpha$ such that $Y \to Y_1$ is a local reversal of the kernel V, and the word Y_1 lies in the same set $\mathscr{K}_\alpha$, $\mathscr{L}_\alpha$ or $\mathscr{M}_\alpha$ as contains both X and Y. Since the occurrence $R*D*S \in \operatorname{ComReg}(\alpha-1, Y)$ is stable in the reversal $Y \to Y_1$, according to Def. III.2.12 we have

$$f_{Y\to Y_1}(R*D*S) = R'*D*S', \tag{16}$$

where either $R' \overset{\circ}{\equiv} R$ or $S' \overset{\circ}{\equiv} S$. Then by III.2.13, we have for every occurrence $W \in \operatorname{Reg}(\alpha, Y)$ contained in $R*D*S$,

$$f_\alpha(W; Y, Y_1) = \phi(W; R*D*S, R'*D*S'). \tag{17}$$

Since $Y \overset{\alpha}{\sim} Y_1$, using 2.18(e), 2.19 and (16) we get the relation

$$f_\alpha(P*E*Q; X, Y_1) = f_\alpha(R*D*S; Y, Y_1) = R'*D*S'.$$

Since $I_\alpha(X, Y_1) = I_\alpha(X, Y) - 1$, by induction there exists a word $Z \in \mathscr{R}_\alpha$ such that $X \overset{\alpha-1}{\approx} Z$, $f_{\alpha-1}(P*E*Q; X, Z) = R_1*D*S_1$, and, for all occurrences $U \in \operatorname{Reg}(\alpha, X)$ contained in $P*E*Q$, the relation

$$f_{\alpha-1}(U; X, Z) = \phi(f_\alpha(U; X, Y_1); R'*D*S', R_1*D*S_1) \tag{18}$$

holds.

In addition, Z lies in the same set $\mathscr{K}_\alpha$, $\mathscr{L}_\alpha$ or $\mathscr{M}_\alpha$ that contains X, Y and Y_1. Setting $W \rightleftharpoons f_\alpha(U; X, Y)$, we have by 2.18(e) that

$$f_\alpha(U; X, Y_1) = f_\alpha(W; Y, Y_1),$$

whence it follows by (18) and (17) that (15) holds.

2.25. Suppose that $X \overset{\alpha}{\sim} Y$, $P*E*Q \in \operatorname{Reg}(\alpha, X)$ and $f_\alpha(P*E*Q; X, Y) = R*D*S$. We shall say that the occurrence $P*E*Q$ *transforms to* $R*D*S$ *locally in rank* α, if the conditions

$$V = T * F * uEQ \Rightarrow f_\alpha(V;\ X,\ Y) = T * F * vDS \tag{19}$$

and

$$V = PEu * H * G \Rightarrow f_\alpha(V;\ X,\ Y) = RDv * H * G \tag{20}$$

hold for all $V \in \operatorname{Reg}(\alpha, X)$. If condition (19) (or (20)) holds, then we shall say that $P * E * Q$ *transforms* to $R * D * S$ *locally from the left* (*from the right* respectively) *in rank* α.

It follows easily from 2.18(e) that, if $P * E * Q$ transforms locally in rank α (locally from the left or right) to $R * D * S$ and $R * D * S$ transforms locally in rank α (locally from the left or right) to $L * M * N$, then it transforms locally in rank α to $L * M * N$ (locally from the left or right respectively).

It follows from 2.11 and 2.18(b) that, if $X \in \mathcal{R}_\alpha$ and the occurrence $P * E * Q \in \operatorname{Reg}(\alpha, X)$ transforms locally in rank $\alpha - 1$ (locally from the left or right) to $R * E * S$, then it transforms locally in rank α (locally from the left or right respectively) to the same $R * E * S$.

If $X, Y \in R_\alpha$, $X \overset{\alpha}{\sim} Y$ *is a local real reversal of a kernel* $V \in \operatorname{Ker}(\alpha, X)$, *and* V *is contained in an occurrence* $W \in \operatorname{Reg}(\alpha, X)$, *then* W *transforms locally in rank* α *to the occurrence* $f_\alpha(W;\ X,\ Y)$.

2.26. *If* $X \overset{\alpha}{\sim} Y$, $P * E * Q \in \operatorname{Reg}(\alpha, X)$ *and* $f_\alpha(P * E * Q;\ X,\ Y) = R * D * S$, *then there is a word* $Z \in \mathcal{R}_\alpha$ *such that* $X \overset{\alpha}{\sim} Z$, $f_\alpha(P * E * Q;\ X,\ Z) = F * D * H$ *and* $P * E * Q$ *transforms to* $F * D * H$ *locally in rank* α. *If in addition* X *and* Y *both lie in the one of the sets* $\mathcal{K}_\alpha$, $\mathcal{L}_\alpha$ *or* $\mathcal{M}_\alpha$, *then* Z *may be taken to lie in that same set.*

For $\alpha = 0$ we have $D \overset{0}{=} E$ and can take $Z \rightleftarrows X$.

For $\alpha > 0$ we go by induction on the number $I_\alpha(P * E * Q;\ X,\ Y)$ of kernels $V \in \operatorname{Ker}(\alpha, X)$ which are contained in $P * E * Q$ and are such that $\neg \operatorname{Rel}(V, f_\alpha(V;\ X,\ Y))$.

If $I_\alpha(P * E * Q;\ X,\ Y) = 0$ then, by 2.24, it is possible to find a word $Z_1 \in \mathcal{R}_\alpha$ such that $X \overset{\alpha-1}{\sim} Z_1$, $f_{\alpha-1}(P * E * Q; X, Z) = R_1 * D * S_1$ and Z_1 belongs to whichever of the sets $\mathcal{K}_\alpha$, $\mathcal{L}_\alpha$ or $\mathcal{M}_\alpha$ it is that contains both words X and Y. Then by the induction assumption, we can find a word $Z \in \mathcal{R}_{\alpha-1}$ such that $X \overset{\alpha-1}{\sim} Z$, $f_{\alpha-1}(P * E * Q;\ X,\ Z) = F * D * H$ and the occurrence $P * E * Q$ transforms to $F * D * H$ locally in rank $\alpha - 1$, and therefore also in rank α. If in addition X and Z_1 belong to one of the sets $\mathcal{K}_\alpha$, $\mathcal{L}_\alpha$ or $\mathcal{M}_\alpha$, then Z belongs to the corresponding set $\mathcal{K}_{\alpha-1}$, $\mathcal{L}_{\alpha-1}$ or $\mathcal{M}_{\alpha-1}$, whence it follows by 1.24 that it belongs to one of $\mathcal{K}_\alpha$, $\mathcal{L}_\alpha$ or $\mathcal{M}_\alpha$.

Suppose that $I_\alpha(P * E * Q;\ X,\ Y) > 0$, that is, for some kernel $V \in \operatorname{Ker}(\alpha, X)$ contained in $P * E * Q$, we have $\neg \operatorname{Rel}(V, f_\alpha(V;\ X,\ Y))$. Then we can find by 2.21 an $X_1 \in \mathcal{R}_\alpha$ such that $X \to X_1$ is a local real reversal of V, and X_1 belongs to that one of the sets $\mathcal{K}_\alpha$, $\mathcal{L}_\alpha$ or $\mathcal{M}_\alpha$ to which X and Y belong. Suppose that $f_\alpha(P * E * Q;\ X,\ X_1) = P_1 * E_1 * Q_1$. Clearly, $X_1 \overset{\alpha}{\sim} Y$ and $f_\alpha(P_1 * E_1 * Q_1;\ X_1,\ Y) = R * D * S$. Since $I_\alpha(P_1 * E_1 * Q_1;\ X_1,\ Y) = I_\alpha(P * E * Q;\ X,\ Y) - 1$, by the second induction hypothesis we can find a word $Z \in \mathcal{R}_\alpha$ such that $X_1 \overset{\alpha}{\sim} Z$, $f_\alpha(P_1 * E_1 * Q_1;\ X_1,\ Z) = F * D * H$, $P_1 * E_1 * Q_1$ transforms locally in rank α into $F * D * H$, and Z belongs to that one of the sets $\mathcal{K}_\alpha$, $\mathcal{L}_\alpha$ or $\mathcal{M}_\alpha$ to which X_1 and Y belong. It then follows from the properties of local transformations set out in 2.25 that $P * E * Q$ transforms locally in rank α into $F * D * H$, that is, Z is the required word.

2.27. *Suppose that* $X \overset{\alpha}{\sim} Y$, $P * E * Q \in \operatorname{Reg}(\alpha, X)$ *and* $R * D * S = f_\alpha(P * E * Q; X, Y)$. *If one of the words E and D is a start or an end of the other, then* $E \overset{\circ}{=} D$ *and for each occurrence* $U \in \operatorname{Reg}(\alpha, X)$ *which is contained in* $P * E * Q$,

$$f_\alpha(U; X, Y) = \phi(U; P * E * Q, R * E * S). \tag{21}$$

This is clear for $\alpha = 0$. If $\alpha > 0$, we first of all consider the case where each kernel $V \in \operatorname{Ker}(\alpha, X)$ which is contained in $P * E * Q$ is related to the kernel $f_\alpha(V; X, Y)$. In this case, by 2.24 it is possible to find a word $Z \in \mathcal{R}_\alpha$ such that $X \overset{\alpha-1}{\sim} Z$, $f_{\alpha-1}(P * E * Q; X, Z) = R_1 * D * S_1$ and, for each occurrence $U \in \operatorname{Reg}(\alpha, X)$ which is contained in $P * E * Q$, relation (15) holds. Then by the induction assumption, $E \overset{\circ}{=} D$ and, for each occurrence $U \in \operatorname{Reg}(\alpha - 1, X)$ which is contained in $P * E * Q$, and in particular for $U \in \operatorname{Reg}(\alpha, X)$,

$$f_{\alpha-1}(U; X, Z) = \phi(U; P * E * Q, R * E * S),$$

from which by 2.18(b), relation (15) and the equality $E \overset{\circ}{=} D$, relation (21) follows.

It remains for us to prove that, if one of the words E and D is a start or end of the other, then each kernel $V \in \operatorname{Ker}(\alpha, X)$ which is contained in $P * E * Q$ is related to the kernel $f_\alpha(V; X, Y)$. Since the 4 possible cases here are analogous, we restrict ourselves to considering one of them. Suppose, for example, that E is a start of D, that is, $D \overset{\circ}{=} EH$.

We assume that some kernel $V \in \operatorname{Ker}(\alpha, X)$ which is contained in $P * E * Q$ is not related to the kernel $f_\alpha(V; X, Y)$. Let $V \rightleftharpoons PF * A * GQ$ be the first of these kernels from the left. Then by 2.9, the kernel $f_\alpha(V; X, Y)$ is the first kernel $U \in \operatorname{Ker}(\alpha, Y)$ contained in $R * EH * S$ which is not related to the kernels $f_\alpha(U; Y, X)$. Suppose that $f_\alpha(V; X, Y) = RT * B * CS$. By 2.14 and 1.3, we have $l_\alpha(A) \geqslant q$ and $l_\alpha(B) \geqslant q$.

We look at the occurrences $U_1 \rightleftharpoons RT * B * CS$ and $V_1 \rightleftharpoons RF * A * GHS$. By II.5.17, it follows from $\neg \operatorname{Rel}(A, B)$ that $\neg \operatorname{Comp}(U_1, V_1)$. Then by II.6.7, either $V_1 < U_1$ or $U_1 < V_1$.

Suppose that $V_1 < U_1$. By 1.4, a kernel $V_2 \in \operatorname{Ker}(\alpha, Y)$ can be found such that $\operatorname{Comp}(V_1, V_2)$, where V_2 intersects V_1 by 1.7. Suppose that W is a start of the occurrence $R * D * S$ which ends with the kernel V_2. Then $W = R * D_1 * D_2S$, where $\partial(D_1) > \partial(F)$. Since $W \ll U_1$ by 1.3, by 2.18(g) we have $f_\alpha(W; Y, X) \ll V$ and hence $f_\alpha(W; Y, X) = P * E_1 * E_2 Q$, where $\partial(E_1) \leqslant \partial(F)$. Clearly, one of the words E_1 and D_1 is a start of the other. In addition each kernel $U \in \operatorname{Ker}(\alpha, Y)$ which is contained in W is related to the kernel $f_\alpha(U; Y, X)$ by hypothesis. Then, by what has been proved above, we must have $D_1 \overset{\circ}{=} E_1$, which is impossible since $\partial(D_1) > \partial(E_1)$. Hence we cannot have $V_1 < U_1$.

If $U_1 < V_1$, the word TB is a proper start of the word FA, that is, $E \overset{\circ}{=} TBE'$ for some E'. We then look at the occurrences $V = PF * A * GQ$ and $U_2 \rightleftharpoons PT * B * E'Q$. Since $\neg \operatorname{Comp}(U_2, V)$ and $U_2 < V$, by analogy with the case $V_1 < U_1$

we again get a contradiction. Hence we must have Rel $(V, f_\alpha(V; X, Y))$ for every kernel $V \in \operatorname{Ker}(\alpha, X)$ contained in $P * E * Q$. We have proved what was required.

2.28. *Suppose that* $X \overset{\alpha}{\sim} Y$ *and* $P * E * Q \in \operatorname{Reg}(\alpha, X)$. *If* $f_\alpha(P * E * Q; X, Y) = P * E * S$, *then for each occurrence* $V \in \operatorname{Reg}(\alpha, X)$ *which is contained in* $* PE * Q$, *we have*

$$f_\alpha(V; X, Y) = \phi(V; * PE * Q, * PE * S) .$$

If $f_\alpha(P * E * Q; X, Y) = R * E * Q$, *then for each occurrence* $V \in \operatorname{Reg}(\alpha, X)$ *which is contained in* $P * EQ *$ *we have*

$$f_\alpha(V; X, Y) = \phi(V; P * EQ *, R * EQ *) .$$

Suppose that $f_\alpha (P * E * Q; X, Y) = P * E * S$ and $P \mathrel{\underline{\underline{\circ}}} FH$, where the occurrence $F * HE * Q$ starts with the first left kernel of rank α of X. Then it follows from 2.18 and 2.27 that $f_\alpha (F * HE * Q; X, Y) = F * HE * S$, and the first assertion of the lemma holds. The second one can prove in an analogous fashion.

2.29. *Suppose that* $X \overset{\alpha}{\sim} Y$ *and* $P * E * Q \in \operatorname{Reg}(\alpha, X)$. *If* $P * E * S \in \operatorname{Reg}(\alpha, Y)$, *then* $f_\alpha (P * E * Q; X, Y) = P * E * S$. *If* $R * E * Q \in \operatorname{Reg}(\alpha, Y)$, *then* $f_\alpha (P * E * Q; X, Y) = R * E * Q$.

By the obvious analogy, it is sufficient to prove the first of these assertions. Suppose that $P * E * S \in \operatorname{Reg}(\alpha, Y)$. For $\alpha = 0$ everything is clear. For $\alpha > 0$ we proceed by a second induction on the parameter $I_\alpha (X, Y)$. If $I_\alpha (X, Y) = 0$, then we have by 2.13 that $X \overset{\alpha-1}{\sim} Y$. Then, by the first inductive hypothesis, we have $P * E * S = f_{\alpha-1}(P * E * Q; X, Y)$, and it only remains to quote 2.18(b).

Suppose that $I_\alpha (X, Y) > 0$, that is, some kernel $V \in \operatorname{Ker}(\alpha, X)$ is not related to the kernel $f_\alpha (V; X, Y) \rightleftharpoons U$. Then by 2.14, U and V are active. By 1.3, either $* PE * Q \ll V$ or V is contained in $* PE * Q$.

If $* PE * Q \ll V$, then by 2.21 we can find a word $Z \in \mathscr{R}_\alpha$ such that the transition $X \to Z$ is a local real reversal of the kernel V, that is, $f_\alpha(P * E * Q; X, Z) = P * E * Q_1$, for some Q_1. Since $Y \overset{\alpha}{\sim} Z$ and $I_\alpha (Y, Z) = I_\alpha(X, Y) - 1$, by the second induction hypothesis we have $f_\alpha (P * E * Q_1; Z, Y) = P * E * S$, and hence $f_\alpha (P * E * Q; X, Y) = P * E * S$.

The case $* PE * S \ll U$ is dealt with similarly.

It therefore remains only to look at the case where some active kernels of rank α are contained in the occurrences $* PE * Q$ and $* PE * S$. Suppose that

$$V_1 \rightleftharpoons F * A * GQ \quad \text{and} \quad U_1 \rightleftharpoons T * B * HS$$

are the first active kernels of rank α from the left of the words X and Y. Then $U_1 = f_\alpha(V_1; X, Y)$, by 2.17, and the words FA and TB are starts of the word PE. Hence one of F and T is a start of the other.

If $F \mathrel{\underline{\underline{\circ}}} T$, then one of A, B is a start of the other, whence $A \mathrel{\underline{\underline{\circ}}} B$ follows by 2.27. Then by 2.18(i), one of the occurrences $F * BH * S$ and $f_\alpha(F * AG * Q; X, Y)$ is a start of the other, and hence by 2.27, $f_\alpha(F * AG * Q;\ X,\ Y) = F * BH * S$. It follows from this by 2.28 that

$$f_\alpha(P * E * Q;\ X,\ Y) = P * E * S\,.$$

We assume that $\neg(F \overset{\circ}{=} T)$. Then, without loss of generality, we may assume that $\partial(T) < \partial(F)$.

By 1.3, we have $l_\alpha(B) \geqslant q$. We look at the occurrence

$$V_2 \rightleftharpoons T * B * HQ = \phi(U_1;\ * PE * S,\ * PE * Q)\,. \tag{22}$$

By II.6.12, a start V_3 of V_2 can be found such that $V_3 \in \text{Norm}(\alpha, X, q-4)$. Since by 1.6 the occurrence V_1 has no proper normalised continuation to the left, it follows from $\partial(T) < \partial(F)$ that $\neg\text{Comp}(V_1, V_3)$ and $V_3 < V_1$. By 1.4, V_3 is compatible with some kernel $V_4 \in \text{Ker}(\alpha, X)$, where in addition $V_4 \ll V_1$ by 1.4.

Suppose that $U_4 \rightleftharpoons f_\alpha(V_4;\ X,\ Y)$ and $V_5 \rightleftharpoons \phi(U_4;\ * PE * S,\ * PE * Q)$. By 2.18(g), it follows from $V_4 \ll V_1$ that $U_4 \ll U_1$, whence $V_5 \ll V_2$ by (22). By II.6.11, we have $V_5 \in \text{Norm}(\alpha,\ X,\ 9)$. Since $\neg\text{Comp}(U_4,\ U_1)$, by II.5.8 we have $\neg\text{Comp}(V_5,\ V_2)$ and hence $\neg\text{Comp}(V_5,\ V_4)$. Since there is no active kernel of rank α in X to the left of V, by III.2.4 the occurrence V_5 is stable in a reversal of an arbitrary active kernel of rank α of X. Hence some normalised continuation V_6 of it is a kernel of rank α. Then $V_6 < V_4$.

We consider the occurrences

$$U_6 \rightleftharpoons f_\alpha(V_6;\ X,\ Y) \quad \text{and} \quad V_7 \rightleftharpoons \phi(U_6;\ * PE * S,\ * PE * Q)\,.$$

By 2.18(f), it follows from $V_6 < V_4$ that $V_7 < V_5$; by II.5.8, it follows from $\neg\text{Comp}(U_6,\ U_4)$ that $\neg\text{Comp}(V_7,\ V_5)$, and from $U_6 \in \text{Norm}(\alpha,\ Y,\ 9)$ it follows by II.7.4 and II.5.5 that $V_7 \in \text{Norm}(\alpha,\ X,\ 9)$. Since by III.2.4 V_7 is stable in a reversal of an arbitrary active kernel of rank α, it is compatible with some kernel $V_8 \in \text{Ker}(\alpha,\ X)$. Then $\neg\text{Comp}(V_8,\ V_6)$, and hence $V_8 < V_6$.

On repeating the argument of the last paragraph $r-4$ times we produce r kernels

$$V_{2r} < V_{2r-2} < \ldots < V_6 < V_4 \ll V_1$$

of rank α of X, where $r = \partial_\alpha(X) + 1$. Thus the assumption that $\neg(F \overset{\circ}{=} T)$ leads to a contradiction.

2.30. *Suppose that* $X \overset{\alpha}{\sim} Y$, $V \in \text{Ker}(\alpha,\ X)$ *and* $\text{Rel}(V, f_\alpha(V;\ X,\ Y))$. *If* $r \geqslant 9$, *the occurrence* $U \in \text{Norm}(\alpha,\ X,\ r)$ *is compatible with the kernel* V *and is stable both in a reversal of some fixed kernel* $W_0 \in \text{Ker}(\alpha,\ X)$ *and in a reversal of any kernel* $W \in \text{Ker}(\alpha, X)$ *which is not compatible with* V *and is such that* $\neg\text{Rel}(W, f_\alpha(W;\ X,\ Y))$, *then there exists an occurrence* $U' \in \text{Norm}(\alpha,\ Y,\ r)$, *compatible with* $f_\alpha(V;\ X,\ Y)$, *which is stable in a reversal of the kernel* $f_\alpha(W_0;\ X,\ Y)$ *and in a reversal of an arbitrary kernel* $W' \in \text{Ker}(\alpha,\ Y)$ *which is not compatible with* U', *where* $\neg\text{Rel}(W', f_\alpha(W';\ Y,\ X))$.

If $X \overset{\alpha-1}{\sim} Y$, then by 2.13, (9) and II.5.16, we can take $U' \rightleftharpoons f_{\alpha-1}(U; X, Y)$. In the contrary case, we have by 2.6 a simple sequence (7) of real reversals of rank α in which only those kernels $W \in \mathrm{Ker}(\alpha, X)$ for which $\neg\mathrm{Rel}(W, f_\alpha(W; X, Y))$ are active. If there is only one reversal in this sequence, then by 1.10, III.2.6 and III.1.2, we can take $U' \rightleftharpoons f_{X\to Y}(U)$, and the general case proceeds by induction on the number of reversals in the sequence (7).

2.31. *Suppose that $FQ \overset{\alpha}{\sim} RQ$, $FT \overset{\alpha}{\sim} RT$, $F_1 * E * F_2Q$ is an active kernel of rank α of the word FQ and that*

$$f_\alpha(F_1 * E * F_2Q;\ FQ,\ RQ) = R_1 * D * R_2Q\,.$$

If $V \in \mathrm{Ker}(\alpha, FT)$ then

$$\mathrm{Comp}(V, F_1 * E * F_2T) \Longleftrightarrow \mathrm{Comp}(f_\alpha(V; FT, RT), R_1 * D * R_2T)\,. \tag{23}$$

The symmetric assertion for $QF \overset{\alpha}{\sim} QR$ and $TF \overset{\alpha}{\sim} TR$ can be proved in an analogous way.

Together with the result as stated, we shall prove by simultaneous induction on the parameter $\partial(F_1R_1)$ the assertion obtained from it by interchanging the words T and Q.

We assume that for some kernel $V \in \mathrm{Ker}(\alpha, FT)$, relation (23) is violated. Suppose, for example, that

$$\mathrm{Comp}(V;\ F_1 * E * F_2T)$$

and

$$\neg\mathrm{Comp}(f_\alpha(V;\ FT,\ RT),\ R_1 * D * R_2T)\,. \tag{24}$$

We choose the parameter $\partial(F_1R_1)$ to have the least possible value for which relation (23), or that obtained from it by interchanging T and Q, is violated.

By 1.3, $F_1 * E * F_2Q \in \mathrm{Norm}(\alpha, FQ, q)$. From II.6.12 and 1.7 it follows that $l_\alpha(V) \geqslant q - 38$. We set

$$U \rightleftharpoons f_\alpha(V;\ FT,\ RT)\,.$$

By 2.12, we have $l_\alpha(U) \geqslant q - 38$. From (24) and II.6.7 it follows that either $R_1 * D * R_2T < U$, or $U < R_1 * D * R_2T$. In the first case we consider a kernel $W \in \mathrm{Ker}(\alpha, RT)$ which is compatible with $R_1 * D * R_2T$. Setting $W_1 \rightleftharpoons f_\alpha(W; RT, FT)$ we have $W_1 < V$, by 2.9. Then $\neg\mathrm{Comp}(W_1, F_1 * E * F_2T)$ and $W_1 < F_1 * E * F_2T$, that is, we have a situation similar to the case $U < R_1 * P * R_2T$.

It is therefore enough to consider the case where $U < R_1 * D * R_2T$.

Since by hypothesis (23) holds for smaller values of $\partial(F_1R_1)$, by 2.14 and II.5.17 we have that, for an arbitrary kernel $W \in \mathrm{Ker}(\alpha, FT)$ contained in $* F_1E * F_2T$,

$$\neg\mathrm{Comp}(W, F_1 * E * F_2T) \;\&\; \neg\mathrm{Rel}(W, f_\alpha(W; FT, RT)) \Rightarrow$$
$$\exists W_1(W_1 \in \mathrm{Ker}(\alpha, FQ) \;\&\; \mathrm{Comp}(W_1, \phi(W; * F * T, * F * Q)) \;\&$$
$$\neg\mathrm{Rel}(W_1, f_\alpha(W_1; FQ, RQ)))\,. \tag{25}$$

Analogous conditions hold for arbitrary kernels of rank α contained in the occurrences $* F_1E * F_2Q$, $* R_1D * R_2T$ and $* R_1D * R_2Q$.

By II.6.12, if V' is an arbitrary start of $F_1 * E * F_2T$, we have $V' \in \mathrm{Norm}(\alpha, FT, q - 4)$, from which it follows by III.2.15, III.1.3 and (25) that V' is stable in a reversal of an arbitrary kernel $W \in \mathrm{Ker}(\alpha, FT)$ provided that $W < V'$, $\neg\mathrm{Comp}(W, V)$ and $\neg\mathrm{Rel}(W, f_\alpha(W; FT, RT))$. Then by III.2.10, some start $\bar{V} \in \mathrm{Norm}\,(\alpha, FT, q - 21)$ of it has the same property on both sides, from which it follows by 2.30 and 1.3 that the kernel U is compatible with some occurrence $\bar{U} \in \mathrm{Norm}(\alpha, RT, q - 21)$ which is stable in a reversal of an arbitrary kernel $W_1 \in \mathrm{Ker}(\alpha, RT)$ which is not compatible with U and not related to $f_\alpha(W_1; RT, FT)$.

Since by (24) $\bar{U}$ is not compatible with $R_1 * D * R_2T$, we have by II.6.7 and II.6.11 that $\phi(\bar{U}; * R * T, * R * Q) \in \mathrm{Norm}(\alpha, RQ, q - 21)$. From the corresponding relation of type (25), it follows from III.2.15 and III.2.10, in an analogous way to what has gone before, that some start $\bar{U}_1 \in \mathrm{Norm}(\alpha, RQ, q - 38)$ of the occurrence $\phi(\bar{U}; * R * T, * R * Q)$ is stable in a reversal of an arbitrary kernel $W \in \mathrm{Ker}(\alpha, RQ)$ such that $\neg\mathrm{Comp}(W, \bar{U}_1)$ and $\neg\mathrm{Rel}(W, f_\alpha(W; RQ, FQ))$, and is also stable in a reversal of the kernel $R_1 * D * R_2Q$.

By 1.4, we can find a kernel $U_1 \in \mathrm{Ker}(\alpha, RQ)$ such that $\mathrm{Comp}(U_1, \bar{U}_1)$. Suppose that $V_1 \rightleftharpoons f_\alpha(U_1; RQ, FQ)$. By 2.30, V_1 is compatible with some occurrence $\bar{V}_1 \in \mathrm{Norm}(\alpha, FQ, q - 38)$, which is stable in a reversal of an arbitrary kernel $W \in \mathrm{Ker}(\alpha, FQ)$ with $\neg\mathrm{Comp}(W, V_1)$ and $\neg\mathrm{Rel}(W, f_\alpha(W; FQ, RQ))$, and also in a reversal of the kernel $F_1 * E * F_2Q$. By III.2.2, we have

$$\bar{U}_1 = R_3 * D_1 * R_4DR_2Q \quad\text{and}\quad \bar{V}_1 = F_3 * E_1 * F_4EF_2Q\,,$$

where $R_3D_1R_4 \mathrel{\underline{\circ}} R_1$ and $F_3E_1F_4 \mathrel{\underline{\circ}} F_1$.

We shall now prove that *if we take any kernel* $V_1 \in \mathrm{Ker}(\alpha, FQ)$ *which is situated to the left of* V *and any occurrence* $\bar{V}_1 \in \mathrm{Norm}(\alpha, FQ, q - 38)$ *compatible with it which has the properties mentioned in the preceding paragraph, then we can find a new kernel* $V_3 \in \mathrm{Ker}(\alpha, FQ)$ *and an occurrence* $\bar{V}_3 \in \mathrm{Norm}(\alpha, FQ, q - 38)$ *compatible with it, which again has these properties and where, in addition,* $V_3 < V_1$.

By II.6.11, the occurrence $\bar{V}_2 \rightleftharpoons F_3 * E_1 * F_4EF_2T$ is normalised. We prove that it is stable in a reversal of an arbitrary kernel $W \in \mathrm{Ker}(\alpha, FT)$ with $\neg\mathrm{Comp}(V_2, W)$ and $\neg\mathrm{Rel}(W, f_\alpha(W; FT, RT))$. If $\mathrm{Comp}(W, F_1 * E * F_2T)$, this follows by III.2.15 and III.1.3 from the stability of $\bar{V}_1$ in a reversal of $F_1 * E * F_2Q$. If $\neg\mathrm{Comp}(W, F_1 * E * F_2T)$, then, if $F_1 * E * F_2T < W$, it follows from III.2.11 and, if $W < F_1 * E * F_2T$, from (25) by III.2.15 and III.1.3. By 1.4, the occurrence $\bar{V}_2$ is compatible with some kernel $V_2 \in \mathrm{Ker}(\alpha, FT)$, where $V_2 < V$. We look at the kernel $U_2 \rightleftharpoons f_\alpha(V_2; FT, RT)$. By 2.9, $U_2 < U$. By 2.30, U_2 is compatible with some occurrence $\bar{U}_2 \in \mathrm{Norm}(\alpha, RT, q - 38)$ which is stable in a reversal of an arbitrary kernel $W \in \operatorname{Ker}(\alpha, RT)$ with $\neg$ Comp (W, U_2) and $\neg$ Rel $(W, f_\alpha(W; RT, FT))$.

Moreover, by analogy with the foregoing, it follows from relation (25) for $* R_1 D * R_2 Q$ that the occurrence $\bar{U}_3 \rightleftharpoons \phi(\bar{U}_2; * R * T, * R * Q)$ is normalised and is stable in a reversal of an arbitrary kernel $W \in \operatorname{Ker}(\alpha, RQ)$ such that $\neg\operatorname{Comp}(\bar{U}_3, W)$ and $\neg\operatorname{Rel}(W, f_\alpha(W; RQ, FQ))$. By 2.4, $\bar{U}_3$ is compatible with some kernel $U_3 \in \operatorname{Ker}(\alpha, RQ)$, where $U_3 < U_1$. Then by 2.30, the kernel $V_3 \rightleftharpoons f_\alpha(U_3; RQ, FQ)$ is compatible with some occurrence $\bar{V}_3 \in \operatorname{Norm}(\alpha, FQ, q-38)$ that is stable in a reversal of an arbitrary kernel $W \in \operatorname{Ker}(\alpha, FQ)$ such that $\neg\operatorname{Comp}(W, V_3)$ and $\neg\operatorname{Rel}(W, f_\alpha(W; FQ, RQ))$. Moreover we have $V_3 < V_1$ by 2.9, and therefore $\bar{V}_3 < \bar{V}_1 \ll F_1 * E * F_2 Q$. Then by III.2.11, $\bar{V}_3$ is stable in a reversal of $F_1 * E * F_2 Q$.

Thus it follows from the assumption that (23) is violated for the kernel V that there is an unbounded number of kernels of rank α of the word FT situated to the left of V. Hence relation (23) holds.

2.32. *Suppose that $F \overset{\alpha}{\sim} R$ and $FQ \overset{\alpha}{\sim} RQ$. If $F_1 * E * F_2 \in \operatorname{Ker}(\alpha, F)$ and $\neg\operatorname{Rel}(F_1 * E * F_2, f_\alpha(F_1 * E * F_2; F, R))$, then the occurrence $F_1 * E * F_2 Q$ is compatible with some kernel $V \in \operatorname{Ker}(\alpha, FQ)$, where one of the occurrences $F_1 * E * F_2 Q$ and V is a start of the other.*

The symmetrical assertion for $QF \overset{\alpha}{\sim} QR$ can be proved in an analogous way.

Suppose that $f_\alpha(F_1 * E * F_2; F, R) = R_1 * D * R_2$. Since $l_\alpha(E) \geqslant q$, by II.6.12 we can find words E_1 and E_2 such that $E \overset{\circ}{=} E_1 E_2$ and $F_1 * E_1 * E_2 F_2 Q \in \operatorname{Norm}(\alpha, FQ, q-4)$. By 1.4, there exists a kernel $V \in \operatorname{Ker}(\alpha, FQ)$ which is compatible with $F_1 * E_1 * E_2 F_2 Q$. From $\neg\operatorname{Rel}(E, D)$ it follows by 2.31 that $\neg\operatorname{Rel}(V, f_\alpha(V; FQ, RQ))$, that is, V is active and hence $l_\alpha(V) \geqslant q$. From 1.14 and III.2.15 it follows that $F_1 * E_1 * E_2 F_2 Q$ is stable in a reversal of an arbitrary active kernel $W \in \operatorname{Ker}(\alpha, FQ)$ which is situated on its left and not compatible with it. Since $F_1 * E * F_2 \in \operatorname{Ker}(\alpha, F)$, then for the same reason, no normalised continuation of $F_1 * E_1 * E_2 F_2 Q$ to the left has these properties. Hence, one of the occurrences $F_1 * E_1 * E_2 F_2 Q$ and V is a start of the other.

2.33. *Suppose that $X \overset{\alpha}{\sim} Y$, $P * E * Q \in \operatorname{Reg}(\alpha, X)$ and $f_\alpha(P * E * Q, X, Y) = R * E * S$. Then $X \overset{\alpha}{\sim} REQ \overset{\alpha}{\sim} PES$,*

$$f_\alpha(P * E * Q;\, X,\, REQ) = R * E * Q \tag{26}$$

and

$$f_\alpha(P * E * Q; X, PES) = P * E * S\,.$$

If, in addition, the words X and Y both belong to one of the sets $\mathscr{K}_\alpha$, $\mathscr{L}_\alpha$ and $\mathscr{M}_\alpha$, then the words REQ and PES belong to that same set.

By 2.18(d), we can restrict ourselves to proving the assertion involving the word REQ.

For $\alpha = 0$, everything is clear. For $\alpha > 0$, we proceed by a second induction on the parameter $I_\alpha(X, Y)$. If $I_\alpha(X, Y) = 0$ then, by 2.13, $X \overset{\alpha-1}{\sim} Y$ and by 2.18(b), $f_{\alpha-1}(P * E * Q; X, Y) = R * E * S$. Then by the first inductive assumption we have that $X \overset{\alpha-1}{\simeq} REQ$ and $f_{\alpha-1}(P * E * Q;\ X,\ REQ) = R * E * Q$. We need only now quote 2.11, 2.18(b) and 1.24.

Suppose that $I_\alpha(X, Y) > 0$, that is, for some kernel $V \in \mathrm{Ker}(\alpha, X)$ we have $\neg\mathrm{Rel}(V, f_\alpha(V;\ X,\ Y))$. Then by 2.27, the kernel V is not contained in $P * E * Q$ whence, by 2.14 and 1.3, it follows that either $V \ll P * E * Q$ or $P * E * Q \ll V$.

By 2.21, we can find a word $Z \in \mathscr{R}_\alpha$ such that $X \to Z$ is a local reversal of the kernel V and Z belongs to whichever set $\mathscr{K}_\alpha$, $\mathscr{L}_\alpha$ or $\mathscr{M}_\alpha$ it is that X and Y belong to. According to definition III.2.12, we have

$$f_\alpha(P * E * Q;\ X,\ Z) = P_1 * E * Q_1\,, \tag{27}$$

where if $V \ll P * E * Q$ we have $Q_1 \overset{\circ}{=} Q$, and if $P * E * Q \ll V$ we have $P_1 \overset{\circ}{=} P$.

Since $I_\alpha(Z, Y) = I_\alpha(X, Y) - 1$, by the second induction assumption we have $P_1EQ_1 \overset{\alpha}{\simeq} REQ_1 \overset{\alpha}{\simeq} RES$,

$$f_\alpha(P_1 * E * Q_1;\ Z,\ REQ_1) = R * E * Q_1 \tag{28}$$

and

$$f_\alpha(R * E * S;\ Y,\ REQ_1) = R * E * Q_1\,,$$

where REQ_1 belongs to the same set as X and Y.

If $Q_1 \overset{\circ}{=} Q$, then the required relation (26) follows from (27) and (28). We may therefore assume that $P_1 \overset{\circ}{=} P$ and that each kernel $U \in \mathrm{Ker}(\alpha, X)$ which is contained in $* P * EQ$ is related to the kernel $f_\alpha(U;\ X,\ Y)$. Then by 2.28, for each kernel $W \in \mathrm{Ker}(\alpha, Z)$ contained in $* P * EQ$, we have $\mathrm{Rel}(W, f_\alpha(W;\ Z,\ REQ_1))$. On the other hand, the same is true by 2.28 for each kernel $W \in \mathrm{Ker}(\alpha, Z)$ contained in $P * EQ_1 *$. Therefore $I_\alpha(Z, REQ_1) = 0$. Then by 2.13, $PEQ_1 \overset{\alpha-1}{\simeq} REQ_1$ and by 2.18(b), it follows from (28) that

$$f_{\alpha-1}(P * E * Q_1;\ Z,\ REQ_1) = R * E * Q_1\,. \tag{29}$$

It follows from (27) by 2.19 and III.2.1 that

$$\mathrm{MutNorm}_{\alpha-1}(P * E * Q, P * E * Q_1)\,.$$

Then by II.7.6 and (29), we can find a word X' such that $X \overset{\alpha-1}{\simeq} X'$ and

$$f_{\alpha-1}(P * E * Q;\ X,\ X') = R * E * Q'\,,$$

whence we obtain by the first induction assumption that $X \overset{\alpha-1}{\simeq} REQ$ and

$$f_{\alpha-1}(P * E * Q;\ X,\ REQ) = R * E * Q'\,.$$

By II.6.13 and 1.24, the word REQ belongs to that set $\mathscr{K}_\alpha$, $\mathscr{L}_\alpha$ or $\mathscr{M}_\alpha$ to which X, Y belong. It now remains for us to quote 2.11 and 2.18(b).

2.34. *If $X \overset{\alpha}{\sim} Y$, $P * E * Q \in \operatorname{Reg}(\alpha, X)$ and $f_\alpha(P * E * Q;\ X,\ Y) = R * D * S$ then words F and H can be found such that $X \overset{\alpha}{\sim} FDQ$, $X \overset{\alpha}{\sim} PDH$ and the occurrence $P * E * Q$ transforms locally in rank α from the left into $F * D * Q$ and locally from the right into $P * D * H$. Furthermore, the words FDQ and PDH can be chosen from that set $\mathscr{K}_\alpha$, $\mathscr{L}_\alpha$ or $\mathscr{M}_\alpha$ to which X and Y belong.*

By 2.26, we can find a word FDH such that $X \overset{\alpha}{\sim} FDH$ and the occurrence $P * E * Q$ transforms into $F * D * H$ locally in rank α from both left and right. It follows from this by 2.33 that $X \overset{\alpha}{\sim} FDS$ and $X \overset{\alpha}{\sim} RDH$, where $f_\alpha(P * E * Q;\ X,\ FDS) = F * D * S$ and $R * D * H = f_\alpha(P * E * Q;\ X,\ RDH)$.

Suppose that $PEQ_1 * T * Q_2 \in \operatorname{Reg}(\alpha, X)$. Then by 2.25, we have

$$f_\alpha(PEQ_1 * T * Q_2;\ X,\ FDH) = FDQ_3 * T * Q_2\,, \tag{30}$$

where $Q_3TQ_2 \overset{0}{=} H$. Since in addition

$$f_\alpha(F * D * H;\ FDH,\ RDH) = R * D * H\,,$$

we have by 2.27 that

$$f_\alpha(F * DQ_3T * Q_2;\ FDH,\ RDH) = R * DQ_3T * Q_2$$

and

$$f_\alpha(FDQ_3 * T * Q_2;\ FDH,\ RDH) = RDQ_3 * T * Q_2\,. \tag{31}$$

From (30) and (31) there follows

$$f_\alpha(PEQ_1 * T * Q_2;\ X,\ RDH) = RDQ_3 * T * Q_2\,.$$

Hence $P * E * Q$ transforms to $R * D * H$ locally from the right in rank α. It may be shown in an analogous way that it transforms to $F * D * S$ locally from the left.

2.35. *Suppose that $X, Y \in \mathscr{K}_\alpha$, $X \overset{\alpha}{\sim} Y$ and $R * E * Q \in \operatorname{Reg}(\alpha, X)$. If $f_\alpha(R * E * Q;\ X,\ Y) = F * E * H$, then $R \overset{\alpha}{\sim} F$ and $Q \overset{\alpha}{\sim} H$.*

For $\alpha = 0$ all is clear. For $\alpha > 0$ we proceed by a second induction on the parameter $I_\alpha(X, Y)$. If $I_\alpha(X, Y) = 0$, then $X \overset{\alpha-1}{\sim} Y$ and $f_{\alpha-1}(R * E * Q;\ X,\ Y) = F * E * H$, whence, by the first induction hypothesis, $R \overset{\alpha-1}{\sim} F$ and $Q \overset{\alpha-1}{\sim} H$. Then by 1.21 and 2.11, we obtain $R \overset{\alpha}{\sim} F$ and $Q \overset{\alpha}{\sim} H$.

Suppose that $I_\alpha(X, Y) > 0$, that is, for some kernel $V \in \operatorname{Ker}(\alpha, X)$ we have $\neg\operatorname{Rel}(V, f_\alpha(V;\ X,\ Y))$. Then by 2.27, V is not contained in $R * E * Q$ and by 1.3, $l_\alpha(V) \geqslant q$ and either $V \ll R * E * Q$ or $R * E * Q \ll V$.

Suppose that $V \ll R * E * Q$. The second case is analogous and we omit it. By 2.21 there exists a word $Z \in \mathscr{K}_\alpha$ such that $X \to Z$ is a local reversal of the kernel V, that is, $X \overset{\alpha}{\sim} Z$ and $f_\alpha(R * E * Q;\ X,\ Z) = R_1 * E * Q$ for some R_1. Then $Z \overset{\alpha}{\sim} Y$ and, by 2.18(e) we have $f_\alpha(R_1 * E * Q;\ Z,\ Y) = F * E * H$. Since $I_\alpha(Z, Y) = I_\alpha(X, Y) - 1$, then, by the second induction hypothesis, we have the relations

$$R_1 \overset{\alpha}{\sim} F \text{ and } Q \overset{\alpha}{\sim} H\,.$$

Hence it remains for us to prove that $R \overset{\alpha}{\sim} R_1$.

By *IV.3.16* and *IV.2.33*, there is a word $X' \in \mathscr{K}_{\alpha-1}$ such that $X \overset{\alpha-1}{\sim} X'$, $f_{\alpha-1}(R * E * Q; X, X') = R' * E * Q$, $R' \in \mathscr{K}_{\alpha-1}$ and the occurrence $f_{\alpha-1}(V; X, X')$ is periodised in rank $\alpha - 1$. Then $X' \in \mathscr{K}_\alpha$ by 1.24, and by III.3.27, the transition $X' \to Z$ is a real reversal of the occurrence $W \rightleftharpoons f_{\alpha-1}(V; X, X')$.

Since W is periodised in rank $\alpha - 1$, then according to II.4.5, we have $W = G * A^t A_1 * TEQ$, where $A \eqcirc A_1 A_2$ is an elementary period of rank α and $GA^t A_1 T \eqcirc R'$. By III.1.10,

$$R_1 EQ \overset{\alpha-1}{\sim} [G, A^{-n+t}, A_1 TEQ]_{\alpha-1}\,. \tag{32}$$

We consider the occurrence $U \rightleftharpoons G * A^{t-1} A_1 * A_2 A^3 A_1 T$. Since by 1.21 $R' \in \mathscr{R}_\alpha$, by II.6.12 we have $U \in \text{Norm}(\alpha, R', q - 4)$, whence by 1.18 and III.1.9 it follows that we can find a q_1-reversal $R' \to R''$ of U. Then by III.1.10, we have

$$R'' \overset{\alpha-1}{\sim} [G, A^{-n+t-4}, A^4 A_1 T]_{\alpha-1}\,. \tag{33}$$

Since $A_1 T, A^4, EQ \in \mathscr{R}_\alpha$ by 1.21, it follows from (32) and (33) by *V.1.8* and *V.1.11* that

$$R_1 \overset{\alpha-1}{\sim} [G, A^{-n+t}, A_1 T]_{\alpha-1} \overset{\alpha-1}{\sim} R''\,.$$

It then follows by 2.11 that $R'' \overset{\alpha}{\sim} R_1$, since $R_1 \in \mathscr{R}_\alpha$.

On the other hand, by the first induction assumption we get that $f_{\alpha-1}(R * E * Q; X, X') = R' * E * Q$ implies $R \overset{\alpha-1}{\sim} R'$, that is, by 2.11 we have $R \overset{\alpha}{\sim} R'$. Finally, by 1.27, we have $R' \overset{\alpha}{\sim} R''$. Therefore, $R \overset{\alpha}{\sim} R_1$. This is what we wanted to prove.

2.36. *If $X \overset{\alpha}{\sim} X^{-1}$, then X is the empty word.*

Suppose that $X \overset{\alpha}{\sim} X^{-1}$. If $\alpha = 0$, then since X is uncancellable, it is empty. For $\alpha > 0$ we proceed by a second induction on the parameter $I_\alpha(X, X^{-1})$. We assume that X is non-empty. Then by the first induction assumption we have $\neg(X \overset{\alpha-1}{\sim} X^{-1})$, whence 2.13 yields that $I_\alpha(X, X^{-1}) > 0$. If $I_\alpha(X, X^{-1}) = 1$, then $X \to X^{-1}$ is a real reversal of rank α. But by III.1.12, this is impossible. Hence, $I_\alpha(X, X^{-1}) \geqslant 2$. Suppose that V is the first left kernel of rank α of X for which we have $\neg\text{Rel}(V, f_\alpha(V; X, X^{-1}))$. Then by the principle of symmetry, V^{-1} is the first right kernel of X^{-1} for which $\neg\text{Rel}(V^{-1}, f_\alpha(V^{-1}; X^{-1}, X))$. Since $I_\alpha(X, X^{-1}) \geqslant 2$, the kernels V^{-1} and $f_\alpha(V; X, X^{-1})$ are distinct. By 2.21, we can find a word $Z \in \mathscr{R}_\alpha$ such that the transition $X \to Z$ is a real reversal of V. Then $Z \overset{\alpha}{\sim} X \overset{\alpha}{\sim} X^{-1}$ and $I_\alpha(Z, X^{-1}) = I_\alpha(X, X^{-1}) - 1$. By the principle of symmetry, the transition $X^{-1} \to Z^{-1}$ is a real reversal of V^{-1}, that is, $Z^{-1} \overset{\alpha}{\sim} X^{-1} \overset{\alpha}{\sim} Z$. Since $\neg\text{Rel}(V^{-1}; f_\alpha(V^{-1}; X^{-1}, Z))$, we have $I_\alpha(Z, Z^{-1}) = I_\alpha\, Z, X^{-1}) - 1 = I_\alpha(X, X^{-1}) - 2$. Since the word Z is non-empty, we have by the second induction assumption that $\neg(Z \overset{\alpha}{\sim} Z^{-1})$. The assumption that $X \overset{\alpha}{\sim} X^{-1}$ for some non-empty X thus leads to a contradiction.

2.37. *If A is an elementary period of rank α and $A^q \in \mathscr{K}_{\alpha-1}$, then*

$$A^{3q} \overset{\alpha}{\sim} A^{-n+3q}.$$

By I.4.3 and II.6.13, the words A^{3q} and A^{-n+3q} belong to $\mathscr{K}_{\alpha-1}$. By II.6.7 and 1.19, they belong to $\mathscr{K}_\alpha$. Hence it will suffice to prove that the transition

$$A^{3q} \to A^{-n+3q}$$

is a real reversal of rank α. But this follows easily from III.3.24, III.1.10 and III.3.27.

§ 3. Mutually Normalised Occurrences

3.1. *If U, V, W are regular occurrences of rank α, then*

$$\mathrm{MutNorm}_\alpha(U, U),$$
$$\mathrm{MutNorm}_\alpha(U, V) \Rightarrow \mathrm{MutNorm}_\alpha(V, U),$$
$$\mathrm{MutNorm}_\alpha(U, V) \,\&\, \mathrm{MutNorm}_\alpha(V, W) \Rightarrow \mathrm{MutNorm}_\alpha(U, W).$$

This follows directly from definition I.4.31.

3.2. *If $X \overset{\alpha}{\sim} Y$ and $V \in \mathrm{Reg}(\alpha, X)$, then* $\mathrm{MutNorm}_\alpha(V, f_\alpha(V; X, Y))$.
This follows from I.4.31, 2.18(d) and 2.18(e).

3.3. *If* $\mathrm{MutNorm}_\alpha(P * E * Q,\ R * E * S)$, *then* $\partial_\alpha(R * E * S) = \partial_\alpha(P * E * Q)$ *and, for an arbitrary occurrence V that is contained in $P * E * Q$, we have*

$$V \in \mathrm{Ker}(\alpha, PEQ) \Longleftrightarrow \phi(V;\ P * E * Q,\ R * E * S) \in \mathrm{Ker}(\alpha, RES), \quad (1)$$

$$V \in \mathrm{Act}(\alpha, PEQ) \Longleftrightarrow \phi(V;\ P * E * Q,\ R * E * S) \in \mathrm{Act}(\alpha, RES). \quad (2)$$

By 2.18(c) and I.4.31, it follows from $\mathrm{MutNorm}_\alpha(P * E * Q,\ R * E * S)$ that we can find a word Z such that $RES \overset{\alpha}{\sim} Z$,

$$f_\alpha(R * E * S;\ RES,\ Z) = R_1 * E * S_1 \quad (3)$$

and the relations obtained from (1) and (2) by replacing $R * E * S$ by $R_1 * E * S_1$ hold. Then by 2.27, 2.17, 2.18(a) and relation (3), we have that (1) and (2) hold. It follows from (1) that $\partial_\alpha(P * E * Q) = \partial_\alpha(R * E * S)$.

3.4. *If* $\mathrm{MutNorm}_\alpha(P * E * Q,\ F * D * H)$, *then* $\partial_\alpha(P * E * Q) = \partial_\alpha(F * D * H)$ *and a word Z can be found such that $FDH \overset{\alpha}{\sim} Z$, $f_\alpha(F * D * H;\ FDH, Z) = R * E * S$ and conditions* (1) *and* (2) *hold for an arbitrary occurrence V contained in $P * E * Q$.*
We find the desired word Z by means of 2.18(c) and I.4.31. Then by 3.2 and 3.1,

we have $\text{MutNorm}_\alpha(P * E * Q,\ R * E * S)$, whence by 3.3 and 2.18(j) it follows that (1) and (2) hold and

$$\partial_\alpha(F * D * H) = \partial_\alpha(R * E * S) = \partial_\alpha(P * E * Q)\,.$$

3.5. *Suppose that* $\text{MutNorm}_\alpha(P * E * Q,\ R * E * S)$, $PEQ \overset{\alpha}{\sim} X$, $RES \overset{\alpha}{\sim} Y$, $f_\alpha(P * E * Q;\ PEQ,\ X) = P_1 * D * Q_1$ *and* $f_\alpha(R * E * S;\ RES,\ Y) = R_1 * D * S_1$. *Then for an arbitrary occurrence* $V \in \text{Reg}(\alpha, PEQ)$ *which is contained in* $P * E * Q$, *we have*

$$f_\alpha(\phi(V;\ P * E * Q,\ R * E * S);\ RES,\ Y) = \phi(f_\alpha(V;\ PEQ,\ X);\ P_1 * D * Q_1,\ R_1 * D * S_1)\,. \tag{4}$$

Since $\text{MutNorm}_\alpha(P_1 * D * Q_1,\ R_1 * D * S_1)$ by 3.2 and 3.1, by 3.3 along with relation (1) we have that, if U is an arbitrary occurrence contained in $P_1 * D * Q_1$, then

$$U \in \text{Ker}(\alpha, X) \Longleftrightarrow \phi(U;\ P_1 * D * Q_1,\ R_1 * D * S_1) \in \text{Ker}(\alpha, Y)\,.$$

Then by 1.2 and 2.18(f), relation (4) holds for an arbitrary kernel $V \in \text{Ker}(\alpha, PEQ)$, and from this it easily follows that it also holds for an arbitrary occurrence $V \in \text{Reg}(\alpha, PEQ)$.

3.6. *If* $\text{MutNorm}_\alpha(P * E * Q,\ R * E * S)$, *then an arbitrary occurrence* V *contained in* $P * E * Q$ *satisfies the condition*

$$V \in \text{Reg}(\alpha, PEQ) \Rightarrow \text{MutNorm}_\alpha(V,\ \phi(V;\ P * E * Q,\ R * E * S))\,.$$

This follows from 3.3 and 3.5.

3.7. *Suppose that* $\text{MutNorm}_\alpha(P * E * Q,\ R * D * S)$. *If* V *is a start (or end) of* $P * E * Q$ *and* $V \in \text{Reg}(\alpha, PEQ)$, *then a start (end, respectively)* U *of* $R * D * S$ *can be found such that* $\text{MutNorm}_\alpha(U, V)$.

By 3.4 we can find a word Z such that $PEQ \overset{\alpha}{\sim} Z$ and $f_\alpha(P * E * Q;\ PEQ,\ Z) = F * D * H$. We set

$$U \rightleftharpoons \phi(f_\alpha(V;\ PEQ,\ Z);\ F * D * H,\ R * D * S)\,.$$

Since $\text{MutNorm}_\alpha(R * D * S,\ F * D * H)$, $\text{MutNorm}_\alpha(U, V)$ follows by 3.2, 3.6 and 3.1. By 2.18(i), U is a start (end, respectively) of the occurrence $R * D * S$.

3.8. *Suppose that* $\text{MutNorm}_\alpha(U, V)$, *where* $U \in \text{Ker}(\alpha, X)$ *and* $V \in \text{Ker}(\alpha, Y)$. *If* $\text{Rel}(U, V)$, *then* $\text{MutNorm}_{\alpha-1}(U, V)$ *and* $l_\alpha(U) = l_\alpha(V)$.

By 3.4, we can find a word Z such that $X \overset{\alpha}{\sim} Z$ and $\text{Bas}(f_\alpha(U;\ X,\ Z)) = \text{Bas}(V)$. Then $\text{Rel}(U, V)$ implies $\text{Rel}(U, f_\alpha(U;\ X,\ Z))$, whence by 2.12 we obtain $\text{MutNorm}_{\alpha-1}(U, f_\alpha(U;\ X,\ Z))$ and $l_\alpha(U) = l_\alpha(f_\alpha(U;\ X,\ Z)) = l_\alpha(V)$. Furthermore, by 1.1 and II.5.5, we have $\text{MutNorm}_{\alpha-1}(V, f_\alpha(U;\ X,\ Z))$. Therefore, $\text{MutNorm}_{\alpha-1}(U, V)$.

3.9. *Suppose that $P * E * Q$ and $R * E * S$ are regular occurrences of rank α. Then*

$$\text{MutNorm}_\alpha(P * E * Q,\ R * E * S) \Rightarrow \text{MutNorm}_{\alpha-1}(P * E * Q,\ R * E * S),$$

and, if these occurrences do not contain active kernels of rank α, then the converse implication holds:

$$\text{MutNorm}_{\alpha-1}(P * E * Q,\ R * E * S) \Rightarrow \text{MutNorm}_\alpha(P * E * Q,\ R * E * S).$$

Suppose that $\text{MutNorm}_\alpha(P * E * Q, R * E * S)$. We consider kernels U and V of rank α, where U is a start and V is an end of $P * E * Q$. By 3.6 and 3.8, we have $\text{MutNorm}_{\alpha-1}(U, \phi(U; P * E * Q, R * E * S))$ and $\text{MutNorm}_{\alpha-1}(V, \phi(P * E * Q, R * E * S))$ whence by 1.1 and II.7.3 there follows $\text{MutNorm}_{\alpha-1}(P * E * Q, R * E * S)$.

Suppose that $\text{MutNorm}_{\alpha-1}(P * E * Q, R * E * S)$ and that the occurrences $P * E * Q$ and $R * E * S$ do not contain active kernels of rank α. We shall prove the relation $\text{MutNorm}_\alpha(P * E * Q, R * E * S)$. Suppose that

$$PEQ \overset{\alpha}{\sim} X \quad \text{and} \quad f_\alpha(P * E * Q;\ PEQ,\ X) = F * D * H. \tag{5}$$

By 2.26 we may assume that $R * E * Q$ transforms locally into $F * D * H$. Then for an arbitrary active kernel $V \in \text{Ker}(\alpha, PEQ)$ which is not contained in $P * E * Q$, we have $\text{Rel}(V, f_\alpha(V; PEQ, X))$, whence by 2.14 and 2.13, $PEQ \overset{\alpha-1}{\sim} X$. By 2.18(b) it follows from (5) that

$$f_{\alpha-1}(P * E * Q;\ PEQ,\ X) = F * D * H.$$

Then since $\text{MutNorm}_{\alpha-1}(P * E * Q, R * E * S)$ a word Y can be found such that $RES \overset{\alpha-1}{\sim} Y$ and

$$f_{\alpha-1}(R * E * S;\ RES,\ Y) = R_1 * D * S_1,$$

whence

$$RES \overset{\alpha}{\sim} Y \text{ and } f_\alpha(R * E * S;\ RES,\ Y) = R_1 * D * S_1 \tag{6}$$

follows by 2.11 and 2.18(b).

If Y is an arbitrary word satisfying (6), we can find a word X which satisfies (5) in an analogous way.

Since by 2.17 $F * D * H$ and $R_1 * D * S_1$ do not contain active kernels of rank α, it remains only to prove that an arbitrary kernel V contained in $F * D * H$ satisfies the condition

$$V \in \text{Ker}(\alpha, X) \Longleftrightarrow \phi(V;\ F * D * H,\ R_1 * D * S_1) \in \text{Ker}(\alpha, Y). \tag{7}$$

By III.2.4, the occurrence $F * D * H$ is stable in a reversal of an arbitrary active kernel $W \in \mathrm{Ker}(\alpha, X)$. Suppose that $V \in \mathrm{Ker}(\alpha, X)$. We set $U \rightleftharpoons \phi(V; F*D*H, R_1*D*S_1)$. By 1.1, $V \in \mathrm{Norm}(\alpha, X, 9)$, and by III.2.3, no normalised continuation of V is contained in $F*D*H$. It follows from this by II.5.15 that $U \in \mathrm{Norm}(\alpha, Y, 9)$, and that no normalised continuation of U is contained in R_1*D*S_1. Since R_1*D*S_1 is stable in a reversal of an arbitrary active kernel $W \in \mathrm{Ker}(\alpha, Y)$, by III.2.3 it follows that U is stable in a reversal of an arbitrary active kernel $W \in \mathrm{Ker}(\alpha, Y)$. We assume that $U \notin \mathrm{Ker}(\alpha, Y)$. Then some normalised continuation $\bar{U}$ of U distinct from U is also stable in a reversal of an arbitrary active kernel $W \in \mathrm{Ker}(\alpha, Y)$. Hence $\bar{U}$ is not contained in R_1*D*S_1, that is, it intersects either $*R_1*DS_1$ or R_1D*S_1*. In the first case, by II.5.7 $\bar{U}$ will be compatible with an initial kernel $U_1 \in \mathrm{Ker}(\alpha, Y)$ of R_1*D*S_1, and their union $\bar{U}_1$, which is distinct from U_1, will be stable in a reversal of an arbitrary active kernel $W \in \mathrm{Ker}(\alpha, Y)$, by III.2.4. By 1.5, this contradicts the hypothesis that $U_1 \in \mathrm{Ker}(\alpha, Y)$. Hence $\bar{U}$ cannot intersect $*R_1*DS_1$. Similarly, $\bar{U}$ cannot intersect R_1D*S_1*. Therefore $U \in \mathrm{Ker}(\alpha\ Y)$.

We have proved that if V is an arbitrary occurrence contained in $F*D*H$, then $V \in \mathrm{Ker}(\alpha, X) \Rightarrow U \in \mathrm{Ker}(\alpha, Y)$. The converse implication for such occurrences V is proved in an analogous fashion. Hence (7) holds for such occurrences V. This proves what was required.

3.10. *If* $\mathrm{MutNorm}_\alpha(P*E*Q, R*D*S)$ *where one of the words* E *and* D *is a start (or end) of the other, then* $E \mathrel{\overset{\circ}{\equiv}} D$.

This follows from 3.4 and 2.27.

3.11. *Suppose that* $U \in \mathrm{Ker}(\alpha, X)$, $V \in \mathrm{Ker}(\alpha, Y)$ *and* $\mathrm{MutNorm}_\alpha(U, V)$. *Then the following conditions hold:*

$$U \in \mathrm{Act}(\alpha, X) \Longleftrightarrow V \in \mathrm{Act}(\alpha, Y),$$
$$\neg\, \mathrm{Rel}(U, V) \Rightarrow (\mathrm{Rel}(U^{-1}, V)\ \&\ U \in \mathrm{Act}(\alpha, X)). \tag{8}$$

By 3.4, a word Z can be found such that $X \overset{\alpha}{\sim} Z$, $\mathrm{Bas}(V) = \mathrm{Bas}(f_\alpha(U; X, Z))$ and

$$f_\alpha(U; X, Z) \in \mathrm{Act}(\alpha, Z) \Longleftrightarrow V \in \mathrm{Act}(\alpha, Y),$$

whence (8) follows by 2.17.

Suppose that $\neg\, \mathrm{Rel}(U, V)$. Then $\neg\, \mathrm{Rel}(U, f_\alpha(U; X, Z))$ and so, by 2.12, $\mathrm{Rel}(U^{-1}, f_\alpha(U; X, Z))$, that is, $\mathrm{Rel}(U^{-1}, V)$. Furthermore, by 2.14 we have $U \in \mathrm{Act}(\alpha, X)$.

3.12. *If* $X \in \mathscr{R}_\alpha$, *then a word* $Y \in \bar{\mathscr{M}}_\alpha$ *can be found such that* $X \overset{\alpha}{\sim} Y$.

By 2.20 we may assume that $X \in \mathscr{M}_\alpha$. For $\alpha = 0$ we have $\bar{\mathscr{M}}_\alpha = \mathscr{R}_\alpha$.

Suppose that $\alpha > 0$. By induction we can find a word $X_1 \in \bar{\mathscr{M}}_{\alpha-1}$ such that $X \overset{\alpha-1}{\sim} X_1$. Then by 2.11, $X \overset{\alpha}{\sim} X_1$ and $X_1 \in \mathscr{M}_\alpha$ by 1.24. We may therefore assume that $X \in \mathscr{M}_\alpha \cap \bar{\mathscr{M}}_{\alpha-1}$.

We list all the kernels of rank α of a word X in the order in which they are situated, from the left to the right. We denote by $\mu_\alpha(X)$ the maximal index i of a kernel

$V_i \in \text{Ker}(\alpha, X)$ for which there exists a $V_j \in \text{Ker}(\alpha, X)$ such that $i < j$ and V_i and V_j violate the relation

$$\text{MutNorm}_\alpha(V_i, V_j) \Rightarrow \text{Rel}(V_i, V_j). \tag{9}$$

If such an i does not exist, then we set $\mu_\alpha(X) = 0$. Since $X \in \mathcal{M}_\alpha \cap \bar{\mathcal{M}}_{\alpha-1}$, because $\mu_\alpha(X) = 0$ it follows by I.4.35 that $X \in \bar{\mathcal{M}}_\alpha$. We proceed by induction on $\mu_\alpha(X)$. Suppose that $\mu_\alpha(X) = r > 0$. Then for some $t > r$, condition (9) is violated for kernels V_r and V_t, that is, we have

$$\text{MutNorm}_\alpha(V_r, V_t) \;\&\; \neg\, \text{Rel}(V_r, V_t), \tag{10}$$

from which it follows by 3.11 that $V_r \in \text{Act}(\alpha, X)$ and

$$\text{Rel}(V_r^{-1}, V_t). \tag{11}$$

By 2.21, a word $X' \in \bar{\mathcal{M}}_{\alpha-1}$ can be found such that the transition $X \to X'$ is a local reversal of the kernel V_r. Then by 3.8 and III.2.6 we have $X' \in \mathcal{R}_\alpha$. Hence, $X \overset{\alpha}{\sim} X'$. We set

$$W_i \rightleftharpoons f_\alpha(V_i\,;\, X, X')$$

for $i = 1, 2, \ldots, \partial_\alpha(X)$. By 3.2, we have $\text{MutNorm}_\alpha(V_i, W_i)$, from which it follows by 3.1 that for arbitrary i, j,

$$\text{MutNorm}_\alpha(V_i, V_j) \Longleftrightarrow \text{MutNorm}_\alpha(W_i, W_j). \tag{12}$$

By III.1.5, we have $\text{Rel}(V_r^{-1}, W_r)$, whence by (11) it follows that $\text{Rel}(W_r, V_t)$. Since in addition $\text{MutNorm}_\alpha(W_r, V_t)$ by (10), we have $l_\alpha(W_r) = l_\alpha(V_t)$ by 3.8. Furthermore, by III.2.6 we have, for $i \neq r$,

$$\text{Rel}(W_i, V_i) \text{ and } l_\alpha(W_i) = l_\alpha(V_i). \tag{13}$$

Therefore $X' \in \mathcal{M}_\alpha$.

Since $r = \mu_\alpha(X)$, it follows from (12) and (13) that, for $i, j > r$,

$$\text{MutNorm}_\alpha(W_i, W_j) \Rightarrow \text{Rel}(W_i, W_j), \tag{14}$$

that is, $\mu_\alpha(X') \leqslant r$. Assume that $\mu_\alpha(X') = r$, that is, for some $k > r$ we have

$$\text{MutNorm}_\alpha(W_r, W_k) \;\&\; \neg\, \text{Rel}(W_r, W_k). \tag{15}$$

Since $\text{Rel}(W_r, V_t)$ and $\text{Rel}(W_t, V_t)$ we have $\text{Rel}(W_r, W_t)$. Then by (15), $\neg\, \text{Rel}(W_t, W_k)$, from which $\neg\, \text{Rel}(V_t, V_k)$ follows by (13). Since $\text{MutNorm}_\alpha(V_r, V_k)$ by (15) and (12), we have $\text{MutNorm}_\alpha(V_t, V_k)$ by (10). Therefore the kernels V_t and V_k violate (9), that is, $\mu_\alpha(X) > r$. Thus the assumption $\mu_\alpha(X') = r$ leads to a contradiction. Hence $\mu_\alpha(X') < r$. Then by the second induction assumption we can find a word $Y \in \bar{\mathcal{M}}_\alpha$ such that $X' \overset{\alpha}{\sim} Y$ that is, $X \overset{\alpha}{\sim} Y$.

3.13. *Suppose that* $X, Y \in \mathcal{R}_\alpha$, $P * E * Q \in \mathrm{Reg}(\alpha, X)$, $R * E * S \in \mathrm{Reg}(\alpha, Y)$ *and* $\mathrm{MutNorm}_\alpha(P * E * Q,\ R * E * S)$.

If $P \mathrel{\overline{\underline{\circ}}} R$ *and there are no active kernels of rank* α *to the left of* $P * E * Q$, *then for every* $\beta \leqslant \alpha$ *and every kernel* $V \in \mathrm{Ker}(\beta, X)$ *contained in* $* PE * Q$, *we have*

$$l_\beta(V) \geqslant \frac{n+1}{2} \Rightarrow \phi(V;\, * PE * Q,\ * PE * S) \in \mathrm{Ker}(\beta, Y)\,.$$

Similarly, if $Q \mathrel{\overline{\underline{\circ}}} S$ *and there are no active kernels of rank* α *to the right of* $P * E * Q$, *then for every* $\beta \leqslant \alpha$ *and every kernel* $V \in \mathrm{Ker}(\beta, X)$ *contained in* $P * EQ *$, *we have*

$$l_\beta(V) \geqslant \frac{n+1}{2} \Rightarrow \phi(V;\ P * EQ *,\, R * EQ *) \in \mathrm{Ker}(\beta, Y)\,.$$

Suppose that $P \mathrel{\overline{\underline{\circ}}} R$, there are no active kernels of rank α to the left of $P * E * Q$, $\beta \leqslant \alpha$, $V \in \mathrm{Ker}(\beta, X)$, $l_\beta(V) \geqslant (n+1)/2$ and that V is contained in $* PE * Q$. We set

$$U \rightleftharpoons \phi(V;\, * PE * Q,\, * PE * S)\,.$$

If $\beta < \alpha$, then by 1.1 we have $P * E * Q \in \mathrm{ComReg}(\beta, X)$ and by 3.9, $\mathrm{MutNorm}_\beta(P * E * Q,\, R * E * S)$, whence $\mathrm{MutNorm}_\beta(V, U)$ follows by II.7.4. Then $U \in \mathrm{Ker}(\beta, Y)$ by 3.3.

Suppose that $\beta = \alpha$. Since $V \in \mathrm{Act}(\alpha, X)$ by 1.20, by hypothesis V is not contained in $* P * EQ$. It is then contained in $P * E * Q$, by 1.3, from which it follows by 3.3 that $U \in \mathrm{Ker}(\alpha, Y)$.

3.14. *Suppose that* $X, Y \in \mathcal{R}_\alpha$, $P * E * Q \in \mathrm{Reg}(\alpha, X)$, $R * E * S \in \mathrm{Reg}(\alpha, Y)$, $\mathrm{MutNorm}_\alpha(P * E * Q,\, R * E * S)$ *and that* Y *belongs to one of the sets* $\mathcal{K}_\alpha$, $\mathcal{L}_\alpha$, $\mathcal{M}_\alpha$ *or* $\bar{\mathcal{M}}_\alpha$. *Then a word* Z *can be found in that same set such that* $X \overset{\alpha}{\sim} Z$ *and*

$$f_\alpha(P * E * Q;\, X, Z) = F * E * H\,.$$

Furthermore, if $P \mathrel{\overline{\underline{\circ}}} R$ *and if there are no active kernels of rank* α *in* $* P * EQ$, *then we can take* $F \mathrel{\overline{\underline{\circ}}} P$. *If* $Q \mathrel{\overline{\underline{\circ}}} S$ *and there are no active kernels of rank* α *in* $PE * Q *$, *we can take* $H \mathrel{\overline{\underline{\circ}}} Q$.

By 2.22, we can find words F and H such that

$$X \overset{\alpha}{\sim} FEQ \overset{\alpha}{\sim} PEH\,, \tag{16}$$

$$f_\alpha(P * E * Q;\ X,\ FEQ) = F * E * Q\,, \tag{17}$$

$$f_\alpha(P * E * Q;\ X,\ PEH) = P * E * H\,, \tag{18}$$

and, for every kernel V of rank β, $0 < \beta \leqslant \alpha$ contained in $* F * EQ$ or $PE * H *$, we have $l_\beta(V) \leqslant (n+1)/2$.

Suppose that $Y \in \mathscr{M}_\alpha$. We note first of all that if $P \mathrel{\overline{\underline{\circ}}} R$ and there are no active kernels of rank α in $* P * EQ$, then we can take REH for the required $Z \in \mathscr{M}_\alpha$. Indeed, by 3.13 it follows in this case from MutNorm_α $(P * E * H,\ R * E * S)$ that the inequality $l_\beta(V) \leqslant (n+1)/2$ is also satisfied for a kernel $V \in \text{Ker}(\beta, REH)$ contained in $* RE * H$. At the same time, each kernel $V \in \text{Ker}(\beta, REH)$ with $l_\beta(V) > (n+1)/2$ must, by 1.20 and 1.3, be contained in one of the occurrences $* RE * H$ and $RE * H *$.

In the case where $Q \mathrel{\overline{\underline{\circ}}} S$ and there are no active kernels of rank α in $PE * Q *$, one can prove analogously that $FEQ \in \mathscr{M}_\alpha$.

In the general case we set $Z \rightleftharpoons FEH$. It follows from (16), (17) and (18), in view of 2.23, that $FEQ \overset{\alpha}{\simeq} Z$ and

$$f_\alpha(P * E * Q;\ X, Z) = f_\alpha(F * E * Q;\ FEQ, Z) = F * E * H\,. \tag{19}$$

It follows by 3.3 and 3.9 from $\text{MutNorm}_\alpha(F * E * H,\ R * E * S)$, that, if V is an arbitrary occurrence contained in $F * E * H$ and $\beta \leqslant \alpha$, then

$$V \in \text{Ker}(\beta, Z) \Longleftrightarrow \phi(V;\ F * E * H,\ R * E * S) \in \text{Ker}(\beta, Y)\,. \tag{20}$$

We assume that for some kernel $U \in \text{Ker}(\beta, Z)$ with $\beta \leqslant \alpha$, we have $l_\beta(U) > (n+1)/2$. Since $Y \in \mathscr{M}_\alpha$, it follows by (20) that U is not contained in $F * E * H$. Assume that U is contained in $* F * EH$. If $\beta = \alpha$, it follows from (19) by 2.28 that

$$\phi(U;\ * F * EH,\ * F * EQ) \in \text{Ker}(\beta, FEQ)\,.$$

If $\beta < \alpha$, by 3.9 we have MutNorm_β $(F * E * H,\ F * E * Q)$, from which we get the same result by II.7.4. But by hypothesis every kernel $V \in \text{Ker}(\beta, FEQ)$, contained in $* F * EQ$ satisfies $l_\beta(V) \leqslant (n+1)/2$. Therefore $Z \in \mathscr{M}_\alpha$.

In the case where Y belongs to $\mathscr{K}_\alpha$ or $\mathscr{L}_\alpha$, the proof that Z belongs to the same set is similar.

It remains to consider the case where $Y \in \bar{\mathscr{M}}_\alpha$. In this case we may assume by what we have already proved that $X \in \mathscr{M}_\alpha$. We shall establish the existence of the required word Z analogously to 3.12.

If $\alpha > 0$, then by the induction assumption we can find a word $X_1 \in \bar{\mathscr{M}}_{\alpha-1}$ such that $X \overset{\alpha-1}{\simeq} X_1$ and

$$f_{\alpha-1}(P * E * Q;\ X, X_1) = F * E * H\,.$$

If $P \mathrel{\overline{\underline{\circ}}} R$, then for the occurrence $P_1 * DE * Q$, which begins with the first active kernel of rank $\alpha - 1$ of X, we have $\text{MutNorm}_{\alpha-1}(P_1 * DE * Q,\ P_1 * DE * S)$, by II.7.4. Therefore we may assume by 2.28 and the induction hypothesis that for $P \mathrel{\overline{\underline{\circ}}} R$ the word F coincides with the word P and, similarly, if $Q \mathrel{\overline{\underline{\circ}}} S$, the word H coincides with S.

By 2.11, $X \overset{\alpha-1}{\simeq} X_1$, by 1.24 $X_1 \in \mathcal{M}_\alpha$, and by 2.18(b), $f_\alpha(P * E * Q, X, X_1) = F * E * H$. Hence, taking X_1 in the place of X, we shall have $X \in \mathcal{M}_\alpha \cap \overline{\mathcal{M}}_{\alpha-1}$.

Listing all kernels of X of rank α in the order in which they appear from left to right, we choose as a paramenter for a second induction $\mu'_\alpha(X)$, the maximal index i of a kernel $V_i \in \operatorname{Ker}(\alpha, X)$, not contained in $P * E * Q$, for which there exists a kernel $V_j \in \operatorname{Ker}(\alpha, X)$ such that V_i and V_j violate (9) and either $i < j$ or V_j is contained in $P * E * Q$. Since $Y \in \overline{\mathcal{M}}_\alpha$, it follows by 3.6 that if $\mu'_\alpha(X)$ is not defined, then $X \in \overline{\mathcal{M}}_\alpha$. In the contrary case, on carrying out a local reversal of the kernel V_r according to 2.21, where $r = \mu'_\alpha(X)$, we obtain as in 3.12 a word $X' \in \mathcal{M}_\alpha \cap \overline{\mathcal{M}}_{\alpha-1}$, where $X \overset{\alpha}{\simeq} X'$ and $\mu'_\alpha(X') < \mu'_\alpha(X)$. Furthermore, since the reversal $X \to X'$ is local, we have $f_\alpha(P * E * Q; X, X') = P_1 * E * Q_1$, where $\operatorname{MutNorm}_\alpha(P_1 * E * Q_1, R * E * S)$ and either $P_1 \overset{\circ}{\equiv} P$ or $Q_1 \overset{\circ}{\equiv} Q$. The required word $Z \in \overline{\mathcal{M}}_\alpha$ is now found by the second induction assumption.

3.15. *Suppose that* $X, Z \in \mathcal{R}_\alpha$, $P * E * uDQ$, $PEu * D * Q \in \operatorname{Reg}(\alpha, X)$, $P' * E * S$, $R * D * Q' \in \operatorname{Reg}(\alpha, Z)$, $\operatorname{MutNorm}_\alpha(P * E * uDQ,\ P' * E * S)$ *and* $\operatorname{MutNorm}_\alpha(PEu * D * Q,\ R * D * Q')$. *If the word* Z *belongs to one of the sets* $\mathcal{K}_\alpha$, $\mathcal{L}_\alpha$, $\mathcal{M}_\alpha$ *or* $\overline{\mathcal{M}}_\alpha$ *then a word* Y *can be found in that same set such that* $X \overset{\alpha}{\simeq} Y$, $F * E * vDH = f_\alpha(P * E * uDQ; X, Y)$ *and* $FEv * D * H = f_\alpha(PEu * D * Q; X, Y)$.

If in addition $P \overset{\circ}{\equiv} P'$ *and there are no active kernels of rank* α *in* $* P * EuDQ$, *then we can take* $F \overset{\circ}{\equiv} P$. *If* $Q \overset{\circ}{\equiv} Q'$ *and there are no active kernels of rank* α *in* $PEuD * Q *$, *we can take* $H \overset{\circ}{\equiv} Q$.

The proof is similar to that of 3.14. First of all, by 2.23 a word v can be found such that $X \overset{\alpha}{\simeq} PEvDQ$,

$$f_\alpha(P * E * uDQ;\ X,\ PEvDQ) = P * E * vDQ\,,$$
$$f_\alpha(PEu * D * Q;\ X,\ PEvDQ) = PEv * D * Q\,,$$

and every kernel V of rank β, $0 < \beta \leqslant \alpha$, contained in $PE * v * DQ$ satisfies the inequality $l_\beta(V) \leqslant (n + 1)/2$. If $P \overset{\circ}{\equiv} P'$, $Q \overset{\circ}{\equiv} Q'$ and there are no active kernels of rank α in the occurrences $* P * EuDQ$ and $PEuD * Q *$ then we have by 1.20, 1.3 and 3.13 that

$$Z \in \mathcal{M}_\alpha \Rightarrow PEvDQ \in \mathcal{M}_\alpha\,,$$
$$Z \in \mathcal{L}_\alpha \Rightarrow PEvDQ \in \mathcal{L}_\alpha\,,$$
$$Z \in \mathcal{K}_\alpha \Rightarrow PEvDQ \in \mathcal{K}_\alpha\,.$$

In the contrary case, we find by 2.22 words F and H such that $PEvDQ \overset{\alpha}{\simeq} FEvDH$,

$$f_\alpha(P * EvD * Q;\ PEvDQ,\ FEvDH) = F * EvD * H\,,$$

and each kernel V of rank β, $0 < \beta \leqslant \alpha$, contained in the occurrences $* F * EvDH$ or $FEvD * H *$ satisfies the inequality $l_\beta(V) \leqslant (n + 1)/2$. Then, as we convinced ourselves in 3.14, since $Z \in \mathcal{M}_\alpha$ ($Z \in \mathcal{K}_\alpha$ or $Z \in \mathcal{L}_\alpha$) it follows that $FEvDH \in \mathcal{M}_\alpha$ ($FEvDH \in \mathcal{K}_\alpha$ or $FEvDH \in \mathcal{L}_\alpha$, respectively).

Since we have in addition $\operatorname{MutNorm}_\alpha(F * E * vDH, P' * E * S)$ and $\operatorname{MutNorm}_\alpha(FEv * D * H, R * D * Q')$, when $Z \in \bar{\mathscr{M}}_\alpha$ we can take the word $FEvDH$ instead of X or, in other words, we may assume that $X \in \mathscr{M}_\alpha$.

Furthermore, as in 3.14, we shall proceed by induction on α, and for $\alpha > 0$ by a second induction on the parameter $\mu'_\alpha(X)$, which is to be the maximal index i of a kernel $V_i \in \operatorname{Ker}(\alpha, X)$ not contained in the occurrences $P * E * uDQ$ and $PEu * D * Q$, and for which there exists a kernel $V_j \in \operatorname{Ker}(\alpha, X)$ such that V_i and V_j violate (9) and either $i < j$ or V_j is contained in one of the occurrences $P * E * uDQ$ and $PEu * D * Q$.

3.16. *Suppose that $X \in \mathscr{K}_\alpha$ and $P * E * Q \in \operatorname{Reg}(\alpha, X)$. Then we can find a word $Y \in \mathscr{K}_\alpha$ such that $X \overset{\alpha}{\sim} Y$, $f_\alpha(P * E * Q; X, Y) = F * E * H$, $F \in \mathscr{K}_\alpha$, $H \in \mathscr{K}_\alpha$ and every occurrence $W \in \operatorname{Reg}(\alpha, Y)$, contained in $* F * EH$ or $FE * H *$ is periodised in rank α.*

If $X \in \mathscr{M}_\alpha$, then we can choose the desired Y in $\mathscr{M}_\alpha$.

By 3.12, we find a word $Z \in \bar{\mathscr{M}}_\alpha$ such that $X \overset{\alpha}{\sim} Z$. Set $f_\alpha(P * E * Q; X, Z) = R * D * S$. Since $f_\alpha(R * D * S; Z, X) = P * E * Q$, by 2.26 we find a word $Y \in \mathscr{K}_\alpha$ ($Y \in \mathscr{M}_\alpha$, if $X \in \mathscr{M}_\alpha$) such that $Z \overset{\alpha}{\sim} Y$, $f_\alpha(R * D * S; Z, Y) = F * E * H$ and the occurrence $R * D * S$ transforms into $F * E * H$ locally in rank α. Then by 3.2, every occurrence $V \in \operatorname{Reg}(\alpha, Y)$ which is contained in either $* F * EH$ or $FE * H *$ is periodised in rank α. By 2.18(e), we have $f_\alpha(P * E * Q; X, Y) = F * E * H$. If $Y \in \mathscr{M}_\alpha$, then $F, H \in \mathscr{K}_\alpha$ by 1.21.

It remains to prove that if $X \in \mathscr{K}_\alpha$, then the words F and H can be chosen in $\mathscr{K}_\alpha$.

By 2.22, we can find an F' such that $Y \overset{\alpha}{\sim} F'EH$, $F' * E * H = f_\alpha(F * E * H; Y, F'EH)$ and, for each kernel V of rank β, $0 < \beta \leqslant \alpha$ contained in $* F' * EH$, we have $l_\beta(V) \leqslant (n+1)/2$. If there are no kernels of rank α in $* F * EH$, then it follows from 1.7 and 1.19 that $F' \in \mathscr{K}_\alpha$.

Suppose that $F_1 * G * uEH$ is a maximal regular occurrence of rank α contained in $* F * EH$. Then by 2.23, there exists a word v such that $F_1GvEH \in \mathscr{K}_\alpha$, $Y \overset{\alpha}{\sim} F_1GvEH$,

$$f_\alpha(F * E * H;\ Y,\ F_1GvEH) = F_1Gv * E * H\,,$$

$$f_\alpha(F_1 * G * uEH;\ Y,\ F_1GvEH) = F_1 * G * vEH\,,$$

and, for every kernel V of rank β, $0 < \beta \leqslant \alpha$, contained in $F_1G * v * EH$, we have $l_\beta(V) \leqslant (n+1)/2$. Then $f_\alpha(F_1 * G * vEH;\ F_1GvEH, Z) = F_1 * G * R_1DS$ by 2.25. In view of 3.2, the occurrence $F_1 * G * vEH$ is periodised in rank α and, by 3.13, $l_\beta(V) \leqslant (n+1)/2$ for every kernel V of rank β, $0 < \beta \leqslant \alpha$, contained in $* F_1Gv * EH$. Thus in this case we may take F_1Gv for the desired word $F' \in \mathscr{K}_\alpha$.

Similarly, there exists a word $H' \in \mathscr{K}_\alpha$ such that $F'EH' \in \mathscr{K}_\alpha$, $F'EH \overset{\alpha}{\sim} F'EH'$, $f_\alpha(F' * E * H;\ F'EH,\ F'EH') = F' * E * H'$, and the maximal regular occurrence of rank α contained in $F'E * H' *$ is periodised in rank α. Then we have $X \overset{\alpha}{\sim} F'EH'$ and $f_\alpha(P * E * Q;\ X,\ F'EH') = F' * E * H'$. This is what we wanted to prove.

3.17. *Suppose that $X, Y \in \mathscr{R}_\alpha$, $P * E * Q \in \operatorname{Reg}(\alpha, X)$, $R * D * S \in \operatorname{Reg}(\alpha, Y)$ and $\operatorname{MutNorm}_\alpha(P * E * Q, R * D * S)$.*

If $X \overset{\alpha}{\sim} X_1$ and $f_\alpha(P * E * Q; X, X_1) = P_1 * G * Q_1$, then there is a word Y_1 such that $Y \overset{\alpha}{\sim} Y_1$, $f_\alpha(R * D * S; Y, Y_1) = R_1 * G * S_1$, and the word Y_1 lies in the same set $\mathscr{K}_\alpha$, $\mathscr{L}_\alpha$, $\mathscr{M}_\alpha$ or $\bar{\mathscr{M}}_\alpha$ as contains X_1.

For given X_1, the desired word Y_1 can be chosen so that the occurrence $R * D * S$ transforms locally in rank α to $R_1 * G * S_1$ and Y_1 lies in the same set $\mathscr{K}_\alpha$, $\mathscr{L}_\alpha$ or $\mathscr{M}_\alpha$ as contains both X_1 and Y.

Suppose that $X \overset{\alpha}{\sim} X_1$ and $f_\alpha(P * E * Q; X, X_1) = P_1 * G * Q_1$. By definition I.4.31, there exists a word Y' such that $Y \overset{\alpha}{\sim} Y'$ and $f_\alpha(R * D * S; Y, Y') = R' * G * S'$. Then, by 3.2 and 3.1, we have $\text{MutNorm}_\alpha(P_1 * G * Q_1, R' * G * S')$. Thus the word Y_1 required in the first assertion exists in view of 3.14.

By 2.26, the second assertion follows from what has been proved already.

3.18. *Suppose that $X, Y \in \mathscr{R}_\alpha$, $P * E * Q \in \text{Reg}(\alpha, X)$, $R * D * S \in \text{Reg}(\alpha, Y)$. If $\text{MutNorm}_\alpha(P * E * Q, R * D * S)$, then there is a word Z such that $X \overset{\alpha}{\sim} Z$, $f_\alpha(P * E * Q; X, Z) = F * D * H$ and Z lies in the same set $\mathscr{K}_\alpha$, $\mathscr{L}_\alpha$, $\mathscr{M}_\alpha$ or $\bar{\mathscr{M}}_\alpha$ as Y. The desired word Z may be chosen so that $P * E * Q$ transforms to $F * D * H$ locally in rank α and Z lies in the same set $\mathscr{K}_\alpha$, $\mathscr{L}_\alpha$ or $\mathscr{M}_\alpha$ as contains both X and Y.*

This follows from 3.17 by virtue of the relations $Y \overset{\alpha}{\sim} Y$ and $R * D * S = f_\alpha(R * D * S; Y, Y)$.

3.19. *Suppose that $X \in \bar{\mathscr{M}}_\alpha$, $P * E * Q \in \text{Reg}(\alpha, X)$ and that $R * D * S \in \text{Reg}(\alpha, X)$. If $\text{MutNorm}_\alpha(P * E * Q, R * D * S)$, then $E \overset{\circ}{=} D$.*

If $\alpha = 0$, we have $E \overset{\circ}{=} D$ by I.4.1. Suppose that $\alpha > 0$. By 3.18, there exists a word $Z \overset{\alpha}{\sim} X$ such that $f_\alpha(P * E * Q; X, Z) = F * D * H$ and $\text{Bas}(f_\alpha(V; X, Z)) \overset{\circ}{=} \text{Bas}(V)$ for every kernel $V \in \text{Ker}(\alpha, X)$ not intersecting $P * E * Q$.

Let us prove that $I_\alpha(X, Z) = 0$. Suppose that the kernel $V \in \text{Ker}(\alpha, X)$ is not related to $f_\alpha(V; X, Z)$. Then V intersects $P * E * Q$, whence it follows from 2.14 and 1.3 that V is contained in $P * E * Q$. Let $U \rightleftharpoons \phi(f_\alpha(V; X, Z); F * D * H, R * D * S)$. By 3.6, it follows from $\text{MutNorm}_\alpha(F * D * H, R * D * S)$ and $\text{MutNorm}_\alpha(V, f_\alpha(V; X, Z))$ that $\text{MutNorm}_\alpha(V, U)$. Then, from the definition of $\bar{\mathscr{M}}_\alpha$, we have $\text{Rel}(V, U)$ and $\text{Rel}(V, f_\alpha(V; X, Z))$. Thus $I_\alpha(X, Z) = 0$.

By 2.13, $X \overset{\alpha-1}{\sim} Z$, and $f_{\alpha-1}(P * E * Q; X, Z) = F * D * H$ by 2.18(b). Then $\text{MutNorm}_{\alpha-1}(P * E * Q, F * D * H)$. Since in addition we have $\text{MutNorm}_{\alpha-1}(F * D * H, R * D * S)$ because of 3.9, it follows that $\text{MutNorm}_{\alpha-1}(P * E * Q, R * D * S)$, whence it follows by the inductive hypothesis that $E \overset{\circ}{=} D$.

3.20. *Suppose that $X, Y \in \mathscr{R}_\alpha$, $P * E * Q \in \text{Reg}(\alpha, X)$, $R * E * S \in \text{Reg}(\alpha, Y)$, $\text{MutNorm}_\alpha(P * E * Q, R * E * S)$, $X \overset{\alpha}{\sim} Z$ and $F * D * H = f_\alpha(P * E * Q; X, Z)$.*

If $P \overset{\circ}{=} R$ and there are no active kernels of rank α in $* P * EQ$, then there is a word T such that

$$Y \overset{\alpha}{\sim} FDT \quad \text{and} \quad f_\alpha(P * E * S; Y, FDT) = F * D * T.$$

If $Q \overset{\circ}{=} S$ and there are no active kernels of rank α in $PE * Q *$, then there is a word T such that $Y \overset{\alpha}{\sim} TDH$ and $T * D * H = f_\alpha(R * E * Q; Y, TDH)$.

If Z lies in one of the sets $\mathscr{K}_\alpha$, $\mathscr{L}_\alpha$, $\mathscr{M}_\alpha$ or $\bar{\mathscr{M}}_\alpha$, then the words TDH and FDT may be chosen from that same set.

By the principle of symmetry, we may restrict ourselves to a proof of the existence of FDT.

Suppose that $P \overset{\circ}{=} R$ and that there are no active kernels of rank α in $* P * EQ$. For $\alpha = 0$ we have $D \overset{\circ}{=} E$ and $\mathcal{M}_0 = \mathcal{R}_0$. For $\alpha > 0$ we proceed by induction on the parameter $I_\alpha(X, Z)$. If $I_\alpha(X,Z) = 0$, by 2.13 and 2.18(b) we have $X \overset{\alpha-1}{\approx} Z$ and $f_{\alpha-1}(P * E * Q;\ X, Z) = F * D * H$, whence by 1.1 and II.7.6 it follows that

$$Y \overset{\alpha-1}{\approx} FDT_1 \quad \text{and} \quad f_{\alpha-1}(R * E * S;\ Y,\ FDT_1) = F * D * T_1$$

for some word T_1. Then, by 2.11 and 2.18(b), we have $Y \overset{\alpha}{\sim} FDT_1$ and $F * D * T_1 = f_\alpha(R * E * S;\ Y,\ FDT_1)$. Since $\mathrm{MutNorm}_\alpha(F * D * T_1, F * D * H)$, the desired word FDT exists by 3.14.

Suppose that $I_\alpha(X, Z) > 0$, that is, we have $\neg \mathrm{Rel}\,(V, f_\alpha(V;\ X, Z))$ for some kernel $V \in \mathrm{Ker}\,(\alpha, X)$. By 2.21, there exists a word $X_1 \in \mathcal{R}_\alpha$ such that $X_1 \overset{\alpha}{\sim} Z$, $I_\alpha(X_1, Z) = I_\alpha(X, Z) - 1$ and the transition $X \to X_1$ is a local real reversal of the kernel V.

If V is contained in $PE * Q *$, then $f_\alpha(P * E * Q;\ X, X_1) = P * E * Q_1$ for some Q_1. Then we have $f_\alpha(P * E * Q;\ X_1,\ Z) = F * D * H$, $\mathrm{MutNorm}_\alpha(P * E * Q_1, P * E * S)$, and the desired word FDT exists by induction.

Suppose that V is not contained in $PE * Q *$. Then by 1.3, it must be contained in $P * E * Q$. Set

$$f_\alpha(P * E * Q;\ X,\ X_1) = P' * E' * Q' \,. \tag{21}$$

Since $X_1 \overset{\alpha}{\sim} Z$, $f_\alpha(P' * E' * Q';\ X_1, Z) = F * D * H$ and $I_\alpha(X_1, Z) = I_\alpha(X, Z) - 1$, then in view of the inductive hypothesis it is enough for us to find a Y_1 such that $Y \overset{\alpha}{\sim} Y_1$ and

$$f_\alpha(P * E * S;\ Y,\ Y_1) = P' * E' * S' \tag{22}$$

for some S'.

By 3.3, the occurrence

$$V' \rightleftharpoons \phi(V;\ P * E * Q,\ P * E * S)$$

is an active kernel of rank α of the word Y.

Assume that V is not an end of the occurrence $P * E * Q$. Then $V = Pu * A * vBQ$, where $PuAv * B * Q \in \mathrm{Ker}\,(\alpha, X)$, and $PuAv * B * S \in \mathrm{Ker}\,(\alpha, Y)$ by 3.3. Since the reversal $X \to X_1$ is local, we have $X_1 \overset{\circ}{=} P_1u_1A_1v_1BQ$, where

$$f_\alpha(PuAv * B * Q;\ X,\ X_1) = P_1u_1A_1v_1 * B * Q \,. \tag{23}$$

We denote the word $P_1u_1A_1v_1BS$ by Y_1. By III.2.17, the transition $Y \to Y_1$ is a local reversal of the kernel V', and

$$f_\alpha(PuAv * B * S;\ Y,\ Y_1) = P_1u_1A_1v_1 * B * S \,. \tag{24}$$

By 2.18(i) it follows from (24) that

$$f_\alpha(P * E * S;\ Y,\ Y_1) = P'' * E'' * S$$

where $P''E'' \overset{\circ}{=} P'E' \overset{\circ}{=} P_1u_1A_1v_1B$. Consequently one of the words E' and E'' is an end of the other. Since $\mathrm{MutNorm}_\alpha\,(P * E * Q,\ P * E * S)$ implies that $\mathrm{MutNorm}_\alpha(P' * E' * Q,\ P'' * E'' * S)$ we have by 3.10 that $E'' \overset{\circ}{=} E'$ and $P'' \overset{\circ}{=} P'$, that is, the desired relation (22) holds.

It remains to consider the case where V is an end of $P * E * Q$, that is $V = PG * A * Q$, where $GA \overset{\circ}{=} E$. Suppose that $X_1 \overset{\circ}{=} P_1A_1Q_1$, where $P_1 * A_1 * Q_1$ is a maximal image of the kernel V in the reversal $X \to X_1$. By III.2.16, there is a local reversal $Y \to P_1A_2S_1$ of the kernel V' such that $P_1 * A_2 * S_1$ is a maximal image of V' and one of the words A_1 and A_2 is a start of the other.

Let

$$f_\alpha(P * E * S;\ Y,\ P_1A_2S_1) = P'' * E'' * S'\,.$$

If the word G is empty, then $V' = P * E * S$, and by assumption there are no active kernels of rank α to the left of $P'' * E'' * S'$, whence by 1.6 it follows that $P'' * E'' * S'$ is a start of the occurrence $P_1 * A_2 * S_1$. In an analogous way it follows from (21) and the fact that $V = P * E * Q$ that $P' * E' * Q'$ is a start of the occurrence $P_1 * A_1 * Q_1$. Thus $P'' \overset{\circ}{=} P_1' \overset{\circ}{=} P'$, and one of the words E' and E'' is a start of the other. Then, by 3.10, it follows that $E'' \overset{\circ}{=} E'$ since $\mathrm{MutNorm}_\alpha(P' * E' * Q',\ P' * E'' * S')$.

If G is non-empty, then $G \overset{\circ}{=} Bu$, where $P * B * uAQ \in \mathrm{Ker}\,(\alpha\ X)$. By 3.3, $P * B * uAS \in \mathrm{Ker}\,(\alpha, Y)$. Since $X \to X_1$ is a local reversal, $X_1 \overset{\circ}{=} PBu_1A_1Q_1$, where $P' * E' * Q'$ is a start of the occurrence $P * Bu_1A_1 * Q_1$. For the same reason, $P'' * E'' * S'$ is a start of $P * Bu_1A_2 * Q_1$. Thus $P' \overset{\circ}{=} P''$, and one of the words E' and E'' is a start of the other, whence by 3.10 it follows that $E'' \overset{\circ}{=} E'$. As in previous cases, the desired word is found by using the inductive hypothesis.

3.21. *Suppose that* $X, Y \in \mathscr{R}_\alpha$, $P * E * Q \in \mathrm{Reg}(\alpha, X)$, $R * E * S \in \mathrm{Reg}(\alpha, Y)$ *and, for every occurrence* V *contained in* $P * E * Q$, *the conditions*

$$V \in \mathrm{Ker}(\alpha, X) \Longleftrightarrow \phi(V;\ P * E * Q,\ R * E * S) \in \mathrm{Ker}(\alpha, Y)\,, \tag{25}$$

$$V \in \mathrm{Ker}(\alpha, X) \Longrightarrow \mathrm{MutNorm}_\alpha(V, \phi(V; P * E * Q, R * E * S)) \tag{26}$$

hold. Then $\mathrm{MutNorm}_\alpha(P * E * Q,\ \ R * E * S)$.

For $\alpha = 0$, all is clear. Suppose that $\alpha > 0$, $X \overset{\alpha}{\simeq} X_1$ and

$$f_\alpha(P * E * Q;\ X,\ X_1) = P_1 * D * Q_1\,.$$

By I.4.31, we must establish the existence of a word Y_1 such that $Y \overset{\alpha}{\simeq} Y_1$, $f_\alpha(R * E * S;\ Y,\ Y_1) = R_1 * D * S_1$ and the conditions

$$V \in \mathrm{Ker}(\alpha, X_1) \Longleftrightarrow \phi(V;\ P_1 * D * Q_1,\ R_1 * D * S_1) \in \mathrm{Ker}(\alpha,\ Y_1), \tag{27}$$

$$V \in \mathrm{Ker}(\alpha\ X_1) \Rightarrow (V \in \mathrm{Act}(\alpha, X_1) \Longleftrightarrow$$
$$\phi(V; P_1 * D * Q_1, R_1 * D * S_1) \in \mathrm{Act}(\alpha, Y_1)) \tag{28}$$

hold for every occurrence V contained in $P_1 * D * Q_1$.

By 2.26, we may assume that $P * E * Q$ transforms to $P_1 * D * Q_1$ locally in rank α.

We prove the existence of the desired word Y_1 by induction on the parameter $I_\alpha(X, X_1)$.

If $I_\alpha(X, X_1) = 0$ then by 2.13 and 2.18(b), we have $X \overset{\alpha-1}{\approx} X_1$ and $f_{\alpha-1}(P * E * Q; X, X_1) = P_1 * D * Q_1$. Since by 3.9 and II.7.3 it follows from (26) that $\mathrm{MutNorm}_{\alpha-1}(P * E * Q, R * E * S)$, there exists a word Y_1 such that $Y \overset{\alpha-1}{\approx} Y_1$ and $f_{\alpha-1}(R * E * S; Y, Y_1) = R_1 * D * S_1$. In addition, by 3.5 we have

$$\phi(f_{\alpha-1}(V; X, X_1); P_1 * D * Q_1, R_1 * D * S_1)$$
$$= f_{\alpha-1}(\phi(V;\ P * E * Q,\ R * E * S);\ Y,\ Y_1) \tag{29}$$

for every occurrence $V \in \mathrm{Reg}(\alpha - 1, X)$ contained in $P * E * Q$. Then, by virtue of 2.11 and 2.18(b) we have $Y \overset{\alpha}{\sim} Y_1$ and

$$f_\alpha(R * E * S;\ Y,\ Y_1) = R_1 * D * S_1 .$$

Condition (27) follows from (25) and (29) in view of 2.18(a) and (b). By 3.3, the condition

$$V \in \mathrm{Ker}(\alpha, X) \Rightarrow$$
$$(V \in \mathrm{Act}(\alpha, X) \Longleftrightarrow \phi(V;\ P * E * Q,\ R * E * Q,\ R * E * S) \in \mathrm{Act}(\alpha, Y)) \tag{30}$$

follows from (26) for every occurrence V contained in $P * E * Q$. Condition (28) follows from (29) and (30) in view of 2.18(a) and III.3.27.

Suppose that $I_\alpha(X, X_1) > 0$, that is, we have $\neg\mathrm{Rel}(U, f_\alpha(U; X, X_1))$ for some kernel $U \in \mathrm{Ker}(\alpha, X)$. Since by assumption $P * E * Q$ transforms to $P_1 * D * Q_1$ locally in rank α, U is contained in $P * E * Q$. By 2.21, there is a word $Z \in \mathscr{R}_\alpha$ such that $X \to Z$ is a local reversal of the kernel U and $I_\alpha(Z, X_1) = I_\alpha(X, X_1) - 1$. The following three cases are possible:

I. U is a start of the occurrence $P * E * Q$,

II. U is an end of $P * E * Q$,

III. U is neither a start nor an end of $P * E * Q$.

Case I. By(26), we may assume that $U \neq P * E * Q$. Then we have $U = P * A * uBQ$, where $PAu * B * Q \in \mathrm{Ker}(\alpha, X)$. In such a case, the word X is of the form $Z \overset{\circ}{=} P'A'vBQ$, where $P' * A' * vBQ$ is a maximal image of the occurrence U in the reversal $X \to Z$. By 3.3 it follows from (26) that $R * A * uBS \in \mathrm{Act}(\alpha, Y)$. Then by III.2.15 there is a local reversal of the kernel $W \rightleftarrows R * A * uBS$,

$$Y \to R'A''vBS ,$$

where $R' * A'' * vBS$ is a maximal image of W and one of the words A' and A'' is an end of the other.

Suppose that

$$f_\alpha(U; X, Z) = P'u_1 * A_1 * v_1vBQ, \; Z_1 \rightleftharpoons R'A''vBs$$

and

$$f_\alpha(W; Y, Z_1) = R'u_2 * A_2 * v_2Bs\,.$$

Clearly, one of the words A_1v_1 and A_2v_2 is an end of the other. Therefore, by 1.6, III.3.31 and III.2.15, it follows from (30) that $v_1 \overline{\underline{\circ}}\, v_2$, and consequently that one of the words A_1, A_2 is an end of the other. Then by 3.2, 3.1 and 3.10, it follows from MutNorm$_\alpha(U, W)$ that $A_1 \overline{\underline{\circ}}\, A_2$. Thus we have

$$f_\alpha(P * E * Q; X, Z) = P'u_1 * A_1v_1vB * Q$$

and

$$f_\alpha(R * E * S; Y, Z_1) = R'u_2 * A_1v_1vB * S\,.$$

In addition, since the reversals $X \to Z$ and $Y \to Z_1$ are local, and in view of 3.2, it follows from (25) and (26) that the relations analogous to (25) and (26) are satisfied by every occurrence V contained in $P'u_1 * A_1v_1vB * Q$. Since $Z \overset{\alpha}{\sim} X_1$, $I_\alpha(Z, X_1) = I_\alpha(X, X_1) - 1$ and $f_\alpha(P'u_1 * A_1v_1vB * Q;\, Z, X_1) = P_1 * D * Q_1$, the inductive hypothesis yields the existence of a word Y_1 such that $Z_1 \overset{\alpha}{\sim} Y_1$, $f_\alpha(R'u_2 * A_1v_1vB * S;\, Z_1, Y_1) = R_1 * D * S_1$ and conditions (27) and (28) hold. Then $Y \overset{\alpha}{\sim} Y_1$ and $f_\alpha(R * E * S; Y, Y_1) = R_1 * D * S_1$, that is, Y_1 is the desired word.

Case II is analogous to Case I.

Case III. By 1.3, we have $U = PCu * A * vBQ$, where $P * C * uAvBQ$ and $PCuAv * B * Q$ are kernels of rank α of X. Then $Z \overline{\underline{\circ}}\, PCu'A'v'BQ$, where $PCu' * A' * v'BQ$ is a maximal image of the kernel U in the reversal $X \to Z$. By III.2.18, the transition

$$Y \to RCu'A'v'BS$$

is a local reversal of the kernel $W \rightleftharpoons RCu * A * vBS$, and $RCu' * A' * v'BS$ is a maximal image of W in this reversal. Set

$$f_\alpha(U; X, Z) = PCu'u_1 * A_1 * v_1v'BQ$$

and

$$f_\alpha(W; Y, Z_1) = RCu'u_2 * A_2 * v_2v'S\,,$$

where $u_1A_1v_1 \overline{\underline{\circ}}\, A' \overline{\underline{\circ}}\, u_2A_2v_2$. Exactly as the equality $v_1 \overline{\underline{\circ}}\, v_2$ was established in

Case I by 1.6 we find here the equation $v_1 \mathrel{\overline{\underline{\circ}}} v_2$ on the right and $u_1 \mathrel{\overline{\underline{\circ}}} u_2$ on the left. Thus $A_1 \mathrel{\overline{\underline{\circ}}} A_2$. Since

$$f_\alpha(P * E * Q;\ X,\ Z) = P * Cu'u_1A_1v_1v'B * Q$$

and

$$f_\alpha(R * E * S;\ Y,\ Z_1) = R * Cu'u_1A_1v_1v'B * S\,,$$

we may repeat the argument of Case I.

3.22. *Suppose that* $X, Y \in P_\alpha$, $P * E * Q \in \mathrm{Ker}(\alpha, X)$, $R * E * S \in \mathrm{Ker}(\alpha, Y)$. *If* $P \mathrel{\overline{\underline{\circ}}} R$ *and the condition*

$$V \in \mathrm{Act}(\alpha, X) \Longleftrightarrow \phi(V;\ * PE * Q,\ * PE * S) \in \mathrm{Act}(\alpha, Y) \tag{31}$$

is satisfied for every occurrence V *contained in* $* PE * Q$, *then every such occurrence also satisfies the condition*

$$V \in \mathrm{Ker}(\alpha, X) \Longleftrightarrow \phi(V;\ * PE * Q,\ * PE * S) \in \mathrm{Ker}(\alpha, Y)\,. \tag{32}$$

Analogously, if $Q \mathrel{\overline{\underline{\circ}}} S$ *and the condition*

$$V \in \mathrm{Act}(\alpha, X) \Longleftrightarrow \phi(V;\ P * EQ *,\ R * EQ *) \in \mathrm{Act}(\alpha, Y) \tag{33}$$

is satisfied for every occurrence V *contained in* $P * EQ *$, *then every such occurrence also satisfies the condition*

$$V \in \mathrm{Ker}(\alpha,\ Y) \Longleftrightarrow \phi(V;\ P * EQ *, R * EQ *) \in \mathrm{Ker}(\alpha,\ Y)\,.$$

Suppose that $P \mathrel{\overline{\underline{\circ}}} R$, and that the condition (31) is satisfied. Suppose that V is contained in $* PE * Q$ and $V \in \mathrm{Ker}(\alpha, X)$. If V is different from $P * E * Q$, then $V < P * E * Q$ and $\neg\ \mathrm{Comp}(V,\ P * E * Q)$. Let W be the maximal normalised continuation of the kernel V. We have $W < P * E * Q$, by virtue of II.5.7. Set

$$V' \rightleftharpoons \phi(V;\ * PE * Q,\ * PE * S)$$

and

$$W' \rightleftharpoons \phi(W;\ \ * PE * Q,\ * PE * S)\,.$$

By II.5.5, we have $\mathrm{MutNorm}_{\alpha-1}(P * E * Q,\ P * E * S)$, whence by II.7.4 and II.5.5 it follows that $V', W' \in \mathrm{Norm}(\alpha, Y, 9)$, and, by II.5.8 and II.7.4, W' is the maximal normalised continuation of the occurrence V'. Clearly, the nearest kernels of rank α to V (or V') on the left or right are contained in $* PE * Q$ (in $* PE * S$ respectively). Thus, by 1.6 and III.2.15, it follows that $V' \in \mathrm{Ker}(\alpha, Y)$ since $V \in \mathrm{Ker}(\alpha, X)$ and (31) holds.

3.23. *Suppose that* $X, Y \in \mathscr{R}_\alpha$, $P * E * Q \in \mathrm{Ker}(\alpha, X)$, $R * E * S \in \mathrm{Ker}(\alpha, Y)$ *and* $\mathrm{MutNorm}_\alpha(P * E * Q, R * E * S)$.

If $P \mathrel{\underline{\underline{\circ}}} R$ *and* (31) *is satisfied for every occurrence* V *contained in* $* PE * Q$, *then every such* V *satisfies also the condition*

$$V \in \mathrm{Reg}(\alpha, X) \Rightarrow \mathrm{MutNorm}_\alpha(V, \phi(V; * PE * Q, * PE * S)). \tag{34}$$

If $Q \mathrel{\underline{\underline{\circ}}} S$ *and relation* (33) *is satisfied for every occurrence* V *contained in* $P * EQ *$, *then every such* V *satisfies also the condition*

$$V \in \mathrm{Reg}(\alpha, X) \Rightarrow \mathrm{MutNorm}_\alpha(V, \phi(V; P * EQ *, R * EQ *)).$$

Suppose that $P \mathrel{\underline{\underline{\circ}}} R$ and that (31) holds. Then, by 3.22, (32) also holds. By 3.21, it suffices to prove the desired condition (34) for $V \in \mathrm{Ker}(\alpha, X)$.

Suppose that V is contained in $* PE * Q$, $V \in \mathrm{Ker}(\alpha, X)$ and $U \rightleftharpoons \phi(V; * PE * Q, * PE * S)$. By (32), we have $U \in \mathrm{Ker}(\alpha, Y)$.

Suppose that $X \overset{\alpha}{\sim} Z$ and $f_\alpha(V; X, Z) = F * D * H$. We must find a word Z_1 such that $Y \overset{\alpha}{\sim} Z_1$ and $f_\alpha(U; Y, Z_1) = F_1 * D * H_1$ for some F_1 and H_1. We may assume that V transforms to $F * D * H$ locally in rank α. Then $I_\alpha(X, Z) \leqslant 1$.

If $I_\alpha(X, Z) = 0$, then we have $X \overset{\alpha-1}{\sim} Z$ and $f_{\alpha-1}(V; X, Z) = F * D * H$ by 2.13 and 2.18(b). Since we have $\mathrm{MutNorm}_{\alpha-1}(V, U)$ by II.7.4, the desired word Z_1 exists in the present case in view of I.4.31, 2.11 and 2.18(b).

Suppose that $I_\alpha(X, Z) = 1$, that is, $\neg \mathrm{Rel}(V, f_\alpha(V; X, Z))$. Then, by 2.21, there exists a word $X_1 \in \mathscr{R}_\alpha$ such that $X \to X_1$ is a local real reversal of the kernel V and $I_\alpha(X_1, Z) = I_\alpha(X, Z) - 1 = 0$. In such a case V does not intersect $P * E * Q$, that is, $V = P_1 * A * uEQ$, where $P_1 A u \mathrel{\underline{\underline{\circ}}} P$. Then $X_1 \mathrel{\underline{\underline{\circ}}} P'A'vEQ$, where $P' * A' * vEQ$ is a maximal image of the kernel V in the reversal $X \to X_1$. Set $Y_1 \rightleftharpoons P'A'vES$. By (31), $U \in \mathrm{Act}(\alpha, Y)$ since $V \in \mathrm{Act}(\alpha, X)$. By III.2.17, the transition $Y \to Y_1$ is a local real reversal of the kernel U, and the occurrence $P' * A' * vES$ is a maximal image of the kernel U in this reversal. Thus we have

$$\mathrm{Comp}(f_\alpha(V; X, X_1), P' * A' * vEQ)$$

and

$$\mathrm{Comp}(f_\alpha(U; Y, Y_1), P' * A' * vES).$$

Suppose that

$$f_\alpha(V; X, X_1) = P'u_1 * A_1 * v_1 vEQ$$

and

$$f_\alpha(U; Y, Y_1) = P'u_2 * A_2 * v_2 vES. \tag{35}$$

If W is the kernel of rank α of the word X_1 that is closest to $f_\alpha(V; X, X_1)$ on the left (or on the right), it is contained in $* P'A'vE * Q$, and by (32) the occurrence $W_1 \rightleftharpoons \phi(W; * P'A'vE * Q, * P'A'vE * S)$ is the kernel of rank α of Y_1 that is closest to $f_\alpha(U; Y, Y_1)$ on the left (on the right respectively). The converse assertion is also true. In addition, by 2.17 it follows from (31) that

$$W \in \mathrm{Act}(\alpha, X_1) \Longleftrightarrow W_1 \in \mathrm{Act}(\alpha, Y_1).$$

Then we have $u_1 \overset{\alpha}{=} u_2$ and $v_1 \overset{\alpha}{=} v_2$ by 1.6 and III.2.15. Thus $A_1 \overset{\alpha}{=} A_2$. Since $Y, X_1 \in \mathcal{R}_\alpha$, it follows that $Y_1 \in \mathcal{R}_\alpha$. By II.5.5 we have

$$\mathrm{MutNorm}_{\alpha-1}(P'u_1 * A_1 * v_1vEQ, P'u_1 * A_1 * v_1vES).$$

Since in addition $f_\alpha(P'u_1 * A_1 * v_1vEQ;\ X_1, Z) = F * D * H$ and $I_\alpha(X_1, Z) = 0$ then, as was proved above, there is a word $Z_1 \in \mathcal{R}_\alpha$ such that $Y_1 \overset{\alpha}{\sim} Z_1$ and $f_\alpha(P'u_1 * A_1 * v_1vES;\ Y_1, Z_1) = F_1 * D * H_1$, whence by (35) it follows that $f_\alpha(U; Y, Z_1) = F_1 * D * H_1$.

In an analogous way, for every word Z_1 with $Y \overset{\alpha}{\sim} Z_1$, there exists a word Z such that $X \overset{\alpha}{\sim} Z$ and $\mathrm{Bas}(f_\alpha(V; X, Z)) \overset{\alpha}{=} \mathrm{Bas}(f_\alpha(U; Y, Z_1))$.

Since $f_\alpha(V; X, Z)$ and $f_\alpha(U; Y, Z_1)$ are active kernels of rank α, condition (c) of I.4.31 is also satisfied. Thus we have $\mathrm{MutNorm}_\alpha(V, U)$. This is what we wanted to prove.

3.24. *Suppose that* $X, Y \in \mathcal{R}_\alpha$,

$$P * A * ECQ,\ PAE * C * Q \in \mathrm{Norm}(\alpha, X, q) \cap \mathrm{Act}(\alpha, X)$$

and

$$R * A * ECS,\ RAE * C * S \in \mathrm{Norm}(\alpha, Y, q) \cap \mathrm{Act}(\alpha, Y).$$

Then, for every occurrence V contained in $PA * E * CQ$, the condition

$$V \in \mathrm{Reg}(\alpha, X) \Rightarrow \mathrm{MutNorm}_\alpha(V;\ \phi(V;\ PA * E * CQ,\ RA * E * CS))$$

is satisfied.

This assertion is true also in the case where the word RPA (or CQS) is empty.

By III.3.30 and III.2.15, the conditions

$$V \in \mathrm{Act}(\alpha, X) \Longleftrightarrow \phi(V; PA * E * CQ, RA * E * CS) \in \mathrm{Act}(\alpha, Y) \tag{36}$$

and

$$V \in \mathrm{Ker}(\alpha, X) \Longleftrightarrow \phi(V;\ PA * E * CQ,\ RA * E * CS) \in \mathrm{Ker}(\alpha, Y) \tag{37}$$

are satisfied for every occurrence V contained in $PA * E * CQ$.

By III.2.10, we may assume that the occurrence $P * A * ECQ$ (or $PAE * C * Q$) is stable in a reversal of the active kernel of rank α that is nearest to it on the right (or left respectively), and that it contains not less than $q - 17$ segments. Then by (36), (37) and III.2.15, the occurrences $R * A * ECS$ and $RAE * C * S$ will have the analogous property. Therefore, for kernels $V \in \mathrm{Ker}(\alpha, X)$ contained in $PA * E * CQ$, the relation $\mathrm{MutNorm}_\alpha(V, \phi(V; PA * E * CQ, RA * E * CS))$ is proved exactly as in 3.23. Here III.2.18 is to be used instead of III.2.17.

3.25. *Suppose that $X, Y \in \mathscr{R}_\alpha$, $P * E * Q \in \mathrm{Reg}(\alpha, X)$, $R * E * S \in \mathrm{Reg}(\alpha, Y)$, U is a start of $P * E * Q$, V is an end of $P * E * Q$, and that every kernel of rank α contained in $P * E * Q$ is contained in one of the occurrences U and V. If*

$$\mathrm{MutNorm}_\alpha(U,\ \phi(U;\ P * E * Q,\ R * E * S))$$

and

$$\mathrm{MutNorm}_\alpha(V,\ \phi(V;\ P * E * Q,\ R * E * S))\,,$$

then $\mathrm{MutNorm}_\alpha(P * E * Q, R * E * S)$.

This follows from 3.3, 3.6 and 3.21.

3.26. *Suppose that $P * A * uCvBQ$, $PAu * C * vBQ$ and $PAuCv * B * Q$ are regular occurrences of rank α in a word $X \in \mathscr{R}_\alpha$, set $Y \rightleftharpoons RAuCvBS$, where $Y \in \mathscr{R}_\alpha$, and suppose that*

$$\mathrm{MutNorm}_\alpha(P * AuCvB * Q,\ \ R * AuCvB * S)\,. \tag{38}$$

If $X \overset{\alpha}{\sim} PAu_1Dv_1BQ$, where the occurrence $PAu * C * vBQ$ transforms to $PAu_1 * D * v_1BQ$ locally in rank α, then $Y \overset{\alpha}{\sim} RAu_1Dv_1BS$, where the occurrence $RAu * C * vBS$ transforms to $RAu_1 * D * v_1BS$ locally in rank α.

We set $X_1 \rightleftharpoons PAu_1Dv_1BQ$. For $\alpha = 0$, everything is clear. For $\alpha > 0$ we proceed by induction on the parameter $I_\alpha(X, X_1)$. If $I_\alpha(X, X_1) = 0$, then $X \overset{\alpha-1}{\sim} X_1$, and we may use the inductive hypothesis.

Suppose that $I_\alpha(X, X_1) > 0$, that is, there is some kernel $V \in \mathrm{Ker}(\alpha, X)$ that is not related to $f_\alpha(V; X, X_1)$. Then V is contained in $PAu * C * vBQ$. Suppose that $X \to X'$ is a local reversal of the kernel V, where

$$f_\alpha(PAu * C * vBQ;\ X,\ X') = PAu' * C' * v'BQ\,. \tag{39}$$

Clearly, the maximal normalised continuation W of the kernel $f_\alpha(V; X, X')$ is contained in $PA * u'C'v' * BQ$. By III.2.18, with $Y' \rightleftharpoons RAu'C'v'BS$, the transition $Y \to Y'$ is a local reversal of the kernel $U \rightleftharpoons \phi(V; PAu * C * vBQ, RAu * C * vBS)$, and the occurrence $\phi(W; PA * u'C'v' * BQ, RA * u'C'v' * BS)$ is the maximal normalised continuation of the kernel $f_\alpha(U; Y, Y')$. By 3.3, it follows from (38) that $U \in \mathrm{Act}\,(\alpha, Y)$, that is, $Y \to Y'$ is a real reversal. Since $\mathrm{MutNorm}_\alpha(U, V)$, it follows by III.1.5 and 3.8 that $Y' \in \mathscr{R}_\alpha$, since $Y, X' \in \mathscr{R}_\alpha$. If the active kernel $V' \in \mathrm{Ker}(\alpha, X')$ that is nearest to $f_\alpha(V; X, X')$ on the left (or the right) is contained in $P * Au'C'v'B * Q$, then by 3.3 and III.3.31 the occurrence

$$\phi(V';\ P * Au'C'v'B * Q,\ R * Au'C'v'B * S)$$

is the active kernel of rank α of Y' that is nearest to $f_\alpha(U;\ Y,\ Y')$ on the left (on the right respectively). The converse assertion is also true. By III.2.15 and III.2.4, it follows from this that

$$f_\alpha(U;\ Y,\ Y') = \phi(f_\alpha(V;\ X,\ X');\ P * Au'C'v'B * Q,\ R * Au'C'v'B * S).$$

By (39) and the fact that both reversals are local, this means that

$$f_\alpha(RAu * C * vBS;\ Y,\ Y') = RAu' * C' * v'BS\,.$$

Finally, since $I_\alpha(X',\ X_1) = I_\alpha(X,\ X_1) - 1$, and since by 2.25 the occurrence $PAu' * C' * v'BQ$ transforms to $PAu_1 * D * v_1BQ$ locally in rank α, we get from the second inductive hypothesis that $Y' \overset{\alpha}{\sim} RAu_1Dv_1BS$, where $RAu' * C' * v'BS$ transforms to $RAu_1 * D * v_1BS$ locally in rank α. This was what we wanted to prove.

Chapter V. Coupling in Rank α

§1. Properties of the Operation of "Coupling"

We shall show in this section that the operation "coupling in rank α", defined for arbitrary pairs of words in $\mathscr{R}_\alpha$, is determined up to equivalence in rank α, and is associative.

1.1. *Suppose that $\alpha > 0$ and $X, Y \in \mathscr{R}_\alpha$.*
If $[X, Y]_{\alpha-1} = Z$ and $Z \in \mathscr{R}_\alpha$, then $[X, Y]_\alpha = Z$.
If $[X, Y]_0 = Z$ and $Z \in \mathscr{R}_\alpha$, then $[X, Y]_\alpha = Z$.
This follows from I.4.36 and IV.2.11.

1.2. We introduce the concept of local coupling in rank α for words $X, Y \in \mathscr{L}_\alpha$.

Take $X, Y \in \mathscr{L}_\alpha$. A word $Z \in \mathscr{K}_\alpha$ is said to be the result of *local coupling of the words X and Y in rank α* if there are words $X_1T \in \mathscr{K}_\alpha$ and $T^{-1}Y_1 \in \mathscr{K}_\alpha$ such that $X \overset{\alpha}{\sim} X_1T$, $Y \overset{\alpha}{\sim} T^{-1}Y_1$, $Z \overset{0}{\equiv} X_1Y_1$, and the following conditions are satisfied for these relations:

(a) If $P * E * Q \in \mathrm{Reg}(\alpha, X_1T)$ is an occurrence contained in $* X_1 * T$, then

$$f_\alpha(P * E * Q; X_1T, X) = P * E * Q_1$$

for some Q_1.

If $P * E * Q \in \mathrm{Reg}(\alpha, T^{-1}Y_1)$ is an occurrence contained in $T^{-1} * Y_1 *$, then

$$f_\alpha(P * E * Q, T^{-1}Y_1, Y) = P_1 * E * Q$$

for some P_1.

(b) If $V \in \mathrm{Norm}(\alpha, X_1T, 9)$ is an occurrence contained in $* X_1 * T$, then $l_\alpha(V) \leqslant (n+1)/2 + 55$. If $V \in \mathrm{Norm}(\alpha, T^{-1}Y_1, 9)$ is contained in $T^{-1} * Y_1 *$, then $l_\alpha(V) \leqslant (n+1)/2 + 55$.

(c) If $X_2 * D * T_1 \in \mathrm{Reg}(\alpha, X)$ and $T_1^{-1} * D^{-1} * Y_2 \in \mathrm{Reg}(\alpha, Y)$, then we have

$$f_\alpha(X_2 * D * T_1;\ X,\ X_1T) = X_3 * D * T_1\,,$$
$$f_\alpha(T_1^{-1} * D^{-1} * Y_2;\ Y,\ T^{-1}Y_1) = T_1^{-1} * D^{-1} * Y_3$$

for some X_3 and Y_3, and thus DT_1 is an end of the word T.

We shall denote local coupling of the words X and Y in rank α by

$$[X, Y]'_\alpha\,.$$

1.3. *If $X, Y \in \mathscr{L}_\alpha$, then there is a word $Z \in \mathscr{K}_\alpha$ such that*

$$Z = [X, Y]'_\alpha .$$

For $\alpha = 0$, all is clear. Suppose that $\alpha > 0$. By the inductive hypothesis there exist words $X'T \in \mathscr{K}_{\alpha-1}$ and $T^{-1}Y' \in \mathscr{K}_{\alpha-1}$ such that

$$X \overset{\alpha-1}{\simeq} X'T, \; Y \overset{\alpha-1}{\simeq} T^{-1}Y', \tag{1}$$

$X'Y' \in \mathscr{K}_{\alpha-1}$, and, for relation (1), the condition obtained by replacing α by $\alpha - 1$ in (a), (b) and (c) is also satisfied. Then, in view of IV.1.24, we have $X'T \in \mathscr{K}_\alpha$ and $T^{-1}Y \in \mathscr{K}_\alpha$, whence it follows by IV.2.11 that

$$X \overset{\alpha}{\simeq} X'T \text{ and } Y \overset{\alpha}{\simeq} T^{-1}Y'. \tag{2}$$

In addition, conditions (a) and (c) in rank α are satisfied in view of IV.1.1 and IV.2.18(b), and condition (b) in view of IV.1.18 and II.5.16. Thus, relations (2) comply with all the necessary conditions, apart from that requiring $X'Y' \in \mathscr{K}_\alpha$.

We choose words $X'T, T^{-1}Y' \in \mathscr{K}_\alpha$ satisfying conditions (2), $X'Y' \in \mathscr{K}_{\alpha-1}$ and conditions (a), (b), (c), and such that the parameter $\partial_\alpha(* X' * T)$ is minimal. We show that $X'Y' \in \mathscr{K}_\alpha$, that is, $X'Y' = [X, Y]'_\alpha$ for such a choice of $X'T$ and $T^{-1}Y'$.

Suppose that $X'Y' \notin \mathscr{K}_\alpha$. Then, by IV.1.19, there exists an occurrence

$$U \in \mathrm{MaxNorm}(\alpha, X'Y', n - 217).$$

Since $n - 217 > (n + 1)/2 + 59$, by condition (b) and II.6.18 we have $U = R * ED * G$, where $RE \overset{\circ}{=} X'$, $DG \overset{\circ}{=} Y'$,

$$l_\alpha(E) \leqslant \frac{n+1}{2} + 59 \text{ and } l_\alpha(D) \leqslant \frac{n+1}{2} + 59 .$$

By II.5.2 we have

$$l_\alpha(E) + l_\alpha(D) \geqslant l_\alpha(ED) - 2 \geqslant n - 219 \geqslant 446 . \tag{3}$$

Since $l_\alpha(E) \leqslant (n + 1)/2 + 59$, it follows from (3) that

$$l_\alpha(D) \geqslant \frac{n-1}{2} - 278 \geqslant 54 . \tag{4}$$

Clearly, without loss of generality we may assume that $l_\alpha(E) \geqslant l_\alpha(D)$. Then $l_\alpha(E) \geqslant 223$, by (3). In view of IV.1.20, the occurrence $R * E * T$ is compatible with some active kernel $U_1 \in \mathrm{Ker}(\alpha, X'T)$. By II.6.12, $E \overset{\circ}{=} E_0 v_0$, where $l_\alpha(E_0) \geqslant l_\alpha(E) - 4$ and the occurrence $R * E_0 * v_0 T$ is normalised and has no proper normalised continuation to the left. Since $R * ED * G$ is a normalised continuation of the occurrence $R * E_0 * v_0DG$ to the right and $T^{-1}DG \in \mathcal{K}_\alpha$, by II.3.10 the maximal normalised continuation of $R * E_0 * v_0T$ to the right is contained in $R * E * T$. Then the kernel U_1 is also contained in $R * E * T$, that is, we have

$$U_1 = R' * E' * T', \quad E \overset{\circ}{=} uE'v, \quad R' \overset{\circ}{=} Ru \quad \text{and} \quad T' \overset{\circ}{=} vT$$

for some R', E', T', u and v. By (a), some Q' satisfies the relation

$$f_\alpha(U_1;\ X'T,\ X) = R' * E' * Q'. \tag{5}$$

In view of IV.1.7, the inequalities

$$l_\alpha(uE') \leqslant l_\alpha(E') + 17, \quad l_\alpha(E'v) \leqslant l_\alpha(E') + 21$$

and

$$l_\alpha(E') \geqslant l_\alpha(E) - 38 \geqslant 185$$

are satisfied.

By II.5.2, the word E' can be represented in the form $E' \overset{\circ}{=} E_1E_2E_3$, where the occurrence $R'E_1 * E_2 * E_3vT$ is normalised, $l_\alpha(E_1) \geqslant 9$, $l_\alpha(E_3) \geqslant 9$ and $l_\alpha(E_2) = l_\alpha(E_1E_2) - 11 = l_\alpha(E') - 22$. Since the occurrences $R' * E' * Q'$ and $R' * E' * vY'$ are also normalised, by II.7.19 there exist words C_1, C_2, C_3 such that the relations

$$X \overset{\alpha-1}{\sim} R'C_1C_2C_3Q', \quad R'C_1C_2C_3Q' \in \mathcal{L}_{\alpha-1},$$
$$X'T \overset{\alpha-1}{\sim} R'C_1C_2C_3vT, \quad R'C_1C_2C_3vT \in \mathcal{K}_{\alpha-1}$$

and

$$X'Y' \overset{\alpha-1}{\sim} R'C_1C_2C_3vY', \quad R'C_1C_2C_3vY' \in \mathcal{K}_{\alpha-1},$$

are satisfied, where the occurrences $R'E_1 * E_2 * E_3Q'$, $R'E_1 * E_2 * E_3vT$ and $R'E_1 * E_2 * E_3vY'$ transform locally in rank $\alpha - 1$ to $R'C_1 * C_2 * C_3Q'$, $R'C_1 * C_2 * C_3vT$ and $R'C_1 * C_2 * C_3vY'$ respectively, which occurrences are periodised in rank $\alpha - 1$. Then, by IV.1.24 we have

$$R'C_1C_2C_3Q' \in \mathcal{L}_\alpha, \quad R'C_1C_2C_3vT \in \mathcal{K}_\alpha$$

and

$$R'C_1C_2C_3vY' \in \mathcal{K}_\alpha \Longleftrightarrow X'Y' \in \mathcal{K}_\alpha. \tag{6}$$

Taking $R'C_1C_2C_3Q'$ as X and $R'C_1C_2C_3vT$ instead of $X'T$, we get the analogous situation, in which it is required to prove the relation $R'C_1C_2C_3vY' \in \mathcal{K}_\alpha$. Therefore, and in view of (6), we may assume with a change of notation that the ocurrence $R'E_1 * E_2 * E_3vT$ is periodised in rank $\alpha - 1$, that is, $E_2 \overset{\circ}{=} A^tA_1$, where $A \overset{\circ}{=} A_1A_2$ is an elementary period of rank α and

$$t \geqslant l_\alpha(E_2) - 1 = l_\alpha(E') - 23 \geqslant 162 .$$

Then, in view of II.6.13, we will have the relations $A^t A_1 \in \mathscr{L}_{\alpha-1}$ and

$$R' E_1 A^t A_1 \in \mathscr{L}_{\alpha-1} .$$

Since $t \leqslant l_\alpha(E_2) + 1 = l_\alpha(E') - 21 \leqslant (n+1)/2 + 38$, it follows that

$$n - t + 20 \geqslant \frac{n-1}{2} - 18 \geqslant 314 . \tag{7}$$

Let F be the maximal start of the elementary word D satisfying the inequality

$$l_\alpha(F) \leqslant \frac{n-1}{2} - 131$$

and terminating with some pseudokernel of rank $\alpha - 1$. Then $D \mathrel{\underset{\circ}{=}} FD_1$, where $l_\alpha(F) \geqslant 54$ by (4), so that by II.5.2 we have

$$72 \leqslant l_\alpha(A^{10} A_1 E_3 vF) \leqslant l_\alpha(A^{10} A_1 E_3 v) + l_\alpha(F) + 2 \leqslant \frac{n-1}{2} - 86 . \tag{8}$$

According to II.5.4, we have

$$R' E_1 A^{t-10} * A^{10} A_1 E_3 vF * D_1 G \in \operatorname{Norm}(\alpha,\ X'Y',\ 72) .$$

By II.7.11, there exists a word N and a number

$$h \geqslant n - t + 20 - l_\alpha(A^{10} A_1 E_3 vF) - 2 , \tag{9}$$

such that

$$A^{10} A_1 E_3 vFNA^h \in \mathscr{K}_{\alpha-1} , \tag{10}$$

$$A^{n-t+20} \overset{\alpha-1}{\sim} A^{10} A_1 E_3 vFNA^h , \tag{11}$$

and, therefore,

$$A^{-n+t-20} \overset{\alpha-1}{\sim} A^{-h} N^{-1} F^{-1} v^{-1} E_3^{-1} A_1^{-1} A^{-10} . \tag{12}$$

From (7), (8) and (9) it follows that $h \geqslant 66$. Since $T^{-1} FD_1 G \in \mathscr{K}_{\alpha-1}$, it follows from (10) by II.6.13 that $T^{-1} FNA^h \in \mathscr{K}_{\alpha-1}$, whence we get from the principle of symmetry that

$$A^{-h} N^{-1} F^{-1} T \in \mathscr{K}_{\alpha-1} . \tag{13}$$

Since the occurrence $R * uE_1A'A_1 * E_3vT$ has no proper normalised continuation to the left and $l_\alpha(uE_1A'A_1) - l_\alpha(A'A_1) \leqslant 28$, by II.7.20 there exist words

$$R_1SA^{10} \in \mathscr{K}_{\alpha-1} \quad \text{and} \quad A^{-10}S^{-1}N_1A^{-r} \in \mathscr{K}_{\alpha-1}, \tag{14}$$

such that the conditions

$$RuE_1A^{10} \overset{\alpha-1}{\simeq} R_1SA^{10}, \tag{15}$$

$$A^{-h} \overset{\alpha-1}{\simeq} A^{-10}S^{-1}N_1A^{-r} \tag{16}$$

and

$$R_1N_1A^{-r} \in \mathscr{K}_{\alpha-1}$$

are satisfied, where $r \geqslant h - 46 \geqslant 20$. Then by II.6.13, it follows from (13) that

$$R_1N_1A^{-r}N^{-1}F^{-1}T \in \mathscr{K}_{\alpha-1}. \tag{17}$$

Suppose that $X'T \to X_1$ is a real reversal of the kernel U_1. Set $X_2 \rightleftharpoons R_1N_1A^{-r}N^{-1}F^{-1}T$. Using III.1.10, relations (15), (12), (16), (17) and the properties *V.1.7*, *V.1.8* and *V.1.1* of coupling in rank $\alpha - 1$, we find

$$\begin{aligned} X_1 &\overset{\alpha-1}{\simeq} [RuE_1A^{10}, A^{-n+t-20}, A^{10}A_1E_3vT]_{\alpha-1} \\ &\overset{\alpha-1}{\simeq} [R_1SA^{10}, A^{-h}, N^{-1}F^{-1}v^{-1}E_3^{-1}A_1^{-1}A^{-10}, A^{10}A_1E_3vT]_{\alpha-1} \\ &\overset{\alpha-1}{\simeq} [R_1N_1A^{-r}, N^{-1}F^{-1}T]_{\alpha-1} \overset{\alpha-1}{\simeq} R_1N_1A^{-r}N^{-1}F^{-1}T \overset{\circ}{=} X_2. \end{aligned}$$

Then $X'T \to X_2$ is also a real reversal of the kernel U_1. By IV.1.26, $X_2 \in \mathscr{K}_\alpha$. Consequently,

$$X'T \overset{\alpha}{\simeq} X_2.$$

Since $l_\alpha(F) \geqslant 54$ by assumption, then on shortening F^{-1} by not more than 4 segments on the left and on the right, we get $F^{-1} \overset{\circ}{=} u'F'v'$ by II.6.6, where the occurrences

$$W_1 \rightleftharpoons G^{-1}D_1^{-1}u' * F' * v'T \quad \text{and} \quad W_2 \rightleftharpoons R_1N_1A^{-r}N^{-1}u' * F' * v'T$$

are normalised, and $l_\alpha(F') \geqslant 46$. Then, by II.5.5, we will have the relation

$$\mathrm{MutNorm}_{\alpha-1}(W_1, W_2). \tag{18}$$

Clearly, the occurrence W_2 is compatible with the kernel $f_\alpha(U_1; X'T, X_2)$.

We consider the kernel $V_1 \in \mathrm{Ker}(\alpha, X'T)$ that is nearest to U_1 on the left. Suppose that $V_1 = P * H * Q$. Then, by condition (a), we have for some Q_1,

$$f_\alpha(V_1;\ X'T,\ X) = P * H * Q_1\,. \tag{19}$$

On the other hand, since V_1 is stable in the reversal $X'T \to X_2$ and the word $R_1 S\dot{A}^{10}$ is obtained because of II.7.20, where $RuE_1A^{10} \in \mathscr{L}_{\alpha-1}$, it follows that PH is a start of R_1, that is, we have $R_1 \,\overline{\underline{\circ}}\, PHR_2$ and the relation

$$\mathrm{MutNorm}_{\alpha-1}(V_1,\quad P * H * R_2N_1A^{-r}N^{-1}F^{-1}T) \tag{20}$$

is satisfied.

We set

$$V_2 \rightleftharpoons P * H * R_2N_1A^{-r}N^{-1}F^{-1}T\,.$$

Since (20) and (19) give $\mathrm{MutNorm}_{\alpha-1}(P * H * Q_1,\ V_2)$ and $X \in \mathscr{L}_{\alpha-1}$, by II.7.6 there exists a word $X_3 \in \mathscr{L}_{\alpha-1}$ such that $X_2 \overset{\alpha-1}{\backsim} X_3$ and

$$f_{\alpha-1}(V_2;\ X_2,\ X_3) = P * H * Q_2 \tag{21}$$

for some Q_2. Suppose that

$$f_{\alpha-1}(W_2;\ X_2,\ X_3) = PHQ_3 * F_1 * T_1\,, \tag{22}$$

where $Q_3F_1T_1 \,\overline{\underline{\circ}}\, Q_2$. By II.7.6, it follows from (18) and (22) that some word $Y_1 \in \mathscr{L}_{\alpha-1}$ satisfies the relation $G^{-1}D_1^{-1}F^{-1}T \overset{\alpha-1}{\backsim} Y_1$ and

$$f_{\alpha-1}(W_1;\ G^{-1}D_1^{-1}F^{-1}T,\ Y_1) = G_1 * F_1 * T_1\,,$$

whence it follows from the principle of symmetry that

$$T^{-1}Y' \overset{\alpha-1}{\backsim} Y_1^{-1} \text{ and } f_{\alpha-1}(W_1^{-1};\ T^{-1}Y', Y_1^{-1}) = T_1^{-1} * F_1^{-1} * G_1^{-1}\,. \tag{23}$$

By the inductive hypothesis, we can accomplish the local coupling $[X_3,\ Y_1^{-1}]'_{\alpha-1}$, that is, there exist words $X''T_2 \in \mathscr{K}_{\alpha-1}$ and $T_2^{-1}Y'' \in \mathscr{K}_{\alpha-1}$ such that

$$X_3 \overset{\alpha-1}{\backsim} X''T_2,\ Y_1^{-1} \overset{\alpha-1}{\backsim} T_2^{-1}Y''\,, \tag{24}$$

$X''Y'' \in \mathscr{K}_{\alpha-1}$, and for (24) the corresponding relations (a), (b) and (c) are satisfied. By IV.1.24, it follows that $X''T_2 \in \mathscr{K}_\alpha$ since $X_2 \in \mathscr{K}_\alpha$ and $T^{-1}Y' \in \mathscr{K}_{\alpha-1}$. By IV.2.11 we have

$$X \overset{\alpha}{\backsim} X''T_2 \text{ and } T_2^{-1}Y'' \overset{\alpha}{\backsim} Y\,. \tag{25}$$

We show that conditions (a), (b) and (c) are also satisfied for relation (25).

Indeed, from condition (c) for relations (24), it follows that F_1T_1 is an end of the word T_2. Therefore condition (c) for relations (25) is satisfied in view of the same condition for relations (2).

Since $T^{-1}Y' \overset{\alpha-1}{\backsim} T_2^{-1}Y''$, conditions (a) and (b) for the relation $Y \overset{\alpha}{\backsim} T_2^{-1}\ Y''$ are satisfied in view of IV.2.18(b), II.5.16 and these same conditions for the relation $Y \overset{\alpha}{\sim} T^{-1}Y'$.

It remains to check conditions (a) and (b) for the first of relations (25).

Since $l_\alpha(F_1) = l_\alpha(F') \geqslant 46$, and the kernel

$$U' \rightleftharpoons f_\alpha(U_1;\, X'T,\, X''T_2) \tag{26}$$

is compatible with $f_\alpha(W_2;\, X_2,\, X''T_2)$, U' is not contained in $* X'' * T_2$ because of IV.1.7. Thus condition (a) needs to be checked for kernels of rank α situated to the left of U'. Of these, the nearest to U' is the kernel

$$V' \rightleftharpoons f_{\alpha-1}(V_2;\, X_2,\, X''T_2)\,.$$

Since $f_\alpha(V';\, X''T_2,\, X) = P * H * Q_1$ by (19), condition (a) for kernels of rank α situated to the left of U' holds by IV.2.28, and condition (a) for the first conditions in (2) and (24).

It follows from IV.2.11 and condition (a) for the first relation in (2) that, just as for all kernels of rank α situated to the left of V', we have the relation

$$l_\alpha(V') = l_\alpha(f_\alpha(V';\, X''T_2,\, X'T)) = l_\alpha(f_\alpha(V';\, X''T_2,\, X))\,,$$

for V' itself, whence by IV.1.7 it follows since $X \in \mathscr{R}_\alpha$ that the inequality

$$l_\alpha(V) \leqslant \frac{n+1}{2} + 55 \tag{27}$$

holds for occurrences $V \in \mathrm{Norm}(\alpha,\, X''T_2,\, 9)$ compatible with it. Thus it remains for us to check inequality (27) for an arbitrary occurrence $V \in \mathrm{Norm}(\alpha,\, X''T_2,\, 9)$ contained in $* X'' * T_2$, situated to the right of V' and not compatible with it.

Let us suppose that V is such an occurrence and $l_\alpha(V) > (n+1)/2 + 55$. Then V is compatible with some active kernel of rank α. Since the kernel U' nearest to V' on the right intersects $X'' * T_2 *$, V must be compatible with U'. Let $\bar{V}$ be the maximal normalised continuation of V. Since V is contained in $* X'' * T_2$, by II.5.2 we have

$$l_\alpha(V) \leqslant l_\alpha(\bar{V}) - l_\alpha(F_1) + 1 \leqslant l_\alpha(\bar{V}) - l_\alpha(F) + 9\,. \tag{28}$$

Since the reversal $X'T \to X''T_2$ is a reversal of the kernel U_1 and $\bar{V}$ is an image of U_1 in this reversal, by III.1.8 we have $l_\alpha(\bar{V}) + l_\alpha(\bar{U}) \leqslant n + 1$, where $\bar{U}$ is the maximal normalised continuation of U_1. Then, since $l_\alpha(\bar{U}) \geqslant l_\alpha(E) - 4$, it follows from (28) that

$$l_\alpha(V) \leqslant n - l_\alpha(\bar{U}) - l_\alpha(F) + 10 \leqslant n - l_\alpha(E) - l_\alpha(F) + 14\,. \tag{29}$$

By assumption, F is the maximal start of the word D such that $l_\alpha(F) \leqslant (n-1)/2 - 131$. Thus either $F \overset{\circ}{=} D$ or $l_\alpha(F) = (n-1)/2 - 131$. In the first case, $l_\alpha(V) \leqslant 233$ because of (29) and (3), and in the second we have $l_\alpha(V) \leqslant (n+1)/2 - 78$ since $l_\alpha(E) \geqslant 223$ and (29) holds.

Thus we have proved that conditions (a), (b) and (c) are satisfied for relations (25). Since U' intersects $X'' * T_2 *$, $\partial_\alpha(*X'' * T_2) < \partial_\alpha(* X' * T)$. But this contradicts the assumption on the choice of the words $X'T$ and $T^{-1}Y'$. Thus $X'Y' \in \mathscr{K}_\alpha$, that is,*

$$X'Y' = [X, Y]_\alpha .$$

1.4. *If $X, Y \in \mathscr{R}_\alpha$, then there are words $X'T \in \mathscr{K}_\alpha$ and $T^{-1}Y' \in \mathscr{K}_\alpha$ such that $X \overset{\alpha}{\sim} X'T$, $Y \overset{\alpha}{\sim} T^{-1}Y'$ and $X'Y' \in \mathscr{K}_\alpha$, that is, $X'Y' = [X, Y]_\alpha$.*

This follows from 1.3 and IV.2.20.

1.5. *Suppose that $XT \overset{\alpha}{\sim} X'T_1$, $T^{-1}Y' \overset{\alpha}{\sim} T_1^{-1}Y'$ and $X'Y' \in \mathscr{R}_\alpha$.*

If $* X * T$ contains no more than $q - 6$ segments of some kernel $V_1 \in \mathrm{Ker}(\alpha, XT)$ that is not related to the kernel $f_\alpha(V_1; XT, X'T_1)$ and is compatible with some occurrence

$$XG * E * Q \in \mathrm{Norm}(\alpha, XT, 9) , \tag{30}$$

where the occurrence $Q^{-1} * E^{-1} * G^{-1}Y$ is normalised and is stable in a reversal of every active kernel of rank α that is not compatible with it and situated to its right, then $Q^{-1} * E^{-1} * G^{-1}Y$ is compatible with some kernel $W_1 \in \mathrm{Ker}(\alpha, T^{-1}Y)$ that is not related to the kernel $f_\alpha(W_1; T^{-1}Y, T_1^{-1}Y')$.

If $T^{-1} * Y *$ contains not more than $q - 6$ segments of some kernel $V_1 \in \mathrm{Ker}(\alpha, T^{-1}Y)$ that is not related to the kernel $f_\alpha(V_1; T^{-1}Y, T_1^{-1}Y')$ and is compatible with some occurrence

$$Q * E * GY \in \mathrm{Norm}(\alpha, T^{-1}Y, 9) ,$$

where the occurrence $XG^{-1} * E^{-1} * Q^{-1}$ is normalised and stable in a reversal of every active kernel of rank α that is not compatible with it and is situated to the left of it, then $XG^{-1} * E^{-1} * Q^{-1}$ is compatbile with some kernel $W_1 \in \mathrm{Ker}(\alpha, XT)$ that is not related to the kernel $f_\alpha(W_1; XT, X'T_1)$.

We prove these two assertions by simultaneous induction on the parameter

$$\eta \rightleftharpoons l_\alpha(XT, X'T_1) + l_\alpha(T^{-1}Y, T_1^{-1}Y'),$$

---

*Since F_1T_1 is an end of the word T_2, $X'' * T_2 *$ is contained in not less than $l_\alpha(F_1)$ segments of the maximal normalised continuation of the kernel $U' \in \mathrm{Ker}(\alpha, X''T_2)$, which is not related to the kernel $f_\alpha(U'; X''T_2, X)$. In addition, by relations (4), (22) and the definition of the word F (p. 210), and having regard to the fact that we shortened it (on p. 211) by not more than 8 segments, we have the relation

$$l_\alpha(F_1) = l_\alpha(F') \geqslant \frac{n-1}{2} - 286 .$$

This means that the local coupling in rank α that we defined satisfies the extra condition:

(d) If γ is the smallest number such that $X \overset{\gamma}{\sim} X_1T_1$, $\gamma > 0$, and the kernel $V \in \mathrm{Ker}(\gamma, X_1T)$ is not related to the kernel $f_1(V; X_1T, X)$, then not less than $(n-1)/2 - 286$ segments of its maximal normalised continuation are contained in $X_1 * T *$.

We do not include this property in definition 1.2, because we need it only in Chapter VII (p. 280).

and, for a given value η of the parameter, by induction on the parameter $\gamma(V_1) \rightleftharpoons \partial(Q)$. Since the assertions to be proved are symmetrical, we shall restrict attention to the first of them.

Suppose that $V_1 \in \operatorname{Ker}(\alpha, XT)$,

$$\neg \operatorname{Rel}(V_1, f_\alpha(V_1; XT, X'T_1)), \tag{31}$$

that the occurrence (30) is compatible with V_1, and that $Q^{-1} * E^{-1} * G^{-1}Y$ is normalised and stable in a reversal of every active kernel of rank α that is not compatible with it and situated to its right. From (31) it follows that $I_\alpha(XT, X'T_1) > 0$.

Without loss of generality, we may assume that the occurrence (30) is an end of the occurrence V_1, that is, $V_1 = P * DGE * Q$, where $PD \overset{\circ}{\equiv} X$ and $P * D * GEQ$ is a maximal start of the kernel V_1 contained in $* X * T$. Of course, the words D and G may be empty. We set

$$V_2 \rightleftharpoons PDG * E * Q \quad \text{and} \quad W_2 \rightleftharpoons Q^{-1} * E^{-1} * G^{-1}Y.$$

Since V_2 and W_2 are normalised by assumption, by II.5.5 we have

$$\operatorname{MutNorm}_{\alpha-1}(V_2, W_2^{-1}). \tag{32}$$

Suppose that $\eta = 1$. Then $I_\alpha(T^{-1}Y, T_1^{-1}Y') = 0$, whence it follows by IV.2.13 that

$$T^{-1}Y \overset{\alpha-1}{\approx} T_1^{-1}Y'. \tag{33}$$

By IV.3.12 we find an $X_1 \in \mathscr{M}_{\alpha-1}$ such that $XT \overset{\alpha-1}{\approx} X_1$. Suppose that

$$f_\alpha(V_2; XT, X_1) = P_1 * A'A_1 * Q_1, \tag{34}$$

where $A \overset{\circ}{\equiv} A_1A_2$ is an elementary period of rank α. By (32), (34) and II.7.6, there exists a word $Y_1 \in \mathscr{M}_{\alpha-1}$ such that $Y^{-1}T \overset{\alpha-1}{\approx} Y_1$ and

$$f_{\alpha-1}(W_2^{-1}; Y^{-1}T, Y_1) = R_1 * A'A_1 * Q_1,$$

whence it follows by the principle of symmetry that

$$T^{-1}Y \overset{\alpha-1}{\approx} Y_1^{-1} \quad \text{and} \quad f_{\alpha-1}(W_2; T^{-1}Y, Y_1^{-1}) = Q_1^{-1} * A_1^{-1}A^{-\prime} * R_1^{-1}. \tag{35}$$

Since $\eta = 1$, it follows from (31) that the transition

$$XT \to X'T_1 \tag{36}$$

is a real reversal of the kernel V_1. By III.1.3, the transition (36) is a reversal of the occurrence V_2, while by III.1.4 the transition $X_1 \to X'T_1$ is a reversal of the occurrence (34). Then, by III.1.10, we have the relation

$$X'T_1 \overset{\alpha-1}{\sim} [P_1A^{10}, A^{-n+t-20}, A^{10}A_1Q_1]_{\alpha-1}\,.$$

It follows from (33) and (35) that

$$T_1^{-1}Y' \overset{\alpha-1}{\sim} Y_1^{-1}\,.$$

Using the last two relations and the property of coupling of rank $\alpha - 1$ formulated in *V.1.1*, *V.1.10*, *V.1.7* and *V.1.8*, we get

$$\begin{aligned} X'Y' &= [X'\,T_1, T_1^{-1}Y']_{\alpha-1} \\ &\overset{\alpha-1}{\sim} [[P_1A^{10}, A^{-n+t-20}, A^{10}A_1Q_1]_{\alpha-1}, Q_1^{-1}A_1^{-1}A^{-t}R_1^{-1}]_{\alpha-1} \\ &\overset{\alpha-1}{\sim} [P_1A^{10}, A^{-n}, A^{-10}R_1^{-1}]_{\alpha-1}\,. \end{aligned} \tag{37}$$

Clearly, we may assume that V_2 is a maximal end of the kernel V_1 contained in $X * T *$, for which the corresponding occurrence W_2 is normalised and stable in a reversal of every kernel of rank α situated to the right and not compatible with it. In such a case, we have by II.6.18 and III.2.10,

$$l_\alpha(E^{-1}G^{-1}) \leqslant l_\alpha(E^{-1}) + 21\,. \tag{38}$$

If V is the maximal normalised continuation to the left of the occurrence V_2, then, by IV.1.7 and II.5.2, it follows from (38) that

$$l_\alpha(V) - l_\alpha(V_2) \leqslant l_\alpha(D) + l_\alpha(GE) + 19 - l_\alpha(E) \leqslant l_\alpha(D) + 40\,.$$

By II.7.20, there exist words $P_2F_1A^{10}$and $A^{-10}F_1^{-1}HA^{-r}$ such that

$$P_1A^{10} \overset{\alpha-1}{\sim} P_2F_1A^{10}, \qquad A^{-n} \overset{\alpha-1}{\sim} A^{-10}F_1^{-1}HA^{-r}, \qquad P_2HA^{-r} \in \mathscr{K}_{\alpha-1} \tag{39}$$

and

$$r \geqslant n - l_\alpha(V) - 18 + l_\alpha(V_2) \geqslant n - l_\alpha(D) - 58\,. \tag{40}$$

Then we have $P_2HA^{-r-10}R_1^{-1} \in \mathscr{R}_{\alpha-1}$ by II.6.13, and it follows from (37) and (39) that

$$X'Y' \overset{\alpha-1}{\sim} P_2HA^{-r-10}R_1^{-1}\,.$$

In addition, since $X'Y' \in \mathscr{R}_\alpha$ it follows by IV.1.24 that $P_2HA^{-r-10}R_1^{-1} \in \mathscr{R}_\alpha$. But this is impossible by IV.1.18 and (40), since $l_\alpha(D) \leqslant q - 6 = 84$ by hypothesis. Hence the case $\eta = 1$ is impossible.

Suppose that $I_\alpha(XT, X'T_1\,) > 1$. Then there exists a kernel $V \in \text{Ker}\,(\alpha, XT)$ such that $\neg$ Rel $(V, f_\alpha(V; XT, X'T_1))$. By IV.2.21, an $X_2 \in \mathscr{R}_\alpha$ can be found such that $XT \to X_2$ is a local reversal of the kernel V, and

$$I_\alpha(X_2, X'T_1) = I_\alpha(XT, X'T_1) - 1\,. \tag{41}$$

If $V \ll V_1$, then for some P' we have $f_\alpha(V;\, XT,\, X_2) = P' * DGE * Q$. Then, on replacing the word XT and the kernel V_1 by the word X_2 and the kernel $P' * DGE * Q$, we can use the induction assumption, by (41).

Suppose that $V_1 \ll V$, that is, $V = PDGEQ_2 * F * Q_3$, where $Q_2FQ_3 \mathrel{\underline{\circ}} Q$. Then for some R we have

$$f_{XT \to X_2}(V_1) = f_\alpha(V_1;\, XT,\, X_2) = P * DGE * R\,. \tag{42}$$

By III.2.4, V_2 is stable in the reversal $XT \to X_2$ and

$$f_{XT \to X_2}(V_2) = PDG * E * R\,.$$

Since $\gamma(V) = \partial(Q_3) < \gamma(V_1)$, by the induction assumption the occurrence $W \rightleftharpoons Q_3^{-1} * F^{-1} * Q_2^{-1}E^{-1}G^{-1}Y$ is compatible with some kernel $W' \in \mathrm{Ker}\,(\alpha,\, T^{-1}Y)$ which is not related to the kernel $f_\alpha(W';\, T^{-1}Y,\, T_1^{-1}Y')$. Then the occurrences W' and W are active. Hence $W^{-1} = Y^{-1}GEQ_2 * F * Q_3$ is active also. Since $W_2 \in \mathrm{ComReg}(\alpha - 1,\, Y^{-1}T)$, then by III.2.17, $W_2^{-1} = Y^{-1}G * E * Q_2FQ_3$ is stable in a reversal of W^{-1}, the transition

$$Y^{-1}T \to Y^{-1}GER \rightleftharpoons Y_2$$

is a local reversal of W^{-1} and hence

$$f_{Y^{-1}T \to Y_2}(W_2^{-1}) = Y^{-1}G * E * R\,.$$

Then by the principle of symmetry the transition

$$T^{-1}Y \to Y_2^{-1}$$

is a local reversal of W and

$$f_{T^{-1}Y \to Y_2^{-1}}(W_2) = R^{-1} * E^{-1} * G^{-1}Y\,. \tag{43}$$

Since $T^{-1}Y \to Y_2^{-1}$ is a real reversal of W' and $\neg\ \mathrm{Rel}\,(W', f_\alpha(W';\, T^{-1}Y,\, T_1^{-1}Y'))$, it follows that

$$I_\alpha(Y_2^{-1},\, T_1^{-1}Y') = I_\alpha(T^{-1}Y,\, T_1^{-1}Y') - 1\,. \tag{44}$$

We replace the words T, XT, $T^{-1}Y$ and the kernel V_1 by the words ER, X_2, Y_2^{-1} and the kernel $V' \rightleftharpoons f_\alpha(V_1;\, XT,\, X_2)$. Since $\neg\ \mathrm{Rel}\,(V', f_\alpha(V';\, X_2,\, X'\,T_1))$ by (31) and (42), we obtain by the induction assumption from (41) and (44) that the occurrence $W_3 \rightleftharpoons R^{-1} * E^{-1} * G^{-1}Y$ is compatible with some kernel $W_0 \in \mathrm{Ker}\,(\alpha,\, Y_2^{-1})$ that is not related to the kernel $f_\alpha(W_0;\, Y_2^{-1},\, T_1^{-1}Y')$. Then by (43) and III.2.6, the occurrence $W_2 = f_{Y_2^{-1} \to T^{-1}Y}(W_3)$ is compatible with $W_1 \rightleftharpoons f_{T_2^{-1} \to T^{-1}Y}(W_0)$, which is clearly not related to its image

$$f_\alpha(W_1;\, T^{-1}Y,\, T_1^{-1}Y') = f_\alpha(W_0;\, Y_2^{-1},\, T_1^{-1}Y')\,.$$

We need only now consider the case where $I_\alpha(XT, X'T_1) = 1$ and $\eta > 1$, that is, $I_\alpha(T^{-1}Y, T_1^{-1}Y') > 0$. In this case some kernel $W \in \text{Ker}(\alpha, T^{-1}Y)$ is not compatible with $f_\alpha(W; T^{-1}Y, T_1^{-1}Y')$. We assume that $\neg$ Comp (W, W_2). Then by II.5.7, we have either $W < W_2$ or $W_2 < W$.

If $W < W_2$, it follows by the induction assumption and from the fact that $\gamma(W) < \gamma(V_1)$ that there is a kernel $V \in \text{Ker}(\alpha, XT)$ situated to the right of V_1 which is not related to $f_\alpha(V; XT, X'T_1)$. Then $I_\alpha(XT, X'T_1) > 1$, that is, we are in a case that we have already considered.

Suppose that $W_2 < W$. Then by hypothesis, W_2 is stable in a reversal of W. On applying the local reversal $T^{-1}Y \to Y_3$ of W, we get the word $Y_3 \mathrel{\underline{\underline{\circ}}} Q^{-1}E^{-1}R_1$, where $f_{T^{-1}Y \to Y_3}(W_2) = Q^{-1} * E^{-1} * R_1$ and $I_\alpha(Y_3, T_1^{-1}Y') = I_\alpha(T^{-1}Y, T_1^{-1}Y') - 1$. Since by III.2.19 the occurrence $W_4 \rightleftharpoons Q^{-1} * E^{-1} * R_1$ is stable in a reversal of an arbitrary kernel of rank α situated on its right and with which it is not compatible, by the induction assumption there exists a kernel $W_5 \in \text{Ker}(\alpha, Y_3)$ such that Comp (W_4, W_5) and $\neg$ Rel $(W_5, f_\alpha(W_5; Y_3, T_1^{-1}Y'))$. Then, by III.2.6, W_2 is compatible with the kernel $W_1 \rightleftharpoons f_{Y \to T^{-1}Y}(W_5)$, which is not related to the kernel $f_\alpha(W_1; T^{-1}Y, T_1^{-1}Y')$.

1.6. *Suppose that $XT \overset{\alpha}{\sim} X'T_1$, $T^{-1}Y \overset{\alpha}{\sim} T_1^{-1}Y'$ and $X'Y' \in \mathscr{R}_\alpha$.*

If $* X * T$ contains no more than $q_1 + 20$ segments of a kernel $V \in$ Ker (α, XT) and $\neg$ Rel$(V, f_\alpha(V; XT, X'T_1))$, then $T \mathrel{\underline{\underline{\circ}}} GEQ$, where $XG * E * Q \in$ Norm $(\alpha, XT, 21)$, Comp $(V, XG * E * Q)$ and the occurrence $Q^{-1} * E^{-1} * G^{-1}Y$ is normalised and compatible with some kernel $W \in$ Ker $(\alpha, T^{-1}Y)$ that is not related to $f_\alpha(W; T^{-1}Y, T_1^{-1}Y')$.

If $T^{-1} * Y *$ contains no more than $q_1 + 20$ segments of a kernel $W \in$ Ker $(\alpha, T^{-1}Y)$ and $\neg$Rel $(W, f_\alpha(W; T^{-1}Y, T_1^{-1}Y'))$, then $T^{-1} \mathrel{\underline{\underline{\circ}}} QEG$, where $Q * E * GY \in$ Norm $(\alpha, T^{-1}Y, 21)$, Comp $(W, Q * E * GY)$ and the occurrence $XG^{-1} * E^{-1} * Q^{-1}$ is normalised and compatible with some kernel $V \in$ Ker (α, XT) that is not related to $f_\alpha(V; XT, X'T_1)$.

Clearly, we need only prove the first assertion. Suppose that $* X * T$ contains no more than $q_1 + 20$ segments of the kernel $V \in \text{Ker}(\alpha, XT)$, where $\neg$ Rel$(V, f_\alpha(V; XT, X'T_1))$. Let $\bar{V}$ be the maximal normalised continuation of V. Then by IV.2.12, we have $l_\alpha(\bar{V}) \geqslant q + 34$, and by IV.1.7, $* X * T$ contains no more than $q_1 + 37$ segments of $\bar{V}$, whence it follows by II.5.2 that if V_1 is a maximal end of $\bar{V}$ contained in $X * T *$, then $l_\alpha(V_1) \geqslant q - q_1 - 5 = 48$. Suppose that $V_1 = XG * DE * Q$, where $XGD * E * Q$ is the end of V_1 beginning with its 22nd left segment. Clearly

$$XGD * E * Q \in \text{Norm}(\alpha, XT, 27).$$

Moreover, it follows from II.6.12 and III.2.9 that the occurrence $Q^{-1} * E^{-1} * D^{-1}G^{-1}Y$ is normalised and stable in a reversal of an arbitrary active kernel of rank α that is situated to its right and is not compatible with it. Then by 1.5, there exists a kernel $W \in \text{Ker}(\alpha, T^{-1}Y)$ such that Comp $(W, Q^{-1} * E^{-1} * D^{-1}G^{-1}Y)$ and W is not related to $f_\alpha(W; T^{-1}Y, T_1^{-1}Y')$.

1.7. We can now prove the uniqueness of coupling in rank α:

If $XT \overset{\alpha}{\sim} X'T_1$, $T^{-1}Y \overset{\alpha}{\sim} T_1^{-1}Y'$, $XY \in \mathcal{R}_\alpha$ and $X'Y' \in \mathcal{R}_\alpha$, then $XY \overset{\alpha}{\sim} X'Y'$.

For $\alpha = 0$ this is clear. For $\alpha > 0$ the proof proceeds by induction on the parameter

$$\eta \rightleftharpoons I_\alpha(XT, X'T_1) + I_\alpha(T^{-1}Y, T_1^{-1}Y').$$

We note that, by 1.4, it is enough to consider the case where

$$X'T_1 \in \mathcal{K}_\alpha, \; T_1^{-1}Y' \in \mathcal{K}_\alpha \quad \text{and} \quad X'T' \in \mathcal{K}_\alpha. \tag{45}$$

Moreover, by 1.4 and 1.7 for rank $\alpha - 1$, we can restrict ourselves to the case where

$$XT \in \mathcal{K}_{\alpha-1}, \; T^{-1}Y \in \mathcal{K}_{\alpha-1} \quad \text{and} \quad XY \in \mathcal{K}_{\alpha-1}. \tag{46}$$

If $\eta = 0$, then by IV.2.13 we have $XT \overset{\alpha-1}{\sim} X'T_1$ and $T^{-1}Y \overset{\alpha-1}{\sim} T_1^{-1}Y'$ whence $XY \overset{\alpha-1}{\sim} X'Y'$ follows by the induction assumption, that is, $XY \overset{\alpha}{\sim} X'Y'$.

Suppose that $\eta > 0$. We assume that $* X * T$ contains no more than $q_1 + 20$ segments of some kernel $V \in \text{Ker}(\alpha, XT)$ that is not related to $f_\alpha(V; XT, X'T_1)$. Then by 1.6, the word T has the form $T \overset{\circ}{=} GEQ$, where

$$XG * E * Q \in \text{Norm}(\alpha, XT, 21), \; \text{Comp}(V, XG * E * Q) \tag{47}$$

and, for some kernel $W \in \text{Ker}(\alpha, T^{-1}Y)$, the relations

$$Q^{-1} * E^{-1} * G^{-1}Y \in \text{Norm}(\alpha, T^{-1}Y, 21), \quad \text{Comp}(W, Q^{-1} * E^{-1} * G^{-1}Y) \tag{48}$$

and

$$\neg \, \text{Rel}(W, f_\alpha(W; T^{-1}Y, T_1^{-1}Y')) \tag{49}$$

hold.

By II.5.2, the word E can be represented in the form $E \overset{\circ}{=} E_1 u E_2$ where the occurrences

$$V_1 \rightleftharpoons XG * E_1 * uE_2Q \quad \text{and} \quad V_2 \rightleftharpoons XGE_1u * E_2 * Q$$

are normalised, $l_\alpha(V_1) \geqslant 9$ and $l_\alpha(V_2) \geqslant 9$. By IV.3.12 for rank $\alpha - 1$, there exists an $X_1 \in \mathcal{M}_{\alpha-1}$ such that $XT \overset{\alpha-1}{\sim} X_1$. Suppose that $f_{\alpha-1}(V_2; XT, X_1) = R * D * S$. Then by IV.2.26 for rank $\alpha - 1$, there exists an $X_2 \in \mathcal{K}_{\alpha-1}$ such that $XT \overset{\alpha-1}{\sim} X_2$,

$$f_{\alpha-1}(V_1; XT, X_2) = XG * E_1 * vDQ_1$$

and

$$f_{\alpha-1}(V_2; XT, X_2) = XGE_1v * D * Q_1. \tag{50}$$

By II.4.5, we have $D \overset{\circ}{=} A'A_1$, where $A \overset{\circ}{=} A_1A_2$ is an elementary period of rank α.

We set $Y_1 \rightleftharpoons Y^{-1}GE_1vDQ_1$. Since $Y^{-1}T \in \mathscr{K}_{\alpha-1}$ and $X_2 \in \mathscr{K}_{\alpha-1}$, it follows by II.6.13 that $Y_1 \in \mathscr{K}_{\alpha-1}$. It then follows from *V.2.1* since $XT \overset{\alpha-1}{\sim} X_2$ that

$$Y^{-1}T \overset{\alpha-1}{\sim} Y_1. \tag{51}$$

We set

$$W_1 \rightleftharpoons Y^{-1}G * E_1 * uE_2Q \quad \text{and} \quad W_2 \rightleftharpoons Y^{-1}GE_1u * E_2 * Q\,.$$

By (48), we have $Y^{-1}G * E_1uE_2 * Q \in \operatorname{Norm}(\alpha,\ Y^{-1}T,\ 21)$. Hence $W_1 \in \operatorname{ComReg}(\alpha - 1,\ Y^{-1}T)$. It follows from (51) by II.7.7 that

$$f_{\alpha-1}(W_1;\ Y^{-1}T,\ Y_1) = Y^{-1}G * E_1 * vDQ_1\,.$$

Then by II.7.8 and (50), we have

$$f_{\alpha-1}(W_2;\ Y^{-1}T,\ Y_1) = Y^{-1}GE_1v * D * Q_1\,,$$

whence, by the principle of symmetry,

$$T^{-1}Y \overset{\alpha-1}{\sim} Y_1^{-1},\ f_{\alpha-1}(W_2^{-1};\ T^{-1}Y,\ Y_1^{-1}) = Q_1^{-1} * D^{-1} * v^{-1}E_1^{-1}G^{-1}Y\,. \tag{52}$$

By IV.2.21, there exist words $X_3 \in \mathscr{R}_\alpha$ and $Y_3 \in \mathscr{R}_\alpha$ such that the transitions

$$XT \to X_3 \quad \text{and} \quad T^{-1}Y \to Y_3 \tag{53}$$

are real reversals of the kernels V and W, where $I_\alpha(X_3,\ X'T_1) = I_\alpha(XT,\ X'T_1) - 1$ and $I_\alpha(Y_3,\ T_1^{-1}Y') = I_\alpha(T^{-1}Y,\ T_1^{-1}Y') - 1$. By 1.4 for rank $\alpha - 1$, there exist words $X''T_2 \in \mathscr{K}_{\alpha-1}$ and $T_2^{-1}Y'' \in \mathscr{K}_{\alpha-1}$ such that

$$X_3 \overset{\alpha-1}{\sim} X''T_2,\ Y_3 \overset{\alpha-1}{\sim} T_2^{-1}Y'' \tag{54}$$

and $X''Y'' \in \mathscr{K}_{\alpha-1}$, that is, $[X_3,\ Y_3]_{\alpha-1} = X''Y''$. By (47) and (48), the transitions (53) can be regarded as reversals of V_2 and W_2^{-1}. Then by III.1.4, (50) and (52), the transitions $X_2 \to X_3$ and $Y_1^{-1} \to Y_3$ are reversals of $XGE_1v * D * Q_1$ and $Q_1^{-1} * D^{-1} * v^{-1}E_1^{-1}G^{-1}Y$, respectively. Hence by III.1.10 we have

$$X_3 \overset{\alpha-1}{\sim} [XGE_1v,\ A^{-n+t},\ A_1Q_1]_{\alpha-1}$$

and

$$Y_3 \overset{\alpha-1}{\sim} [Q_1^{-1}A_1^{-1},\ A^{n-t},\ v^{-1}E_1^{-1}G^{-1}Y]_{\alpha-1}\,.$$

From this, using (54), II.6.13 and the properties of coupling of rank $\alpha - 1$ we obtain

$$X''Y'' \overset{\alpha-1}{\simeq} [X_3, Y_3]_{\alpha-1}$$
$$\overset{\alpha-1}{\simeq} [[XGE_1v, A^{-n+t}, A_1Q_1]_{\alpha-1}, [Q_1^{-1}A_1^{-1}, A^{n-t}, v^{-1}E_1^{-1}G^{-1}Y]_{\alpha-1}]_{\alpha-1}$$
$$\overset{\alpha-1}{\simeq} [XGE_1v, v^{-1}E_1^{-1}G^{-1}Y]_{\alpha-1} \overset{\alpha-1}{\simeq} XY.$$

Since $XY \in \mathcal{R}_\alpha$, it follows by IV.2.11 that $X''Y'' \in \mathcal{R}_\alpha$ and $XY \overset{\alpha}{\simeq} X''Y''$. Since $I_\alpha(X''T_2, X'T_1) + I_\alpha(T_2^{-1}Y'', T_1^{-1}Y') = \eta - 2$, we have by the induction hypothesis that $X''Y'' \overset{\alpha}{\simeq} X'Y'$. Hence $XY \overset{\alpha}{\simeq} X'Y'$.

We have thus dealt with the case where $* X * T$ contains no more than $q_1 + 20$ segments of some kernel $V \in \mathrm{Ker}(\alpha, XT)$, where $\neg\,\mathrm{Rel}(V, f_\alpha(V; XT, X'T_1))$. The case where $T^{-1} * Y *$ contains no more than $q_1 + 20$ segments of some kernel $W \in \mathrm{Ker}(\alpha, T^{-1}Y)$, where $\neg\,\mathrm{Rel}(W, f_\alpha(W; T^{-1}Y, T_1^{-1}Y'))$ is treated in an analogous way. As a result, we may assume in what follows that $* X * T$ contains not less than $q_1 + 21$ segments of each kernel $V \in \mathrm{Ker}(\alpha, XT)$, where $\neg\,\mathrm{Rel}(V, f_\alpha(V; XT, X'T_1))$ and $T^{-1} * Y *$ contains not less than $q_1 + 21$ segments of each $W \in \mathrm{Ker}(\alpha, T^{-1}Y)$ that is not related to $f_\alpha(W; T^{-1}Y, T_1^{-1}Y')$.

Suppose that $I_\alpha(XT, X'T_1) > 1$. Then there exist at least two kernels $V \in \mathrm{Ker}(\alpha, XT)$, which are not related to $f_\alpha(V; XT, X'T_1)$. Suppose that V_1 and V_2 are kernels of this sort such that $V_2 < V_1$. By hypothesis, each of them contains not less than $q_1 + 21$ segments of the occurrence $* X * T$. Since they are active kernels, $V_2 \ll V_1$.

Suppose that $X \overset{\circ}{=} PDuEv$, where $V_2 = P * D * uEvT$ and $V' \rightleftharpoons PDu * E * vT$ is a maximal start of V_1 contained in $* X * T$. Then $l_\alpha(V') \geqslant q_1 + 21$.

We consider the occurrences

$$U_2 \rightleftharpoons P * D * uEvY \quad \text{and} \quad U' \rightleftharpoons PDu * E * vY.$$

In general, U' need not be normalised. However, by II.6.12, we may make it normalised by shortening E by 4 segments on the right. Therefore we may assume that U' is normalised and $l_\alpha(U') \geqslant q_1 + 17$.

Let $\bar{V}_2$ be the maximal normalised continuation of V_2. Since $X'T_1 \in \mathcal{K}_\alpha$, it follows by IV.2.12 that $l_\alpha(\bar{V}_2) \geqslant 166 = q + 76$. Since $\bar{V}_2$ does not intersect V', the occurrence

$$\bar{U}_2 \rightleftharpoons \phi(\bar{V}_2; * PDu * EvT, * PDu * EvY)$$

is normalised, by II.6.11. By II.5.8, we have $\mathrm{Comp}(\bar{U}_2, U_2)$. Since $l_\alpha(\bar{U}_2) \geqslant q + 76$, it follows by IV.1.20 that U_2 is active.

By IV.2.21, there exists $X_4 \in \mathcal{R}_\alpha$ such that the transition

$$XT \to X_4$$

is a local reversal of V_2, $I_\alpha(X_4, X'T_1) = I_\alpha(XT, X'T_1) - 1$ and $X_4 \in \mathcal{K}_{\alpha-1}$. Suppose that $X_4 \overset{\circ}{=} P_1D_1u_1EvT$, where the occurrence $V_3 \rightleftharpoons P_1 * D_1 * U_1EvT$ is a maximal image of V_2 in the reversal $XT \to X_4$. Since U_2 is active and $\mathrm{MutNorm}_{\alpha-1}(V', U')$ by II.5.5, the transition

$$XY \to P_1 D_1 u_1 E v Y \tag{55}$$

is by III.2.17 a real reversal of U_2, where U' is stable and $U_3 \rightleftharpoons P_1 * D_1 * u_1 E v Y$ is a maximal image of U_2.

We set $Z \rightleftharpoons P_1 D_1 u_1 E v Y$. Since $X_4 \in \mathscr{K}_{\alpha-1}$ and $XY \in \mathscr{K}_{\alpha-1}$, it follows by II.6.13 that $Z \in \mathscr{K}_{\alpha-1}$. By IV.1.25, $Z \in \mathscr{P}_\alpha$. Now $l_\alpha(U_3) = l_\alpha(V_3) \leqslant l_\alpha(f_\alpha(V_2; XT, X_4)) + 34$ and, by IV.2.12, $l_\alpha(f_\alpha(V_2; XT, X_4)) = l_\alpha(f_\alpha(V_2; XT, X'T_1))$, so that $l_\alpha(U_3) \leqslant n - 184$ since $X'T_1 \in \mathscr{K}_\alpha$. Then by IV.1.13, $Z \in \mathscr{R}_\alpha$, that is, $XY \overset{\alpha}{\sim} Z$. Moreover, since $I_\alpha(X_4, X'T_1) = I_\alpha(XT, X'T_1) - 1$, it follows by the induction assumption that $Z \overset{\alpha}{\sim} X'Y'$. Therefore $XY \overset{\alpha}{\sim} X'Y'$.

We have dealt with the case $I_\alpha(XT, X'T_1) > 1$. The case $I_\alpha(T^{-1}Y, T_1^{-1}Y') > 1$ is treated similarly. We may therefore assume that $I_\alpha(XT, X'T_1) \leqslant 1$ and $I_\alpha(T^{-1}Y, T_1^{-1}Y') \leqslant 1$. Suppose that $I_\alpha(T^{-1}Y, T_1^{-1}Y') = 0$. Then $I_\alpha(XT, X'T_1) = 1$, since $\eta > 0$, that is, there exists a unique kernel $V_1 \in \operatorname{Ker}(\alpha, XT)$ such that $\neg \operatorname{Rel}(V_1, f_\alpha(V_1; XT, X'T_1))$. Then $XT \to X'T_1$ is a reversal of V_1. Suppose that $V' \rightleftharpoons P * E * vT$ is a maximal start of V_1 contained in $* X * T$. By hypothesis, $l_\alpha(V') \geqslant q_1 + 21$. We may assume that

$$P * E * vY \in \operatorname{Norm}(\alpha, XY, q_1 + 17)\,.$$

Suppose that $E \,\overline{\underline{\circ}}\, E_1 u E_2$, where the occurrences $P * E_1 * u E_2 v T$ and $P E_1 u * E_2 * v T$ are normalised, $l_\alpha(E_1) \geqslant q_1$ and $l_\alpha(E_2) \geqslant 9$. We shall first of all prove that we may assume that the occurrence $P * E_1 * u E_2 v T$ is periodised in rank $\alpha - 1$. Indeed, let $U_1 \rightleftharpoons P * E_1 * u E_2 v T$ and $U_2 \rightleftharpoons P E_1 u * E_2 * v T$. By IV.3.12, there exists an $X_1 \in \mathscr{M}_{\alpha-1}$ such that $XT \overset{\alpha-1}{\sim} X_1$. Then $f_{\alpha-1}(U_1; XT, X_1) = R * A' A_1 * S$, where A is an elementary period of rank α. By IV.2.26, there exists an $X_2 \in \mathscr{K}_{\alpha-1}$ such that $XT \overset{\alpha-1}{\sim} X_2$ and

$$f_{\alpha-1}(U_1;\ XT,\ X_2) = P_1 * A' A_1 * u_1 E_2 v T\,.$$

Since $X_2 \in \mathscr{K}_{\alpha-1}$ and $XY \in \mathscr{K}_{\alpha-1}$, it follows that $P_1 A' A_1 u_1 E_2 v Y \in \mathscr{K}_{\alpha-1}$. Since $XT \overset{\alpha-1}{\sim} X_2$ it follows by *V.2.1* that $XY \overset{\alpha-1}{\sim} P_1 A' A_1 u_1 E_2 v Y$. We then have $X_2 \in \mathscr{R}_\alpha$ and $XY \overset{\alpha}{\sim} P_1 A' A_1 u_1 E_2 v Y$ by IV.2.11, since $XT, XY \in \mathscr{R}_\alpha$. It is therefore sufficient to prove that $P_1 A' A_1 u_1 E_2 v Y \overset{\alpha}{\sim} X'Y'$.

Since $I_\alpha(X_2, X'T_1) = I_\alpha(XT, X'T_1)$, then changing the notation, we may assume that $P * E_1 * u E_2 v T$ is periodised in rank $\alpha - 1$. In this case we have, by III.1.10,

$$X'T_1 \overset{\alpha-1}{\sim} [P, A^{-n+t}, A_1 u E_2 v T]_{\alpha-1}\,. \tag{56}$$

By III.1.9, there exists a q_1-reversal

$$XY \to Z$$

of the occurrence $P * E * vY$. Since this is a reversal of $P * E_1 * u E_2 v Y$, we have by III.1.10 that

$$Z \overset{\alpha-1}{\sim} [P, A^{-n+t}, A_1uE_2vY]_{\alpha-1}\,. \tag{57}$$

Since $T^{-1}Y \overset{\alpha-1}{\sim} T_1^{-1}Y'$ using (56) and (57) together with the properties of coupling in rank $\alpha - 1$, we get

$$X'Y' \overset{\alpha-1}{\sim} [X'T_1, T_1^{-1}Y']_{\alpha-1} \overset{\alpha-1}{\sim} [P, A^{-n+t}, A_1uE_2vT, T^{-1}Y]_{\alpha-1}$$
$$\overset{\alpha-1}{\sim} [P, A^{-n+t}, A_1uE_2vY]_{\alpha-1} \overset{\alpha-1}{\sim} Z\,.$$

Then $XY \to X'Y'$ is also a q_1-reversal of $P * E * vY$, whence $XY \overset{\alpha}{\sim} X'Y'$ by IV.1.27.

The case $I_\alpha(XT, X'T_1) = 0$ is dealt with in an analogous fashion. We need only consider, therefore, the case where

$$I_\alpha(XT, X'T_1) = I_\alpha(T^{-1}Y, T_1^{-1}Y') = 1.$$

Suppose that $V_1 \in \mathrm{Ker}(\alpha, XT)$, $\neg\, \mathrm{Rel}(V_1, f_\alpha(V_1; XT, X'T_1))$, $W_1 \in \mathrm{Ker}(\alpha, T^{-1}Y)$ and $\neg\, \mathrm{Rel}(W_1, f_\alpha(W_1; T^{-1}Y, T_1^{-1}Y'))$. Then the transitions

$$XT \to X'T_1 \quad \text{and} \quad T^{-1}Y \to T_1^{-1}Y' \tag{58}$$

are reversals of these kernels. By hypothesis, $* X * T$ and $T^{-1} * Y *$ contain not less than $q_1 + 21$ segments of the kernels V_1 and W_1, that is, we have occurrences

$$V_2 \rightleftharpoons P * E * FT \text{ and } W_2 \rightleftharpoons T^{-1}G * D * Q\,,$$

where $l_\alpha(V_2) \geqslant q_1 + 21$, $l_\alpha(W_2) \geqslant q_1 + 21$, $\mathrm{Comp}(V_1, V_2)$ and $\mathrm{Comp}(W_1, W_2)$. By shortening E and D by not more than 4 segments, we obtain that the occurrences

$$V_3 \rightleftharpoons P * E * FGDQ \text{ and } W_3 \rightleftharpoons PEFG * D * Q$$

are normalised. Then $l_\alpha(V_3) \geqslant q_1 + 17$ and $l_\alpha(W_3) \geqslant q_1 + 17$.

Suppose that $E \mathrel{\bar{\bar{\circ}}} E_1uE_2$ and $D \mathrel{\bar{\bar{\circ}}} D_1vD_2$, where the occurrences

$$V_2' \rightleftharpoons P * E_1 * uE_2FT, \quad V_2'' \rightleftharpoons PE_1u * E_2 * FT\,,$$
$$W_2' \rightleftharpoons T^{-1}G * D_1 * vD_2Q, \; W_2'' \rightleftharpoons T^{-1}GD_1v * D_2 * Q$$

are normalised and satisfy the inequalities

$$l_\alpha(V_2') \geqslant q_1, l_\alpha(V_2'') \geqslant 9, l_\alpha(W_2') \geqslant 9, l_\alpha(W_2'') \geqslant q_1\,.$$

Then, clearly, the occurrences

$$V_3' \rightleftharpoons P * E_1 * uE_2FGDQ \quad \text{and} \quad W_3'' \rightleftharpoons PEFGD_1v * D_2 * Q$$

are also normalised, that is,

$$V_3', W_3'' \in \text{Norm}(\alpha; XY, q_1).$$

As established in the preceding case, we may assume that $P * E_1 * uE_2FT$ is periodised in rank $\alpha - 1$. Furthermore, if we employ assertions IV.3.18 and IV.3.14 in these arguments instead of IV.3.12, we shall convince ourselves that we may assume in addition that E_1 is equal to the base of an arbitrary given occurrence U' which is periodised in rank $\alpha - 1$ and which satisfies $\text{MutNorm}_{\alpha-1}(V_2', U')$. We may similarly assume that W_2'' is periodised in rank $\alpha - 1$ and D_2 is equal to the base of an arbitrary given occurrence U'', periodised in rank $\alpha - 1$, which satisfies $\text{MutNorm}_{\alpha-1}(W_2'', U'')$. By IV.3.12, we first of all find a word $Z' \in \mathscr{M}_{\alpha-1}$ such that $XY \overset{\alpha-1}{\sim} Z'$, and we choose the occurrences U' and U'' to be

$$U' \rightleftharpoons f_{\alpha-1}(V_3'; XY, Z') \text{ and } U'' \rightleftharpoons f_{\alpha-1}(W_3''; XY, Z'). \tag{59}$$

We can do this, since by II.5.5 we have $\text{MutNorm}_{\alpha-1}(V_2', V_3')$ and $\text{MutNorm}_{\alpha-1}(W_2'', W_3'')$. Thus we let

$$E_1 \rightleftharpoons \text{Bas}(U') \text{ and } D_2 \rightleftharpoons \text{Bas}(U'').$$

We shall then have $E_1 \overset{\circ}{=} A^t A_1$ and $D_2 \overset{\circ}{=} B^r B_1$, where $A \overset{\circ}{=} A_1 A_2$ and $B \overset{\circ}{=} B_1 B_2$ are elementary periods of rank α.

Since the transitions (58) are reversals of V_2' and W_2'', we have by III.1.10 that

$$X'T_1 \overset{\alpha-1}{\sim} [PA^{10}, A^{-n+t-20}, A^{10}A_1uE_2FT]_{\alpha-1} \tag{60}$$

and

$$T_1^{-1}Y' \overset{\alpha-1}{\sim} [T^{-1}GD_1vB^{10}, B^{-n+r-20}, B^{10}B_1Q]_{\alpha-1}. \tag{61}$$

Since V_2' is a start of V_1, by IV.1.7 and II.7.20 there exist words P_1HA^{10} and $A^{-10}H^{-1}SA^{-k}$ such that

$$PA^{10} \overset{\alpha-1}{\sim} P_1HA^{10}, A^{-10}H^{-1}SA^{-k} \overset{\alpha-1}{\sim} A^{-n+t-20}, P_1SA^{-k} \in \mathscr{K}_{\alpha-1}$$

and $k \geqslant n - t + 20 - 17 - 18 > q$. It then follows from (60) that

$$X'T_1 \overset{\alpha-1}{\sim} [P_1SA^{-10}, A^{-k+10}, A^{10}A_1uE_2FT]_{\alpha-1}. \tag{62}$$

We shall prove that V_3 and W_3 are not compatible. Assume that they are. In this case, E_1 and D_2 may be chosen such that $\text{MutNorm}_{\alpha-1}(V_3', W_3'')$. By II.2.2, for this to occur it is enough that V_3' and W_3'' correspond to one another in phase relative to the original elementary period for the word $EFGD$, and by II.2.9 for this it is sufficient that their initial pseudokernels of rank $\alpha - 1$ correspond to one another in phase and their final pseudokernels of rank $\alpha - 1$ also correspond in phase and the relation $l_\alpha(V_3') = l_\alpha(W_3'')$ holds. We may then deduce from (59) that

$\mathrm{MutNorm}_{\alpha-1}(U', U'')$, from which we get $\mathrm{Bas}(U') \overset{\circ}{\equiv} \mathrm{Bas}(U'')$ by IV.3.19, that is, $A'A_1 \overset{\circ}{\equiv} B'B_1$. Since A and B are simple words we may assume by I.2.9 that

$$A \overset{\circ}{\equiv} B, \quad t = r \quad \text{and} \quad A_1 \overset{\circ}{\equiv} B_1 . \tag{63}$$

Since $XY \in \mathscr{R}_\alpha$, it follows by IV.1.18 that $l_\alpha(A'A_1uE_2FGD_1vA'A_1) \leqslant n - 142$. Then

$$l_\alpha(A^{10}A_1uE_2FGD_1vA^{10}) \leqslant n - 2t - 122 .$$

Since $k \geqslant n - t - 15$, by II.7.11 there exists a word H_1 such that

$$A^{k-10} \overset{\alpha-1}{\sim} A^{10}A_1uE_2FGD_1vA^{10}H_1A^h , \tag{64}$$

where $A^{10}A_1uE_2FGD_1vA^{10}H_1A^h \in \mathscr{K}_{\alpha-1}$ and

$$h \geqslant k - (n - 2t - 122) - 2 \geqslant t + 105 .$$

By the principle of symmetry, it follows from (64) that

$$A^{-k+10} \overset{\alpha-1}{\sim} A^{-h}H_1^{-1}A^{-10}v^{-1}D_1^{-1}G^{-1}F^{-1}E_2^{-1}u^{-1}A_1^{-1}A^{-10} . \tag{65}$$

Further, since W_2'' is an end of the kernel W_2, by (63), IV.1.7 and II.7.20 we can find, in an analogous manner to the foregoing, a word $A^{-10}S_1Q_1 \in \mathscr{K}_{\alpha-1}$ and a number $s \geqslant n - t - 16$ such that

$$[A^{-n+r-20}, A^{10}A_1Q]_{\alpha-1} \overset{\alpha-1}{\sim} A^{-s}S_1Q_1 \overset{\alpha-1}{\sim} [A^{-s+10}, A^{-10}S_1Q_1]_{\alpha-1} ,$$

whence by (61) and (63),

$$T_1^{-1}Y' \overset{\alpha-1}{\sim} [T^{-1}GD_1vA^{10}, A^{-s+10}, A^{-10}S_1Q_1]_{\alpha-1} . \tag{66}$$

Using now the relations (62), (66) and (65) and the properties of coupling of rank $\alpha - 1$, we obtain

$$\begin{aligned} X'Y' &\overset{\alpha-1}{\sim} [X'T_1, T_1^{-1}Y']_{\alpha-1} \\ &\overset{\alpha-1}{\sim} [[P_1SA^{-10}, A^{-k+10}, A^{10}A_1uE_2FT]_{\alpha-1}, [T^{-1}GD_1vA^{10}, A^{-s+10}, A^{-10}S_1Q_1]_{\alpha-1}]_{\alpha-1} \\ &\overset{\alpha-1}{\sim} [P_1SA^{-10}, [A^{-k+10}, A^{10}A_1uE_2FGD_1vA^{10}]_{\alpha-1}, A^{-s}S_1Q_1]_{\alpha-1} \\ &\overset{\alpha-1}{\sim} [P_1SA^{-10}, A^{-h}H_1^{-1}, A^{-s}S_1Q_1]_{\alpha-1} \overset{\alpha-1}{\sim} P_1SA^{-h-10}H_1^{-1}A^{-s}S_1Q_1 . \end{aligned}$$

Since $X'Y' \in \mathscr{R}_\alpha$, it follows from this by IV.1.24 that

$$P_1SA^{-h-10}H_1^{-1}A^{-s}S_1Q_1 \in \mathscr{R}_\alpha .$$

But this is impossible, by IV.1.18, since

$$l_\alpha(A^{-h-10}H_1^{-1}A^{-s}) > h + s > t + 105 + n - t - 16 > n\,.$$

Therefore V_3 and W_3 are not compatible.

By III.1.9 and III.2.14, there exists a local q_1-reversal $XY \to Z$ of V_3'. Since $XY \in \mathscr{K}_{\alpha-1}$, we may assume by IV.3.15 that $Z \in \mathscr{K}_{\alpha-1}$. By III.2.9 and III.2.4, the occurrence W_3'' is stable in the reversal $XY \to Z$. Therefore for some R' we have

$$f_{XY\to Z}(W_3'') = R' * B'B_1 * Q\,.$$

By III.1.10, the condition

$$Z \overset{\alpha-1}{\simeq} [P,\ A^{-n+t},\ A_1uE_2FGD_1vB'B_1Q]_{\alpha-1}$$

holds, from which there follows by *V.1.7*, *V.1.8* and *V.1.9* that

$$R' \overset{\alpha-1}{\simeq} [P,\ A^{-n+t},\ A_1uE_2FGD_1v]_{\alpha-1}\,. \tag{67}$$

Clearly, the occurrence $W_4 \rightleftharpoons R' * B'B_1 * Q$ is normalised. Suppose that $\bar{W}_4$ is its maximal normalised continuation. Since $XY \in \mathscr{R}_\alpha$, it follows from IV.1.18 and III.2.10 that

$$l_\alpha(\bar{W}_4) \leqslant n - 142 + 17\,.$$

Then by III.1.9, there exists a q_1-reversal $Z \to Z_1$ of W_4. Moreover III.1.10 yields

$$Z_1 \overset{\alpha-1}{\simeq} [R',\ B^{-n+r},\ B_1Q]_{\alpha-1}\,,$$

from which we obtain, on using (67), (60) and (61), that

$$\begin{aligned} Z_1 &\overset{\alpha-1}{\simeq} [P,\ A^{-n+t},\ A_1uE_2FGD_1v,\ B^{-n+r},\ B_1Q]_{\alpha-1} \\ &\overset{\alpha-1}{\simeq} [P,\ A^{-n+t},\ [A_1uE_2FT,\ T^{-1}GD_1v]_{\alpha-1},\ B^{-n+r},\ B_1Q]_{\alpha-1} \\ &\overset{\alpha-1}{\simeq} [X'T_1,\ T_1^{-1}Y']_{\alpha-1} \overset{\alpha-1}{\simeq} X'Y'\,. \end{aligned}$$

Then by III.1.4, the transition $Z \to X'Y'$ is also a q_1-reversal of W_4.

Suppose that U is a maximal image of V_3 in the reversal $XY \to Z$ and U_1 is a maximal start of U which is stable in the reversal $Z \to X'Y'$. Since $X'Y' \in \mathscr{K}_{\alpha-1}$, it follows by IV.1.18 and III.2.6 that

$$l_\alpha(U_1) = l_\alpha(f_{Z\to Y'X'}(U_1)) \leqslant n - 184\,,$$

so that we have $l_\alpha(U) \leqslant n - 167$ by III.2.10. Then if $\bar{V}_3$ is the maximal normalised continuation of V_3, we have by III.1.8 that

$$l_\alpha(\bar{V}_3) \geqslant n - 18 - l_\alpha(U) \geqslant 149 > q + 34\,,$$

and it follows from this by IV.1.20 that V_3 is active. That W_4 is also active is proved in an analogous way. Therefore we have a sequence

$$XY \to Z \to X'Y'$$

of real reversals of rank α, that is, $XY \overset{\alpha}{\sim} X'Y'$.

1.8. *For arbitrary words X, Y, $Z \in \mathscr{R}_\alpha$, the relation*

$$[[X, Y]_\alpha, Z]_\alpha \overset{\alpha}{\sim} [X, [Y, Z]_\alpha]_\alpha$$

holds.

The proof goes by induction on the length of Y. We may assume by IV.2.20 and 1.7 that $X, Z \in \mathscr{M}_\alpha$. If Y is empty, then we get what we want from the self-evident equalities

$$[X, \Lambda]_\alpha = X = [\Lambda, X]_\alpha\,.$$

Suppose that $\partial(Y) = 1$, that is, Y is some letter a. Then by IV.1.22, we have

$$[X, a]_0 \in \mathscr{K}_\alpha \text{ and } [a, Z]_0 \in \mathscr{K}_\alpha\,,$$

from which we find by 1.1 that

$$[X, a]_\alpha = [X, a]_0 \quad \text{and} \quad [a, Z]_\alpha = [a, Z]_0\,.$$

In view of 1.7, we need only prove that

$$[[X, a]_0, Z]_\alpha \overset{\alpha}{\sim} [X, [a, Z]_0]_\alpha\,. \tag{68}$$

By 1.4, there exist words $X_1 \in \mathscr{K}_\alpha$ and $Z_1 \in \mathscr{K}_\alpha$ which satisfy

$$[X, a]_0 \overset{\alpha}{\sim} X_1,\ Z \overset{\alpha}{\sim} Z_1 \tag{69}$$

and $[X_1, Z_1]_0 \in \mathscr{K}_\alpha$, that is,

$$[[X, a]_0, Z]_\alpha = [X_1, Z_1]_0\,. \tag{70}$$

Since $[X_1, a^{-1}]_0 \in \mathscr{R}_\alpha$ and $[a, Z_1]_0 \in \mathscr{R}_\alpha$ by IV.1.22, it follows from (69) by 1.1 and 1.7 that

$$X = [[X, a]_0, a^{-1}]_0 \overset{\alpha}{\sim} [X_1, a^{-1}]_0$$

and

$$[a, Z]_0 \overset{\alpha}{\sim} [a, Z_1]_0 .$$

We then have by 1.7 that

$$[X, [a, Z]_0]_\alpha \overset{\alpha}{\sim} [[X_1, a^{-1}]_0, [a, Z_1]_0]_\alpha = [X_1, Z_1]_0 . \tag{71}$$

The required relation (68) now follows from (70) and (71).

Suppose that $\partial(Y) > 1$. By IV.1.23, there then exist words $R \in \mathscr{R}_\alpha$ and $Q \in \mathscr{R}_\alpha$ such that $Y \overset{\circ}{=} RQ$. Since $\partial(R) < \partial(Y)$ and $\partial(Q) < \partial(Y)$, application of the induction assumption and the coupling properties 1.1 and 1.7 yields

$$\begin{aligned}[][[X, Y]_\alpha, Z]_\alpha &\overset{\alpha}{\sim} [[X, [R, Q]_\alpha]_\alpha, Z]_\alpha \\ &\overset{\alpha}{\sim} [[[X, R]_\alpha, Q]_\alpha, Z]_\alpha \overset{\alpha}{\sim} [[X, R]_\alpha, [Q, Z]_\alpha]_\alpha \\ &\overset{\alpha}{\sim} [X, [R, [Q, Z]_\alpha]_\alpha]_\alpha \overset{\alpha}{\sim} [X, [[R, Q]_\alpha, Z]_\alpha]_\alpha \overset{\alpha}{\sim} [X, [Y, Z]_\alpha]_\alpha .\end{aligned}$$

1.9. *If $X, Y, Z \in \mathscr{R}_\alpha$, then*

$$[[X, Y]_\alpha, [Y^{-1}, Z]_\alpha]_\alpha \overset{\alpha}{\sim} [X, Z]_\alpha .$$

This follows from 1.8 and 1.1.

1.10. *If the words XT and $T^{-1}Y$ occur in some words in $\mathscr{K}_\alpha$, then*

$$[XT, T^{-1}Y]_\alpha \overset{\alpha}{\sim} [X, Y]_\alpha .$$

This follows from 1.1 and 1.9, since $X, Y, T \in \mathscr{R}_\alpha$ by IV.1.21.

1.11. *If $X, Y, Z, T \in \mathscr{R}_\alpha$, then*

$$[TX, Y]_\alpha \overset{\alpha}{\sim} TZ \Rightarrow [X, Y]_\alpha \overset{\alpha}{\sim} Z$$

and

$$[Y, XT]_\alpha \overset{\alpha}{\sim} ZT \Rightarrow [Y, X]_\alpha \overset{\alpha}{\sim} Z .$$

Suppose that $[TX, Y]_\alpha \overset{\alpha}{\sim} TZ$. Then by 1.1, we have $[[T, X]_\alpha, Y]_\alpha \overset{\alpha}{\sim} [T, Z]_\alpha$, from which it follows by 1.8 and 1.7 that $[T^{-1}, [T, Z]_\alpha]_\alpha \overset{\alpha}{\sim} [T^{-1}, [T, [X, Y]_\alpha]_\alpha]_\alpha$. Furthermore, we get $Z \overset{\alpha}{\sim} [X, Y]_\alpha$ by 1.9. The second relation is symmetric to the one we have just proved.

§2. Further Properties of Equivalent Words

2.1. *Suppose that $X, Y, T \in \mathscr{R}_\alpha$.*

If $XT, YT \in \mathscr{R}_\alpha$, then $X \overset{\alpha}{\sim} Y \Longleftrightarrow XT \overset{\alpha}{\sim} YT$.

If $TX, TY \in \mathscr{R}_\alpha$, *then* $X \overset{\alpha}{\sim} Y \Longleftrightarrow TX \overset{\alpha}{\sim} TY$.

Suppose that $XT, YT \in \mathscr{R}_\alpha$. If $X \overset{\alpha}{\sim} Y$, then we have by 1.1 and 1.7 that $XT \overset{\alpha}{\sim} [X, T]_\alpha \overset{\alpha}{\sim} [Y, T]_\alpha \overset{\alpha}{\sim} YT$. Since $XT \overset{\alpha}{\sim} YT$, it follows by 1.11 that $X \overset{\alpha}{\sim} Y$.

2.2. *Suppose that* $X, Y \in \mathscr{K}_\alpha$, $X \overset{\alpha}{\sim} Y$, $P * E * FDQ \in \mathrm{Reg}(\alpha, X)$ *and* $PEF * D * Q \in \mathrm{Reg}(\alpha, X)$. *If*

$$f_\alpha(P * E * FDQ;\ X,\ Y) = R * E * HDS$$

and

$$f_\alpha(PEF * D * Q;\ X,\ Y) = REH * D * S\,,$$

then $F \overset{\alpha}{\sim} H$.

By IV.2.35 we have

$$PEF \overset{\alpha}{\sim} REH \quad \text{and} \quad P \overset{\alpha}{\sim} R\,,$$

whence $PE \overset{\alpha}{\sim} RE$ by IV.1.21 and 2.1. Application of 1.1 and 1.7 then yields

$$H \overset{\alpha}{\sim} [E^{-1}R^{-1},\ REH]_\alpha \overset{\alpha}{\sim} [E^{-1}P^{-1},\ PEF]_\alpha \overset{\alpha}{\sim} F\,.$$

2.3. *Suppose that* $X \overset{\alpha-1}{\approx} X'T$, $Y \overset{\alpha-1}{\approx} T^{-1}Y$, $X'Y' \in \mathscr{R}_{\alpha-1}$, $P * E * Q \in \mathrm{ComReg}\ (\alpha - 1, X)$, $Q^{-1} * E^{-1} * R^{-1} \in \mathrm{ComReg}\ (\alpha - 1, Y)$ *and* $\mathrm{MutNorm}_{\alpha-1}(P * E * Q, R * E * Q)$.

If $f_{\alpha-1}(P * E * Q, X, X'T) = F * D * H$, *then* DH *is an end of the word* T, *and for some* F_1,

$$f_{\alpha-1}(Q^{-1} * E^{-1} * R^{-1};\ Y,\ T^{-1}Y') = H^{-1} * D^{-1} * F_1\,.$$

Suppose that $f_\alpha(Q^{-1} * E^{-1} * R^{-1};\ Y,\ T^{-1}Y') = H_1 * D_1 * F_1$. By II.7.6, there exists a word Z such that $Y^{-1} \overset{\alpha-1}{\approx} Z$ and for some G, $f_{\alpha-1}(R * E * Q;\ Y^{-1}, Z) = G * D * H$. Then $Y \overset{\alpha-1}{\approx} Z^{-1}$ and

$$f_{\alpha-1}(H^{-1} * D^{-1} * G^{-1};\ Z^{-1},\ T^{-1}Y') = H_1 * D_1 * F_1\,.$$

By induction on α, we shall prove that $H_1 \overset{\circ}{=} H^{-1}$ and $D_1 \overset{\circ}{=} D^{-1}$. This is evident for $\alpha = 1$. For $\alpha > 1$, we proceed by induction on the parameter $I_{\alpha-1}(Z^{-1}, T^{-1}Y')$. If $I_{\alpha-1}(Z^{-1}, T^{-1}Y') = 0$, then

$$Z \overset{\alpha-2}{\approx} T^{-1}Y',\ f_{\alpha-2}(H^{-1} * D^{-1} * G^{-1};\ Z^{-1},\ T^{-1}Y') = H_1 * D_1 * F_1\,,$$

and therefore we can use the induction hypothesis.

Suppose that $I_{\alpha-1}(Z^{-1}, T^{-1}Y') > 0$, that is, some kernel $V \in \mathrm{Ker}(\alpha - 1, Z^{-1})$ is not related to the kernel $f_{\alpha-1}(V;\ Z^{-1}, Z^{-1}Y')$. Assume that V is contained in $* H^{-1}D^{-1} * G^{-1}$. Then by II.7.4, the occurrence $U \rightleftharpoons \phi(V^{-1};\ G * DH *, F * DH *)$

is a kernel of rank $\alpha - 1$ of the word $X'T$. Since $f_{\alpha-1}(U; X'T, X'T) = U$ and $X'Y' \in \mathscr{R}_{\alpha-1}$, it follows by *V.1.6* that V cannot be contained in $* H^{-1}D^{-1} * G^{-1}$. Then V is contained in $H^{-1}D^{-1} * G^{-1} *$. Applying a local reversal of it we can find a word Z_1 such that $Z^{-1} \overset{\alpha-1}{\sim} Z_1$ and

$$f_{\alpha-1}(H^{-1} * D^{-1} * G^{-1};\ Z^{-1},\ Z_1) = H^{-1} * D^{-1} * G_1$$

for some G_1. Therefore, $Z_1 \overset{\alpha-1}{\sim} T^{-1}Y'$ and

$$f_{\alpha-1}(H^{-1} * D^{-1} * G_1;\ Z_1,\ T^{-1}Y') = H_1 * D_1 * F_1\,.$$

Since in addition $I_{\alpha-1}(Z_1, T^{-1}Y') = I_{\alpha-1}(Z^{-1}, T^{-1}Y') - 1$, the required relations $H_1 \overset{0}{=} H^{-1}$ and $D_1 \overset{0}{=} D^{-1}$ follow from the second induction hypothesis.

2.4. *Suppose that* $X, Y \in \mathscr{X}_\alpha$, $X \overset{\alpha}{\sim} Y$, $P * E * Q \in \operatorname{Reg}(\alpha, X)$ *and* $f_\alpha(P * E * Q; X, Y) = R * G * S$. *If the word* $[EQ, S^{-1}G^{-1}]_0$ *(or* $[G^{-1}R^{-1}, PE]_0$*) is non-empty, then it does not occur in any word in the set* $\mathscr{R}_\alpha$.

We shall establish this for the word $F \rightleftharpoons [EQ, S^{-1}G^{-1}]_0$. For the second word the proof is analogous.

If $\alpha = 0$, F is empty. Suppose that $\alpha > 0$. If $X \overset{\alpha-1}{\sim} Y$, then by IV.2.18(b), we can use the induction assumption. In the contrary case we have $I_\alpha(X, Y) > 0$ by IV.2.13, that is, some kernel $V \in \operatorname{Ker}(\alpha, X)$ is not related to $f_\alpha(V; X, Y)$. If $V \ll P * E * Q$, then on applying a local reversal of V we get $X' \in \mathscr{X}_\alpha$, where $X \overset{\alpha}{\sim} X'$, $f_\alpha(P * E * Q; X, X') = P_1 * E * Q$ and $I_\alpha(X', X) = I_\alpha(X, Y) - 1$. In the contrary case, V must be contained in $P * EQ *$. We shall prove that in such a case the word F is non-empty and does not occur in any word in $\mathscr{R}_\alpha$. It is clearly sufficient to prove this in the case where $V = P * E * Q$ and $I_\alpha(X, Y) = 1$. In this case, the transition $X \to Y$ is a real reversal of V, that is, we have words $X_1, Y_1 \in \mathscr{X}_{\alpha-1}$ such that $X \overset{\alpha-1}{\sim} X_1$, $Y \overset{\alpha-1}{\sim} Y_1$ and the transition $X_1 \to Y_1$ is a simple q-reversal of the occurrence $f_{\alpha-1}(P * E * Q; X, X_1)$. According to definition I.4.16, we have

$$P_1T * A^tA_1 * T_1Q_1 \in \operatorname{MaxNorm}(\alpha, X_1, q)\,, \tag{1}$$

$$P_1H * D^hD_1 * H_1Q_1 \in \operatorname{MaxNorm}(\alpha, Y_1, q)\,, \tag{2}$$

where $A \overset{0}{=} A_1A_2$ is an elementary period of rank α, the word $D \overset{0}{=} D_1D_2$ is a cyclic shift of A^{-1}, the occurrence (1) is periodised in rank $\alpha - 1$, the relation

$$A^{n+20} \overset{\alpha-1}{\sim} A^{t+10}A_1T_1H_1^{-1}D_1^{-1}D^{-h}H^{-1}TA^{10}\,, \tag{3}$$

holds, where both left and right hand sides belong to $\mathscr{X}_{\alpha-1}$ and the occurrences $f_{\alpha-1}(P * E * Q; X, X_1)$ and $f_{\alpha-1}(R * G * S; Y, Y_1)$ are contained in the occurrences (1) and (2) and are compatible with them.

Suppose that

$$E \overset{0}{=} E_1uE_2 \quad\text{and}\quad G \overset{0}{=} G_1vG_2\,,$$

where $l_\alpha(E_1) = l_\alpha(G_1) = 9$,

$$l_\alpha(E_2) = l_\alpha(E) - 11, \; l_\alpha(G_2) = l_\alpha(G) - 11 \tag{4}$$

and that the occurrences

$$V_1 \rightleftharpoons P * E_1 * uE_2Q, \quad V_2 \rightleftharpoons PE_1u * E_2 * Q\,,$$
$$W_1 \rightleftharpoons R * G_1 * vG_2S, \quad W_2 \rightleftharpoons RG_1v * G_2 * S$$

are normalised. Suppose that

$$f_{\alpha-1}(V_2;\; X,\; X_1) = P_3 * E_3 * Q_3 \tag{5}$$

and

$$f_{\alpha-1}(W_2;\; Y,\; Y_1) = R_3 * G_3 * S_3\,. \tag{6}$$

Since occurrences (5) and (6) are contained in (1) and (2), we have $E_3Q_3 \overset{\circ}{=} A_4A'A_1T_1Q_1$ and $G_3S_3 \overset{\circ}{=} D_4D^kD_1H_1Q_1$, where A_4 and D_4 are ends of the periods A and D satisfying the inequalities

$$l_\alpha(E_3) \leqslant l_\alpha(A_4A'A_1) \text{ and } l_\alpha(G_3) \leqslant l_\alpha(D_4D^kD_1)\,. \tag{7}$$

It follows from (5) by IV.2.34 that, for some $X_2 \in \mathscr{K}_{\alpha-1}$, we have $X \overset{\alpha-1}{\simeq} X_2$ and $f_{\alpha-1}(P * E_1 * uE_2Q;\; X,\; X_2) = P * E_1 * u_1E_3Q_3$, whence it follows by IV.2.35 and 2.1 that

$$EQ \overset{\alpha-1}{\simeq} E_1u_1A_4A'A_1T_1Q_1\,. \tag{8}$$

We find similarly from (6) that

$$GS \overset{\alpha-1}{\simeq} G_1v_1D_4D^kD_1H_1Q_1 \tag{9}$$

for some v_1. From (3) one can see that the word $B \rightleftharpoons A_4A'A_1T_1H_1^{-1}D_1^{-1}D^{-k}D_4^{-1}$ is an elementary word of rank α. By IV.1.21, it belongs to $\mathscr{R}_{\alpha-1}$. Exactly the same holds for the right hand sides of (8) and (9). Therefore $E_1u_1Bv_1^{-1}G_1^{-1} \in \mathscr{R}_{\alpha-1}$, that is, we have

$$[EQ,\; S^{-1}G^{-1}]_{\alpha-1} \overset{\alpha-1}{\simeq} E_1u_1Bv_1^{-1}G_1^{-1}\,. \tag{10}$$

Assume that F occurs in a word $Z \in \mathscr{R}_\alpha$. Suppose that $F \overset{\circ}{=} F_1F_2$, where $EQ \overset{\circ}{=} F_1N$ and $S^{-1}G^{-1} \overset{\circ}{=} N^{-1}F_2$. Then it follows easily from 2.3 that E_1 is a start of F_1 and G_1^{-1} is an end of F_2. Since they occur in the words X, $Y^{-1} \in \mathscr{K}_\alpha$, then by IV.1.21 and II.6.13, we have $F \in \mathscr{R}_{\alpha-1}$. Consequently

$$[EQ,\; S^{-1}G^{-1}]_{\alpha-1} \overset{\alpha-1}{\simeq} F\,. \tag{11}$$

It follows from (10) and (11) that

$$F \overset{\alpha-1}{\simeq} E_1 u_1 B v_1^{-1} G_1^{-1} . \tag{12}$$

Application of II.5.2 and (7), (6), (5) and (4) yields

$$\begin{aligned} l_\alpha(B) \geqslant l_\alpha(E_3) + l_\alpha(G_3) - 1 &= l_\alpha(E_2) + l_\alpha(G_2) - 1 \\ &= l_\alpha(E) + l_\alpha(G) - 23. \end{aligned}$$

In addition, $l_\alpha(E) + l_\alpha(G) \geqslant n - 86$ by IV.1.13. Hence $l_\alpha(B) \geqslant n - 109$. It then follows by II.6.6 and II.5.16 from (12) that some elementary $(n-117)$-power of rank α occurs in F. By IV.1.18, this means that F cannot occur in a word $Z \in \mathscr{R}_\alpha$.

2.5. *Suppose that* $\alpha > 0$, $X \in \mathscr{K}_\alpha$, $Y \in \mathscr{K}_\alpha$, $X \overset{\alpha}{\sim} Y$, $P * E * Q \in \operatorname{Reg}(\alpha, X)$ *and*

$$f_\alpha(P * E * Q;\ X,\ Y) = R * D * S .$$

Then there exist words $F, F_1 \in \mathscr{R}_\alpha$ *such that*

$$E \overset{\alpha}{\sim} [F,\ D,\ F_1]_\alpha , \tag{13}$$

$$R \overset{\alpha}{\sim} [P,\ F]_\alpha,\ S \overset{\alpha}{\sim} [F_1,\ Q]_\alpha \tag{14}$$

and each of F *and* F_1 *is the result of a coupling of the form*

$$[G,\ N,\ G']_{\alpha-1} ,$$

where $G, G' \in \mathscr{A}_\alpha$, *the word* $N \in \mathscr{R}_{\alpha-1}$ *occurs in some elementary word of rank* α, *and* $l_\alpha(N) \leqslant 44$.

In addition, if $P * E * Q$ *does not contain active kernels of rank* α, *then* $F, F_1 \in \mathscr{A}_\alpha$ *and* α *can be replaced by* $\alpha - 1$ *in* (13).

By IV.1.21, the words P, Q, R, S, E, D belong to $\mathscr{R}_\alpha$.

By IV.2.26 and IV.2.35, we may assume that $P * E * Q$ transforms into $R * D * S$ locally in rank α. Then an arbitrary kernel $V \in \operatorname{Ker}(\alpha, X)$ not contained in $P * E * Q$ is related to the occurrence $f_\alpha(V;\ X,\ Y)$. In this case, the proof proceeds by induction on the parameter $I_\alpha(X,\ Y)$.

If $I_\alpha(X,\ Y) = 0$, then by IV.2.13 and IV.2.18(b), we have $X \overset{\alpha-1}{\simeq} Y$ and $f_{\alpha-1}(P * E * Q;\ X,\ Y) = R * D * S$, whence it follows by II.7.14 that for some $F, F_1 \in \mathscr{A}_\alpha$ the relations (13) and (14) hold with α replaced by $\alpha - 1$.

Suppose that $I_\alpha(X,\ Y) > 0$, that is, some kernel $V \in \operatorname{Ker}(\alpha, X)$ is not related to $f_\alpha(V;\ X,\ Y)$. Then V is contained in $P * E * Q$ by hypothesis.

We consider first of all the case where $V = P * E * Q$. Then $I_\alpha(X,\ Y) = 1$ and the transition $X \to Y$ is a real reversal of V, that is, there exist words $X_1 \in \mathscr{K}_\alpha$, $Y_1 \in \mathscr{K}_\alpha$ and an elementary period $A \overset{\circ}{=} A_1 A_2$ of rank α such that $X \overset{\alpha-1}{\simeq} X_1$, $Y \overset{\alpha-1}{\simeq} Y_1$,

$$X_1 \overset{\circ}{\equiv} P_1TA^tA_1T_1Q_1, \quad Y_1 \overset{\circ}{\equiv} P_1HB^hB_1H_1Q_1$$

and the transition $X_1 \to Y_1$ is a simple reversal of $f_\alpha(V; X, X_1)$. Then by definition I.4.16, we have the relation

$$A^{n+20} \overset{\alpha-1}{\simeq} A^{t+10}A_1T_1H_1^{-1}B_1^{-1}B^{-h}H^{-1}TA^{10}, \tag{15}$$

the left and right hand sides of which belong to $\mathscr{K}_{\alpha-1}$, and where B is a cyclic shift of A^{-1}. In addition, by III.1.6 the occurrences $f_{\alpha-1}(P * E * Q;\ X,\ X_1)$ and $f_{\alpha-1}(R * D * S;\ Y,\ Y_1)$ are contained in the occurrences $P_1T * A^tA_1 * T_1Q_1$ and $P_1H * B^hB_1 * H_1Q_1$, respectively. Suppose that

$$f_{\alpha-1}(P * E * Q;\ X,\ X_1) = P_1TM * E_1 * M_1T_1Q_1 \tag{16}$$

and

$$f_{\alpha-1}(R * D * S;\ Y,\ Y_1) = P_1HL * D_1 * L_1H_1Q_1, \tag{17}$$

where $ME_1M_1 \overset{\circ}{\equiv} A^tA_1$ and $LD_1L_1 \overset{\circ}{\equiv} B^hB_1$. We set

$$N \rightleftharpoons M^{-1}T^{-1}HL \text{ and } N_1 \rightleftharpoons L_1H_1T_1^{-1}M_1^{-1}.$$

One can see from (15) that N and N_1 occur in an elementary word of rank α. They belong to $\mathscr{R}_{\alpha-1}$, by IV.1.21. By IV.1.7 and II.5.2, we have $l_\alpha(N), l_\alpha(N_1) \leqslant 44$.

By II.7.14, there exist words $G, G_1, G', G_1' \in \mathscr{A}_\alpha$ such that

$$E \overset{\alpha-1}{\simeq} [G, E_1, G_1]_{\alpha-1}, \ D_1 \overset{\alpha-1}{\simeq} [G', D, G_1']_{\alpha-1}, \tag{18}$$

$$[P, G]_{\alpha-1} \overset{\alpha-1}{\simeq} P_1TM, \quad [P_1HL, G']_{\alpha-1} \overset{\alpha-1}{\simeq} R, \tag{19}$$

$$[G_1, Q]_{\alpha-1} \overset{\alpha-1}{\simeq} M_1T_1Q_1, [G_1', L_1H_1Q_1]_{\alpha-1} \overset{\alpha-1}{\simeq} S. \tag{20}$$

We set

$$F \rightleftharpoons [G, N, G']_{\alpha-1} \text{ and } F_1 \rightleftharpoons [G_1', N_1, G_1]_{\alpha-1}. \tag{21}$$

Since P_1, TM, $HL \in \mathscr{R}_{\alpha-1}$, on applying properties of coupling in rank $\alpha - 1$, we obtain from (19) and (21) that

$$R \overset{\alpha-1}{\simeq} [P_1, HL, G']_{\alpha-1} \overset{\alpha-1}{\simeq} [P, G, M^{-1}T^{-1}, HL, G']_{\alpha-1} \overset{\alpha-1}{\simeq} [P, F]_{\alpha-1}. \tag{22}$$

By II.7.12 and IV.1.18, since $l_\alpha(N) \leqslant 44$ it follows that $N, F \in \mathscr{R}_\alpha$. We then get from (22) by 1.1 that

$$R \overset{\alpha}{\simeq} [P, F]_\alpha.$$

The second relation in (14) is obtained in the same way as (20), (21).

It only remains to prove (13). By 1.1, (18) and (21), in order to do this it is sufficient to prove that

$$E_1 \overset{\alpha}{\sim} [N, D_1, N_1]_\alpha .$$

Clearly, we may assume without loss of generality that $l_\alpha(D_1) \leqslant l_\alpha(E_1)$. In the contrary case we interchange the rôles of these words. Since $l_\alpha(D_1) \leqslant (n+1)/2$, it follows by II.7.12 and IV.1.18 that $[N, D_1, N_1]_{\alpha-1} \in \mathscr{R}_\alpha$. By 1.1 and IV.1.27, we need then only prove that the transition

$$E_1 \to [N, D_1, N_1]_{\alpha-1} \tag{23}$$

is a q_1-reversal of rank α. Since $A^t A_1 \overset{\circ}{=} M E_1 M_1$ and $l_\alpha(E_1) \geqslant q$, it follows from II.6.6, III.1.9 and III.1.10 that the transition

$$E_1 \to [M^{-1}, A^{-n+t+1} A_2^{-1}, M_1^{-1}]_0$$

is a q_1-reversal of rank α. In addition, it follows from (15) by 2.1 that

$$A^{-n+t+1} A_2^{-1} \overset{\alpha-1}{\sim} T^{-1} H L D_1 L_1 H_1 T_1^{-1} ,$$

that is,

$$[M^{-1}, A^{-n+t+1} A_2^{-1}, M_1^{-1}]_0 \overset{\alpha-1}{\sim} [N, D_1, N_1]_{\alpha-1} .$$

Then by III.1.4, (23) is also a q_1-reversal of rank α.

We consider the case where V is a proper start of the occurrence $P * E * Q$. We then have $E \overset{\circ}{=} E_1 u E_2$ by IV.1.3, where $V = P * E_1 * u E_2 Q$ and $P E_1 u * E_2 * Q \in \text{Reg}(\alpha, X)$. Suppose that $D \overset{\circ}{=} D_1 v D_2$, where

$$f_\alpha(V; X, Y) = R * D_1 * v D_2 S$$

and

$$f_\alpha(P E_1 u * E_2 * Q; X, Y) = R D_1 v * D_2 * S .$$

By IV.2.34, there exists a $Z \in \mathscr{K}_\alpha$ such that

$$X \overset{\alpha}{\sim} Z \overset{\alpha}{\sim} Y, \quad I_\alpha(X, Z) = 1, \quad I_\alpha(Z, Y) = I_\alpha(X, Y) - 1 ,$$

$$f_\alpha(V; X, Z) = R * D_1 * u_1 E_2 Q$$

and

$$f_\alpha(P E_1 u * E_2 * Q; X, Z) = R D_1 u_1 * E_2 * Q .$$

Then

$$f_\alpha(RD_1u_1 * E_2 * Q;\ Z,\ Y) = RD_1v * D_2 * S\,.$$

By what we have proved, there exist words $F, F' \in \mathscr{R}_\alpha$ such that

$$E_1 \overset{\alpha}{\sim} [F, D_1, F']_\alpha\,, \tag{24}$$
$$R \overset{\alpha}{\sim} [P, F]_\alpha$$

and

$$u_1E_2Q \overset{\alpha}{\sim} [F', uE_2Q]_\alpha\,. \tag{25}$$

Furthermore, by the induction assumption, there exist words F_0, $F_1 \in \mathscr{R}_\alpha$ such that

$$E_2 \overset{\alpha}{\sim} [F_0, D_2, F_1]_\alpha\,, \tag{26}$$

$$RD_1v \overset{\alpha}{\sim} [RD_1u_1, F_0]_\alpha \tag{27}$$

and

$$S \overset{\alpha}{\sim} [F_1, Q]_\alpha\,.$$

It follows from (24) and (27) by 1.11 that

$$u_1 \overset{\alpha}{\sim} [F', u]_\alpha \quad\text{and}\quad v \overset{\alpha}{\sim} [u_1, F_0]_\alpha\,. \tag{28}$$

Further, using (24), (26) and (28) we get

$$E \overset{\circ}{=} E_1uE_2 \overset{\alpha}{\sim} [F, D_1, F', u, F_0, D_2, F_1]_\alpha$$
$$\overset{\alpha}{\sim} [F, D_1, v, D_2, F_1]_\alpha \overset{\alpha}{\sim} [F, D, F_1]_\alpha\,.$$

The case where V is an end of $P * E * Q$ is treated in an analogous fashion.

It remains to consider the case where $E \overset{\circ}{=} E_1uE_2vE_3$, where $V = PE_1u * E_2 * vE_3Q$, $P * E_1 * uE_2vE_3Q \in \mathrm{Reg}(\alpha, X)$ and $PE_1uE_2v * E_3 * Q \in \mathrm{Reg}(\alpha, X)$.

In this case, application of a local reversal of the kernel V yields a word $Z \in \mathscr{K}_\alpha$ such that $X \overset{\alpha}{\sim} Z \overset{\alpha}{\sim} Y$, $I_\alpha(Z, Y) = I_\alpha(X, Y) - 1$ and

$$f_\alpha(P * E * Q;\ X,\ Z) = P * E_1CE_3 * Q$$

for some C. We then have by 2.1 that

$$E \overset{\alpha}{\sim} E_1CE_3\,. \tag{29}$$

Since, in addition, $f_\alpha(P * E_1CE_3 * Q;\ Z,\ Y) = R * D * S$, by the induction assumption there exist words F, $F_1 \in \mathscr{R}_\alpha$ such that (14) holds and $E_1CE_3 \overset{\alpha}{\sim} [F, D, F_1]_\alpha$, whence the required relation (13) follows by (29).

With the proof of 2.5, *we have completed the complicated induction which we began in Chapter* II. *As a result, all of the assertions mentioned in Chapters* II, III, IV *and* V *have been proved for arbitrary values of the parameter* α.

The propositions which follow in this section are necessary to justify the principle of effectiveness (I.5.4).

2.6. *For arbitrary words* X, Y *and* $\alpha \geqslant 0$, *the following holds*:

$$X \overset{\alpha}{\sim} Y \Rightarrow \partial(Y) \leqslant n^{8^\alpha}\, \partial(X)\,. \tag{30}$$

Induction on α. Suppose that $X \overset{\alpha}{\sim} Y$. If $\alpha = 0$, then $\partial(Y) = \partial(X)$. Suppose that $\alpha > 0$. If $I_\alpha(X, Y) = 0$, then by IV.2.13, $X \overset{\alpha-1}{\approx} Y$, and we can use the induction assumption.

Suppose that $I_\alpha(X, Y) = r > 0$. By IV.3.12, there exists an $X_1 \in \mathscr{M}_{\alpha-1}$ such that $X \overset{\alpha-1}{\approx} X_1$. Suppose that

$$X_1 \overset{\circ}{=} u_1 E_1 u_2 E_2 \ldots u_r E_r u_{r+1}\,,$$

where, for $1 \leqslant i \leqslant r$, the occurrences

$$V_i \rightleftarrows u_1 E_1 u_2 E_2 \ldots u_i * E_i * u_{i+1} \ldots E_r u_{r+1}$$

are all the kernels of rank α of the words X_1 which are not related to $f_\alpha(V_i;\, X_1, Y)$. By IV.2.6 and IV.2.5, there exists a simple sequence of real reversals

$$X_1 \to X_2 \to X_3 \to \ldots \to X_i \to X_{i+1} \to \ldots \to X_r \to X_{r+1}\,, \tag{31}$$

where $X_i \to X_{i+1}$ is a reversal of the kernel $f_\alpha(V_i;\, X_1, X_i)$, and $X_{r+1} \overset{\alpha-1}{\approx} Y$. Then

$$\partial(Y) \leqslant n^{8^{\alpha-1}}\, \partial(X_{r+1})\,. \tag{32}$$

By IV.2.21, we may assume that all the reversals in (31) are local, and all the X_i lie in $\mathscr{M}_{\alpha-1}$. Then for every $i > 1$, we have, for some v_i and D_i,

$$X_i \overset{\circ}{=} v_1 D_1 v_2 D_2 \ldots v_{i-1} D_{i-1} u_i' E_i \ldots u_r E_r u_{r+1}\,, \tag{33}$$

where

$$f_\alpha(V_{i-1};\, X_1, X_i) = v_1 D_1 v_2 \ldots v_{i-1} * D_{i-1} * u_i' E_i \ldots u_r E_r u_{r+r}\,,$$
$$f_\alpha(V_{i-2};\, X_1, X_i) = v_1 D_1 v_2 \ldots v_{i-2} * D_{i-2} * v_{i-1} D_{i-1} u_i' \ldots u_r E_r u_{r+1}$$

and

$$f_\alpha(V_i;\, X_1, X_i) = v_1 D_1 v_2 \ldots v_{i-1} D_{i-1} u_i' * E_i * u_{i+1} \ldots u_r E_r u_{r+1}\,.$$

We consider any one of the reversals $X_i \to X_{i+1}$ in the sequence (31). Since the occurrences V_i and $f_\alpha(V_i;\, X_1, X_i)$ are periodised in rank $\alpha - 1$, by II.4.5 we have

$E_i \overset{\circ}{=} A^t A_1$, where A is an elementary period of rank α and A_1 is a start of A. By III.1.10, the relation

$$X_{i+1} \overset{\alpha-1}{\sim} [v_1 D_1 v_2 \ldots v_{i-1} D_{i-1} u_i', A^{-n+t}, A_1 u_{i+1} E_{i+1} \ldots u_r E_r u_{r+1}]_{\alpha-1}$$

holds, whence it follows from (33) and 1.11 that

$$v_i D_i u_{i+1}' \overset{\alpha-1}{\sim} [u_i', [A^{-n+t}, A_1 u_{i+1}]_{\alpha-1}]_{\alpha-1} . \tag{34}$$

Here, for $i = 1$, we understand u_1' as u_1. It follows from (34) and the inductive hypothesis that

$$\begin{aligned} &\partial(v_i D_i u_{i+1}') \\ &\quad \leqslant n^{8^{\alpha-1}}(\partial(u_i')n^{8^{\alpha-1}} + n^{8^{\alpha-1}}(n^{8^{\alpha-1}} n\partial(E_i) + n^{8^{\alpha-1}}\partial(u_{i+1}))) \\ &\quad \leqslant n^{3\cdot 8^{\alpha-1}+1}\, \partial(E_i u_{i+1}) + n^{2\cdot 8^{\alpha-1}}\, \partial(u_i') . \end{aligned} \tag{35}$$

If $i > 1$, then to bound $\partial(u_i')$ we firstly use III.2.15 to find a local reversal of the kernel V_{i-1},

$$X_1 \to u_1 E_1 u_2 \ldots E_{i-2} u_{i-1}'' D_{i-1} u_i' E_i u_{i+1} \ldots E_r u_{r+1} ,$$

whence, in a fashion analogous to (35), we get

$$\partial(u_{i+1}'' D_{i-1} u_i') \leqslant n^{3\cdot 8^{\alpha-1}+1}\, \partial(E_{i-1} u_i) + n^{2\cdot 8^{\alpha-1}} \partial(u_{i-1}) . \tag{36}$$

Further, using relations (32), (35), (36) and $X \overset{\alpha-1}{\sim} X_1$, we get

$$\begin{aligned} \partial(Y) &\leqslant n^{8^{\alpha-1}}\partial(X_{r+1}) \leqslant n^{8^{\alpha-1}} \sum_{i=1}^{r} \partial(v_i D_i u_{i+1}') \\ &\leqslant n^{4\cdot 8^{\alpha-1}+1} \sum_{i=1}^{r} \partial(E_i u_{i+1}) + n^{5\cdot 8^{\alpha-1}} \sum_{i=2}^{r} \partial(u_{i-1}) + n^{6\cdot 8^{\alpha-1}+1} \sum_{i=2}^{r} \partial(E_{i-1} u_i) \\ &\leqslant 3n^{6\cdot 8^{\alpha-1}+1}\, \partial(X_1) \leqslant n^{8^\alpha}\, \partial(X) . \end{aligned}$$

2.7. Relation (30) gives a rough upper bound for the length of any word Y equivalent in rank α to a given word $X \in \mathscr{R}_\alpha$. It can be shown that a more exact bound holds:

$$X \overset{\alpha}{\sim} Y \Rightarrow \partial(Y) \leqslant n^\alpha \partial(X) .$$

However, we shall not need this.

2.8. *If* $P * F * Q \in \mathrm{Reg}(\alpha, X)$, $X \overset{\alpha}{\sim} Y$ *and*

$$f_\alpha(P * F * Q;\, X, Y) = R * G * S ,$$

then $\partial(G) \leqslant n^{8^\alpha} \partial(F)$.

This is proved by induction on α. By IV.2.26, for $\alpha > 0$ we may assume that every kernel $V \in \mathrm{Ker}(\alpha, X)$ that is not related to $f_\alpha(V; X, Y)$ is contained in $P * E * Q$. In a case like this we repeat the arguments of 2.6. Firstly, by IV.3.12 we find an $X_1 \in \mathscr{M}_{\alpha-1}$ such that $X \overset{\alpha-1}{\simeq} X_1$. Set $f_{\alpha-1}(P * F * Q; X, X_1) = P_1 * F_1 * Q_1$. Then $P_1 * F_1 * Q_1 \in \mathrm{Reg}(\alpha, X_1)$ and

$$f_\alpha(P_1 * F_1 * Q_1; X_1, Y) = R * G * S.$$

Next, we select all kernels $V_i \in \mathrm{Ker}(\alpha, X_1)$ that are not related to $f_\alpha(V_i; X_1, Y)$, and find a simple sequence (31) of local reversals. Clearly, the initial (final) kernel of rank α of the occurrence $P_1 * F_1 * Q_1$ either coincides with V_1 (with V_r respectively) or does not intersect it. In the latter case, having added the final kernel, we may assume that $P_1 * F_1 * Q_1$ begins with V_1 and ends with V_r, assuming that there cannot be reversals with fewer kernels in (31). Further, it remains for us to repeat the calculation of length of words performed in 2.6, ignoring the words u_1, v_1, u_{r+1} and v_{r+1}, which in our case coincide with P_1, R, Q_1 and S respectively.

2.9. *If $P * E * Q$ is a normal generating occurrence of rank α in an integral word $Y \in \mathrm{Int}(X, \alpha, A)$ and $l_{\alpha,A}(E) \geqslant 3$, then there is a generating occurrence $R * E * S$ in an integral word $Y_1 \in \mathrm{Int}(X_1, \alpha, A)$ such that the inequality*

$$\partial(X_1) < n^{8\alpha}\partial(E) \tag{37}$$

is satisfied.

Set $f_{\alpha-1}(P * E * Q; Y, X) = P_1 * E_1 * Q_1$. Then by 2.8, we have

$$\partial(E_1) \leqslant n^{8\alpha-1}\partial(E). \tag{38}$$

Clearly, $\partial(E_1) > 2\partial(A)$. Suppose that $X \overset{\circ}{=} A_1 A' A_2$, where A_1 is an end of A and A_2 is a start of A. We consider a word $X_1 \rightleftharpoons A_1 A^r A_2$ such that

$$\partial(E_1) + n\partial(A) \leqslant \partial(X_1) \leqslant n\partial(E_1). \tag{39}$$

By II.7.9, $X_1 \in \mathrm{Per}(\alpha, A)$. Further, (37) follows from (38) and (39). Let $R_1 * E_1 * S_1$ be an occurrence in the word X_1 that is interior with respect to the period A and corresponds in phase to $P_1 * E_1 * Q_1$ in the sense of I.2.6. Then, by II.1.8 we have

$$\mathrm{MutNorm}_{\alpha-1}(P_1 * E_1 * Q_1,\ R_1 * E_1 * S_1),$$

whence it follows by I.4.31 that there exists a word Y_1 such that $X_1 \overset{\alpha-1}{\simeq} Y_1$ and $f_{\alpha-1}(R_1 * E_1 * S_1; X_1, Y_1) = R * E * S$ for some R and S.

2.10. By IV.3.4 and 2.8, the relation

$$\mathrm{MutNorm}_\alpha(P * E * Q,\ R * D * S) \Rightarrow \partial(D) \leqslant n^{8\alpha}\partial(E)$$

holds.

Chapter VI. Periodic Groups of Odd Exponent

In Chapters I-V we constructed a theory of transformations of periodic words in a group alphabet

$$a_1, a_2, \ldots, a_m, a_1^{-1}, a_2^{-1}, \ldots a_m^{-1}. \tag{1}$$

This theory is essentially linked to the choice of a fixed odd natural parameter $n \geqslant 665$. In this chapter we shall describe a representation of free periodic groups of odd exponent $n \geqslant 665$, using the concepts that were considered in the preceding chapters. This representation allows us to investigate various properties of these groups and their subgroups.

§1. Existence of Infinite Periodic Groups of Odd Exponent

1.1. In this section we note some simple properties of absolutely reduced words in the alphabet (1), which were defined in I.4.34.

It follows from II.2.11 and IV.1.19 that the relation

$$\mathscr{A}_i \subset \mathscr{A}_{i+1} \tag{2}$$

holds for arbitrary $i > 0$.

Using IV.1.21, we get the condition

$$PEQ \in \mathscr{A}_i \cap \mathscr{K}_{i-1} \Rightarrow E \in \mathscr{A}_{i+1}. \tag{3}$$

It follows easily from IV.2.11, IV.2.15 and II.5.16 that

$$(X \in \mathscr{A}_j \;\&\; X \overset{i}{\sim} Y) \Rightarrow Y \in \mathscr{A}_j \tag{4}$$

for arbitrary i, j.

It follows from II.7.13 with empty E that

$$(X \in \mathscr{A}_i \;\&\; Y \in \mathscr{A}_i \;\&\; Z = [X, Y]_{i-1}) \Rightarrow Z \in \mathscr{A}_{i+1}. \tag{5}$$

If $X \in \mathscr{A}_\alpha$, then $X \in \mathscr{R}_i$ for $i \leqslant \alpha$ by IV.1.17, and by IV.1.19 and II.2.11 we have the same for $i > \alpha$. Conversely, if $X \in \mathscr{R}_i$ for all $i \geqslant 1$, then, by II.5.24, $X \in \mathscr{A}_\alpha$ for some $\alpha < \log_5 \partial(X)$, that is, $X \in \mathscr{A}$. Thus the relation

$$\mathscr{A} = \bigcup_{i=1}^{\infty} \mathscr{A}_i = \bigcap_{i=0}^{\infty} \mathscr{R}_i \tag{6}$$

holds.

1.2. *For every nonempty word* $X \in \mathscr{A}_\alpha$, *there is a word* $S \in \mathscr{A}_{\alpha+2}$ *and an elementary period* A *of rank* $\gamma \leqslant \alpha + 1$ *such that* Norm$(\alpha + 1, S, 26)$ *is empty,* A^q *occurs in some word of the class* $\mathscr{M}_{\gamma-1}$, *and the relation*

$$X \overset{\alpha}{\simeq} [S, A^r, S^{-1}]_\alpha \tag{7}$$

is satisfied for some r, $0 < r \leqslant (n + 1)/2 + 46$.

In addition, there is no occurrence of an elementary $((n + 1)/2 + 46)$*-power of any rank in the word* A^r.

We consider any word $Z \in \mathscr{K}_\alpha$ such that $X \overset{\alpha-1}{\simeq} Z$. We say that an occurrence $T * Y * T^{-1}$ in the word Z is *bounded in rank* α if it contains no more than $(n + 1)/2$ segments of an arbitrary kernel of rank i, $0 < i \leqslant \alpha$. By IV.2.20, there exists a $Z_0 \in \mathscr{M}_{\alpha-1}$ such that $X \overset{\alpha-1}{\simeq} Z_0$. Clearly, $Z_0 \in \mathscr{K}_\alpha$ and every occurrence of the form $T * Y * T^{-1}$ in this word Z_0 is bounded in rank $\alpha + 1$.

We shall call the sequence $(r_\alpha, r_{\alpha-1}, \ldots, r_2, r_1, r_0)$ the *index in rank* α of an arbitrary occurrence $P * E * Q$ in a word $Z \in \mathscr{R}_\alpha$, where $r_i = \partial_i (P * E * Q)$. We order these sequences lexicographically, that is, we suppose

$$(r_\alpha, r_{\alpha-1}, \ldots, r_2, r_1, r_0) < (s_\alpha, s_{\alpha-1}, \ldots, s_2, s_1, s_0)$$

if and only if $r_k < s_k$ for some $k \leqslant \alpha$ and $r_i = s_i$ for $k < i \leqslant \alpha$.

We choose an occurrence $T * Y * T^{-1}$ in a word $Z \in \mathscr{K}_\alpha$ that is bounded in rank $\alpha + 1$, where $X \overset{\alpha-1}{\simeq} TYT^{-1}$, and that has the smallest index in rank α relative to the lexicographical ordering in the class of all occurrences bounded in rank $\alpha + 1$. Then the word Y^q is uncancellable, that is, $Y^q \in \mathscr{K}_0$.

First we analyse the case where $Y^q \notin \mathscr{K}_{\alpha+1}$. Let γ be the smallest number such that $Y^q \notin \mathscr{K}_\gamma$. Then $0 < \gamma \leqslant \alpha + 1$, and, by IV.1.19, there exists an occurrence

$$P * H * Q \in \text{MaxNorm}(\gamma, Y^q, n - 217). \tag{8}$$

Since by (4), $Z \in \mathscr{A}_\alpha$ as $X \in \mathscr{A}_\alpha$, it follows that $\gamma < \alpha$.

By IV.1.7, there is no occurrence of an elementary $((n + 1)/2 + 43)$-power of rank γ in the word Y. Thus H does not occur in Y.

Let us show that Y occurs in H. Assume that this is not so. Then $H \mathrel{\overline{\underline{\circ}}} ED$, where $Y \mathrel{\overline{\underline{\circ}}} RE \mathrel{\overline{\underline{\circ}}} DG$ for some R and G. Since $l_\gamma(E), l_\gamma(D) \leqslant (n + 1)/2 + 42$, it follows from II.5.2 that

$$l_\gamma(D), l_\gamma(E) \geqslant \frac{n-1}{2} - 261 \geqslant 71 .$$

Without loss of generality, we may assume that $l_\gamma(E) \geqslant l_\gamma(D)$. Then

$$l_\gamma(E) \geqslant \frac{n-1}{2} - 109 \geqslant 223\,. \tag{9}$$

By IV.1.4, there exist kernels $U, V \in \operatorname{Ker}(\gamma, Y)$ such that

$$\operatorname{Comp}(U, R * E *) \quad \text{and} \quad \operatorname{Comp}(V, * D * G)\,.$$

Since Y does not occur in H, $\neg \operatorname{Comp}(* D * G, R * E *)$, so that $* D * G < R * E *$. By II.6.7 and II.6.14, we have

$$TR * ED * GT^{-1} \in \operatorname{MaxNorm}(\gamma, TYYT^{-1}, n - 217)\,. \tag{10}$$

By II.6.12, it follows from (10) that some start $TR * E_0 * u_0 T^{-1}$ of the occurrence $TR * E * T^{-1}$, where $l_\gamma(E_0) \geqslant l_\gamma(E) - 4 \geqslant 219$, is normalised and has no proper normalised continuation to the left. Let U_1 be a kernel of rank γ that is compatible with $TR * E * T^{-1}$. By II.3.10, it follows from (10) that the maximal normalised continuation $\bar{U}_1$ of the kernel U_1 is contained in $TR * E * T^{-1}$. Thus $l_\gamma(U_1) \leqslant (n + 1)/2$. Similarly, we can convince ourselves that a kernel $V_1 \in \operatorname{Ker}(\gamma, Z)$ compatible with $T * D * GT^{-1}$ is contained in it and that $l_\gamma(V_1) \leqslant (n + 1)/2$.

It follows from (9) that the kernel U_1 is active. By IV.2.21, there is a local reversal

$$Z \to Z_1 \tag{11}$$

of U_1 such that $Z_1 \in \mathscr{X}_{\gamma-1}$. Then $Z_1 \in \mathscr{X}_\gamma$ by IV.1.26, whence by IV.2.11 and IV.1.24 it follows that $Z \overset{\alpha-1}{\sim} Z_1$ and $Z_1 \in \mathscr{X}_\alpha$. By shortening the word D on the right by not more than 17 of its segments, we can arrange that the kernel V_1 is an end of the occurrence $* TD * GT^{-1}$. Then, since the reversal (11) is local, we will have $Z_1 \overset{\circ}{=} TDG_1$, where $f_{Z \to Z_1}(V_1)$ is an end of the occurrence $* TD * G_1$. In V.1.3, in an analogous situation, we proved in detail using II.7.11 and II.7.20 that the word Z_1 in the local reversal (11) may be chosen so that it has $F^{-1}T^{-1}$ as an end, where F is a maximal start of the word D satisfying the inequality

$$l_\gamma(F) \leqslant \frac{n-1}{2} - 131 \tag{12}$$

and ending with some pseudokernel of rank $\alpha - 1$. Then $G_1 \overset{\circ}{=} G_2 F^{-1} T^{-1}$ for some G_2, $D \overset{\circ}{=} FD_1$ for some D_1 and, shortening F on the left by 4 segments, we get

$$Tu * F_1 * D_1 GT^{-1} \in \operatorname{Norm}(\gamma, Z, 50)$$

in view of II.6.12, where $uF_1 \overset{\circ}{=} F$. In addition the occurrences

$$Tu * F_1 * D_1 G_2 F_1^{-1} u^{-1} T^{-1} \quad \text{and} \quad TuF_1 D_1 G_2 * F_1^{-1} * u^{-1} T^{-1}$$

will also be normalised; the first of them is an image of the occurrence $Tu * F_1 * D_1 GT^{-1}$ in the reversal (11), and the second is compatible with the kernel $f_\gamma(U_1; Z, Z_1)$.

By IV.2.23 and III.1.4, we may assume that every kernel of rank β of Z_1 contained in $TF * D_1G_2 * F^{-1}T^{-1}$, $0 < \beta < \gamma$, contains not more than $(n+1)/2$ segments. As was established in V.1.3, it follows easily from $l_\gamma(ED) \geqslant n - 217$ and the conditions on the choice of F that $TE * D_1G_2 * F^{-1}T^{-1}$ contains not more than $(n+1)/2$ segments of the kernel $f_\gamma(U_1; Z, Z_1)$. The analogous condition for the remaining kernels of rank γ of Z_1 follows from III.2.6 and the boundedness of the occurrence $T * Y * T^{-1}$ in rank $\alpha + 1$. The same is true for kernels of rank i for $\gamma < i \leqslant \alpha + 1$, by IV.2.11. Thus the occurrence $TF * D_1G_2 * F^{-1}T^{-1}$ is also bounded in rank $\alpha + 1$. On the other hand, the kernels V_1 and U_1 are contained in $T * Y * T^{-1}$, while their images $f_\gamma(V_1;\ Z,\ Z_1)$ and $f_\gamma(U_1;\ Z,\ Z_1)$ are not contained in $TF * D_1G_2 * F^{-1}T^{-1}$, that is, $\partial_\gamma(TF * D_1G_2 * F^{-1}T^{-1}) < \partial_\gamma(T * Y * T^{-1})$. Moreover, since $\gamma < i \leqslant \alpha$, we have $\partial_i(TF * D_1G_2 * F^{-1}T^{-1}) \leqslant \partial_i(T * Y * T^{-1})$ by IV.2.18. Consequently the index in rank α of the occurrence $TF * D_1G_2 * F^{-1}T^{-1}$ is less than the index of $T * Y * T^{-1}$, and this contradicts the choice of the occurrence $T * Y * T^{-1}$.

Thus, we have proved that Y occurs in H if $Y^q \notin \mathscr{X}_\gamma$. Since $l_\gamma(Y) \leqslant (n+1)/2 + 42$, it follows in view of II.5.2 and II.6.7 that the occurrence $P * H * Q$ is compatible with its shift to the left by a period Y whenever $\partial(P) \geqslant \partial(Y)$. Similarly on the right. By assumption, $Y^q \in \mathscr{X}_{\gamma-1}$. Since pseudokernels of rank $\gamma - 1$ of occurrence (8) are kernels of rank $\gamma - 1$ and $l_\gamma(H) > 4q$, there exists a kernel $V \in \mathrm{Ker}\,(\gamma - 1,\ Y^q)$ that is interior relative to the period Y and such that $\partial(\mathrm{Bas}(V)) < 2\partial(V)$. Consequently, $Y^q \in \mathrm{Per}(\gamma,\ Y)$. Let us consider a normal generating occurrence $P_1 * H_1 * Q_1$ of rank γ in the word Y^q with $\partial(H_1) > 10\partial(Y)$. Suppose that $H_1 \overset{\circ}{=} B^t B_1$, where $\partial(B) = \partial(Y)$ and B_1 is a start of B. Clearly, we may assume that $P_1 * H_1 * Q_1$ is contained in $P * H * Q$. Then the word H_1 is an elementary word of rank γ, just like H, obtained from some original elementary period, which may in general be different from Y. Since $\partial(H_1) > 10\partial(Y)$, it follows that

$$P_1 * H_1 * Q_1 \in \mathrm{Norm}(\gamma,\ Y^q,\ 9)\,. \tag{13}$$

By IV.3.12, there exists a word $Z' \in \mathscr{M}_{\gamma-1}$ such that $Y^q \overset{\alpha-1}{\sim} Z'$. Set

$$f_{\gamma-1}(P_1 * H_1 * Q_1;\ Y^q,\ Z') = P' * H' * Q'\,. \tag{14}$$

Since $Z' \in \mathrm{Int}(Y^q,\ \gamma,\ Y)$ and $l_{\gamma,Y}(H_1) \geqslant 9$, we have $H' \overset{\circ}{=} D^t D_1$ by II.4.1, where D is an image of the period B in the occurrence $P' * H' * Q'$ and D_1 is a start of D. Suppose that $D \overset{\circ}{=} A^r$, where $r > 0$ and A is a simple word. Then $H' \overset{\circ}{=} A^s A_1$, where A_1 is a start of A. On the other hand, since (13) and (14) give that $P' * H' * Q' \in \mathrm{Norm}(\gamma,\ Z',\ 9)$, we have $H' \overset{\circ}{=} C^h C_1$ by II.4.5, where C is an elementary period of rank γ. Then C is a simple word, and consequently $A \overset{\circ}{=} C$, that is, A is an elementary period of rank γ. By II.4.3, we have $X' \in \mathscr{X}_{\gamma-1}$ for some generating occurrence $R' * H' * S'$ in some word $X' \in \mathrm{Per}\,(\gamma,\ D)$, and

$$\mathrm{MutNorm}_{\gamma-1}(P' * H' * Q',\ R' * H' * S')\,.$$

Then, by II.7.15, there exists a word $T_1 \in \mathscr{A}_\gamma$ such that

$$B \overset{\gamma-1}{\simeq} [T_1, A', T_1^{-1}]_{\gamma-1}\,. \tag{15}$$

The word B is a cyclic shift of Y, that is, $B \overset{\circ}{\equiv} Y_2Y_1$, where $Y \overset{\circ}{\equiv} Y_1Y_2$. Using II.5.16 and II.5.2, we have

$$l_\gamma(A') = l_\gamma(B) \leqslant l_\gamma(Y_2) + l_\gamma(Y_1) + 2 \leqslant l_\gamma(Y) + 3 \leqslant \frac{n+1}{2} + 45\,. \tag{16}$$

Since $Z' \in \mathscr{M}_{\gamma-1}$, it follows from IV.2.12 that $A' \in \mathscr{K}_{\gamma-1}$, whence it follows from (16) and IV.1.19 that $A' \in \mathscr{K}_\gamma$. Then $A' \in \mathscr{A}_{\gamma+1}$ by II.2.12. Similarly, since $Y^q \in \mathscr{K}_{\gamma-1}$, it follows that $B \in \mathscr{R}_\gamma$ and $B \in \mathscr{A}_{\gamma+1}$. Since $\gamma < \alpha$, by (2) we have

$$A', B \in \mathscr{A}_{\alpha+1}\,.$$

By (16), II.7.12 and IV.1.19, $[T_1, A']_{\gamma-1} \in \mathscr{K}_i$ for all $i \geqslant \gamma$. Then, by V.1.1,

$$[T_1, A']_{\gamma-1} = [T_1, A']_\alpha$$

and by IV.2.11, relation (15) can be rewritten as

$$B \overset{\alpha}{\simeq} [T_1, A', T_1^{-1}]_\alpha\,. \tag{17}$$

We make the definition

$$S \rightleftharpoons [TY_1, T_1]_\alpha\,. \tag{18}$$

Since $TYT^{-1} \in \mathscr{A}_\alpha$ and $T_1 \in \mathscr{A}_\gamma$, it follows that $TY_1, T_1 \in \mathscr{A}_{\alpha+1}$. Then $S \in \mathscr{A}_{\alpha+2}$ by II.7.13, and, by II.7.12, Norm $(\alpha+1, S, 26)$ is empty. Further, using (17), (18) and the properties of coupling, we get

$$\begin{aligned} X \overset{\alpha}{\simeq} TYT^{-1} &\overset{\alpha}{\simeq} [TY_1, Y_2T^{-1}]_\alpha \overset{\alpha}{\simeq} [TY_1, [B, Y_1^{-1}T^{-1}]_\alpha]_\alpha \\ &\overset{\alpha}{\simeq} [TY_1, [[T_1, A', T_1^{-1}]_\alpha, Y_1^{-1}T^{-1}]_\alpha]_\alpha \overset{\alpha}{\simeq} [S, A', S^{-1}]_\alpha\,. \end{aligned}$$

The desired relation $r \leqslant (n+1)/2 + 46$ follows from (16). Since $Z' \in \mathscr{M}_{\gamma-1}$, it follows from II.7.10 that the word A^q occurs in some word in $\overline{\mathscr{M}}_{\gamma-1}$. Finally, utilizing IV.1.7, II.2.12 and the condition $Z' \in \mathscr{M}_{\gamma-1}$, we get easily that there are no occurrences of elemenatry $((n+1)/2 + 46)$-powers of any rank in A^q.

Suppose now that $Y^q \in \mathscr{K}_{\alpha+1}$. Since $Z \in \mathscr{A}_\alpha$, it follows that there is no occurrence of an elementary 3-power of rank $\alpha+1$ in Y. Thus there cannot exist an occurrence $V \in \mathrm{Norm}(\alpha+1, Y^q, 9)$ that is interior relative to the period Y and such that $\partial(\mathrm{Bas}(V)) < 2\partial(Y)$, that is, $Y^q \notin \mathrm{Per}\,(\alpha+2, Y)$. We denote by γ the largest number such that $Y^q \in \mathrm{Per}\,(\gamma, Y)$. By II.1.2, $\gamma \leqslant \alpha+1$. We find a word $Z' \in \mathscr{M}_{\gamma-1}$ such that $Y^q \overset{\gamma-1}{\simeq} Z'$ and

$$f_{\gamma-1}(P_1 * H_1 * Q_1;\, Y^q, Z') = P' * H' * Q'\,,$$

where $P_1 * H_1 * Q_1$ is a normal generating occurrence of rank γ of the word Y^q, $H_1 \overset{\circ}{=} B'B_1$ and B is a cyclic shift of the period Y. Then $H' \overset{\circ}{=} D'D_1$, where D is an image of the period D in the occurrence $P' * H' * Q'$. Suppose that $D \overset{\circ}{=} A'$, where A is a simple word.

Then A is a minimal period of rank γ. Let us assume that A is not elementary in rank γ. Then $A^q \in \text{Per}(\gamma + 1, A)$ by II.1.16, whence it follows easily that $P' * H' * Q'$ contains a kernel $U \in \text{Ker}(\gamma, Z')$ such that $\partial(\text{Bas}(U)) < 2\partial(A)$. Then $f_{\gamma-1}(U; Z', Y^q) \in \text{Ker}(\gamma, Y^q)$ and $\partial(\text{Bas}(f_{\gamma-1}(U; Z, Y^q))) < 2\partial(Y)$, that is, $Y^q \in \text{Per}(\gamma + 1, Y)$. But this contradicts the choice of the number γ. Consequently, A is elementary in rank γ. In the way analogous to the case considered above, we can now find the desired word $S \in \mathscr{A}_{\alpha+2}$ and prove the relations $r \leqslant (n + 1)/2 + 46$ and (7).

1.3. The *relation* $\sim$ is defined on the set $\mathscr{A}$ in the following way:

$$X \sim Y \Longleftrightarrow \exists i(X \overset{i}{\sim} Y).$$

Clearly, this relation is reflexive and symmetric. By (6) and IV.2.11, it is transitive. By IV.2.16, the condition

$$X \in \mathscr{A}_1 \Rightarrow (X \sim Y \Longleftrightarrow X \overset{\circ}{=} Y) \tag{19}$$

holds.

Let $\mathscr{B}$ be the set of all equivalence classes of $\mathscr{A}$ under $\sim$. If $X \in \mathscr{A}$, then we denote the element of $\mathscr{B}$ containing X by $\{X\}$.

1.4. We define the *binary operation* $\circ$ on $\mathscr{B}$ as follows:

$$x \circ y = z \Longleftrightarrow \exists\, XYZ\, \exists i(X \in x \,\&\, Y \in y \,\&\, Z \in z \,\&\, Z = [X, Y]_{i^.}). \tag{20}$$

Lemma. *The set $\mathscr{B}$ is a group under $\circ$. We denote it by $\Gamma(m, n)$.*

Suppose that $x, y \in \mathscr{B}$, $X \in x$ and $Y \in y$. By (2), we may assume that $X, Y \in \mathscr{A}_i$ for some $i \geqslant 1$. Set $Z = [X, Y]_{i-1}$. Then $Z \in \mathscr{A}_{i+1}$ in view of (5), that is, $x \circ y = \{Z\}$. If $[X, Y]_j = Z_1$ for some $Z_1 \in \mathscr{A}$, then by (6), V.1.1 and V.1.7 we have $Z \overset{\alpha}{\sim} Z_1$ when $\alpha = \max(i - 1, j)$, that is, $Z_1 \in \{Z\}$. This means that $\circ$ is uniquely defined on $\mathscr{B}$. By V.1.8, it is associative. Finally, by V.1.1, the class $\{\Lambda\}$ of the empty word Λ is the identity element of the semigroup thus defined, and, for arbitrary $X \in \mathscr{A}$, the element $\{X^{-1}\}$ is a left and right inverse for $\{X\}$.

1.5. Theorem. *For every $m > 1$ and every odd $n \geqslant 665$, there is an infinite group on m generators satisfying the identical relation*

$$x^n = 1. \tag{21}$$

We shall prove that the group $\Gamma(m, n)$ that we have constructed is the one we want. Suppose that $x \in \mathscr{B}$. By IV.2.20 and (4), there exists an $X \in x$ such that $X \in \mathscr{A}_i \cap \mathscr{M}_{i-1}$ for some $i > 0$. Then, by (3), every start of the word X lies in $\mathscr{A}$. If

$$X \doteq a_{i_1}^{\sigma_1} a_{i_2}^{\sigma_2} \dots a_{i_r}^{\sigma_r},$$

where $\sigma_j = \pm 1$, then by (20) we have

$$x = \{a_{i_1}^{\sigma_1}\} \circ \{a_{i_2}^{\sigma_2}\} \circ \dots \circ \{a_{i_r}^{\sigma_r}\}.$$

Thus $\boldsymbol{\Gamma}(m, n)$ is generated by the elements

$$\{a_1\}, \{a_2\}, \dots, \{a_m\}. \tag{22}$$

By I.3.2, the set $\mathscr{A}_1$ is infinite, and so it follows from (19) that $\boldsymbol{\Gamma}(m, n)$ is infinite.

Take $x \in \mathscr{B}$ and $X \in x$. Then $X \in \mathscr{A}_\alpha$ for some $\alpha > 0$. By 1.2, there exists a word $S \in \mathscr{A}_{\alpha+2}$ for which Norm$(\alpha + 1, S, 26)$ is empty, and an elementary period A of rank $\gamma \leqslant \alpha + 1$ such that the word A^q occurs in some word in $\bar{\mathscr{M}}_{\gamma-1}$, and relation (7) holds for some natural number $r \leqslant (n + 1)/2 + 46$. Then $[S, A^r]_\alpha \in \mathscr{R}_{\alpha+1}$ by II.7.12 and IV.1.19, that is, $[S, A^r]_\alpha = [S, A^r]_{\alpha+1}$. Since $A^r \in \mathscr{A}_{\gamma+1}$, we have $[S, A^r]_\alpha \in \mathscr{A}_{\alpha+3}$ by (2) and (5). Further, we get from (7) in view of (20) and (4) that

$$x = \{S\} \circ \{A^r\} \circ \{S\}^{-1}.$$

Thus it suffices for us to prove relation (21) with $x = \{A^r\}$. Since $A^i \in \mathscr{A}_{\gamma+1}$ for $0 < i \leqslant n - 3q$, it follows from IV.2.37 that

$$\{A\}^{3q} = \{A^{3q}\} = \{A^{-n+3q}\} = \{A\}^{-n+3q},$$

that is, $\{A\}^n = 1$, and consequently $\{A^r\}^n = 1$.

1.6. The theorem proved in 1.5 is a stronger version of the main result of [5]. Instead of the old upper bound 4381 on the exponent n, we have the bound 665. It seems that, using our method, it will be possible to reduce the upper bound on n even further. However, the proof of the existence of an infinite periodic group with exponent n less than 100 may require a more essential change of method. The problems for exponents 5,7,8 and 9 also remain open, but our methods do not apply in these cases.

Still greater interest resides in the open question concerning the existence of infinite groups of exponent 2^k. It is to be expected that the method set out in this book will be useful for a solution of Burnside's problem for such exponents. But there are fundamental difficulties in the way. The fact of the matter is that some of the lemmas proved above by simultaneous induction are false for even n. Thus, for example, in the contradiction of Lemma III.1.12 for $n = 2k$, the transition $a_1^k \to a_1^{-k}$ is a reversal of rank 1. For even n, Lemmas IV.2.13, IV.2.35 and IV.2.36 are also false. Moreover, since the lemmas proved in Chapters II-V are mutually dependent, the whole system of concepts collapses for even n. On the other hand, it follows trivially from Theorem 1.5 that there exist infinite groups of exponents of the form $2^k n$, where n is odd and greater than 663.

To investigate groups of exponent 2^k, we must seek new concepts with suitable properties in place of those that fail to work for even n. In particular this is so in respect of the relation Rel (V, W), which for odd n allowed us to formulate Lemma IV.2.13, which was very important for the inductive step.

§2. Systems of Defining Relations for a Free Group of Finite Exponent

We let $\mathbf{B}(m, n)$ denote the free group of exponent n on m generators. It is given by a set of generators

$$a_1, a_2, \ldots, a_m \tag{23}$$

with the identical relation (21) *, which is equivalent to a system of defining relations of the form $A^n = 1$, where A runs over all words in the alphabet (1). In this section we shall show that, for $m > 1$ and odd $n \geqslant 665$, the correspondence

$$a_i \longleftrightarrow \{a_i\} \tag{24}$$

between the generators (23) and (22) gives an isomorphism of $\mathbf{B}(m, n)$ onto $\boldsymbol{\Gamma}(m, n)$, and we shall consider the question of whether $\mathbf{B}(m, n)$ can be defined by a finite number of relations.

At the end of the section we shall show that $\mathbf{B}(m, n)$ has exponential growth for $m > 1$ and $n \geqslant 665$.

2.1. By IV.3.12, II.4.5, II.5.16 and II.7.10, for every elementary 9-power E of rank $\alpha > 0$ in the alphabet (1), there exists an elementary period A of rank α such that Rel(E, A^n) and $PA^nQ \in \mathscr{M}_{\alpha-1}$ for some P and Q.

For every $\alpha > 0$, *we choose the set* $\mathscr{E}_\alpha$ *consisting of periodic words of rank* α *of the form* A^n *with elementary period* A, *such that the following conditions are satisfied:*

(a) For every elementary word E of rank α there exists one and only one word A^n such that

$$A^n \in \mathscr{E}_\alpha \;\&\; (\mathrm{Rel}(E, A^n) \vee \mathrm{Rel}(E, A^{-n}))\,.$$

(b) If $A^n \in \mathscr{E}_\alpha$, then, for some P and Q,

$$PA^nQ \in \mathscr{M}_{\alpha-1}\,.$$

Let $\mathscr{E}$ denote the set $\bigcup_{i=1}^{\infty} \mathscr{E}_i$.

---

*For convenience of reference we shall use a uniform numbering of formulae for all sections of this chapter.

2.2. Let $\mathbf{B}(m, n, 0)$ be the free group with generators (23). For each $\alpha > 0$, we denote by $\mathbf{B}(m, n, \alpha)$ the group on the same generators (23), with defining relations

$$A^n = 1 \ (A^n \in \bigcup_{i=1}^{\alpha} \mathscr{E}_i)\,. \tag{25}$$

We shall prove below that the identical relation (21) is equivalent to the system of defining relations

$$A^n = 1 \ (A^n \in \mathscr{E})\,, \tag{26}$$

that is, $\mathbf{B}(m, n)$ is given by the generators (23) with defining relations (26).

2.3. *If $X \overset{\alpha}{\sim} Y$, then $X = Y$ in $\mathbf{B}(m, n, \alpha)$.*

Proceed by induction on α. Suppose that this is true for $\alpha < \gamma$, and that $X \overset{\gamma}{\sim} Y$. If $\gamma = 0$ or $X \overset{\gamma-1}{\sim} Y$, then everything is clear. In the contrary case, we have a sequence of real reversals of rank γ transforming X to Y. Thus it is enough to consider the case where the transition $X \to Y$ is a real reversal of rank γ. Suppose that this is a reversal of the occurrence

$$R * E * S \in \mathrm{Norm}(\gamma, X, 9)\,. \tag{27}$$

According to definition 2.1, there exists an elementary period A of rank γ such that $A^n \in \mathscr{E}_\gamma$, $PA^nQ \in \overline{\mathscr{M}}_{\gamma-1}$ for some P and Q, and either $\mathrm{Rel}(E, A^n)$, or else $\mathrm{Rel}(E, A^{-n})$. Then the relations

$$A^n = 1 = A^{-n} = C^{-n}\,, \tag{28}$$

hold in $\mathbf{B}(m, n, \gamma)$, where C is an arbitrary cyclic shift of A or A^{-1}. Let us consider the case when $\mathrm{Rel}(E, A^n)$. The case $\mathrm{Rel}(E, A^{-n})$ is analogous.

Let $P_1 * E * Q_1$ and $P_2 * A^n * Q_2$ be generating occurrences in integral words $Y_1 \in \mathrm{Int}(X_1, \gamma, B)$ and $Y_2 \in \mathrm{Int}(X_2, \gamma, B)$ respectively. By II.5.5, it follows from (27) that

$$\mathrm{MutNorm}_{\gamma-1}(R * E * S,\ P_1 * E * Q_1)\,. \tag{29}$$

Set

$$f_{\gamma-1}(P_1 * E * Q_1;\ Y_1, X_1) = P' * D * Q'\,. \tag{30}$$

By III.1.8, $l_\gamma(E) = l_{\gamma,B}(D) \leqslant n - q + 1$. Since $l_{\gamma,B}(A^n) \geqslant n - 2$, there exists a generating occurrence $P'' * D * Q''$ in a word X_2 that corresponds in phase to $P' * D * Q'$ in the sense of I.2.6, such that the occurrence $f_{\gamma-1}(P'' * D * Q''; X_2, Y_2)$ is contained in $P_2 * A^n * Q_2$. Set

$$f_{\gamma-1}(P'' * D * Q'';\ X_2, Y_2) = P_3 * F * Q_3\,. \tag{31}$$

Then $F \in \mathrm{Per}(A)$, that is, $F \overset{\circ}{=} C'C_1$, where C is a cyclic shift of A and C_1 is a start of C. Since we have $\mathrm{MutNorm}_{\gamma-1}(P' * D * Q', P'' * D * Q'')$ by II.1.8, it follows from (29) and (30) that $\mathrm{MutNorm}_{\gamma-1}(R * E * S, P'' * D * Q'')$. Then, by (31) there exists a word $Z \overset{\gamma-1}{\sim} X$ such that

$$f_{\gamma-1}(R * E * S;\ X,\ Z) = R_1 * C'C_1 * S_1\,. \tag{32}$$

By III.3.27, the transition $Z \to Y$ is a real reversal of the occurrence (32), and by III.1.10,

$$Y \overset{\gamma-1}{\sim} [R_1,\ C^{-n+t},\ C_1S_1]_{\gamma-1}\,.$$

Then, by (28) and the inductive hypothesis, the relations

$$X = Z = R_1C^tC_1S_1 = R_1C^{-n+t}C_1S_1 = Y$$

hold in $\mathbf{B}(m, n, \gamma)$, that is, $X = Y$ in $\mathbf{B}(m, n, \gamma)$.

2.4. *For every $\alpha \geqslant 0$ and every word X in the alphabet* (1), *there exists a word $Y \in \mathscr{X}_\alpha$ such that $X = Y$ in $\mathbf{B}(m, n, \alpha)$. If $\alpha \geqslant \partial(X)$, then we can choose such a Y from $\mathscr{A}_{\alpha+1}$.*

Proceed by induction on $\partial(X)$. For $\partial(X) = 0$, all is clear. Suppose that $X \overset{\circ}{\equiv} X_1a$, where a is a letter in the alphabet (1). By the inductive hypothesis, there exists a word $Y_1 \in \mathscr{X}_\alpha$ such that $X_1 = Y_1$ in $\mathbf{B}(m, n, \alpha)$. By V.1.4, there exists a $Y \in \mathscr{X}_\alpha$ such that $Y = [Y_1, a]_\alpha$, that is, $Y = [Y_2, a]_0$ for some $Y_2 \overset{\alpha}{\sim} Y_1$. Then, in view of 2.3, $Y_2 = Y_1$ in $\mathbf{B}(m, n, \alpha)$, so that we have

$$X = X_1a = Y_1a = Y_2a = Y$$

in $\mathbf{B}(m, n, \alpha)$.

If $\alpha \geqslant \partial(X)$, then $\alpha - 1 \geqslant \partial(X_1)$, and by induction we may assume that $Y_1 \in \mathscr{A}_\alpha$. Then $Y \in \mathscr{A}_{\alpha+1}$ by (5).

2.5. *For every word X in the alphabet* (1) *that is not equal to* 1 *in $\mathbf{B}(m, n)$, there exist words T and E such that $E^n \in \mathscr{E}$ and $X = TE^rT^{-1}$ in $\mathbf{B}(m, n)$ for some integer r such that $|r| \leqslant (n + 1)/2 + 46$.*

By 2.4, we may assume that $X \in \mathscr{A}_\alpha$ for some α. Then, by 1.2 and 1.3, there is a word S and an elementary period A of rank $\gamma \leqslant \alpha + 1$ such that the word A^q occurs in some word of the class $\mathscr{M}_{\gamma-1}$, and the relation $X = SA^rS^{-1}$ is satisfied in $\mathbf{B}(m, n)$ for some r, $0 < r \leqslant (n + 1)/2 + 46$. If some cyclic shift of the word A^n or of A^{-n} lies in $\mathscr{E}$, then everything is clear. In every case, according to 2.1 there exists an elementary word $E^n \in \mathscr{E}_\gamma$ such that either $\mathrm{Rel}(E^n, A^n)$, or else $\mathrm{Rel}(E^n, A^{-n})$. By II.7.22 and 2.3, in the first case there exists a word H such that $A = HEH^{-1}$ in $\mathbf{B}(m, n)$. Then $X = TE^rT^{-1}$, where $T \rightleftharpoons SH$. In the second case, we get similarly that $A = HE^{-1}H^{-1}$, so that $X = SHE^{-r}H^{-1}S^{-1}$ in $\mathbf{B}(m, n)$. That was what we wanted to prove.

2.6. Theorem. *The correspondence* (24) *generates an isomorphism of* $\mathbf{B}(m, n)$ *onto* $\boldsymbol{\Gamma}(m, n)$.

It follows immediately from (3) and (20) that, if $X \in \mathscr{A}_\alpha \cap \mathscr{K}_{\alpha-1}$ and $X \doteq d_1 d_2 \ldots d_r$, where the d_i are letters in the alphabet (1), then $\{d_1\} \circ \{d_2\} \circ \ldots \circ \{d_r\} = \{X\}$. Consequently, by 2.4 it is enough to prove that the condition

$$X = Y \text{ in } \mathbf{B}(m, n) \Longleftrightarrow \{X\} = \{Y\} \text{ in } \boldsymbol{\Gamma}(m, n) \tag{33}$$

holds for arbitrary $X, Y \in \mathscr{A}$. In one direction this follows from the fact that identity (21) holds in $\boldsymbol{\Gamma}(m, n)$, and in the other direction it follows from 1.3 and 2.3.

2.7. Let $\mathscr{B}_\alpha$ be the set of all equivalence classes of words that $\mathscr{R}_\alpha$ splits into under the relation $\overset{\alpha}{\sim}$. Denoting by $\{X\}_\alpha$ the class of $\mathscr{B}_\alpha$ containing the arbitrary word $X \in \mathscr{R}_\alpha$, we get analogously to 1.4 that *the set* $\mathscr{B}_\alpha$ *is a group under the operation*

$$\{X\}_\alpha \circ \{Y\}_\alpha \rightleftharpoons \{[X, Y]_\alpha\}_\alpha, \tag{34}$$

generated by the elements

$$\{a_1\}_\alpha, \{a_2\}_\alpha, \ldots, \{a_m\}_\alpha.$$

We shall denote this group by $\boldsymbol{\Gamma}(m, n, \alpha)$.

It is easy to convince oneself that, for arbitrary $\alpha \geqslant 0$, the correspondence

$$a_i \longleftrightarrow \{a_i\}_\alpha$$

generates an isomorphism of $\mathbf{B}(m, n, \alpha)$ onto $\boldsymbol{\Gamma}(m, n, \alpha)$. Indeed, 2.4 tells us that it is enough to prove the following assertion.

2.8. *For arbitrary words* $X, Y \in \mathscr{R}_\alpha$, *the following condition is satisfied:*

$$X = Y \textit{ in } \mathbf{B}(m, n, \alpha) \Longleftrightarrow X \overset{\alpha}{\sim} Y. \tag{35}$$

By 2.3, it remains to prove that $X \overset{\alpha}{\sim} Y$ whenever $X = Y$ in $\mathbf{B}(m, n, \alpha)$.

Suppose that $A^n = 1$ is an arbitrary defining relation of $\mathbf{B}(n, m, \alpha)$, that is, $A^n \in \mathscr{E}_\gamma$ for some $\gamma \leqslant \alpha$. Since, according to 2.1, A is an elementary period of rank γ and the word A^n occurs in some word in $\mathscr{M}_{\gamma-1}$, it follows that $A^r \in \mathscr{K}_\alpha$ for some r, $|r| \leqslant n - 3q$, in view of IV.1.21, IV.1.19 and II.2.12. Then $A^{3q} \overset{\gamma}{\sim} A^{-n+3q}$ by IV.2.37, so that $A^{3q} \overset{\alpha}{\sim} A^{-n+3q}$. Further, by (34), the relation

$$\{A\}_\alpha^{3q} = \{A^{3q}\}_\alpha = \{A^{-n+3q}\}_\alpha = \{A_\alpha\}^{-n+3q}$$

is satisfied in $\boldsymbol{\Gamma}(m, n, \alpha)$, that is, $\{A\}_\alpha^n = 1$ in $\boldsymbol{\Gamma}(m, n, \alpha)$. Thus all the defining relations of $\mathbf{B}(m, n, \alpha)$ hold in $\boldsymbol{\Gamma}(m, n, \alpha)$. So $\{X\}_\alpha = \{Y\}_\alpha$ in $\boldsymbol{\Gamma}(m, n, \alpha)$, that is, $X \overset{\alpha}{\sim} Y$, whenever $X = Y$ in $\mathbf{B}(m, n, \alpha)$.

2.9. Theorem. *The relations* (26) *form a system of defining relations for* $\mathbf{B}(m, n)$.

Let $\mathbf{B}'(m, n)$ be the group given by the generators (23) and defining relations (26). Clearly, for every word X in the alphabet (1), $X = 1$ in $\mathbf{B}(m, n)$ whenever $X = 1$ in $\mathbf{B}'(m, n)$. We shall show that the converse of this is also true. Let X be a word in the alphabet (1) such that $X = 1$ in $\mathbf{B}(m, n)$, and set $\alpha = \partial(X)$. By 2.4, there exists a word $Y \in \mathscr{K}_\alpha \cap \mathscr{A}_{\alpha+1}$ such that $X = Y$ in $\mathbf{B}(m, n, \alpha)$. Then $X = Y$ in $\mathbf{B}'(m, n)$ and in $\mathbf{B}(m, n)$, that is, it is enough to prove that $Y = 1$ in $\mathbf{B}'(m, n)$. But if $Y = 1$ in $\mathbf{B}(m, n)$, it follows from (33) that $Y \overset{i}{\sim} \Lambda$ for some $i \geqslant 0$. Since in addition $Y \in \mathscr{R}_i$, $Y = 1$ in $\mathbf{B}(m, n, i)$ by (35), and this means that the same is true in $\mathbf{B}'(m, n)$

2.10. Theorem. *There exists an algorithm for solving the word problem in* $\mathbf{B}(m, n)$, *for odd* $n \geqslant 665$.

According to the principle of effectiveness and by 2.4, there is an algorithm that determines, for an arbitrary word X in the alphabet (1), a word $Y \in \mathscr{A}$ such that $X = Y$ in $\mathbf{B}(m, n)$. Then from (33), (19) and the fact that $\Lambda \in \mathscr{A}_1$, it follows that

$$X = 1 \text{ in } B(m, n) \Longleftrightarrow Y \overset{0}{\sim} \Lambda .$$

2.11. Up to this point, we have not been interested in the question of the existence of elementary words of rank α for arbitrary $\alpha > 0$. Now, for arbitrary $\alpha > 0$, we shall indicate such elementary periodic words of rank α. We shall need this construction in the proof of Theorem 2.13.

Let

$$t_0, t_1, t_2, \ldots, t_i, \ldots \tag{36}$$

be a sequence consisting of the numbers 1 and 2, where $t_0 = 1$, $t_1 = 2$ and no word $t_0t_1t_2 \ldots t_i$ can be represented in the form $RZZZQ$ with non-empty Z (see §3, Ch.I). We define words B_i, $i = 0,1,2, \ldots$, by induction on i:

$$B_0 \rightleftharpoons \Lambda ,$$

$$B_{i+1} \rightleftharpoons (B_i a_{t_i})^{q_1} ,$$

where t_i is the i-th term of the sequence (36).

2.12. *For every* $i \geqslant 0$, B_{i+1} *is an elementary word of rank* $i + 1$ *with period* $B_i a_{t_i}$, *and there is no occurrence in it of an elementary q-power of rank* $\leqslant i$.

Proceed by induction on i. For $i = 0$ everything is clear, since a_1 is an elementary period of rank 1. Suppose that $j > 0$ and our assertions are true for $i < j$. Let us prove them for $i = j$.

Suppose that some elementary q-power E of rank $r \leqslant j$ does occur in B_{j+1}. Suppose that $B_{j+1} \overset{0}{=} PEQ$. We may assume that no elementary q-powers of rank $< r$ occur in B_{j+1}. Since $B_{j+1} \in \mathscr{R}_0$, we have $B_{j+1} \in \mathscr{R}_{r-1}$ by IV.1.19. By the inductive hypothesis, B_j is an elementary word of rank j with period $B_{j-1}\, a_{t_{j-1}}$, and E cannot occur in B_j for $r < j$. For $r = j$, E does not occur in B_j; this by II.6.7. Consequently some one-letter occurrence

$$V \rightleftharpoons (B_j a_{t_j})^k B_j * a_{t_j} * (B_j a_{t_j})^{q_1-k-1} \tag{37}$$

is contained in $P * E * Q$, that is, $E \overset{0}{=} E_1 a_{t_j} E_2$, where E_1 is an end of $(B_j a_{t_j})^k B_j$, and E_2 is a start of $(B_j a_{t_j})^{q_1-k-1}$. Since $1 \leqslant r \leqslant j$, in view of the definition of the words B_i we have that B_r is a start of B_j and $B_j a_{t_j} \overset{\circ}{\equiv} D B_r F$ for some D and F, where

$$F \overset{\circ}{\equiv} a_{t_r} a_{t_{r+1}} \ldots a_{t_j} .$$

Because of the choice of the sequence (31), no word of the form Z^3 occurs in the word F with non-empty Z. The word B_{j+1} can be rewritten in the form

$$B_{j+1} \overset{\circ}{\equiv} R B_r F B_r S ,$$

where the occurrence (37) is an end of $RB_r * F * B_r S$, and if the word R(or S) is non-empty, then $B_r a_{t_r}$ is an end of R(or $a_{t_r} B_r$ is a start of S, respectively). If the occurrences $R * B_r * FB_r S$ and $RB_r F * B_r * S$ are compatible, then $B_r F B_r$ is an elementary word of rank r, and, since by assumption no elementary q-power of rank $< r$ occurs in $B_r F B_r$, it follows by II.4.8 that $B_r F B_r$ is a periodic word. Its period must coincide with the period of the word B_r. For $r = 1$ that is impossible, since F starts with the letter $a_{t_1} \overset{\circ}{\equiv} a_2$, and for $r > 1$ it would mean that the word $B_1 \overset{\circ}{\equiv} a_1^{q_1}$ occurs in F. Thus

$$\neg \mathrm{Comp}\,(R * B_r * FB_r S,\ RB_r F * B_r * S) . \tag{38}$$

Then the occurrence $P * E * Q$ is not compatible with at least one of these occurrences. Suppose that

$$\neg \mathrm{Comp}\,(P * E * Q,\ R * B_r * FB_r S) . \tag{39}$$

The second case is to be considered in the analogous way, and we shall omit it. By II.6.7, it follows from (39) that $E_1 a_{t_j}$ is an end of the word $B_r F$. Moreover, since by assumption no periodic 4-power of rank 1 occurs in F, we have $l_r(E_1 a_{t_j}) \leqslant 17 + 3 + 2 = 22$, by II.6.7 and II.5.2. Consequently, $l_r(E_2) \geqslant q - 26$. Then $P * E * Q$ is compatible with $RB_r F * B_r * S$, the word S is not empty, and $S \overset{\circ}{\equiv} a_{t_r} B_r S_1$ for some S_1, as was remarked above. Since $l_r(B_r) = q_1$, in view of II.5.2 and II.6.7 the occurrence $P * E * Q$ is compatible also with $RB_r F B_r a_{t_r} * B_r * S_1$. But this is false, because of the transitivity of the relation Comp (U, V) and relation (38) with $F \overset{\circ}{\equiv} a_{t_r}$. Thus no elementary q-power of rank $\leqslant j$ occurs in B_{j+1}. Then $B_{j+1} \in \mathscr{M}_j$, by IV. 1.19.

We show now that B_{j+1} is an elementary word of rank $j + 1$ with period $B_j a_{t_j}$. As is clear from (38) (with $r = j$), the occurrences $V_s \rightleftharpoons (B_j a_{t_j})^s * B_j * a_{t_j}(B_j a_{t_j})^{q_1 - s - 1}$, $s = 0,1,2, \ldots, q_1 - 1$, are pairwise incompatible. Let W be the maximal continuation of the occurrence

$$(B_j a_{t_j})^9 * B_j * a_{t_j}(B_j a_{t_j})^{q_1 - 10} .$$

Since W is not compatible with V_8 and V_{10}, by II.6.7 it is contained in $(B_j a_{t_j})^8 * B_j a_{t_j} B_j a_{t_j} B_j * a_{t_j}(B_j a_{t_j})^{q_1 - 11}$, that is, $W \in \mathrm{Inn}(B_{j+1}, B_j a_{t_j})$. By II.1.15, $B_{j+1} \in \mathrm{Per}(j + 1, B_j a_{t_j})$. Since B_{j+1} does not have active kernels of ranks $\leqslant j$, it follows that $Y \overset{\circ}{\equiv} B_{j+1}$ whenever $Y \angle B_{j+1}$, for arbitrary $Y \in \mathscr{R}_j$. Then every semi-integral

9-power of rank $j+1$ contained in B_{j+1} is a periodic word, and, as $B_j a_{t_j}$ is a simple word, by I.4.9 we get that $B_j a_{t_j}$ is a minimal period of rank $j+1$ and, by I.4.10, this period is elementary in rank $j+1$.

2.13. Theorem. *For $m > 1$ and odd $n \geqslant 665$, $\mathbf{B}(m, n)$ cannot be defined by a finite number of defining relations.*

Suppose that $\mathbf{B}(m, n)$ can be defined by finitely many relations. Then it is isomorphic with the group $\mathbf{G}$ given by the generators (23) and a finite set of defining relations

$$D_i = 1 \quad (i = 1, 2, \ldots, k) \tag{40}$$

where D_i is a word in the alphabet (1). This means that, for an arbitrary word X in the alphabet (1),

$$X = 1 \text{ in } \mathbf{B}(m, n) \Longleftrightarrow X = 1 \text{ in } \mathbf{G}. \tag{41}$$

Set $\alpha \rightleftharpoons \max_{1 \leq i \leq k} \partial(D_i) + 1$. By 2.4, there exist words $C_i \in \mathscr{A}_\alpha$ such that $C_i = D_i$ in $\mathbf{B}(m, n, \alpha)$. Since $C_i = 1$ in $\mathbf{B}(m, n)$, by (33) we have $C_i \overset{\gamma_i}{\sim} \Lambda$ for some γ_i. Then $C_i \,\overline{\overline{\circ}}\, \Lambda$ in view of IV.2.16, that is, $C_i = 1$ in $\mathbf{B}(m, n, \alpha)$. Consequently, all the defining relations (40) of $\mathbf{G}$ are satisfied in $\mathbf{B}(m, n, \alpha)$. By (41), this means that we have, for an arbitrary word X in the alphabet (1),

$$X = 1 \text{ in } \mathbf{B}(m, n) \Rightarrow X = 1 \text{ in } \mathbf{B}(m, n, \alpha). \tag{42}$$

Set $X \rightleftharpoons (B_\alpha a_{t_\alpha})^n$. By (42), it follows that $X = 1$ in $\mathbf{B}(m, n, \alpha)$ if $X = 1$ in $\mathbf{B}(m, n)$. By 2.12 and IV.1.19, $B_{\alpha+1} \in \mathscr{R}_\alpha$. Then $X \overset{\alpha}{\sim} 1$ by II.6.13 and (35), whence in view of IV.2.16 it follows that $X \,\overline{\overline{\circ}}\, \Lambda$. We have therefore obtained a contradiction.

2.14. The *growth function* of an infinite group $\mathbf{G}$, in which a system $a_1, a_2, \ldots, a_m$ of generators has been chosen, is the function $\gamma(s)$ of a natural argument s defined as the number of different elements of $\mathbf{G}$ representable in the form of a product of not more than s components of the form $a_i^{\pm 1}$. By $\bar{\gamma}(s)$ we will mean the number of elements that have exactly s such components. Obviously, we have for any group $\mathbf{G}$

$$\gamma(s) \geqslant \bar{\gamma}(s).$$

For an absolutely free group on two generators, we have the obvious equation

$$\bar{\gamma}(s) = 4 \cdot 3^{s-1}$$

for $s > 0$.

We shall say that a function $\gamma(s)$ is *equivalent* to a function $\gamma'(s)$, if there exist c and c' such that

$$\gamma(s) \leqslant \gamma'(cs) \quad \text{and} \quad \gamma'(s) \leqslant \gamma(c's)$$

for all $s > 0$.

The equivalence class containing the growth function of a given finitely generated group does not depend on the choice of the generating set. J. Milnor posed the following question in [19]:

"*Is the growth function of every finitely generated group equivalent either to some power function* s^i *or to the exponential function* 2^s?".

In [20, 21, 22] it was shown that the growth function of every soluble-by-finite group is a power function if it is nilpotent-by-finite, and an exponential function otherwise.

2.15. Theorem. *For* $m > 1$ *and odd* $n \geqslant 665$, $\mathbf{B}(m, n)$ *has exponential growth.*

Clearly, we may restrict to the case $m = 2$. Let $\gamma(s)$ be the growth function of $\mathbf{B}(2, n)$. For $s \geqslant 0$ we denote by $\beta(s)$ the number of uncancellable words of length s in the alphabet (1) that do not contain subwords of the form A^8. Since all such words are contained in $\mathscr{A}_1$, by relations (19) and (33) we have

$$\bar{\gamma}(s) \geqslant \beta(s). \tag{43}$$

For $s > 0$, every uncancellable word of length $s + 1$ is obtained from a word of length s, by adding on the right one of three letters from the alphabet (1), different from b^{-1}, where b is the last letter of the said word of length s.

For $s \geqslant 8i$, we denote by $\delta_i(s)$ the number of uncancellable words of the form XB^8, where $i = \partial(B)$, $s = \partial(XB^8)$ and no word of the form A^8 occurs in X. We have

$$\delta_i(s) \leqslant \beta(s - 8i) \cdot 4^i.$$

If no word of the form A^8 occurs in Y, then such words can occur only at the end of Ya, where a is any letter. Thus

$$\begin{aligned} \beta(s + 1) &\geqslant 3\beta(s) - \sum_{i=1}^{[(s+1)/8]} \delta_i(s + 1) \\ &\geqslant 3\beta(s) - \sum_{i=1}^{[(s+1)/8]} \beta(s - 8i) \cdot 4^i, \end{aligned} \tag{44}$$

where $[(s + 1)/8]$ is the integer part of $(s + 1)/8$.

By induction on s, we can prove the inequality

$$\beta(s + 1) > (2.9)\, \beta(s). \tag{45}$$

For $0 < s < 7$, this inequality is obvious. Assume that it is true for $s < t$. Then for $1 \leqslant i \leqslant [(t + 1)/8]$ we have

$$\beta(t) > 2^{8i-1}\beta(t + 1 - 8i).$$

Replacing s by t in (44), we get

$$\beta(t + 1) \geqslant 3\beta(t) - \sum_{i=1}^{[(t+1)/8]} \frac{\beta(t) \cdot 4^i}{2^{8i-1}} \geqslant \beta(t) \left(3 - \sum_{i=1}^{\infty} \frac{1}{2^{6i-1}}\right) > (2.9) \cdot \beta(t).$$

Since $\beta(1) = 4$, it follows from (45) that $\beta(s) \geqslant 4\cdot(2.9)^{s-1}$ for arbitrary $s > 0$, whence by (43) we get

$$\bar{\gamma}(s) \geqslant 4\cdot(2.9)^{s-1} .$$

This is what we wanted to prove.

2.16. It follows by the arguments of 2.15 that, for an arbitrary natural number k, we can find an odd number N such that the function $\gamma(s)$ for the group $\mathbf{B}(2, N)$ satisfies the inequality

$$\bar{\gamma}(s) \geqslant 4\left(3 - \frac{1}{k}\right)^s .$$

Indeed, if we take $N = 665(2k + 1)$, then we can increase the values of all numerical parameters k times, that is, we can take $p = 9k$. If we then take a word of the form XB^{9k-1} instead of one of the form XB^8, we get in place of (45) the inequality

$$\beta(s + 1) > \left(3 - \frac{1}{k}\right) \beta(s) .$$

In connection with theorem 2.15, it is likely that the answer to the question posed by Kesten in [23] (p. 353) on the reflexivity of a random walk on the periodic group $\mathbf{B}(m, n)$ for $m > 1$ and odd $n \geqslant 665$ is in the negative.

§ 3. Subgroups of a Free Group of Finite Exponent

Working with the representation of the groups $\mathbf{B}(m, n)$ obtained in §2 for odd $n \geqslant 665$, we shall prove, in this section, that the abelian subgroups of these groups are just the cyclic subgroups and that $\mathbf{B}(3, n)$ is a subgroup of $\mathbf{B}(2, n)$ for the stated values of n.

3.1. *For each element X, if $X \neq 1$ in $\mathbf{B}(m, n)$ there exists an element Z of order n such that $X = Z^r$ in $\mathbf{B}(m, n)$ for some r, and each element Y which commutes with X is equal in $\mathbf{B}(m, n)$ to some power of Z.*

Suppose that X is an arbitrary word in the alphabet (1) and $X \neq 1$ in $\mathbf{B}(m, n)$. By 2.5 and 2.1, there exist words S and A such that A is an elementary period of rank γ, A^n occurs in some word in $\mathscr{M}_{\gamma-1}$ and $X = SA^rS^{-1}$ in $\mathbf{B}(m, n)$ for some r such that $0 < r \leqslant (n + 1)/2 + 46$. By 2.4, we may assert that $S \in \mathscr{A}$.

We shall prove that the word SAS^{-1} defines the required element Z. If the order of SAS^{-1} were less than n, then for some i satisfying $0 < |i| < (n + 1)/2$, we would get $A^i = 1$ in $\mathbf{B}(m, n)$. Since $A^i \in \mathscr{A}_{\gamma+1}$, by (33) and IV.2.16, $A^i \mathrel{\overset{\circ}{=}} A$, that is $i = 0$. Therefore SAS^{-1} has order n in $\mathbf{B}(m, n)$.

Suppose that $XY = YX$ in $\mathbf{B}(m, n)$. Then in $\mathbf{B}(m, n)$,

$$A^r = HA^rH^{-1}, \tag{46}$$

where $H \rightleftharpoons S^{-1}YS$. By 2.4, IV.2.20 and (4), we may assert that $H \in \mathscr{M}_\alpha \cap \mathscr{A}_{\alpha+1}$ for some α. Then $H \in \mathscr{M}_i$ for an arbitrary i. By IV.1.21, $A^j \in \mathscr{L}_{\gamma-1}$ for every $j \leqslant n$.

Since for any i, $X^iY = YX^i$ and $A^n = 1$ in $\mathbf{B}(m, n)$, we may assume that r satisfies the condition

$$\frac{n}{3} \leqslant r < \frac{n+1}{2}. \tag{47}$$

Since $A^r, H \in \mathscr{A}$, it follows from (46) by 2.6 that the relation

$$\{A^r\} = \{H\} \circ \{A^r\} \circ \{H\}^{-1}$$

is satisfied in $\boldsymbol{\Gamma}(m, n)$, that is, we have for some k and j that

$$A^r \overset{j}{\sim} [H, [A^r, H^{-1}]_k]_k, \tag{48}$$

where we may assert, in view of V.1.1, that $k \geqslant \gamma$.

By V.1.3 we can find a word $Z_1 \in \mathscr{K}_{\gamma-1}$ such that $Z_1 = [H, A^n]'_{\gamma-1}$, that is, there exist words

$$H_1T \in \mathscr{K}_{\gamma-1} \quad \text{and} \quad T^{-1}F \in \mathscr{K}_{\gamma-1}$$

such that

$$H \overset{\gamma-1}{\rightleftarrows} H_1T, \quad A^n \overset{\gamma-1}{\rightleftarrows} T^{-1}F \quad \text{and} \quad Z_1 \overset{\circ}{=} H_1F, \tag{49}$$

where the hypotheses of V.1.2(a), (b) and (c) hold.

Suppose that $A \overset{\circ}{=} A_1A_2 \overset{\circ}{=} A_3A_4$, where

$$A^9A_1 * A_2A^9A_3 * A_4A^{n-20} \in \mathrm{Reg}(\gamma - 1, A^n).$$

Then, for an arbitrary $0 < i \leqslant n - 10$, the occurrence

$$V_i \rightleftharpoons A^4A_1 * A_2A^iA_3 * A_4A^{n-i-6}$$

is normalised. Let t be the largest integer for which the occurrence

$$f_{\gamma-1}(V_t; A^n, T^{-1}F) \tag{50}$$

is contained in $* \, T^{-1} * F$. Since $H \in \mathscr{M}_\gamma$, it follows that there are no elementary $((n+1)/2 + 43)$-powers of rank γ occurring in T, that is, we have

$$t \leqslant \frac{n+1}{2} + 42 . \tag{51}$$

We consider first of all the case where $t \geqslant 13$. By II.7.6, V.1.3 and condition V.1.2(c), we may assert that, if $T^{-1} \overset{\circ}{\equiv} PEQ$, where

$$P * E * QF \in \mathrm{Norm}(\gamma, T^{-1}F, 9)$$

and

$$H_1 Q^{-1} * E^{-1} * P^{-1} \in \mathrm{Norm}(\gamma, H_1 T, 9) ,$$

then, for some Q_1,

$$f_{\gamma-1}(P * E * QF; T^{-1}F, A^n) = P * E * Q_1 .$$

Then by II.6.12, the word A^{t+1} is a start of T^{-1} for $t \geqslant 13$, that is, for some T_1 we have

$$T^{-1} \overset{\circ}{\equiv} A^{t+1} T_1 .$$

In addition

$$l_\gamma(A^{t+1} T_1) \leqslant t + 5 \tag{52}$$

and

$$H^{-1} \overset{\gamma-1}{\eqsim} T^{-1} H_1^{-1} \overset{\circ}{\equiv} A^{t+1} T_1 H_1^{-1} . \tag{53}$$

Set

$$U \rightleftharpoons A^{t+7} A_1 * A_2 A^9 A_3 * A_4 A^{n-t-18} .$$

It follows from the definition of t that $f_{\gamma-1}(U; A^n, T^{-1}F)$ is contained in $T^{-1} * F *$, whence V.1.2(a) yields that A^{n-t-8} is an end of F. Suppose that

$$F \overset{\circ}{\equiv} F_1 A^{n-t-8} .$$

Since $A^n \overset{\gamma-1}{\eqsim} T^{-1}F$, it then follows by V.2.1 that

$$A^{t+23} \overset{\gamma-1}{\eqsim} T^{-1} F_1 A^{15} . \tag{54}$$

Clearly, $F_1 A^{r-7} T_1$ is an elementary word of rank γ with the original period A. Since the occurrence (50) is contained in $* \, T^{-1} * F$ we have

$$l_\gamma(F) = l_\gamma(F_1 A^{n-t-8}) \leqslant n - t - 5 , \tag{55}$$

and hence

$$l_\gamma(F_1A^{r-7}) \leqslant r-4,$$

from which we get by (52) that

$$l_\gamma(F_1A^{r-7}T_1) \leqslant r < \frac{n+1}{2}. \qquad (56)$$

Since $H_1F \in \mathscr{K}_{\gamma-1}$ and $T^{-1}H_1^{-1} \in \mathscr{K}_{\gamma-1}$, it follows by II.6.13 that

$$H_1F_1A^{r-7}T_1H_1^{-1} \in \mathscr{K}_{\gamma-1}.$$

We assume that for some i,

$$H_1F_1A^{r-7}T_1H_1^{-1} \notin \mathscr{K}_i.$$

Choose i minimal with this property. Then $i \geqslant \gamma$ and by IV.1.19 there exists an occurrence

$$V \in \mathrm{Norm}(i,\ H_1F_1A^{r-7}T_1H_1^{-1},\ n-217).$$

Since $H \in \mathscr{M}_i$, it follows by (49) that there are no elementary $((n+1)/2+43)$-powers of rank i occurring in H_1. Since by II.2.12 there are no elementary 3-powers of rank $> \gamma$ occurring in the word $F_1A^{r-7}T_1$, it follows by II.5.2 and II.5.21 that there are no elementary $(((n+1)/2+43)+3+2+9+2)$-powers of ranks $> \gamma$ occurring in the word $H_1F_1A^{r-7}T_1H_1^{-1}$, that is, i cannot be greater than γ. If $i = \gamma$, then by II.6.7, V is compatible with the occurrence

$$W \rightleftharpoons H_1F_1 * A^{r-7} * T_1H_1^{-1}.$$

On the other hand, since $13 \leqslant t \leqslant (n+1)/2+42$ and $T^{-1}F \in \mathscr{K}_{\gamma-1}$, it follows by II.3.10 that the maximal normalised occurrence compatible with $H_1T_1^{-1} * A^{-t-1} *$ (or with $H_1F_1 * A^{n-t-8} *$) does not intersect $* H_1 * T_1^{-1}A^{-t-1}$ ($* H_1 * F_1A^{n-t-8}$, respectively). Therefore the maximal normalised continuation of W is contained in $H_1 * F_1A^{r-7}T_1 * H_1^{-1}$. Then by (56), V is not compatible with W. Hence

$$H_1F_1A^{r-7}T_1H_1^{-1} \in \mathscr{K}_i$$

for arbitrary i. By IV.1.21 and relations (49) and (6), the words

$$H_1F_1A^{15},\ A^{r-22}T_1H_1^{-1},\ H_1T \quad \text{and} \quad A^{t+23}$$

belong to $\mathscr{R}_k$. Since $k \geqslant \gamma$, using (53), (49), (54) and the properties of coupling of rank k, we find

$$\begin{aligned}[H, [A^r, H^{-1}]_k]_k &= [H, [A^r, A^{t+1}T_1H_1^{-1}]_k]_k \\ &= [H_1T, [A^{t+23}, A^{r-22}T_1H_1^{-1}]_k]_k \\ &= [[H_1T, T^{-1}F_1A^{15}]_k, A^{r-22}T_1H_1^{-1}]_k \\ &= [H_1F_1A^{15}, A^{r-22}T_1H_1^{-1}]_k = H_1F_1A^{r-7}T_1H_1^{-1},\end{aligned}$$

from which it follows by (48) that $A^r \overset{j}{\sim} H_1F_1A^{r-7}T_1H_1^{-1}$. Since there are no active kernels of ranks $> \gamma$ in A^r, by IV.2.15 we may assert that $j = \gamma$. By (47), III.3.24 and IV.1.5, the occurrence $H_1F_1 * A^{r-7} * T_1H_1^{-1}$ is compatible with some active kernel U of rank γ which is related to its image

$$f_\gamma(U; H_1F_1A^{r-7}T_1H_1^{-1}, A^r).$$

Therefore, we have by IV.2.13 that

$$A^r \overset{\gamma}{\rightleftarrows} H_1F_1A^{r-7}T_1H_1^{-1}. \tag{57}$$

By II.7.10, the occurrence

$$W_1 \rightleftharpoons H_1F_1A^{15}A_1 * A_2A^{r-28}A_3 * A_4A^4T_1H_1^{-1}$$

is periodised in rank $\gamma - 1$, and by II.7.21 we have for some s

$$f_{\gamma-1}(W_1;\ H_1F_1A^{r-7}T_1H_1^{-1},\ A^r) = A^sA_1 * A_2A^{r-28}A_3 * A_4A^{26-s},$$

from which there follows by IV.2.35 that

$$H_1F_1A^{15}A_1 \overset{\gamma-1}{\rightleftarrows} A^sA_1.$$

We then have by 2.3, (53) and (54) the equalities

$$H = H_1T = H_1F_1A^{-t-8} = A^{s-t-23}$$

in $\mathbf{B}(m, n)$. Then

$$Y = SA^{s-t-23}S^{-1} = (SAS^{-1})^{s-t-23},$$

that is, SAS^{-1} is the required Z.

We have thus proved our assertion for $t \geqslant 13$.

We now consider the coupling $Z = [A^n, H^{-1}]_{\gamma-1}'$, that is, we find words $GR^{-1} \in \mathscr{K}_{\gamma-1}$ and $RH_2 \in \mathscr{K}_{\gamma-1}$ such that

$$A^n \overset{\gamma-1}{\rightleftarrows} GR^{-1}, \quad H^{-1} \overset{\gamma-1}{\rightleftarrows} RH_2, \quad Z_2 \overset{\circ}{=} GH_2$$

and conditions V.1.2(a), (b), (c) are satisfied. Suppose that

$$U_t \rightleftharpoons A^{n-t-6}A_1 * A_2A^tA_3 * A_4A^4$$

for $0 < i \leqslant n-1$, and let t' be the largest number for which

$$f_{\gamma-1}(U_{t'};\ A^n,\ GR^{-1})$$

is contained in $G * R^{-1} *$. Then, by analogy with (51) and (54), we obtain $t' \leqslant (n+1)/2 + 42$ and

$$A^{t'+23} \overset{\gamma-1}{\rightleftharpoons} A^{15}G_1R^{-1},$$

where $G \overset{\circ}{=} A^{n-t'-8}G_1$. If in addition $t' \geqslant 13$, by symmetry we may repeat the arguments adduced for the case $t \geqslant 13$ in their entirety.

Suppose that $t < 13$ and $t' < 13$ simultaneously.

Since $H_1F,\ GH_2 \in \mathscr{K}_{\gamma-1}$, it follows by II.6.13 that

$$H_1F_1A^{r-t-t'-16}G_1H_2 \in \mathscr{K}_{\gamma-1}\,. \tag{58}$$

It is clear that $F_1A^{r-t-t'-16}G_1$ is an elementary word of rank γ with initial period A, where by (55) and the analogous relation for G_1, we have

$$l_\gamma(F_1A^{r-t-t'-16}G_1) \leqslant r - t - t' - 10 < \frac{n}{2}\,. \tag{59}$$

Assume that $H_1F_1A^{r-t-t'-16}G_1H_2 \notin \mathscr{K}_i$ for some i. We choose i minimal with this property. Then by IV.1.19, there exists an occurrence

$$V \in \operatorname{MaxNorm}(i,\ H_1F_1A^{r-t-t'-16}G_1H_2,\ n-217)\,. \tag{60}$$

Assume that $i > \gamma$. Since by hypothesis there are no elementary $((n+1)/2+43)$-powers of rank i occurring in H_1 and H_2, by II.2.12 and II.5.2 we have

$$V = H' * D_1F_1A^{r-t-t'-16}G_1D_2 * H''\,,$$

where $H'D_1 \overset{\circ}{=} H_1$, $D_2H'' \overset{\circ}{=} H_2$,

$$l_i(D_1) \geqslant n - 217 - l_i(D_2) - 3 - 4 \geqslant n - 224 - \left(\frac{n+1}{2} + 42\right) > 60\,,$$

and, analogously, $l_i(D_2) > 60$. Then, on the one hand, we have $\operatorname{Rel}(D_1, D_2)$ by II.5.17, and on the other hand, since

$$RD_2H'' \overset{i-1}{\rightleftharpoons} H^{-1} \overset{\circ}{=} T^{-1}D_1^{-1}(H')^{-1}\,,$$

we get $\operatorname{Rel}(D_2, D_1^{-1})$ by II.5.16, II.6.7 and II.5.17. This is impossible, by II.5.21.

In view of (58), it remains to consider the case $i = \gamma$. In this case, V is compatible with the occurrence $H_1 * F_1A^{r-t-t'-16}G_1 * H_2$, by II.6.7. Assume that there are not less than 50 segments of the occurrence (60) contained in $* H_1 * F_1A^{r-t-t'-16}G_1H_2$. Then $H_1 \overset{\circ}{=} H'D$, where the occurrence

$$W_1 \rightleftharpoons H' * DF_1A^{r-50}A_3 * A_4A^{33-t-t'}G_1H_2$$

is normalised and $l_\gamma(W_1) \geqslant r$. Since the occurrence

$$W_2 \rightleftharpoons H'DF_1A^{15}A_1 * A_2A^{r-66}A_3 * A_4A^{33-t-t'}G_1H_2$$

is periodised in rank $\gamma - 1$, by IV.3.14 there exists a word $Z' \in \mathscr{M}_{\gamma-1}$ such that $H_1F_1A^{r-t-t'-16}G_1H_2 \overset{\gamma-1}{\sim} Z'$ and, for some P and Q,

$$f_{\gamma-1}(W_2;\ H_1F_1A^{r-t-t'-16}G_1H_2,\ Z') = P * A_2A^{r-66}A_3 * Q\,. \tag{61}$$

We then have

$$f_{\gamma-1}(W_1;\ H_1F_1A^{r-t-t'-16}G_1H_2,\ Z') = P' * A_6A^hA_3 * Q\,,$$

where $P'A_6A^{h-r+66}A_1 \overset{\circ}{=} P$, A_6 is an end of A and

$$h \geqslant l_\gamma(W_1) - 1 \geqslant r - 1\,. \tag{62}$$

It follows from (61) by IV.2.35 that

$$H'DF_1A^{15}A_1 \overset{\gamma-1}{\sim} P'A_6A^{h-r+66}A_1\,. \tag{63}$$

Further, use of (54), (62), (63) and the properties of coupling yields

$$\begin{aligned} H &\overset{\gamma-1}{\sim} H'DT \overset{\gamma-1}{\sim} [[H'DT,\ T^{-1}F_1A^{15}]_{\gamma-1},\ A^{-t-23}]_{\gamma-1} \\ &\overset{\gamma-1}{\sim} [H'DF_1A^{15},\ A^{-t-23}]_{\gamma-1} \overset{\gamma-1}{\sim} [P'A_6A^{h-r+66},\ A^{-t-23}]_{\gamma-1} \\ &\overset{\gamma-1}{\sim} P'A_6A^{h+43-r-t}\,, \end{aligned}$$

from which it follows by the principle of symmetry that

$$H^{-1} \overset{\gamma-1}{\sim} A^{-h-43+r+t}A_6^{-1}(P')^{-1}\,.$$

Since $h+43-r-t \geqslant 30$, by V.2.3 the word $A^{-20}A_2^{-1}$ must be a start of R, and this contradicts the condition $t' < 13$. Hence $* H_1 * F_1A^{r-t-t'-16}G_1H_2$ contains less than 50 segments of (60). It is proved similarly that $H_1F_1A^{r-t-t'-16}G_1 * H_2 *$ contains less than 50 segments of (60). But this contradicts condition (59), by II.5.2. Hence, for an arbitrary i,

$$H_1F_1A^{r-t-t'-16}G_1H_2 \in \mathscr{K}_i\,.$$

Furthermore, we deduce the relation

$$A^r \overset{\gamma-1}{\approx} H_1F_1A^{r-t-t'-16}G_1H_2 \tag{64}$$

from (48), and then convince ourselves using II.7.21 and IV.2.35 that H is equal to some power of A in $\mathbf{B}(m, n)$.

There follows at once from 3.1:

3.2. Theorem. *The centraliser of an arbitrary non-identity element in* $\mathbf{B}(m, n)$ *is a cyclic group of order n, for odd* $n \geqslant 665$.

3.3. Theorem. *Every commutative subgroup of* $\mathbf{B}(m, n)$ *is cyclic of order dividing n, for odd* $n \geqslant 665$.

This follows from 3.2.

3.4. Theorem. *For* $m > 1$ *and odd* $n \geqslant 665$, *the centre of* $\mathbf{B}(m, n)$ *is trivial.*

By (19) and (33), a single letter word a_1 is not equal in $\mathbf{B}(m, n)$ to any element of the set $\mathscr{A}$ distinct from it. Hence, only the powers a_1^i commute with a_1. Similarly, only the powers of a_2 commute with it. Suppose that, for some $|i| \leqslant (n+1)/2$ and $|j| \leqslant (n+1)/2$, $a_1^i = a_2^j$ in $\mathbf{B}(m, n)$. Then $a_1^i a_2^{-j} = 1$ in $\mathbf{B}(m, n)$, and $a_1^i a_2^{-j} \in \mathscr{A}_2$, whence $a_1^i a_2^{-j} \overset{0}{=} \Lambda$ by (33) and (19). Hence there are no non-identity elements in $\mathbf{B}(m, n)$ which commute with a_1 and a_2.

3.5. Theorem. *There is an algorithm solving the conjugacy problem for the group* $\mathbf{B}(m, n)$, *for odd* $n \geqslant 665$.

We have to find an algorithm which checks whether two given words X, Y in the alphabet (1) are conjugate in $\mathbf{B}(m, n)$ or not. By 2.10, we may assume that they are not equal to 1 in $\mathbf{B}(m, n)$. By 2.4 and the principle of effectiveness, we may assume that $X, Y \in \mathscr{A}$. By 1.2 and 2.3, we can find a word S and an elementary period A of some rank γ such that

$$X = SA^rS^{-1}$$

in $\mathbf{B}(m, n)$, for some $r > 0$. Similarly, for the word Y we find a word S_1 and an elementary period B of some rank α such that for some $h > 0$ the equality

$$Y = S_1B^hS_1^{-1}$$

holds in $\mathbf{B}(m, n)$. Without loss of generality we may assume that

$$\alpha \leqslant \gamma. \tag{65}$$

It is clear that X is conjugate to Y in $\mathbf{B}(m, n)$ if and only if for some H in $\mathbf{B}(m, n)$, the equality

$$B^h = HA^rH^{-1} \tag{66}$$

holds. Since $A^n = B^n = 1$ in $\mathbf{B}(m, n)$, by raising both sides of (66) to the appropriate power and if necessary replacing A by A^{-1} (or B by B^{-1}) we may arrange that r and h satisfy

$$\frac{n}{3} \leqslant r < \frac{n+1}{2}, \quad 0 < h < \frac{n+1}{2}. \tag{67}$$

We shall prove that (66) implies that A and B are conjugate in $\mathbf{B}(m, n)$ and $h = r$.

In order to do this, we repeat the analysis of the local couplings $[H, A^r]'_{\gamma-1}$ and $[A^r, H^{-1}]'_{\gamma-1}$ given in 3.1, and we retain the notation introduced there.

If $t \geqslant 13$ then, just as (57) was deduced from (46), we obtain here the relation

$$B^h \overset{\gamma-1}{\rightleftarrows} H_1F_1A^{r-7}T_1H_1^{-1} \tag{68}$$

from (66), where the left and right hand sides belong to $\mathscr{K}_{\gamma-1}$. Moreover, the occurrence

$$W_1 \rightleftarrows H_1F_1A^{15}A_1 * A_2A^{r-28}A_3 * A_4A^4T_1H_1^{-1}$$

is normalised. Set

$$f_{\gamma-1}(W_1;\ H_1F_1A^{r-7}T_1H_1^{-1}, B^h) = P_1 * E_1 * Q_1 .$$

By II.5.16, II.2.12 and (65), we have $\alpha = \gamma$. Clearly, $E_1 \mathrel{\underset{\circ}{=}} C^{r-28}C_1$, where C is a cyclic shift of the period B. By (67), II.7.15 and 2.3, the left periods of the bases of W_1 and $P_1 * E_1 * Q_1$ are conjugate in $\mathbf{B}(m, n)$. Hence the periods A and B are themselves conjugate. In this case, having chosen an appropriate image of the word H, we can write the relation (66) in the form

$$A^h = HA^rH^{-1},$$

whence, as was shown in 3.1, H is equal in $\mathbf{B}(m, n)$ to some power of A. Then $A^{r-h} = 1$ in $\mathbf{B}(m, n)$, whence $A^{r-h} \mathrel{\underset{\circ}{=}} A$ by (33) and (19), that is, $r = h$.

The case $t' \geqslant 13$ follows by symmetry. In the case where we have both $t < 13$ and $t' < 13$, we proceed by analogous arguments, with only this difference: that instead of (68), we use a relation analogous to (64), namely

$$B^h \overset{\gamma-1}{\rightleftarrows} H_1F_1A^{r-t-t'-16}G_1H_2, \tag{69}$$

where the right hand side also belongs to $\mathscr{K}_{\gamma-1}$.

Thus we have proved that if X is conjugate to Y in $\mathbf{B}(m, n)$, then $h = r$ and B is conjugate to A. The converse is clear. Hence we need only check whether, if $\alpha = \gamma$ and $h = r$, the elementary periods A and B of rank γ are conjugate in $\mathbf{B}(m, n)$ or not. But by II.7.15 and (68) and (69), this reduces to verifying the following assertion:

There exists a word $Y_1 \in \mathscr{K}_{\gamma-1}$ and a generating occurrence U of rank γ in the word B^h such that $B^h \overset{\gamma-1}{\rightleftarrows} Y_1$, $l_\gamma(U) \geqslant 9$ and $\operatorname{Bas}(f_{\gamma-1}(U;\ B^h, Y_1)) \in \operatorname{Per}(A)$.

By V.2.6 and the principle of effectiveness, this assertion can be verified by checking through all words $Y_1 \in \mathscr{K}_{\gamma-1}$ which are equivalent to B^h in rank $\gamma - 1$

and by determining $f_{\gamma-1}(U; B^h, Y_1)$ for each of them, where U is a normal generating occurrence in B^h with $l_\gamma(U) \geqslant 9$.

3.6. In 3.1 and 3.5, we have actually described all solutions of the simple equations

$$Ax = xA \text{ and } Ax = xB$$

in $\mathbf{B}(m, n)$.

The method of studying the groups $\mathbf{B}(m, n)$ presented in this book are likely to be helpful for the recognition and description of the set of all solutions of wider classes of equations in $\mathbf{B}(m, n)$.

3.7. Theorem. *For odd* $n \geqslant 665$, $\mathbf{B}(3, n)$ *is isomorphically embedded in* $\mathbf{B}(2, n)$.

By 2.6, it is sufficient to prove that $\mathbf{B}(2, n)$ contains a subgroup isomorphic with $\boldsymbol{\Gamma}(3, n)$. $\mathbf{B}(2, n)$ is isomorphic with $\boldsymbol{\Gamma}(2, n)$.

The construction of $\boldsymbol{\Gamma}(2, n)$ is based, according to 1.4, on the system of concepts which were defined and studied in the first five chapters. We look at this system of concepts in connection with the group alphabet

$$a, b, a^{-1}, b^{-1}. \tag{70}$$

The letters a and a^{-1} are called a-letters, b and b^{-1} are called b-letters.

Side-by-side with the original system of concepts for words in the alphabet (70), we look at the modified system which is obtained from it by restricting the notion of a supporting kernel of rank 0 for periodic words of rank 1. According to definitions I.4.4 and I.4.1, the supporting kernels of rank 0 of a periodic word $X \in$ Per(1, A) are all the occurrences in a word X with single-letter base that are inner relative to the period A.

Suppose that $X \in$ Per(1, A). If some a-letter occurs in A, then only those kernels $V \in$ Ker(0, X) whose bases are a-letters and which are inner relative to the period A are said to be *supporting kernels of rank* 0 *of* X *in the restricted sense*. In the contrary case, that is, where $X \,\overline{\underline{\circ}}\, b^i$ for some integer i, we shall call an arbitrary kernel $V \in$ Ker(0, X) with $V \in$ Inn(X, A) a *supporting kernel of rank* 0 *of* X *in the restricted sense*.

The definitions of all other concepts remain unchanged. Of course, the restriction mentioned is reflected in the contents of most of the concepts considered. Retaining the original name for all of the concepts, in order to avoid confusion we shall append to them the words "*in the restricted sense*".

First of all, as a result of our restriction there is a change in the notion of the number of segments of rank 1 of generating occurrences of rank 1 in words $X \in$ Per(1, A), if the period A contains both an a-letter and a b-letter. For example, the generating occurrence $(ab^2)^9ab * b(ab^2)^{13} * (ab^2)^{12}$ in the word $(ab^2)^{35}$ is not a normal generating occurrence of rank 1 in the restricted sense, and the number of segments of it of rank 1 in the restricted sense with respect to the period (ab^2) is not 14, but 13. The maximal normal generating occurrence of rank 1 in the restricted sense contained in this occurrence is

$$(ab^2)^{10} * (ab^2)^{12}a * b^2(ab^2)^{12} .$$

Normal elementary words of rank 1 in the restricted sense either start and end with a-letters or are powers of b-letters.

Our restriction is reflected in the contents of some ideas not only in rank 1, but also in all ranks $\alpha \geqslant 1$. A kernel of rank α in the original sense may not be a kernel of rank α in the restricted sense, it may be lengthened or shortened according to whether an adjacent active occurrence of rank α fails to be such in the restricted sense or, conversely, an active occurrence of rank α in the restricted sense shows up which is not active in the original sense. Some reduced words of rank α are not reduced in the restricted sense, new words of rank α turn up which are reduced in the restricted sense, and so on. We shall not make a detailed analysis of how each of the concepts under consideration changes as compared with its original version. We note only the following property, which is essential for the proof of our theorem.

Since the formal definitions of all remaining concepts remain unchanged, then all the assertions considered in the preceding chapters, which characterise one or other of the concepts and their interrelationships, remain in force if understood in the restricted sense.

The reader who is familiar with the contents of the foregoing chapters will have no difficulty in verifying this fact, since the proof of each individual assertion taken in the restricted sense is the same as the proof of the corresponding assertion in the old sense.

For the sets of words in the alphabet (70) and the sets of occurrences in such words which arise additionally in the restricted sense, we shall use the corresponding old notation embellished with dashes. For example, $\mathscr{R}'_\alpha$ is the set of reduced words of rank α in the restricted sense in the alphabet (70), $\mathrm{Ker}'(\alpha, X)$ is the set of kernels of rank α of the word X in the restricted sense, and so on.

It follows from what has been said that, on the basis of the concepts considered in the restricted sense we can construct, analogous to $\boldsymbol{\Gamma}(2, n)$, a group $\boldsymbol{\Gamma}'(2, n)$ which is also isomorphic to $\mathbf{B}(2, n)$.

It is therefore sufficient to prove that $\boldsymbol{\Gamma}(3, n)$ is isomorphic with some subgroup of $\boldsymbol{\Gamma}'(2, n)$. We shall prove that it is isomorphic to the subgroup of $\boldsymbol{\Gamma}'(2, n)$ generated by the words

$$bab^{-1},\ b^2ab^{-2} \text{ and } b^4ab^{-4} .$$

In order to do this, we shall define by induction on the length of the word X a lexicographic function $\tau(X)$ on the set $\mathscr{R}_0$ of uncancellable words in the alphabet

$$a_1, a_2, a_3, a_1^{-1}, a_2^{-1}, a_3^{-1} . \tag{71}$$

We set

$$\tau(\Lambda) = \Lambda,\ \tau(a_1^\sigma) = ba^\sigma b^{-1},\ \tau(a_2^\sigma) = b^2a^\sigma b^{-2},\ \tau(a_3^\sigma) = b^4a^\sigma b^{-4}$$

and

$$\tau(Xa_i^\sigma) = [\tau(X), \tau(a_i^\sigma)]_0 ,$$

where $\sigma = \pm 1$; $i = 1,2,3$. Clearly, the function $Y = \tau(X)$ is a one-to-one mapping of the set $\mathscr{R}_0$ onto some subset of the set $\mathscr{R}_0'$ of all uncancellable words in the alphabet (70). The following properties of this function may also be taken as evident.

1) $\tau(X^{-1}) = (\tau(X))^{-1}$.

2) $[X, Y]_0 = Z \iff [\tau(X), \tau(Y)]_0 = \tau(Z)$.

3) If $Y = \tau(X)$, then b^5 does not occur in Y and the number of occurrences of a-letters in Y is $\partial(X)$.

4) $X \mathrel{\overline{\underline{\circ}}} a_{i_1}^{\sigma_1} a_{i_2}^{\sigma_2} \dots a_{i_t}^{\sigma_t} \iff \exists k \exists s\, (\tau(X) \mathrel{\overline{\underline{\circ}}} b^k a^{\sigma_1} b^{r_1} a^{\sigma_2} b^{r_2} \dots b^{r_{t-1}} a^{\sigma_t} b^s)$,

where the σ_t are integers, $i_j = 1,2,3$ and

$$r_j = 0 \iff i_j = i_{j+1} .$$

We shall say that an uncancellable word Y in the alphabet (70) is *regular* if it begins and ends with a-letters. The *significant part* of a word Y in the alphabet (70) is by definition the regular word E such that $Y \mathrel{\overline{\underline{\circ}}} PEQ$, where there are no a-letters occurring in PQ. We denote the significant part of Y by $\bar{Y}$. If E is a regular word, then $\bar{E} \mathrel{\overline{\underline{\circ}}} E$. An occurrence $P * E * Q$ in a word $Y \in \mathscr{R}_0'$ is said to be *regular* if E is a regular word.

5) $X \in \mathrm{Per}(1, A) \iff \overline{\tau(X)} \in \mathrm{Per}(1, B)$, where B is the result of a cyclic cancellation of $\tau(A)$.

If $PEQ \in \mathscr{R}_0$, where E is non-empty, then $\tau(P * E * Q)$ denotes the occurrence $R * D * S$ in $\tau(PEQ)$ such that $D \mathrel{\overline{\underline{\circ}}} \overline{\tau(E)}$ and $R \mathrel{\overline{\underline{\circ}}} \overline{\tau(P)}$.

6) For each $X \in \mathscr{R}_0$, the function $W = \tau(V)$ is a one-to-one mapping of the set of all occurrences of non-empty words in X onto the set of all occurrences of regular words in $\tau(X)$.

7) $V \in \mathrm{Norm}(1, X, r) \iff \tau(V) \in \mathrm{Norm}'(1, \tau(X), r)$.

8) $X \to Y$ is an r-reversal of rank 1 of $V \iff \tau(X) \to \tau(Y)$ is an r-reversal of $\tau(V)$.

On looking over all of the concepts under consideration one by one in the order in which they appear and are studied in the foregoing chapters, we can prove by simultaneous induction on the rank α that the mappings $\tau(X)$ and $\tau(V)$ transfer each of the concepts considered relating to words $X \in \mathscr{R}_\alpha$ in the alphabet (71) and to occurrences V in such words, into their corresponding analogues in the restricted sense, which relate to words $\tau(X)$ and occurrences $\tau(V)$.

There is no need to give the corresponding arguments in detail here, as they present no difficulty to the reader familiar with the contents of the foregoing chapters. Together with other logical equivalences, we obtain in addition the following relations, in which the right hand sides are to be taken in the restricted sense.

$$V \in \mathrm{Norm}(\alpha, X, r) \iff \tau(V) \in \mathrm{Norm}'(\alpha, \tau(X), r) ;$$

$$V \in \mathrm{Ker}(\alpha, X) \Longleftrightarrow \tau(V) \in \mathrm{Ker}'(\alpha, \tau(X)) ;$$
$$X \in \mathscr{R}_\alpha \Longleftrightarrow \tau(X) \in \mathscr{R}'_\alpha ;$$
$$X \overset{\alpha}{\sim} Y \Longleftrightarrow \tau(X) \overset{\alpha}{\sim} \tau(Y) ;$$
$$Z = [X,\ Y]_\alpha \Longleftrightarrow \tau(Z) = [\tau(X), \tau(Y)]_\alpha .$$

By definitions I.4.34, VI.1.3 and IV.1.4, we obtain from these relations the following logical equivalences:

$$X \in \mathscr{A}_\alpha \Longleftrightarrow \tau(X) \in \mathscr{A}'_\alpha ,$$
$$X \sim Y \Longleftrightarrow \tau(X) \sim \tau(Y) ,$$
$$\{Z\} = \{X\} \circ \{Y\} \Longleftrightarrow \{\tau(Z)\} = \{\tau(X)\} \circ \{\tau(Y)\} ,$$

where also the right hand sides must be understood in the restricted sense. But this means that $\Gamma(3, n)$ is isomorphic with the subgroup of $\Gamma'(2, n)$ generated by the elements

$$\{bab^{-1}\},\ \{b^2ab^{-2}\} \text{ and } \{b^4ab^{-4}\} .$$

We have proved the required result.

3.8. Theorem 3.7 gives a negative answer to problem 17 in [15]. It follows at once from it that, for $m > 1$ and odd $n \geqslant 665$, $\mathbf{B}(m, n)$ does not satisfy the minimal condition, that is, there exists an infinite descending chain of subgroups

$$\mathbf{B}(m, n) \supset \mathbf{G}_1 \supset \mathbf{G}_2 \supset \ldots \supset \mathbf{G}_i \supset \ldots .$$

Clearly each subgroup $\mathbf{G}_i$ in this chain can be chosen to be isomorphic with $\mathbf{B}(m, n)$.

3.9. Theorem. *For $m > 1$ and composite odd $n = ks$, where $k \geqslant 665$ and $s > 1$, the group $\mathbf{B}(m, n)$ does not satisfy either the minimal or the maximal conditions for normal subgroups.*

Suppose that $n = ks$, where $k \geqslant 665$, $s > 1$ and n is odd.

We look at an infinite sequence of simple words in an alphabet of two letters a_1 and a_2:

$$Q_1, Q_2, \ldots, Q_i, \ldots ,$$

where $\partial(Q_{i+1}) > \partial(Q_i)$ and no non-empty word of the form EEE (see I.3.4) occurs in any Q_i. Clearly, each Q_i is an elementary period of rank 1.

We denote by $\mathbf{G}_i$ the normal subgroups of $\mathbf{B}(m, n)$ generated by the words

$$Q_1^k, Q_2^k, \ldots, Q_i^k ,$$

and by $\mathbf{F}_i$ the normal subgroups of $\mathbf{B}(m, n)$ generated by

$$Q_{i+1}^k, Q_{i+2}^k, \ldots, Q_j^k, \ldots .$$

We have two infinite sequences of normal subgroups of $\mathbf{B}(m, n)$,

$$\mathbf{G}_1 \subset \mathbf{G}_2 \subset \ldots \subset \mathbf{G}_i \subset \mathbf{G}_{i+1} \subset \ldots$$

and

$$\mathbf{B}(2, n) \supset \mathbf{F}_1 \supset \mathbf{F}_2 \supset \ldots \supset \mathbf{F}_i \supset \mathbf{F}_{i+1} \supset \ldots .$$

We need only prove that $Q_{i+1}^k \notin \mathbf{G}_i$ and $Q_i^k \notin \mathbf{F}_{i+1}$ for each $i \geqslant 1$. In order to do this, we alter the theory of transformations of words in the alphabet (1) which served earlier as a basis for setting up the interpretation $\Gamma(m, n)$ of $\mathbf{B}(m, n)$. The alteration concerns real reversals of rank 1.

Each real reversal of rank 1 has the form

$$PA^tA_1Q \rightarrow PA^{-n+t+1}A_2^{-1}Q\,, \tag{72}$$

where $A \overset{\circ}{=} A_1A_2$ is an elementary period of rank 1. In just those cases where the period A is a cyclic shift on one of the words Q_j or Q_j^{-1} for $1 \leqslant j \leqslant i$, let us agree to consider, instead of the reversal (72), the reversal

$$PA^tA_1Q \rightarrow PA^{-k+t+1}A_2^{-1}Q\,, \tag{73}$$

assuming in addition that $t < k$. Correspondingly, in the definitions of the sets $\mathcal{N}_1, \mathcal{P}_1, \mathcal{R}_1, \mathcal{K}_1, \mathcal{L}_1, \mathcal{M}_1$, in all limitations on the number of segments of occurrences of elementary words of rank 1 or of kernels of rank 1, we must take k instead of the parameter n in just those cases where our concern is about elementary periods that are cyclic shifts of Q_j or Q_j^{-1} for $1 \leqslant j \leqslant i$. Maintaining all other definitions we obtain, instead of $\Gamma(2, m)$, a group $\Gamma^{(i)}(2, m)$ in which the relations

$$Q_j^k = 1 \; (1 \leqslant j \leqslant i)$$

hold. Since k is a divisor of n, $\Gamma^{(i)}(2, n)$ also satisfies the identical relation $x^n = 1$.

Assume that $Q_{i+1}^k \in \mathbf{G}_i$. Then in $\Gamma^{(i)}(2, n)$ we must have $Q_{i+1}^k = 1$, so that $Q_{i+1}^{2q} = Q_{i+1}^{-k+2q}$, that is, $Q_{i+1}^{2q} \sim Q_{i+1}^{-k+2q}$. Since Q_{i+1}^{2q} is an elementary word of rank 1, we have by IV.2.13 that $Q_{i+1}^{2q} \overset{1}{\sim} Q_{i+1}^{-k+2q}$. Since the period Q_{i+1} is not a cyclic shift of Q_j or Q_j^{-1} for $1 \leqslant j \leqslant i$, we have that

$$Q_{i+1}^{2q} \rightarrow Q_{i+1}^{-n+2q}$$

is a real reversal of rank 1. Hence $Q_{i+1}^{2q-k} \overset{1}{\sim} Q_{i+1}^{2q-n}$. Since the unique kernel $V \in \operatorname{Ker}(1, Q_{i+1}^{-k+2q})$ is related to its image $f_1(V; Q_{i+1}^{-k+2q}, Q_{i+1}^{-n+2q})$, it follows by IV.2.13 that $Q_{i+1}^{-k+2q} \overset{0}{\sim} Q_{i+1}^{-n+2q}$, that is, $k = n$. But this contradicts the hypothesis that $s > 1$. Therefore $Q_{i+1}^k \notin \mathbf{G}_i$.

The fact that $Q_i^k \notin \mathbf{F}_{i+1}$ is proved in an analogous way. In order to do this, we need to modify the theory, namely instead of the reversals (72) we consider the reversals (73), in just those cases where A is a cyclic shift of one of the words Q_j or Q_j^{-1} for $j > i$. It is clear, in addition, that the definitions of the sets $\mathscr{N}_1$, $\mathscr{P}_1$, $\mathscr{R}_1$, $\mathscr{K}_1$, $\mathscr{L}_1$ and $\mathscr{M}_1$ are to be altered.

3.10. We remark in conclusion that the interesting open question remains as to whether the groups $\mathbf{B}(m, n)$ satisfy the maximal or minimal conditions on normal subgroups, for prime $n \geqslant 665$.

Note added in proof. When the book was already in print, the author succeeded in finding infinite descending (and ascending) chains of normal subgroups of $\mathbf{B}(m, n)$ for arbitrary odd $n \geqslant 665$ and $m \geqslant n$.

Chapter VII. Further Applications of the Method

§1. Non-Abelian Groups in which Every Pair of Cyclic Subgroups Intersect Non-Trivially

As is well-known, the additive group of rational numbers is torsion-free, and any two cyclic subgroups have non-trivial intersection. The question which naturally arises out of this concerning the existence of non-abelian groups with the same properties (see [14], problem 1.63) remained open for a long time. According to Kargapolov, who first drew the author's attention to this problem, it was posed by Kontorovič as long ago as the forties. Examples of non-abelian groups of this sort were constructed by the author in [11]. We shall give the proof of this result below.

1.1. Suppose that we fix an integer $m > 1$ and odd $n \geqslant 665$. We denote by $\mathbf{A}(m, n)$ the group given by generators

$$a_1, a_2, \ldots, a_m, d \tag{1}$$

and the system of defining relations

$$a_i d = d a_i \ (i = 1, 2, \ldots, m), \tag{2}$$

$$A^n = d \ (A^n \in \bigcup_{i=1}^{\infty} \mathscr{E}_i), \tag{3}$$

where $\mathscr{E}_i$ is the class of elementary words of rank i which were defined in VI.2.1.

It is clear that d is contained in the centre of $\mathbf{A}(m, n)$. By VI.2.29, $\mathbf{B}(m, n)$ is the factor group of $\mathbf{A}(m, n)$ by the subgroup generated by d and, by VI.3.4, this subgroup is the centre of $\mathbf{A}(m, n)$.

The following result is also clear.

1.2. *If X and Y are arbitrary words in the alphabet*

$$a_1, a_2, \ldots, a_m, a_1^{-1}, a_2^{-1}, \ldots, a_m^{-1}, \tag{4}$$

then the equality $X = Y$ holds in the group $\mathbf{B}(m, n)$ if and only if there exists an integer i such that $X = Yd^i$ in $\mathbf{A}(m, n)$.

1.3. Theorem. *If x is an arbitrary non-identity element of $\mathbf{A}(m, n)$, then there exists an integer $s \neq 0$ such that $x^n = d^s$ in $\mathbf{A}(m, n)$.*

Let x be a non-identity element in $\mathbf{A}(m, n)$. By relations (2), the element x is equal to some word of the form Xd^j, where X is a word in the alphabet (4).

Since $X^n = 1$ in $\mathbf{B}(m, n)$ and $\mathbf{B}(m, n) = \mathbf{A}(m, n)/\langle d\rangle$, there exists r such that $X^n = d^r$ in $\mathbf{A}(m, n)$. Then we have $x^n = d^{r+jn}$ in $\mathbf{A}(m, n)$. Since $x \neq 1$ in $\mathbf{A}(m, n)$, it follows that $r + jn \neq 0$.

1.4. In order to prove that there are no elements of finite order in $\mathbf{A}(m, n)$, we have to develop generalisations of the concepts studied in Chapter I-V for words in the alphabet of the group $\mathbf{A}(m, n)$.

We denote by $\mathscr{R}_\alpha^d$ the set of all words of the form Qd^i, where i is any integer and Q is a word in the alphabet (4) that belongs to $\mathscr{R}_\alpha$. Similarly $\mathscr{N}_\alpha^d$, $\mathscr{P}_\alpha^d$, $\mathscr{K}_\alpha^d$, $\mathscr{L}_\alpha^d$, $\mathscr{M}_\alpha^d$, $\bar{\mathscr{M}}_\alpha^d$ and $\mathscr{A}_\alpha^d$ will denote the set of words of form Qd^i, where Q belongs to the sets $\mathscr{N}_\alpha$, $\mathscr{P}_\alpha$, $\mathscr{K}_\alpha$, $\mathscr{L}_\alpha$, $\mathscr{M}_\alpha$, $\bar{\mathscr{M}}_\alpha$, $\mathscr{A}_\alpha$ respectively.

We shall only consider those occurrences in the words Qd^i whose bases occur in Q, that is, they are words in the alphabet (4). Furthermore, if V is an occurrence in the word Q, then Vd^i will denote the occurrence which results from writing the word d^i to the right of V.

The concepts of periodic, integral, semi-integral and elementary words of rank α, of generating occurrence of rank α and of supporting kernel of rank α (more precisely, all concepts which were defined in I.4.3-I.4.10) remain unaltered. They will, as before, relate only to words in the alphabet (4).

We shall study the occurrences of elementary words of rank α in words of the form Qd^i, where $Q \in \mathscr{R}_{\alpha-1}$. All the concepts defined in I.4.11-I.4.15 extend uniquely to such occurrences. In particular, we have in addition that for any integers i and j,

$$\mathrm{Comp}(V, W) \Longleftrightarrow \mathrm{Comp}(Vd^i, Wd^i),$$

$$\mathrm{Rel}(V, W) \Longleftrightarrow \mathrm{Rel}(Vd^i, Wd^j),$$

$$V \in \mathrm{Norm}(\alpha, Q, r) \Longleftrightarrow Vd^i \in \mathrm{Norm}(\alpha, Qd^i, r).$$

We extend the concepts of kernel of rank α and the relation $\mathrm{MutNorm}_\alpha(V, W)$ to occurrences in words of the form Qd^i, so that the relations

$$V \in \mathrm{Ker}(\alpha, Q) \Longleftrightarrow Vd^i \in \mathrm{Ker}(\alpha, Qd^i),$$

$$V \in \mathrm{Reg}(\alpha, Q) \Longleftrightarrow Vd^i \in \mathrm{Reg}(\alpha, Qd^i),$$

$$\mathrm{MutNorm}_\alpha(V, W) \Longleftrightarrow \mathrm{MutNorm}_\alpha(Vd^i, Wd^j)$$

hold.

Further, by simultaneous induction on the rank α we define on $\mathscr{R}_\alpha^d$ the relation $\overset{\alpha}{\approx}$ *of generalised equivalence in rank* α *and the operation of generalised coupling in rank* α,

$$[X, Y]_\alpha^d = Z.$$

These new concepts are not generalisations of the old in the literal sense of that word, but are closely connected with their old analogues. In particular, we shall prove in passing that the relation $\overset{\alpha}{\approx}$ satisfies the following conditions:

$$P \overset{\alpha}{\simeq} Q \Longleftrightarrow \exists i \forall j (Pd^j \overset{\alpha}{\approx} Qd^{j+i}), \tag{5}$$

$$Qd^i \overset{\alpha}{\approx} Qd^j \Rightarrow i = j, \tag{6}$$

where P, Q are words in the alphabet (4) and i, j are integers. The relation (6) means that, for given words P and Q with $P \overset{\alpha}{\simeq} Q$, there exists a unique integer i which satisfies the right hand side of (5).

For arbitrary $P, Q \in \mathscr{R}_0$ we put by definition

$$Pd^i \overset{0}{\approx} Qd^j \Longleftrightarrow (P \overset{0}{\simeq} Q \;\&\; i = j) \Longleftrightarrow Pd^i \overset{0}{\equiv} Qd^j,$$

and

$$[Pd^i, Qd^j]_0^d \rightleftharpoons [P, Q]_0 d^{i+j}. \tag{7}$$

Clearly, (5) and (6) hold if $\alpha = 0$.

Suppose that $\alpha > 0$ and we have already defined the relation $\overset{\alpha-1}{\approx}$ on $\mathscr{R}_{\alpha-1}^d$ satisfying (5) and (6), and the operation $[X, Y]_{\alpha-1}^d$ satisfying

$$[Pd^i, Qd^j]_{\alpha-1}^d \overset{\alpha-1}{\approx} [P, Q]_{\alpha-1}^d \, d^{i+j} \tag{8}$$

and the relations obtained from (16) and (17) below by replacing α by $\alpha - 1$. The existence of such an operation, defined everywhere on $\mathscr{R}_{\alpha-1}^d$, and the facts that it is associative and well defined to within generalised equivalence in rank $\alpha - 1$, are given us by the induction assumption. For $\alpha = 1$, these properties of the operation $[X, Y]_{\alpha-1}^d$ follow trivially from (7).

In order to define $\overset{\alpha}{\approx}$, we now need the concept of a generalised reversal of rank α which is connected with the usual concept of reversal of rank α.

Suppose that $X \in \mathscr{R}_{\alpha-1}$ and the transition $X \to X_1$ is a real reversal of rank α of the occurrence $R * E * S \in \text{Norm}(\alpha, X, 9)$. By VI.2.1, there exists an element A^n in $\mathscr{E}_\alpha$ such that either $\text{Rel}(E, A^n)$ or $\text{Rel}(E, A^{-n})$. As was shown in VI.2.3 (deduction of (32)), in this case we can find a word Z such that $X \overset{\alpha-1}{\approx} Z$ and

$$f_{\alpha-1}(R * E * S; X, Z) = R_1 * C^t C_1 * S_1, \tag{9}$$

where one of the words C^n or C^{-n} is a cyclic shift of the element A^n of $\mathscr{E}_\alpha$. For an arbitrary integer i, there exists by (5) an integer r such that $Xd^i \overset{\alpha-1}{\approx} Zd^r$. We shall then define, for an arbitrary integer i, a *generalised real reversal of rank α of the occurrence $R * E * Sd^i$* to be any transition of the form

$$RESd^i \to Y, \tag{10}$$

where

$$Y \overset{\alpha-1}{\approx} [R_1, [C^{-n+t} d^\sigma, C_1 S_1 d^r]_{\alpha-1}^d]_{\alpha-1}^d \tag{11}$$

and $\sigma = 1$ or $\sigma = -1$ according to whichever of C^{-n} or C^n is a cyclic shift of the word $A^n \in \mathscr{E}_\alpha$.

We shall say that the words $X, Y \in \mathscr{P}_\alpha^d$ are *generalised equivalent in rank* α if either $X \overset{\alpha-1}{\approx} Y$, or there is a sequence of generalised real reversals of rank α which transforms X into Y. If in addition $X, Y \in \mathscr{R}_\alpha^d$, then we shall write $X \overset{\alpha}{\approx} Y$.

It follows from relation (5) for rank $\alpha - 1$, the definition of generalised reversal of rank α and relations (16), (17) below for coupling of rank $\alpha - 1$, that *for any* $P, Q \in \mathscr{R}_{\alpha-1}$ *the transition* $P \to Q$ *is a real reversal of rank* α *of the occurrence* $R * E * S$ *in the word* P *if and only if there exists an integer i such that for an arbitrary integer j the transition* $Pd^j \to Qd^{i+j}$ *is a generalised real reversal of rank* α *of the occurrence* $R * E * Sd^j$.

It follows easily from this that (5) holds also for rank α.

The situation with the proof of (6) for rank α is somewhat more complicated.

Making use of (5), we define a function $f_\alpha^d(V; X, Y)$ which maps $\mathrm{Ker}(\alpha, X)$ onto $\mathrm{Ker}(\alpha, Y)$, where $X \overset{\alpha}{\approx} Y$. Namely, if $Pd^i \overset{\alpha}{\approx} Qd^j$, where $P \overset{\alpha}{\sim} Q$, then for an arbitrary kernel $R * F * Sd^i \in \mathrm{Ker}(\alpha, Pd^i)$, we set

$$f_\alpha^d(R * E * Sd^i;\ Pd^i, Qd^j) \rightleftharpoons f_\alpha(R * E * S;\ P, Q)d^j .$$

In a similar way, we also carry over to generalised reversals of rank α the concepts of stability of an occurrence in a given reversal, its image in the reversal and also the concept of the image of an occurrence $R * E * Sd^i \in \mathrm{Norm}\,(\alpha, X, 9)$ in a reversal of rank α of it. Clearly, for an arbitrary generalised real reversal $X \to Y$ of rank α, where $X, Y \in \mathscr{P}_\alpha^d$, we also have the mapping $f_\alpha^d(V; X, Y)$.

The properties of generalised real reversals of rank α are analogous to the properties of the ordinary real reversals of rank α. The majority of them follow directly from the corresponding properties of the usual real reversals of rank α, and the remainder are proved in an analogous way to the corresponding properties of the ordinary real reversals treated in Chapter III.

In order to prove the symmetry of the relation $\overset{\alpha}{\approx}$, we need to show that, for an arbitrary active occurrence $R * E * S$ of rank α in the word $X \in \mathscr{R}_\alpha$, the above definition of a generalised reversal of rank α of the occurrence $R * E * Sd^i$ does not depend on the choice of the intermediate word Z for which the relations $X \overset{\alpha-1}{\approx} Z$ and (9) hold. Suppose that we have another word Z_1 such that $X \overset{\alpha-1}{\approx} Z_1$ and

$$f_{\alpha-1}(R * E * S;\ X, Z_1) = R_2 * D^h D_1 * S_2 , \tag{12}$$

where one of the words D^n or D^{-n} is a cyclic shift of some element B^n from the set $\mathscr{E}_\alpha$. Then, together with the reversal (10), we now have a further generalised real reversal of rank α of the occurrence $R * E * Sd^i$, namely:

$$RESd^i \to Y_1 ,$$

where

$$Y_1 \overset{\alpha-1}{\approx} [R_2,\ [D^{-n+h}d^{\sigma_1},\ D_1 S_2 d^{r_1}]_{\alpha-1}^d]_{\alpha-1}^d , \tag{13}$$

$Xd^i \overset{\alpha-1}{\approx} Z_1 d^{r_1}$, and $\sigma_1 = 1$ or $\sigma_1 = -1$ according to whether D^{-n} or D^n is a cyclic shift of B^n. Since in addition $\mathrm{Rel}(D^h D_1, C^t C_1)$ follows from (9) and (12), we have by definition VI.2.1 that $A \mathbin{\overset{\circ}{=}} B$. Thus $\sigma_1 = \sigma$. It follows from (9) and (12) that

$$f_{\alpha-1}(R_1 * C^t C_1 * S_1;\, Z, Z_1) = R_2 * D^h D_1 * S_2 \,. \tag{14}$$

It follows from VI.2.1(b) that the occurrence $R_1 * C^t C_1 * S_1$ is periodised in rank $\alpha - 1$. Then, by II.7.21, it follows from (14) that $C^t C_1 \overset{\circ}{\equiv} D^h D_1$, that is, $C \overset{\circ}{\equiv} D$, $C_1 \overset{\circ}{\equiv} D_1$ and $t = h$. By IV.2.35, we have the relations

$$R_1 \overset{\alpha-1}{\simeq} R_2 \text{ and } S_1 \overset{\alpha-1}{\simeq} S_2 \,.$$

Then, in view of (5) there exist integers k and s such that

$$R_1 \overset{\alpha-1}{\approxeq} R_2 d^k \text{ and } S_1 \overset{\alpha-1}{\approxeq} S_2 d^s \,. \tag{15}$$

By the inductive hypothesis, coupling in rank $\alpha - 1$ has all the properties analogous to the properties of ordinary coupling in rank α. Therefore, it follows from (15) that

$$R_1 C^t C_1 S_1 d^r \overset{\alpha-1}{\approxeq} R_2 D^t D_1 S_2 d^{r+k+s} \,.$$

Then

$$Z_1 d^{r_1} \overset{\alpha-1}{\approxeq} X d^l \overset{\alpha-1}{\approxeq} Z d^r \overset{\alpha-1}{\approxeq} Z_1 d^{r+k+s} \,,$$

whence, by relation (6) for rank $\alpha - 1$, it follows that $r_1 = r + k + s$. Further, using associativity and the single-valuedness of generalised coupling in rank $\alpha - 1$, as well as relation (11), (8) and (13), we have

$$\begin{aligned} Y &\overset{\alpha-1}{\approxeq} [R_1, [C^{-n+t} d^{\sigma}, C_1 S_1 d^r]^d_{\alpha-1}]^d_{\alpha-1} \\ &\overset{\alpha-1}{\approxeq} [R_2 d^k, [D^{-n+h} d^{\sigma_1}, D_1 S_2 d^{k+s}]^d_{\alpha-1}]^d_{\alpha-1} \\ &\overset{\alpha-1}{\approxeq} [R_2, [D^{-n+h} d^{\sigma_1}, D_1 S_2 d^{r+k+s}]^d_{\alpha-1}]^d_{\alpha-1} \overset{\alpha-1}{\approxeq} Y_1 \,. \end{aligned}$$

Thus we have proved that the results of any two generalised real reversals of rank α of a given occurrence $R * E * Sd^i$ are generalised equivalent in rank $\alpha - 1$. It follows easily from this that, if $X \to Y$ is a generalised real reversal of a kernel $V \in \mathrm{Ker}(\alpha, X)$, then $Y \to X$ is a generalised real reversal of rank α of the kernel $f^d_\alpha(V;\, X, Y)$. Thus the relation $\overset{\alpha-1}{\approxeq}$ is symmetric. Transitivity follows immediately from the definition.

Further, using the analogues of assertions IV.2.1 and IV.2.5 for generalised real reversals of rank α, we prove the analogue of IV.2.6: if $X \overset{\alpha}{\approxeq} Y$, then either $X \overset{\alpha-1}{\approxeq} Y$, or there exists a simple sequence of generalised real reversals of rank α transforming X to Y. From this it is not hard to deduce relation (6) for rank α.

Assume that $Qd^r \overset{\alpha}{\approxeq} Qd^s$, where $r \neq s$. Since by the induction assumption (6) is true for rank $\alpha - 1$, we have $\neg(Qd^r \overset{\alpha-1}{\approxeq} Qd^s)$. Consequently, there exists a simple sequence

$$Qd^r \overset{\circ}{\equiv} X_1 \to X_2 \to \ldots \to X_l \to X_{l+1} \to \ldots \to X_t \overset{\circ}{\equiv} Qd^s$$

of generalised real reversals of rank α. Then, with Y_t denoting the result of deleting the letters $d^{\pm 1}$ from X_t, we get a simple sequence

$$Q \overset{\circ}{=} Y_1 \to Y_2 \to \ldots \to Y_t \to Y_{t+1} \to \ldots \to Y_t \overset{\circ}{=} Q$$

of ordinary real reversals of rank α. But this is impossible by III.1.5 and IV.2.4, since $I_\alpha(Q, Q) = 0$. Thus relation (6) is true for rank α as well.

On the basis of relations (5) and (6), we can prove the analogue for $\overset{\alpha}{\approx}$ of each of the assertions that were proved for $\overset{\alpha}{\sim}$ in Chapter IV, without significant difficulty.

The *operation of generalised coupling in rank* α is defined for arbitrary words Pd^i, $Qd^j \in \mathscr{R}_\alpha^d$ by the equations

$$[Pd^i, Qd^j]_\alpha^d \rightleftharpoons [P_1 d^r, Q_1 d^s]_0^d = [P_1, Q_1]_0 d^{r+s}, \tag{16}$$

under the assumption that the following conditions are satisfied:

$$Pd^i \overset{\alpha}{\approx} P_1 d^r,\ Qd^j \overset{\alpha}{\approx} Q_1 d^s \quad \text{and} \quad [P_1, Q_1]_0 \in \mathscr{R}_\alpha. \tag{17}$$

The existence of an operation like this, defined everywhere on $\mathscr{R}_\alpha^d$, follows immediately from (5) and the existence of ordinary coupling in rank α. The fact that it is associative and single-valued is proved just as the corresponding properties of ordinary coupling were proved in Chapter V; use of these latter simplifies the proof of their analogues.

1.5. We return now to a study of the group $\mathbf{A}(m, n)$. For $\alpha \geqslant 0$, we denote by $\mathbf{A}(m, n, \alpha)$ the group on generators (1) with defining relations (2) together with

$$A^n = d \quad (A^n \in \bigcup_{i=1}^{\alpha} \mathscr{E}_i). \tag{18}$$

For $\alpha = 0$, relation (18) is absent.

We consider also the group $\Gamma^d(m, n, \alpha)$, whose elements are the equivalence classes $\{X\}_\alpha^d$ of $\mathscr{R}_\alpha^d$ under $\overset{\alpha}{\approx}$, and the group operation is defined in terms of generalised coupling in rank α.

It is easy to convince oneself that the condition

$$X = Y \text{ in } \mathbf{A}(m, n, \alpha) \Longleftrightarrow X \overset{\alpha}{\approx} Y \tag{19}$$

holds for arbitrary words X, $Y \in \mathscr{R}_\alpha^d$. The implication from right to left is proved in exactly the same way as the analogous fact was proved for $X \overset{\alpha}{\sim} Y$ in VI.2.3.

On the other hand, it follows immediately from the definition of generalised real reversals of rank α that the transition

$$A^{3q} \to A^{-n+3q} d$$

is a generalised real reversal for every word $A^n \in \mathscr{E}_\alpha$, that is, that $A^{3q} \overset{\alpha}{\approx} A^{-n+3q} d$.

Thus all relations (18) hold in $\Gamma^d(m, n, \alpha)$. Clearly, we have also $[d, a_i]^d_\alpha = a_i d = [a_i, d]^d_\alpha$, that is, relation (2) is also satisfied in $\Gamma^d(m, n, \alpha)$. Consequently, for any $X, Y \in \mathcal{R}^d_\alpha$, it follows that $\{X\}^d_\alpha = \{Y\}^d_\alpha$ in $\Gamma^d(m, n, \alpha)$, that is $X \overset{\alpha}{\approx} Y$, whenever $X = Y$ in $\mathbf{A}(m, n, \alpha)$.

1.6. Theorem. $\mathbf{A}(m, n)$ *is torsion free, that is, if* $x^t = 1$ *in* $\mathbf{A}(m, n)$ *for some* $t \neq 0$, *then* $x = 1$ *in* $\mathbf{A}(m, n)$.

Suppose that $x = Xd^j$, where X is a word in the alphabet (4). If $X \neq 1$ in $\mathbf{B}(m, n)$, then by VI.2.5 there exist words T and E such that $E^n \in \mathscr{E}$ and $X = TE^rT^{-1}$ in $\mathbf{B}(m, n)$ for some integer r. If $X = 1$ in $\mathbf{B}(m, n)$, then $X = TE^rT^{-1}$ in $\mathbf{B}(m, n)$ with $r = 0$, $T = \Lambda$ and $E \in \mathscr{E}$.

By 1.2, there exists an integer i such that $X = TE^rT^{-1}d^i$ in $\mathbf{A}(m, n)$, so that

$$x^t = TE^{rt}T^{-1}d^{(i+j)t} = 1\,.$$

Then $E^{rt}d^{it+jt} = 1$ in $\mathbf{A}(m, n)$, by relations (2). Suppose that $rt = kn + s$, where $|s| \leqslant (n+1)/2$. Then we have $E^{rt} = E^s d^k$ since $E^n = d$. Thus $E^s d^{k+it+jt} = 1$ in $\mathbf{A}(m, n)$. Since only finitely many of the relations (3) are involved in the deduction of this equation, it follows that $\mathbf{A}(m, n, \gamma)$ satisfies the relation

$$E^s d^{k+it+jt} = 1$$

for some $\gamma > 0$, whence by (19) it follows that

$$E^s d^{k+it+jt} \overset{\gamma}{\approx} \Lambda\,.$$

Then $E^s \overset{\gamma}{\sim} \Lambda$ because of relation (5), whence by IV.2.16 it follows that $E^s \overset{\circ}{\equiv} \Lambda$, that is, $s = 0$. This means that $rt = kn$. On the other hand, by relation (6) it follows that $k + it + jt = 0$, since $d^{k+it+jt} \overset{\gamma}{\approx} \Lambda = d^0$, that is, $k = -(i+j)t$. Consequently, $rt = -n(i+j)t$, whence we get that $r = -n(i+j)$. Then in $\mathbf{A}(m, n)$ we have the equation

$$x = Xd^j = TE^rT^{-1}d^{i+j} = TD^{-i-j}T^{-1}d^{i+j} = 1\,.$$

This is what we wanted to prove.

1.7. It follows from Theorems 1.3 and 1.6 that the *intersection of any two infinite cyclic subgroups of* $\mathbf{A}(m, n)$ *is an infinite cyclic group. Thus every two non-identity subgroups of* $\mathbf{A}(m, n)$ *have non-trivial intersection.*

Up to the publication of [11], in which a description of the groups $\mathbf{A}(m, n)$ was first given, only commutative groups with the property indicated were known. Clearly, every commutative group with these properties is locally cyclic. In contrast with the commutative case, the groups $\mathbf{A}(m, n)$ that we have constructed can have any number $m > 1$ of generators.

The construction introduced in this section allows us to establish yet another property of the free groups $\mathbf{B}(m, n)$.

1.8. Theorem. *For odd* $n \geqslant 665$, *the finite subgroups of* $\mathbf{B}(m, n)$ *are all cyclic.*

Let **G** be a finite subgroup of $\mathbf{B}(m, n)$ generated by the words

$$X_1, X_2, \ldots, X_k.$$

Consider the subgroup **F** of $\mathbf{A}(m, n)$ generated by the words

$$X_1, X_2, \ldots, X_k, d.$$

Since d is contained in the centre of **F**, the central factor group of **F** is finite.

But the finiteness of the central factor group implies that the commutator subgroup is finite (see [18], p.163, for example). Thus the commutator subgroup of **F** is finite. Since the only finite subgroup of $\mathbf{A}(m, n)$ is the identity subgroup, **F** is abelian. Thus, by VI.3.3, G is a cyclic subgroup of $\mathbf{B}(m, n)$, as required.

1.9. Kargapolov has observed that $\mathbf{A}(m, n)$ may be used to settle another question, which was first formulated in [14]. It is known that, if a Sylow subgroup of a locally finite group is central, then it splits off as a direct factor. The question naturally arises: is this true for all periodic groups? ([14], Problem 2.13).

Consider the group $\bar{\mathbf{A}}(m, n)$ obtained from $\mathbf{A}(m, n)$ by adding the extra relation $d^2 = 1$. Clearly, $\bar{\mathbf{A}}(m, n)$ has exponent $2n$, the subgroup $\langle d \rangle$ lies in the centre of $\bar{\mathbf{A}}(m, n)$, and $\langle d \rangle$ is a Sylow 2-subgroup $\bar{\mathbf{A}}_2(m, n)$, since n is odd. Since a_1^n, a_2^n and $(a_1 a_2^{-1})^n$ are elementary words of rank 1, and are not related to any elementary word of rank 1 with another period, it follows trivially from (3) and definition VI.2.1 that each of these three words is equal to d or d^{-1} in $\mathbf{A}(m, n)$. Suppose that $\mathbf{A}(m, n)$ has a decomposition as a direct product:

$$\bar{\mathbf{A}}(m, n) = \mathbf{N} \times \langle d \rangle. \tag{20}$$

Then $a_1 \notin \mathbf{N}$ since $a_1^n = d$, that is, $a_1 \in \mathbf{N}d$. Similarly, $a_2 \in \mathrm{N}d$ since $a_2^n = d$. Thus $a_1 a_2^{-1} \in \mathrm{N}$. Then $d \in \mathbf{N}$ since $(a_1 a_2^{-1})^n = d$, which contradicts (20).

§2. Infinite Independent Systems of Group Identities

One of the very difficult problems on group theory is that of the mutual dependence of identical relations in groups. That is why the finite basis problem for varieties of groups, which is connected with this problem, remained open for a considerable time. The question here was: is every system of group identities equivalent to a finite system? In other words, does every variety of groups have a basis consisting of finitely many identities? ([15], p.39). The negative answer to this question was obtained by Ol'šanskiĭ [16] in an implicit form. Shortly afterwards the author [8] constructed an explicit infinite system of group identities, none of which follows from the others. Vaughan-Lee [17] gave another such system.

We provide here a proof of the result in [8], based on an interpretation for group identities of the method developed in the preceding chapters. The systems of identities that we consider contain only 2 variables, and the varieties they generate are not locally finite, unlike the varieties considered in articles [16] and [17], which are conceptually similar.

2.1. Theorem. *For every odd* $n \geqslant 1003$, *the set of all identities of the form*

$$(x^{rn} y^{rn} x^{-rn} y^{-rn})^n = 1\,, \tag{1}$$

where the parameter r runs over all prime numbers, is an independent system of group identities, that is, none of them follows from the others.

To prove Theorem 2.1, we start with an arbitrary odd $n \geqslant 1003$ and an arbitrary set Π of prime numbers, and construct a group $\Gamma(n, \Pi)$ on 2 generators such that the relation (1) is satisfied if and only if $r \in \Pi$ (see 2.15). To this end, we will find it convenient to have yet another modification of the theory of transformation of words developed in Chapters I-V.

2.2. We shall consider words in a group alphabet

$$a,\ b,\ a^{-1},\ b^{-1}\,. \tag{2}$$

For the sequel we fix an odd $n \geqslant 1003$ and an arbitrary set Π of prime numbers.

As before, we shall use the following notations for auxiliary numerical parameters:

$$q_1 = 37,\ q_2 = q_1 + 17 = 54,\ q = 90\,.$$

For words in the alphabet (2), we define by simultaneous induction on α the concept of *distinguished* period in rank $\alpha - 1$ and analogues of all the concepts that were defined in §4 of Chapter I. Moreover, on the basis of this concept of distinguished period, we alter the content of all the old concepts.

For rank 0, everything remains without change.

Suppose that all the concepts enumerated in 1.4 have been defined in rank $\alpha - 1$. If $X \in \mathscr{R}_{\alpha-1}$, then for integral $k > 1$ we denote by $[X]^k_{\alpha-1}$ the k-fold coupling in rank $\alpha - 1$:

$$[[\ldots[[X, X]_{\alpha-1}, \ldots, X]_{\alpha-1}, X]_{\alpha-1}\,.$$

We shall say that a word C is *distinguished in rank* $\alpha - 1$ if there are words $X, Y \in \mathscr{R}_{\alpha-1}$ and a number $r \in \Pi$ such that

$$[[X]^{rn}_{\alpha-1}, [Y]^{rn}_{\alpha-1}, [X^{-1}]^{rn}_{\alpha-1}, [Y^{-1}]^{rn}_{\alpha-1}]_{\alpha-1} \overset{\alpha-1}{\approx} C^j \tag{3}$$

for some $j \geqslant 1$.

We maintain everything defined in I.4.1–I.4.20 without change.

An elementary period A of rank α is said to be a *distinguished elementary period of rank* α if it is distinguished in some rank less than α. In the contrary case, we shall say that A is a non-distinguished elementary period of rank α. An elementary word of rank α is said to be a *distinguished elementary word in rank* α if and only if it obtained from a distinguished elementary period. We shall say that an occurrence of a distinguished elementary word of rank α is a *distinguished occurrence of rank* α.

Using this concept of distinguished elementary word of rank α, we alter the definitions of some of the concepts from 1.4 in an essential way for ranks $\alpha > 0$.

Indeed, *in definitions* I.4.21–I.4.23 *we shall assume that, in all references to normalised occurrences of elementary words of rank* α, *we have in mind distinguished elementary words of rank* α. *In definition* I.4.26, *all quantifiers governing* W *will also be assumed to be restricted by the condition that* W *is a distinguished occurrence of rank* α.

We maintain all the other definitions in §4 of Chapter I, in Chapters II–V, and in §1 of Chapter VI without any formal change. However, since the concepts under consideration are mutually dependent, the restriction we have introduced is reflected in the content of all the concepts, including those where the definition remains without any formal change. Moreover, we shall maintain the names and notations for the classes of words and relations that we are using to replace the old ones.

The set $\mathscr{B}$ *thus obtained* (see VI.1.3), *with the operation* $\circ$ (see VI.1.4), *is the desired group* $\Gamma(n, \Pi)$.

2.3. The basic properties of the concept of being distinguished in rank $\alpha - 1$ follow easily from the corresponding definitions.

It follows from the relation (11) introduced in II.5.16 that: *if* $V \in \text{Norm}(\alpha, X, 9)$ *and* $X \overset{\alpha-1}{\sim} Y$, *then the elementary words* $\text{Bas}(V)$ *and* $\text{Bas}(f_{\alpha-1}(V; X, Y))$ *are either both distinguished or both non-distinguished in rank* $\alpha - 1$.

If the period A of rank α is distinguished in rank $\alpha - 1$, then A^{-1} is likewise distinguished in rank $\alpha - 1$.

If B *and* C *are periods of rank* α *and*

$$B \overset{\alpha-1}{\sim} [T, C, T^{-1}]_{\alpha-1},$$

then they are either both distinguished or both non-distinguished in rank $\alpha - 1$.

Indeed, if C satisfies relation (3), then we can choose $[T, X, T^{-1}]_{\alpha-1}$ and $[T, Y, T^{-1}]_{\alpha-1}$ as the desired X and Y for period B.

In particular, every cyclic shift of a period A of rank α that is distinguished in rank $\alpha - 1$ is likewise distinguished in rank $\alpha - 1$.

With the notations of II.4.3, it follows from II.7.15 that, *if the period* A *is distinguished in rank* $\alpha - 1$, *then the period* D *is also distinguished in rank* $\alpha - 1$.

It follows from II.7.22 and II.4.3 that, *if* E *and* D *are elementary* 9*-powers of rank* α *such that* $\text{Rel}(E, D)$, *then they are either both distinguished or both non-distinguished.*

From this it follows by II.5.17 that, *if* $\text{Comp}(U, V)$, *then* $\text{Bas}(U)$ *and* $\text{Bas}(V)$ *are either both distinguished or both non-distinguished.*

2.4. The formulations of an overwhelming proportion of the original assertions remain without change. We shall indicate below all the assertions where it is necessary to make some change in formulation. It should be noted that the proofs of all the assertions in Chapter II–V remain essentially just as they were, independently of whether the statements are altered or not. In these proofs we need only take into account the fact that now only occurrences of distinguished elementary words of rank α can be active kernels of rank α, and that the restrictions on the number of factors in words from $\mathscr{P}_\alpha$, $\mathscr{R}_\alpha$, $\mathscr{K}_\alpha$, $\mathscr{L}_\alpha$ and $\mathscr{M}_\alpha$ extend only to occurrences of distinguished elementary words of rank $\leqslant \alpha$. In particular, certain words in these sets may have kernels of rank $\beta \leqslant \alpha$ containing more than n segments, but the bases of such kernels will not be distinguished elementary words, and the kernels themselves will not be active.

In Chapter II, to which reference should be made for all properties mentioned in 2.3 of distinguished periods and elementary words, we have to alter the formulation of assertion II.4.8, which in the new edition must have the following form:

If no distinguished elementary q-power of rank $< \alpha$ *occurs in an elementary word* E *such that* $l_\alpha(E) \geqslant 3$, *then* $E \in \mathrm{Per}(A)$, *where* A *is an elementary period of rank* α.

In I.4.21 we defined the relation "*occurrences U and V adjoin one another*". Because of the restrictions imposed in 2.2, this definition continues to make sense only for distinguished occurrences U and V, so that, if the occurrences U and V adjoin one another, they must be distinguished. In this connection, we have to include in III.3.1 the condition that U and V be distinguished. For the same reason, according to Def. I.4.22 all elements of a left (or right) cascade of rank α, as well as the head, are distinguished occurrences. According to Def. I.4.23, only distinguished occurrences can be completely stable and active occurrences of rank α. For this reason, we must add the condition that the occurrence W be distinguished in assertions III.3.17, III.3.21, III.3.22 and III.3.24.

In the statements of assertions IV.1.3, IV.1.18, IV.1.19, IV.1.20 and IV.1.27, we must add the condition that the occurrence V be distinguished, and in IV.1.16 the condition that E be distinguished. Assertion IV.2.37 is true only for a distinguished elementary period A, while in assertions IV.2.22 and IV.2.23 we must add the restriction that (14) holds for the corresponding distinguished kernels of rank β. Finally, in the new edition, assertion IV.2.16 will have the following form:

If no distinguished elementary periodic q-power of rank $\leqslant \alpha$ *occurs in* X, *then*

$$X \overset{\alpha}{\sim} Y \Rightarrow X \overline{\underline{\circ}} Y .$$

2.5. All assertions of Chapter V and their proofs remain good. But we shall need the concept of *bounded local left* (or *right*) *coupling of rank* α, which was considered in [5], [8], but is absent from Chapter V of this book. The fact is that, for the proof of the existence of such a coupling, we have to raise the bound on n to $n \geqslant 1003$. We give now the definition of this concept.

Suppose that $X, Y \in \mathscr{L}_\alpha$. A word $Z \in \mathscr{K}_\alpha$ is said to be *the result of a bounded right local coupling of rank* α of the words X and Y if there is a word $X_1T \in \mathscr{K}_\alpha$ such that

$$X \overset{\alpha}{\sim} X_1T,\ Y \overline{\underline{\circ}} T^{-1}Y_1,\ Z \overline{\underline{\circ}} X_1Y_1$$

and the following conditions are satisfied:

(a) If the occurrence $P * E * Q \in \mathrm{Reg}(\alpha, X_1T)$ is contained in $* X_1 * T$, then, for some Q_1,

$$f_\alpha(P * E * Q; X_1T, X) = P * E * Q_1 .$$

(b) If $0 < \beta \leqslant \alpha$ and the occurrence $V \in \mathrm{Norm}(\beta, X_1T, 9)$ is distinguished and contained in $* X_1 * T$, then $l_\beta(V) \leqslant (n+1)/2 + 55$; if in addition $X \in \mathscr{M}_\alpha$, then $l_\beta(V) \leqslant (n+1)/2 + 34$.

(c) If $X_2 * D * T_1 \in \mathrm{Reg}(\alpha, X)$ and $T_1^{-1} * D^{-1} * Y_2 \in \mathrm{Reg}(\alpha, Y)$, then the word DT_1 is an end of T.

(d) If γ is the smallest number such that $X \not\mathscr{L} X_1T$, $\gamma > 0$ and the kernel $V \in \mathrm{Ker}(\gamma, X_1T)$ is not related to the kernel $f_\gamma(V; X_1 T, X)$, then $\geqslant 211$ segments of its maximal normalised continuation are contained in the occurrence $X_1 * T *$.

In this case we shall write

$$Z = [X, \underline{Y}]'_\alpha .$$

Bounded left local coupling of rank α of words X and Y is defined in an analogous way and is denoted by $[\underline{X}, Y]'_\alpha$.

2.6. *If $n \geqslant 1003$ and $X, Y \in \mathscr{L}_\alpha$, then there are words $Z \in \mathscr{K}_\alpha$ and $Z' \in \mathscr{K}_\alpha$ such that*

$$Z = [X, \underline{Y}]'_\alpha \quad and \quad Z' = [\underline{X}, Y]'_\alpha .$$

Clearly, it is enough to prove the first assertion. To do that, it is enough to repeat the arguments that were employed in V.1.3, under the additional condition: *no transformation is made on the word Y*. We clarify here why it is that we can restrict to transforming X, and not touch Y. The fact is that, with the notation of V.1.3, we have $l_\alpha(D) \leqslant (n+1)/2 + 59$ since $Y \in \mathscr{L}_\alpha$, and hence, on account of the condition $n \geqslant 1003$ and the relation (3), it follows from V.1.3 that $l_\alpha(E) \geqslant (n-1)/2 - 278 \geqslant 223$, that is, in every case the occurrence $R * E * T$ is compatible with some active kernel $U_1 \in \mathrm{Ker}(\alpha, X_1T)$. Earlier, with $n \geqslant 665$, we could prove only that at least one of the occurrences $R * E * T$ and $T^{-1} * D * S$ has this property, while for $n \geqslant 1003$ we can perform a real reversal of either of these two occurrences.

The bound $l_\beta(V) \leqslant (n+1)/2 + 34$ in condition (b) is established for $X \in \mathscr{M}_\alpha$ essentially as it was in V.1.3.

Finally, the desired condition (d) holds because of the remark made in the footnote on p. 214, since $(n-1)/2 - 286 \geqslant 215$.

As was remarked in the footnote on p. 16, using the existence of such a coupling, we can replace the concept of simple r-reversal of rank α that was introduced at I.4.16 by the simpler concept defined in I.4.18. This also allows us to simplify the proofs of some of the assertions in Chapters III-V. But we shall not do this, rather we use the existence of a bounded left (right) local coupling of rank α to prove a new lemma 2.10, which is needed in the proof of Theorem 2.1.

2.7. In the new theory, the following assertion is the analogue of VI.1.2, and it is proved just as VI.1.2 was proved.

For every non-empty word $X \in \mathscr{A}_\alpha$ *there exists a word* $S \in \mathscr{A}_{\alpha+2}$ *such that* Norm(α, S, 26) *is empty, and an elementary period of some rank* $\gamma \leqslant \alpha + 1$ *such that the word* A^q *occurs in some word in* $\mathscr{M}_{\alpha-1}$, *and the relation*

$$X \overset{\alpha}{\sim} [S, A^r, S^{-1}]_\alpha$$

holds for some $r > 0$, *where there is no occurrence of a distinguished elementary* $((n+1)/2 + 46)$*-power of any rank in the word* A^r.

2.8. *For every nonempty word* $X \in \mathscr{R}_\alpha$, *there exist words* $S \in \mathscr{K}_\alpha$ *and* A *such that the following conditions hold for some* $r > 0$:

(a) $X \overset{\alpha}{\sim} [S, A^r, S^{-1}]_\alpha$,

(b) *A is either an elementary period of some rank* $\gamma \leqslant \alpha$, *or else a minimal period of rank* $\alpha + 1$; *and also the word* A^q *occurs in some word* Z *in* $\mathscr{M}_\alpha$.

The proof of this assertion is also obtained by repeating the arguments of VI.1.2, where the condition $X \in \mathscr{A}_\alpha$ (which is stronger than $X \in \mathscr{R}_\alpha$) is used only in proving the additional properties of the word S and the fact that the period A is elementary in rank $\alpha + 1$.

2.9. *Suppose that* A^q *is a non-distinguished elementary word of rank* $\gamma \leqslant \alpha$ *with period* A (*or that* A *is a minimal period of rank* $\alpha + 1$). *If there is no occurrence of a distinguished elementary* $((n+1)/2 + 67)$*-power of rank* $\leqslant \alpha$ *in the word* T, *then the condition* $TA^qT^{-1} \in \mathscr{K}_{\gamma-1}$ (*or* $TA^qT^{-1} \in \mathscr{K}_\alpha$ *respectively*) *implies that* $TA^iT^{-1} \in \mathscr{R}_\alpha$ *for all* $i \geqslant 1$, *and* $TA^iT^{-1} \in \mathscr{K}_\alpha$ *for all* $i > 10$.

We consider first the case where A^q is a non-distinguished elementary word of rank $\gamma \leqslant \alpha$.

Assume that $TA^iT^{-1} \notin \mathscr{K}_\alpha$ for some $i > 10$. Let β be the smallest number such that $TA^iT^{-1} \notin \mathscr{K}_\beta$. By II.6.13, we have $\beta \geqslant \gamma$. By I.4.26, there exists a kernel $W \in \text{Ker}(\beta, TA^iT^{-1})$ such that W is distinguished and $l_\beta(W) \geqslant n - 217$.

We shall show that one of the occurrences $* T * A^iT^{-1}$ and $TA^i * T^{-1} *$ contains $\leqslant 9$ segments of the occurrence W. Assume that this is not so, that is, $T \doteq T_1C$ and $T^{-1} \doteq DT_2$, where $l_\beta(C) \geqslant 10$, $l_\beta(D) \geqslant 10$, Comp(W, $T_1 * C * A^iT^{-1}$) and Comp(W, $TA^i * D * T_2$). Suppose, for example, that $\partial(C) \geqslant \partial(D)$. Then D^{-1} is an end of C, that is, $C \doteq C_1D^{-1}$. By II.6.7, the occurrence $T_1C_1 * D^{-1} * A^iDT_2$ is compatible with W, and consequently with $TA^i * D * T_2$. But this is impossible, since $\neg$Rel(D, D^{-1}) by II.5.21.

If $\beta > \gamma$, then because of II.2.12, less than three segments of W are contained in $T * A^i * T^{-1}$, while, if $\beta = \gamma$, by II.6.7 there are $\leqslant 16$ segments of W in $T * A^i * T^{-1}$, since W is a distinguished occurrence. By II.5.2, it follows from this that some distinguished elementary $(l_\beta(W) - 11 - 18)$-power of rank $\beta \leqslant \alpha$ must occur in the word T, which is impossible by assumption.

Assume that $TA^iT^{-1} \notin \mathscr{R}_\alpha$ for some $i \geqslant 1$. Then $i \leqslant 10$. Let β be the least number such that $TA^iT^{-1} \notin \mathscr{R}_\beta$. By I.4.26, there exists a kernel $W \in \text{Ker}(\beta, TA^iT^{-1})$ such that W is distinguished and $l_\beta(W) \geqslant n - 175$. If $\beta \geqslant \gamma$, then as was proved above, we get a contradiction. It remains to consider the case $\beta < \gamma$. We can assert that $\leqslant 9$ segments of W occur in $TA^i * T^{-1} *$. Let V be a maximal start of W contained in $* TA^i * T^{-1}$. Since $TA^iT^{-1} \in \mathscr{P}_\beta$, by IV.1.16 there exists a kernel $U \in \operatorname{Ker}(\beta_j, TA^qT^{-1})$ that is compatible with the occurrence $\phi(V; * TA^i * T^{-1}, * TA^i * A^{q-i}T^{-1})$ and such that $l_\beta(U) \geqslant l_\beta(V) - 21 \geqslant n - 267$. But this contradicts the condition $TA^qT^{-1} \in \mathscr{K}_\alpha$.

We come now to the case where A is a minimal period of rank $\alpha + 1$ and $TA^qT^{-1} \in \mathscr{K}_\alpha$. If in addition $i > 10$, then we have $TA^iT^{-1} \in \mathscr{R}_\alpha$ by II.6.13. If $TA^iT^{-1} \notin \mathscr{K}_\alpha$ for some $i \leqslant 10$, then, as was proved at the end of the preceding paragraph, we obtain a contradiction in view of IV.1.16.

2.10. *Suppose that $Y \in \mathscr{L}_\alpha$ and A is either a minimal period of rank $\alpha + 1$ or a non-distinguished elementary period of rank $\gamma \leqslant \alpha$, where the word A^q occurs in some word in the class $\mathscr{M}_\alpha$. Then there is a cyclic shift A_1 of A and a word $TE \in \mathscr{K}_\alpha$ such that $Y \overset{\alpha}{\sim} TE$ and the following conditions are satisfied:*

1) *$[Y, A^i, Y^{-1}]_\alpha \overset{\alpha}{\sim} TA_1^iT^{-1}$ for all $i \geqslant 1$,*

2) *$TA_1^iT^{-1} \in \mathscr{K}_\alpha$ for all $i > 10$,*

3) *there is no occurrence in the word T of a distinguished elementary $((n + 1)/2 + 64)$-power of rank $\leqslant \alpha$, and if $Y \in \mathscr{M}_\alpha$ there is no occurrence in T of a distinguished elementary $((n + 1)/2 + 43)$-power of rank α,*

4) *the word E is an end of A^j for some j, which may be negative,*

5) *for every occurrence $P * Q * TE \in \operatorname{Reg}(\alpha, TE)$ contained in $* T * E$, there exists a word Y' such that*

$$f_\alpha(P * Q * RE; TE, Y) = P * Q * Y'.$$

We note first that $A^i \in \mathscr{L}_\alpha$ for all $i \geqslant q$. This follows from II.6.13 and IV.1.19 since, if A^q is a non-distinguished elementary word of rank $\gamma \leqslant \alpha$, then by II.2.12 and II.6.7, no distinguished elementary q-power of rank $\geqslant \gamma$ can occur in it.

By 2.6, for every $j \geqslant q$ we can perform a bounded right local coupling $[Y, \underline{A^j}]_\alpha'$, that is, there exists a word $TE \in \mathscr{K}_\alpha$ such that

$$Y \overset{\alpha}{\sim} TE, \; A^j \mathrel{\underline{\underline{\circ}}} E^{-1}D, \; TD \in \mathscr{K}_\alpha$$

and the conditions (a), (b), (c) and (d) mentioned in 2.5 are fulfilled. We choose the number j in such a way that $j \geqslant q$ and $\partial(D) \geqslant 11\partial(A)$. Let λ be the smallest number such that $Y \overset{\lambda}{\sim} TE$. Then $\lambda \leqslant \alpha$.

Clearly, we may assume that the word $[Y, \underline{A^j}]_{\lambda-1}'$ does not lie in $\mathscr{K}_\alpha$ if $\lambda > 0$.

It follows from II.6.6 and property (b) that no distinguished elementary $((n + 1)/2 + 64)$-power of rank $\leqslant \alpha$ occurs in the word T, that is, the desired condition 3) is satisfied. Clearly, E is an end of A^j. The desired condition 5) follows from property (d). Let A_1 be a cyclic shift of the period A such that $DE^{-1} \mathrel{\underline{\underline{\circ}}} A_1^j$. Then, by V.1.1, condition 1) will follow from 2), if in addition we have $TA_1^iT^{-1} \in \mathscr{R}_\alpha$ for all $i \geqslant 1$.

Thus, it remains only for us to prove that 2) holds and that $TA_1^iT^{-1} \in \mathscr{R}_\alpha$ for all $i \geqslant 1$.

We carry out also a bounded left local coupling $[\underline{A^j}, Y^{-1}]_\alpha'$, that is, we find a word $E_1^{-1}T_1 \in \mathscr{K}_\alpha$ such that

$$Y^{-1} \overset{\alpha}{\sim} E_1^{-1}T_1, \; A^j \mathrel{\underline{\underline{\circ}}} D_1E_1, \; D_1T_1 \in \mathscr{K}_\alpha$$

and the corresponding properties (a), (b), (c) and (d) are satisfied. We denote by δ the smallest number such that $Y^{-1} \overset{\delta}{\sim} E_1^{-1}T_1$. Clearly, we may assume that $\partial(D_1) \geqslant 11\partial(A)$. It will follow easily from the properties (a), (b) and (d) indicated that conditions 3), 4), 5) hold for the relation $Y \overset{\alpha}{\sim} T_1^{-1}E_1$.

By symmetry, we may restrict to a detailed consideration of the case $\lambda \geqslant \delta$.

Suppose that $\lambda = \delta = 0$. Since A^q is cyclically uncancellable, one of the words E and E_1 is empty. Suppose, for example, that E_1 is empty, that is, $D_1 Y^{-1} \in \mathscr{K}_\alpha$. We have $Y \underset{\circ}{\equiv} TE$, $A^j \underset{\circ}{\equiv} E^{-1}D$ and $TD \in \mathscr{K}_\alpha$. By assumption, A^{11} occurs in the words D and D_1.

By II.6.13, we have $TDE^{-1}T^{-1} \in \mathscr{K}_\alpha$ if A is a minimal period of rank $\alpha + 1$, and $TDE^{-1}T^{-1} \in \mathscr{K}_{\gamma-1}$ if A is an elementary period of rank $\gamma \leqslant \alpha$. Since in addition II.6.6 and condition 2.5(b) give that no distinguished elementary $((n+1)/2+64)$-power of rank $\leqslant \alpha$ occurs in T, we have by 2.9 the desired relations $TA_1^i T^{-1} \in \mathscr{R}_\alpha$ for $i \geqslant 1$, and $TA_1^iT^{-1} \in \mathscr{K}_\alpha$ for $i > 10$.

We come now to the case $\lambda > 0$. By condition 2.5(d), some distinguished elementary 211-power of rank λ occurs in the word E. Then A^j is a periodic word of rank $\lambda + 1$ with period A minimal in rank $\lambda + 1$. By II.6.13, $DE^{-1}T^{-1} \in \mathscr{K}_{\lambda-1}$.

Assume that $DE^{-1}T^{-1} \notin \mathscr{K}_\lambda$. Then there is a distinguished kernel $W \in \mathrm{Ker}(\lambda, DE^{-1}T^{-1})$ such that $l_\lambda(W) \geqslant n-217$. Since by assumption DE^{-1} occurs in some word in $\mathscr{M}_\alpha$, it follows that no distinguished elementary $((n + 1)/2 + 43)$-power of rank λ occurs in it. Thus, by II.5.2, $DE^{-1} * T^{-1} *$ contains $\geqslant (n - 1)/2 - 261 \geqslant 240$ segments of the kernel W. It follows in an analogous way from condition 3) above that $* DE^{-1} * T^{-1}$ contains $\geqslant (n - 1)/2 - 282 \geqslant 219$ segments of W.

Let $DE^{-1} * F^{-1} * R^{-1}$ be an end of W contained in $DE^{-1} * T^{-1} *$. The occurrence $R * F * E$ is compatible with some active kernel $V \in \mathrm{Ker}(\lambda, TE)$. Assume that some kernel $V_1 \in \mathrm{Ker}(\lambda, TE)$ different from V is not related to the kernel $f_\lambda(V_1;\, TE,\, Y)$. Then $V \ll V_1$ by 2.5(d), and by IV.2.12, it follows that $l_\lambda(V_1) \geqslant 134$ since $Y \in K_\gamma$. Set $V_1 = RFE_1 * Q_1 * E_2$. Then, by III.3.24, the occurrence $DE_2^{-1} * Q_1^{-1} * E_1^{-1}F^{-1}R^{-1}$ in the word $DE^{-1}T^{-1}$ is compatible with some active occurrence of rank λ, situated strictly to the left of W. By IV.1.14, it follows from this that

$$\phi(W^{-1};\, * TE * D^{-1},\, * TE *) \in \mathrm{Ker}\,(\lambda, TE)\,,$$

which contradicts the assumption that $TE \in \mathscr{K}_\alpha$. This V is the unique kernel of rank λ of the word TE that is not related to the kernel $f_\lambda(V;\, TE,\, Y)$. Then the transition

$$TE \to Y \tag{4}$$

is a real reversal of the kernel V. Since $l_\lambda(F) \geqslant 240$, by II.6.6 the occurrence $R * F * D$ is compatible with some occurrence $U \in \mathrm{Norm}(\lambda, TD, 232)$. By III.3.24, $U \in \mathrm{Act}(\lambda, TD)$. Suppose that the transition

$$TD \to Z \tag{5}$$

is a reversal of the occurrence U such that $Z \in \mathscr{K}_{\lambda-1}$. Then $Z \in \mathscr{K}_\lambda$, by IV.1.26. Using II.7.19 and III.1.10, we get easily from (4) and (5) that

$$Z \overset{\lambda-1}{\leftrightharpoons} [Y, E^{-1}D]_{\lambda-1}\,. \tag{6}$$

Consider a bounded right local coupling of rank $\lambda - 1$,

$$[Y, \underline{E^{-1}D}]'_{\lambda-1} \rightleftharpoons Z_1\,. \tag{7}$$

It follows from (6) and (7) that $Z \overset{\lambda-1}{\leftrightharpoons} Z_1$. By IV.1.24, we have $Z_1 \in \mathscr{K}_\lambda$. Further, $Z_1 \in \mathscr{K}_\alpha$ since $TD \overset{\lambda}{\sim} Z_1$ and $TD \in \mathscr{K}_\alpha$. But this contradicts the minimality condition in the choice of λ. Thus $DE^{-1}T^{-1} \in \mathscr{K}_\lambda$.

Then $TDE^{-1}T^{-1} \in \mathscr{K}_\lambda$ and $E^{-1}DE^{-1}T^{-1} \in \mathscr{K}_\lambda$ by II.6.13. Using properties of coupling, we find that

$$E^{-1}DE^{-1}T^{-1} = [A^j, E^{-1}T^{-1}]_\lambda \overset{\lambda}{\sim} [A^j, Y^{-1}]_\lambda \overset{\lambda}{\sim} [\underline{A}^j, Y^{-1}]'_\delta\,.$$

Since $[\underline{A}^j, Y^{-1}]'_\delta \in \mathscr{K}_\alpha$, we have $E^{-1}DE^{-1}T^{-1} \in \mathscr{K}_\alpha$ by IV.2.11. From this and the fact that $TD \in \mathscr{K}_\alpha$ it follows by II.6.13 that $TDE^{-1}T^{-1} \in \mathscr{K}_\alpha$, when A is a minimal period of rank $\alpha + 1$, and that $TDE^{-1}T^{-1} \in \mathscr{K}_{\gamma-1}$ when A is an elementary period of rank $\gamma \leqslant \alpha$. Since $DE^{-1} \overset{0}{=} A_1^j$, where A_1 is a cyclic shift of the period A, it remains only for us to use Lemma 2.9.

2.11. *Let D be an elementary period of some rank δ, suppose that D^q occurs in some word in the class $\mathscr{M}_{\delta-1}$, and that no distinguished elementary 17-power of rank α occurs in D^9. If the period D is distinguished in rank α, then it is distinguished in some rank less than α.*

Suppose that D is distinguished in rank α, that is, there exist words $X, Y \in \mathscr{R}_\alpha$ and a number $r \in \Pi$ such that the condition

$$[[X]^{rn}_\alpha, [Y]^{rn}_\alpha, [X^{-1}]^{rn}_\alpha, [Y^{-1}]^{rn}_\alpha]_\alpha \overset{\alpha}{\sim} D^t \tag{8}$$

holds for some $t \geqslant 1$.

By 2.8 and IV.2.20, we can find words $S \in \mathscr{M}_\alpha$ and A such that

$$X \overset{\alpha}{\sim} [S, A^s, S^{-1}]_\alpha \tag{9}$$

for some $s \geqslant 1$, the word A^q occurs in some word in $\mathscr{M}_\alpha$, and A is either an elementary period of some rank $\leqslant \alpha$, or else it is a minimal period of rank $\alpha + 1$. If in addition A is a distinguished elementary period of rank $\leqslant \alpha$, then by IV.2.37 we have $A^{3q} \overset{\alpha}{\sim} A^{-n+3q}$. Then $[A]^n_\alpha \overset{\alpha}{\sim} A$, whence it follows by (9) that $[X]^{rn}_\alpha \overset{\alpha}{\sim} A$, and further that $A \overset{\alpha}{\sim} D^t$ in view of (8). This contradicts the analogue of assertion IV.2.16 that we indicated at the end of 2.4, since D is non-empty. Thus A is not a distinguished elementary period of rank $\leqslant \alpha$. Then, by 2.10, there is a cyclic shift A_1 of the period A and a word $T \in \mathscr{K}_\alpha$ such that

$$[S, A^i, S^{-1}]_\alpha \overset{\alpha}{\sim} TA_1^iT^{-1} \tag{10}$$

for all $i \geqslant 1$, and $TA_1^iT^{-1} \in \mathscr{K}_\alpha$ for arbitrary $i > 10$. In particular we have $TA_1^{srn}T^{-1} \in \mathscr{K}_\alpha$, and hence

$$[X]_{\alpha}^{rn} \overset{\alpha}{\sim} TA_1^{srn}T^{-1}$$

by (9) and (10).

In a similar way we find for the word A words $R \in \mathscr{K}_\alpha$ and B such that $RB^iR^{-1} \in \mathscr{K}_\alpha$ for any $i \geqslant 10$, B is either an non-distinguished elementary period of rank $\leqslant \alpha$ or a minimal period of rank $\alpha + 1$ and, for some $k \geqslant 1$,

$$[Y]_{\alpha}^{rn} \overset{\alpha}{\sim} RB^{krn}R^{-1}.$$

Hence we have

$$[TA_1^{srn}T^{-1}, RB^{krn}R^{-1}, TA_1^{-srn}T^{-1}, RB^{-krn}R^{-1}]_\alpha \overset{\alpha}{\sim} D^t. \tag{11}$$

Assume that some distinguished elementary 25-power of rank α occurs in A_1^n and, if B^n also has this property, then suppose that $\rho_{\alpha+1,A}(A_1^n) \geqslant \rho_{\alpha+1,B}(B^n)$. We shall prove below that this assumption leads to a contradiction.

Suppose that $T_1 \overset{\alpha}{\sim} [R^{-1}, T]_\alpha$ and $T_1 \in \mathscr{M}_\alpha$. By 2.10, there is a word $T_2 \in \mathscr{K}_\alpha$ and a cyclic shift A_2 of the period A_1 such that

$$[T_1, A_1^{srn}, T_1^{-1}]_\alpha \overset{\alpha}{\sim} T_2A_2^{srn}T_2^{-1},$$

where $T_2A_2^{srn}T_2^{-1} \in \mathscr{K}_\alpha$. Then

$$[T_2A_2^{srn}T_2^{-1}, B^{krn}, T_2A_2^{-srn}T_2^{-1}, B^{-krn}]_\alpha \overset{\alpha}{\sim} [R^{-1}, D^t, R]_\alpha \tag{12}$$

follows from relation (11).

We consider the word $A_2^qT_2^{-1}$. Since $A_2^p \in \mathscr{L}_\alpha$, by IV.3.14 there exists a word $A_2^{q-9}T_3 \in \mathscr{L}_\alpha$ such that

$$A_2^qT_2^{-1} \overset{\alpha}{\sim} A_2^{q-9}T_3. \tag{13}$$

Again by 2.10, there is a word $ZE \in \mathscr{K}_\alpha$ such that the conditions

$$A_2^{q-9}T_3 \overset{\alpha}{\sim} ZE, \quad [A_2^qT_2^{-1}, B^{krn}, T_2A_2^{-2}]_\alpha \overset{\alpha}{\sim} ZB_1^{krn}Z^{-1} \tag{14}$$

hold, and $ZB_1^{krn}Z^{-1} \in \mathscr{K}_\alpha$, where B_1 is a cyclic shift of B, the word E is an end of B^j for some integer j, and for every occurrence $R_1 * C * R_2E \in \text{Reg}(\alpha, ZE)$ contained in $* Z * E$ we can find a Z' such that

$$f_\alpha(R_1 * C * R_2E; ZE, A_2^{q-9}T_3) = R_1 * C * Z'. \tag{15}$$

We may assume without loss of generality that $j \geqslant 0$ and that the occurrence $A_2^{30} * A_2^{40} * A_2^{q-79}T_3$ starts with some supporting kernel of rank α. Suppose that $A_2 \overset{\circ}{=} A'u$ and that

$$A_2^{30} * A_2^{40}A' * uA_2^{q-80}T_3 \in \text{Reg}(\alpha, A_2^{q-9}T_3).$$

We shall prove first of all that the occurrence

$$W \rightleftharpoons f_\alpha(A_2^{30} * A_2^{40}A' * uA_2^{q-80}T_3;\ A_2^{q-9}T_3,\ ZE)$$

must intersect the occurrence $* Z * E$. Suppose that this is not so. Then W is contained in $Z * E *$, and since E occurs in B^j then, since $\rho_{\alpha+1,A}(A_1^n) \geqslant \rho_{\alpha+1,B}(B^n)$ and the period A is minimal, it follows that $\rho_{\alpha+1,A}(A_1^n) = \rho_{\alpha+1,B}(B^n)$, that is, $W = ZH_1 * B_0^{40}B' * H_2$, where the period B_0 is a cyclic shift of B and an image of A_2 in W. Then by II.7.15, there exists a word $S \in \mathscr{A}_{\alpha+1}$ for which the conditions

$$A_2 \overset{\alpha}{\sim} [S,\ B_0,\ S^{-1}]_\alpha\,, \tag{16}$$

$$ZH_1 \overset{\alpha}{\sim} [A_2^{30},\ S]_\alpha \text{ and } B'H_2 \overset{\alpha}{\sim} [S^{-1},\ A_2^{q-79}T_3]_\alpha \tag{17}$$

hold.

The word $B'H_2$ is an end of E which starts with B_0. Suppose that $B'H_2 \overset{\circ}{=} P(QP)^t$, where $B_0 \overset{\circ}{=} PQ$ and $B \overset{\circ}{=} QP$. From (13) it follows that $T_2 \overset{\alpha}{\sim} T_3^{-1}A_2^q$. We therefore find, using (16) and (17), that

$$\begin{aligned} T_2A_2^{srn}T_2^{-1} = [T_2,\ A_2^{srn},\ T_2^{-1}]_\alpha &\overset{\alpha}{\sim} [T_3^{-1},\ A_2^{srn},\ T_3]_\alpha \\ &\overset{\alpha}{\sim} [[(B'H_2)^{-1},\ S^{-1},\ A_2^{q-79}]_\alpha,\ [A_2^{srn},\ A_2^{-q+79},\ S,\ B'H_2]_\alpha]_\alpha \\ &\overset{\alpha}{\sim} [B^{-t}P^{-1},\ B_0^{srn},\ PB^t]_\alpha = B^{srn}\,. \end{aligned}$$

Then substitution in (12) gives

$$A \overset{\alpha}{\sim} [R^{-1},\ D^t,\ R]_\alpha\,,$$

which contradicts the fact that D is non-empty.

Hence W intersects $* Z * E$. Then the occurrence

$$f_\alpha(A_2^{10} * A_2^{10}A' * uA_2^{q-30}T_3;\ A_2^{q-9}T_3,\ ZE)$$

must be contained in $* Z * E$ and hence, by (15), A_2^{20} is a start of Z.

In a way analogous to that in which we found the word Z satisfying (14) we can find, again by 2.10, a word $Z_1E_1 \in \mathscr{K}_\alpha$ such that

$$A_2^{-q}T_2^{-1} \overset{\alpha}{\sim} Z_1E_1$$

and

$$[A_2^{-q}T_2^{-1},\ B^{krn},\ T_2A_2^q]_\alpha \overset{\alpha}{\sim} Z_1B^{krn}Z_1^{-1}\,, \tag{18}$$

where $Z_1B_2^{krn}Z_1^{-1} \in \mathscr{K}_\alpha$ and B_2 is a cyclic shift of B, and we again convince ourselves that A_2^{-20} is a start of the word Z_1. By II.6.13, we then have the relation

$$A_2^{srn-3q}ZB_1^{krn}Z^{-1}A_2^{-srn+2q}Z_1B_2^{-krn}Z_1^{-1}A_2^q \in \mathscr{K}_\alpha\,. \tag{19}$$

If D is a distinguished elementary period of rank $\delta \leqslant \alpha$, then D is distinguished in some rank $\leqslant \delta - 1 \leqslant \alpha - 1$. By II.4.9, we may assume that either D is a non-distinguished elementary period of rank $\delta \leqslant \alpha$ or D is a minimal period of rank $\alpha + 1$ and $\alpha + 1 \leqslant \delta$. Suppose that

$$F = [A_2^{-q} T_2^{-1}, R^{-1}]_\alpha \tag{20}$$

and $F \in \mathcal{M}_\alpha$. By 2.10, there exists a word F_1 and a cyclic shift D_1 of the period D such that

$$[F, D^t, F^{-1}]_\alpha \overset{\alpha}{\sim} F_1 D_1^t F_1^{-1}.$$

Then it follows from relations (20), (12), (14), (18) and (19), that

$$A_2^{srn-3q} Z B_1^{krn} Z^{-1} A_2^{-srn+2q} Z_1 B_2^{-krn} Z_1^{-1} A_2^q \overset{\alpha}{\sim} F_1 D_1^t F_1^{-1}.$$

We denote

$$G \rightleftharpoons A_2^{srn-3q} Z B_1^{krn} Z^{-1} A^{-srn+2q} Z_1 B_2^{-krn} Z_1^{-1} A_2^q.$$

We may assume that the occurrences

$$V_1 \rightleftharpoons A_2^9 * A_2^{q-10} A' * u A_2^{srn-4q} Z B_1^{krn} Z^{-1} A_2^{-srn+2q} Z_1 B_2^{-krn} Z_1^{-1} A_2^q$$

and

$$V_2 \rightleftharpoons A_2^{srn+3q} Z B_1^{krn} Z^{-1} A_2^{-srn+2q} Z_1 B_2^{-krn} Z_1^{-1} * A_2^{q-10} A' * u A_2^9$$

are regular occurrences of rank α. From the condition that there are no distinguished elementary 17-powers of rank α occurring in D^9, it follows easily that no distinguished elementary 17-powers of rank α occur in D_1^t. Hence

$$f_\alpha(V_1;\, G,\, F_1 D_1^t F_1^{-1}) = P_1 * C_1 * Q_1 D_1^t F_1^{-1}$$

and

$$f_\alpha(V_2;\, G,\, F_1 D_1^t F_1^{-1}) = F_1 D_1^t Q_2 * C_2 * P_2,$$

where C_1 and C_2 are semi-integral $(q-10)$-powers with original period A_2. Moreover, no semi-integral 25-powers of degree $\alpha + 1$ can occur in the words P_1 and Q_2. Therefore one of the occurrences $P_1 * C_1 * Q_1$ and $P_2^{-1} * C_2^{-1} * Q_2^{-1}$ contains not less than $q - 36$ segments of the other. Since $\rho_{\alpha+1,A}(C_1) = \rho_{\alpha+1,A^{-1}}(C_2)$, the common part of the occurrences $P_1 * C_1 * Q_1$ and $Q_2^{-1} * C_2^{-1} * P_2^{-1}$ contains not less than $q - 36$ segments of each of them. But this is impossible, since $\neg$ Rel (C_1, C_2^{-1}) by II.5.21.

We have thus proved that there are no distinguished 25-powers of rank α in the words A^n and B^n.

Suppose that

$$Y_1 \overset{\alpha}{\sim} [T^{-1}, R]_\alpha .$$

Then the relation

$$[A_1^{srn}, Y_1, B^{krn}, Y_1^{-1}, A_1^{-srn}, Y_1, B^{-krn}, Y_1^{-1}]_\alpha \overset{\alpha}{\sim} [T^{-1}, D^t, T]_\alpha \tag{21}$$

follows from (11).

By IV.2.20, we may assume that $T, Y_1 \in \mathscr{M}_\alpha$.

By 2.10 for rank $\alpha - 1$, there exists a word Y_2E_2 and a cyclic shift B_3 of the period B such that

$$Y_1 \overset{\alpha-1}{\approx} Y_2E_2 \tag{22}$$

and

$$[Y_1, B^{krn}, Y_1^{-1}]_{\alpha-1} \overset{\alpha-1}{\approx} Y_2B_3^{krn}Y_2^{-1}, \tag{23}$$

where $Y_2B_3^{krn}Y_2^{-1} \in \mathscr{K}_{\alpha-1}$, E_2 is an end of B^i for some (possibly negative) i and, for every occurrence

$$P_1 * Q * P_2E_2 \in \operatorname{Reg}(\alpha - 1, Y_2E_2),$$

a word Y' can be found such that

$$f_{\alpha-1}(P_1 * Q * P_2E_2;\ Y_2E_2, Y_1) = P_1 * Q * Y' . \tag{24}$$

Since by IV.1.7 there is no distinguished elementary $((n + 1)/2 + 43)$-power of rank α in Y_2, we have $Y_2B_3^{krn}Y_2^{-1} \in \mathscr{K}_\alpha$, from which it follows by (23) and V.1.1 that

$$[Y_1, B^{krn}, Y_1^{-1}]_\alpha \overset{\alpha}{\sim} Y_2B_3^{krn}Y_2^{-1} . \tag{25}$$

By 2.10, there is a word F_2 and a cyclic shift D_2 of the period D such that

$$[T^{-1}, D^t, T]_{\alpha-1} \overset{\alpha-1}{\approx} F_2D_2^tF_2^{-1}, \tag{26}$$

where there is no distinguished elementary $((n + 1)/2 + 43)$-power of rank α in F_2. Then by II.5.2 and II.5.21, there is no distinguished elementary $((n + 1)/2 + 43 + 18 + 11)$-power of rank α in $F_2D_2^tF_2^{-1}$, from which it follows by IV.1.19 that $F_2D_2^tF_2^{-1} \in \mathscr{R}_\alpha$. Furthermore, it follows from (26) by V.1.1 that

$$[T^{-1}, D^t, T]_\alpha \overset{\alpha}{\sim} F_2D_2^tF_2^{-1} . \tag{27}$$

From (21) we obtain using (27) and (25) that

$$[A_1^{srn}, Y_2B_3^{krn}Y_2^{-1}, A_1^{-srn}, Y_2B_3^{-krn}Y_2^{-1}]_\alpha \overset{\alpha}{\sim} F_2D_2^tF_2^{-1} . \tag{28}$$

We first of all consider the case where the set Norm(α, Y_1, 71) contains no distinguished occurrences. Then by (22) and II.6.6, no distinguished elementary 79-power of rank α occurs in Y_2, from which it follows that no distinguished elementary 116-power of rank α occurs in $Y_2B_3^{krn}Y_2^{-1}$. Then, on setting

$$G_1 \rightleftharpoons [A_1^{srn},\ Y_2B_3^{krn}Y_2^{-1},\ A_1^{-srn},\ Y_2B_3^{-krn}Y_2^{-1}]_{\alpha-1}\,, \tag{29}$$

we get by II.7.12 that Norm(α, G_1, 303) contains no distinguished occurrence. This means that $G_1 \in \mathscr{R}_\alpha$. It then follows from (28) that

$$G_1 \overset{\alpha}{\sim} F_2D_2'F_2^{-1}\,.$$

For any distinguished kernel $V \in$ Ker (α, G_1), we have

$$l_\alpha(V) + l_\alpha(f_\alpha(V;\,G_1,\ F_2D_2'F_2^{-1})) < 303 + \frac{n+1}{2} + 72$$

$$= \frac{n+1}{2} + 375 < n - 86\,,$$

from which it follows by IV.2.12 that $I_\alpha(G_1,\ F_2D_2'F_2^{-1}) = 0$. We then have $G_1 \overset{\alpha-1}{\sim} F_2D_2'F_2^{-1}$ by IV.2.13, and by (29) and the fact that $r \in \Pi$, this means that the period D_2 (and hence the period D) is distinguished in rank $\alpha - 1$.

It now only remains to consider the case where there is some distinguished occurrence $W \in$ Norm(α, Y_1, 71). Since there are no distinguished elementary 25-powers of rank α occurring in the word E_2, it follows from II.5.16 and II.5.2 that $* \ Y_2 * E_2$ contains $\geqslant 44$ segments of the occurrence $f_{\alpha-1}(W;\,Y_1,\ Y_2E_2)$. From this it follows by II.6.12 and II.5.5 that there exists some distinguished occurrence

$$P_1 * Q * P_2 \in \text{Norm}(\alpha,\ Y_2,\ 40) \tag{30}$$

such that

$$\text{MutNorm}_{\alpha-1}(P_1 * Q * P_2,\ P_1 * Q * P_2E_2)\,.$$

Then by (24) we have

$$\text{MutNorm}_{\alpha-1}(P_1 * Q * P_2,\ P_1 * Q * Y')\,,$$

where $P_1QY' \overset{\circ}{=} Y_1$. Since $Y_1 \in \mathscr{M}_{\alpha-1}$, by II.7.6 there exists a word $Y_3 \in \mathscr{M}_{\alpha-1}$ such that $Y_2 \overset{\alpha-1}{\sim} Y_3$ and

$$f_{\alpha-1}(P_1 * Q * P_2;\ Y_2,\ Y_3) = P_1 * Q * P_3\,. \tag{31}$$

Now by 2.10, there exists a word Y_4E_3 and a cyclic shift A_3 of the period A_1 such that

$$Y_3^{-1} \overset{\alpha-1}{\simeq} Y_4 E_3 \tag{32}$$

and

$$[Y_3^{-1}, A_1^{srn}, Y_3]_{\alpha-1} \overset{\alpha-1}{\simeq} Y_4 A_3^{srn} Y_4^{-1}, \tag{33}$$

where $Y_4 A_3^{srn} Y_4^{-1} \in \mathscr{K}_{\alpha-1}$, E_3 is an end of A^i for some (possibly negative) i, and for every occurrence

$$P_4 * Q_1 * P_5 E_3 \in \operatorname{Reg}(\alpha - 1, Y_4 E_3)$$

there is a Y'' such that the relation

$$f_{\alpha-1}(P_4 * Q_1 * P_5 E_3; Y_4 E_3, Y_3^{-1}) = P_4 * Q_1 * Y'' \tag{34}$$

holds.

It follows from (30) and (31) that

$$f_{\alpha-1}(P_3^{-1} * Q^{-1} * P_1^{-1}; Y_3^{-1}, Y_4 E_3) \in \operatorname{Norm}(\alpha, Y_4 E_3, 40). \tag{35}$$

Since 25 segments of the occurrence (35) cannot occur in E_3, by II.5.2 and II.6.12 some start $P_4 * Q_1 * P_5 E_3$ of (35) contains $\geqslant 10$ segments of it and is normalised. Then (34) holds, from which it follows that $P_4 * Q_1 * Y''$ is a start of the occurrence $P_3^{-1} * Q^{-1} * P_1^{-1}$, that is, $P_4 \mathrel{\underline{\underline{\circ}}} P_3^{-1}$ and Q_1 is a start of Q^{-1}.

On the other hand, we get $P_2 \overset{\alpha-1}{\simeq} P_3$ from (31) by IV.2.35, that is, $P_4 \overset{\alpha-1}{\simeq} P_2^{-1}$. Since $Y_2^{-1} \in \mathscr{K}_{\alpha-1}$ and $Y_4 \in \mathscr{K}_{\alpha-1}$, it follows by II.6.13 that $P_2^{-1} Q_1 P_5 \in \mathscr{K}_{\alpha-1}$. Let

$$Z \rightleftharpoons P_2^{-1} Q_1 P_5 .$$

Since $Z \in \mathscr{K}_{\alpha 1}$ and $Y_4 A_3^{srn} Y_4^{-1} \in \mathscr{K}_{\alpha-1}$, it follows that

$$Z A_3^{srn} Z^{-1} \in \mathscr{K}_{\alpha-1} . \tag{36}$$

Since $Y_4 \overset{\alpha-1}{\simeq} Z$, we find from (33) and $Y_2 \overset{\alpha-1}{\simeq} Y_3$ that

$$[Y_2^{-1}, A_1^{srn}, Y_2]_{\alpha-1} \overset{\alpha-1}{\simeq} Z A_3^{srn} Z^{-1} .$$

Since $P_2^{-1} Q_1$ is a start of Y_2^{-1}, from (36) and the fact that $Y_2 B_3^{krn} Y_2^{-1} \in \mathscr{K}_{\alpha-1}$ it follows that

$$Z A_3^{srn} Z^{-1} B_3^{krn} Z A_3^{-srn} Z^{-1} B_3^{-krn} \in \mathscr{K}_{\alpha-1} . \tag{37}$$

From (22), (32), $Y_2 \overset{\alpha-1}{\simeq} Y_3$ and $Y_4 \overset{\alpha-1}{\simeq} Z$ there follows

$$Y_1^{-1} \overset{\alpha-1}{\simeq} E_2^{-1} Y_2{}^{1} \overset{\alpha-1}{\simeq} E_2^{-1} Y_4 E_3 \overset{\alpha-1}{\simeq} E_2^{-1} Z E_3 .$$

Since by hypothesis $Y_1^{-1} \in \mathcal{M}_\alpha$, by IV.1.7 no distinguished $((n+1)/2+43)$-powers of rank α occur in Z. There are then no distinguished $((n+1)/2+43+26+11)$-powers of rank α occurring in $ZA_3^{srn}Z^{-1}$, and no distinguished $((n+1)/2+80+26+11)$-powers of rank α occurring in $ZA_3^{srn}Z^{-1}B_3^{krn}ZA_3^{-srn}Z^{-1}$. Hence no distinguished $((n+1)/2+143)$-powers of rank α occur in the word

$$G_2 \rightleftharpoons ZA_3^{srn}Z^{-1}B_3^{krn}ZA_3^{-srn}Z^{-1}B_3^{-krn}. \tag{38}$$

We then have $G_2 \in \mathcal{K}_\alpha$ by (37). In addition,

$$G_2 \overset{\alpha}{\sim} [[Y_2^{-1}, F_2]_\alpha, D_2^l, [F_2^{-1}, Y_2]_\alpha]_\alpha \tag{39}$$

by (28). By IV.2.20 and 2.10, we find a new word F_3 and a new cyclic shift D_3 of the period D such that

$$[[Y_2^{-1}, F_2]_\alpha, D_2^l, [F_2^{-1}, Y_2]_\alpha]_\alpha \overset{\alpha}{\sim} F_3D_3^lF_3^{-1}, \tag{40}$$

where no distinguished $((n+1)/2+43)$-powers of rank α occur in F_3. Then no distinguished $((n+1)/2+72)$-powers of rank α occur in $F_3D_3^lF_3^{-1}$. From (39), (38) and (40) we get

$$ZA_3^{srn}Z^{-1}B_3^{krn}ZA_3^{srn}Z^{-1}B_3^{-krn} \overset{\alpha}{\sim} F_3D_3^lF_3^{-1}. \tag{41}$$

Assume that $I_\alpha(G_2, F_3D_3^lF_3^{-1}) > 0$, that is, some kernel $V \in \operatorname{Ker}(\alpha, G_2)$ is not related to $f_\alpha(V; G_2, F_3D_2^lF_3^{-1})$. Then we have

$$l_\alpha(V) \geqslant n - 86 - l_\alpha(f_\alpha(V; G_2, F_3D_3^lF_3^{-1})) \geqslant \frac{n-1}{2} + 157 \geqslant 344,$$

by IV.2.12. Since no distinguished 25-powers of rank α occur in A_3^{srn} and B_3^{krn}, it follows from II.5.16 and II.5.21 that $\geqslant 318$ segments of V occur in the word $ZA_3^{srn}Z^{-1}B_3^{krn}ZA_3^{-srn}Z^{-1}$. By II.6.7 and II.5.21, V does not have $\geqslant 10$ segments in each of the words $ZA_3^{srn}Z^{-1}$ and $ZA_3^{-srn}Z^{-1}$. Then there are $\geqslant 281$ segments of the kernel V in one of these words, from which it follows similarly that one of Z or Z^{-1} contains $\geqslant 244$ segments of V, that is,

$$Z \overset{\circ}{=} uCv,$$

where C is a distinguished elementary 244-power of rank α. Then

$$G_2 \overset{\circ}{=} uCvA_3^{srn}v^{-1}C^{-1}u^{-1}B_3^{krn}uCvA_3^{-srn}v^{-1}C^{-1}u^{-1}B_3^{-krn},$$

where the occurrences shown of the words C and C^{-1} are compatible with some kernels of rank α:

$$V_1 < V_2 < V_3 < V_4.$$

By IV.1.7, we have $l_\alpha(V_i) \geqslant 202$. Set

$$U_i \rightleftharpoons f_\alpha(V_i;\, G_2,\, F_3D_3^tF_3^{-1}) \quad (i = 1,2,3,4)\,. \tag{42}$$

Since $l_\alpha(V_i) \leqslant (n+1)/2 + 143$,

$$l_\alpha(U_i) \geqslant 273 \tag{43}$$

by IV.2.12.

Clearly, we may assume that $\geqslant 234$ segments of the kernel U_2 are contained in $* F_3 * D_3^tF_3^{-1}$. Suppose that $F_3 \overset{\circ}{=} u_3C_3v_3$, where $l_\alpha(C_3) \geqslant 234$ and $\mathrm{Comp}(u_3 * C_3 * v_3D_3^tF_3^{-1}, U_2)$. By II.6.12, we may assume that both the occurrences $u_3 * C_3 * v_3D_3^tF_3^{-1}$ and $u_3 * C_3 * v_3D_3^{-t}F_3^{-1}$ are normalised and $l_\alpha(C_3) \geqslant 230$. By III.3.24, both these occurrences are active.

From (42) it follows that

$$f_\alpha(V_4^{-1};\, G_2^{-1},\, F_3D_3^{-t}F_3^{-1}) = U_4^{-1}\,. \tag{44}$$

By IV.1.14 we have

$$\phi(U_4^{-1};\, * u_3 * C_3v_3D_3^{-t}F_3^{-1},\, * u_3 * C_3v_3D_3^tF_3^{-1}) \in \mathrm{Ker}(\alpha,\, F_3D_3^tF_3^{-1})\,.$$

We set

$$U_5 \rightleftharpoons \phi(U_4^{-1};\, * u_3 * C_3v_3D_3^{-t}F_3^{-1},\, * u_3 * C_3v_3D_3^tF_3^{-1})$$

and

$$V_5 \rightleftharpoons f_\alpha(U_5;\, F_3D_3^tF_3^{-1},\, G_2)\,.$$

We assume that $U_5 \ll U_1$. Then $V_5 \ll V_1$, that is, V_5 is contained in $* Z * A_3^{srn}Z^{-1}B_3^{krn}ZA_3^{-srn}Z^{-1}B_3^{-krn}$. Since $l_\alpha(V_5) = l_\alpha(U_5) = l_\alpha(U_4) \geqslant 273$, the occurrence

$$\begin{aligned}\phi(V_5;\, * Z * A_3^{srn}Z^{-1}B_3^{krn}ZA_3^{-srn}Z^{-1}B_3^{-krn}, \\ B_3^{krn} * Z * A_3^{srn}Z^{-1}B_3^{-krn}ZA_3^{-srn}Z^{-1})\end{aligned} \tag{45}$$

is compatible with some kernel $V_6 \in \mathrm{Ker}\,(\alpha,\, G_2^{-1})$, where the kernel V_6 is active by II.6.12 and III.3.24.

Since the kernels V_5, V_1 and V_4^{-1} are also active, by IV.1.6, IV.1.15 and IV.1.7 it follows from (43) that

$$l_\alpha(V_6) \geqslant l_\alpha(V_5) - 21 \geqslant 252\,. \tag{46}$$

Since $V_6 \ll V_4^{-1}$, we have $f_\alpha(V_6;\, G_2^{-1},\, F_3B_3^{-t}F_3^{-1}) \ll U_4^{-1}$.

We denote

$$U_6 \rightleftharpoons f_\alpha(V_6;\, G_2^{-1},\, F_3 D_3^{-\iota} F_3^{-1})$$

and

$$U_7 \rightleftharpoons \phi(U_6;\, * u_3 * C_3 v_3 D_3^{-\iota} F_3^{-1},\, * u_3 * C_3 v_3 D_3^{\iota} F_3^{-1})\,.$$

Since $U_6 \in \mathrm{Ker}(\alpha, F_3 D_3^{-\iota} F_3^{-1})$, it follows by IV.1.14 that

$$U_7 \in \mathrm{Ker}(\alpha,\, F_3 D_3^{\iota} F_3^{-1})\,.$$

Thus, proceeding from the relation $U_5 \ll U_1$, we have found a new active kernel U_7 such that $U_7 \ll U_5$ and $l_\alpha(U_7) = l_\alpha(U_6) = l_\alpha(V_6) \geqslant 252$.

It is clear that if we apply the same argument once more to the inequality $U_7 \ll U_5$, we can find yet another kernel $U_9 \in \mathrm{Ker}(\alpha, F_3 D_3^{\iota} F_3^{-1})$ such that $U_9 \ll U_7$ and $l_\alpha(U_9) \geqslant l_\alpha(U_7) - 21 \geqslant 231$. Moreover, since in this case there is an active kernel

$$V_7 \rightleftharpoons f_\alpha(U_7;\, F_3 D_3^{\iota} F_3^{-1},\, G_2)$$

situated to the left of V_5, then by IV.1.15 the kernel V_6 coincides with the occurrence (45), that is, instead of inequality (46) we have

$$l_\alpha(V_6) = l_\alpha(V_5) \geqslant 273\,.$$

Consequently the kernel U_9 thus found again satisfies the relation

$$l_\alpha(U_9) \geqslant l_\alpha(U_7) - 21 = l_\alpha(V_6) - 21 \geqslant 252\,.$$

On applying our argument a sufficient number of times, we obtain the fact that there are more than $\partial(F_3)$ kernels of rank α to the left of U_1. But this is impossible. Hence the assumption that $U_5 \ll U_1$ leads to a contradiction. Similarly, the assumption that $U_1 \ll U_5$ leads to a contradiction. It follows from this by IV.1.2 and IV.1.3 that

$$U_1 = U_5 = \phi(U_4^{-1};\, * F_3 * D_3^{-\iota} F_3^{-1},\, * F_3 * D_3^{\iota} F_3^{-1})\,.$$

By IV.3.24, we have

$$\mathrm{MutNorm}_\alpha(U_1,\, U_4^{-1})\,, \tag{47}$$

whence by (42) there follows

$$\mathrm{MutNorm}_\alpha(V_1,\, V_4^{-1})\,. \tag{48}$$

Clearly, V_1 is contained in the occurrence

$$* ZA_3^{srn}Z^{-1} * B_3^{krn}ZA_3^{-srn}Z^{-1}B_3^{-krn},$$

that is, for some u_1, C_1 and v_1 we have

$$V_1 = u_1 * C_1 * v_1B_3^{krn}ZA_3^{-srn}Z^{-1}B^{-krn}$$

and $u_1C_1v_1 \overset{\circ}{=} ZA_3^{srn}Z^{-1}$. If the occurrence

$$V' \rightleftharpoons u_1C_1v_1B_3^{krn}v_1^{-1} * C_1^{-1} * u_1^{-1}B_3^{-krn}$$

is not compatible with V_4, then one of the relations

$$V' \ll V_4 \quad \text{or} \quad V_4 \ll V'$$

holds. But in each of these cases, we obtain a contradiction in exactly the same way as we did earlier in the case when $U_5 \ll U_1$. Hence we have Comp(V_4, V'), from which it follows by IV.1.6 and IV.1.15 that one of these occurrences is a start of the other. It then follows by IV.3.10 from (48) that $\mathrm{Bas}(V_4^{-1}) \overset{\circ}{=} \mathrm{Bas}(V_1) \overset{\circ}{=} C_1$, that is, $V' = V_4^{-1}$.

If there are no active kernels of rank α to the left of U_1, it then follows from (47) and (44) by IV.3.20 that, for some word $u_1C_1v_1G' \in \mathscr{K}_\alpha$, the relations

$$F_3D_3'F_3^{-1} \overset{\alpha}{\sim} B_3^{krn}u_1C_1v_1G' \tag{49}$$

and

$$f_\alpha(U_1, F_3D_3'F_3^{-1}; B_3^{krn}u_1C_1v_1G') = B_3^{krn}u_1 * C_1 * v_1G' \tag{50}$$

hold.

A word $u_1C_1v_1G'$ which satisfies relations (49) and (50) can be found even in the case when there are active kernels of rank α to the left of U_1. Indeed, in that case, associating with U_1 and U_4^{-1} all the kernels of rank α situated to the left of them, we get from (47) by IV.3.24 and IV.3.25 that once again the occurrences obtained which have one and the same base are also mutually normalised in rank α. Again, applying IV.3.20 to these occurrences and making use of assertion IV.3.5, we get relation (50).

From (42) for $i = 1$ and (50), it follows that

$$f_\alpha(V_1; G_2, B_3^{krn}u_1C_1v_1G') = B_3^{krn}u_1 * C_1 * v_1G',$$

whence we have

$$u_1 \overset{\alpha}{\sim} B_3^{krn}u_1,$$

by IV.2.35, that is, $B_3^{krn} \overset{\alpha}{\sim} \Lambda$. But this is impossible, since $B_3^{krn} \in \mathscr{R}_\alpha$ and the word B_3 is non-empty. This contradiction comes from the assumption that $I_\alpha(G_2, F_3D_3' F_3^{-1}) > 0$. Hence $I_\alpha(G_2, F_3D_3'F_3^{-1}) = 0$. We then have $G_2 \overset{\alpha-1}{\sim} F_3D_3' F_3^{-1}$ by IV.2.13, from which it follows easily that the period D_3 (and hence also the period D) is distinguished in rank $\alpha - 1$. We have proved the desired result.

Before proceeding to the proof of theorem 2.1 we shall prove an auxiliary result, 2.14, on periods which are distinguished in rank 0. In order to do this, we need the following simple lemma.

2.12. *If $D' \overset{\circ}{=} PAQA^{-1}R$, where A is non-empty and D' is uncancellable, then* $\partial(D) > 2\partial(A)$.

If $\partial(D) \leqslant 2\partial(A)$, then there exist two cyclic shifts TE and ET of the word D, the first of which is a start of $AQA^{-1}R$ and the second is an end of $PAQA^{-1}$. We may assume in addition that T is non-empty. It then follows that either $T \overset{\circ}{=} T^{-1}$ or $T \overset{\circ}{=} AH \overset{\circ}{=} H_1A^{-1}$, where $\partial(H) \leqslant \partial(A)$, since $\partial(T) \leqslant 2\partial(A)$. Both of these cases are clearly impossible.

2.13. If X is a word in the alphabet (2), we shall denote by $l(X)$ the least number k for which the word X can be expressed in the form

$$X \overset{\circ}{=} d_1^{r_1} d_2^{r_2} \dots d_k^{r_k}, \tag{51}$$

where the r_i are integers and each d_i is one of the letters of the alphabet (2). If $l(X) = k$, then powers of b-letters will stand alongside powers of a-letters in the expression (51).

2.14. *Let X, Y be words in the alphabet* (2), *and suppose that the equality*

$$X^{rn}Y^{rn}X^{-rn}Y^{-rn} = D^j, \tag{52}$$

where $r > 1$, holds in the free group, where D is non-empty and cyclically uncancellable. If $l(D) < 5$, then D is a cyclic shift of some word of the form

$$a^{trn}b^{srn}a^{-trn}b^{-srn},$$

where t and s are integers.

Suppose that $X = TAT^{-1}$ and $Y = RBR^{-1}$, where the words A and B are cyclically uncancellable. Clearly, the word $T^{-1}D^jT$ is equal in the free group to some uncancellable word $T_1^{-1}D_1^jT_1$, where D_1 is a cyclic shift of D. Then $\partial(D_1) = \partial(D)$ and $l(D_1) < 6$. Set

$$(T^{-1}RB^{rn}R^{-1}T)A^{-rn}(T^{-1}RB^{-rn}R^{-1}T) = ZA_1^{-rn}Z^{-1}, \tag{53}$$

where $ZA_1^{-rn}Z^{-1}$ is uncancellable and A_1 is a cyclic shift of the word A. Then the equation

$$[A^{rn}, ZA_1^{-rn}Z^{-1}]_0 \overset{\circ}{=} T_1^{-1}D_1^jT_1 \tag{54}$$

follows from (52).

By induction on $\partial(Z)$, we shall prove that (54), as well as the equality

$$[ZA_1^{-rn}Z^{-1}, A^{rn}]_0 \,\overline{\overline{\circ}}\, T_1^{-1}D_1^jT_1 ,$$

where A_1 and D_1 are cyclic shifts of A and D, yield the desired assertion about the word D. By symmetry, we can confine our attention to (54).

Assume that Z is empty. Since D_1 is non-empty, the left-hand side of relation (54) has the form $A^{rn-3}HQA_1^{-rn+3}$, from which it follows easily that $\partial(T_1) < 3\partial(A)$. Then, by 2.12, $\partial(D_1) > 2\partial(A^{rn-7})$, and since by hypothesis $l(D) < 5$, we have $l(D_1) < 6$ and hence $A_1 \,\overline{\overline{\circ}}\, A$. But this is impossible, since D_1 is non-empty.

Suppose that $\partial(Z) > 0$. Since D_1 is non-empty, ZA_1^{-3} cannot be a start of the word A^{-rn}. Hence for some H, E and F we have the equality

$$A^{rn} \,\overline{\overline{\circ}}\, EH, \;\; ZA_1^{-3} \,\overline{\overline{\circ}}\, H^{-1}F \tag{55}$$

and

$$[A^{rn}, ZA_1^{-rn}Z^{-1}]_0 \,\overline{\overline{\circ}}\, EFA_1^{-rn+3}Z^{-1} ,$$

where F is non-empty. Then (54) can be written in the form

$$EFA_1^{-rn+3}Z^{-1} \,\overline{\overline{\circ}}\, T_1D_1^jT_1 . \tag{56}$$

Assume that the word E is empty. Then $ZA_1^{-3} \,\overline{\overline{\circ}}\, A^{-rn}F$, where $\partial(F) < \partial(Z)$. Suppose that $[F, A_1^{-rn}, F^{-1}] \,\overline{\overline{\circ}}\, Z_1A_2^{-rn}Z_1^{-1}$, where A_2 is a cyclic shift of A_1. Then $\partial(Z_1) \leqslant \partial(F) < \partial(Z)$, and by (56) we have the equality

$$[Z_1A_2^{-rn}Z_1^{-1}, A^{rn}]_0 \,\overline{\overline{\circ}}\, T_1^{-1}D_1^jT_1 ,$$

from which it follows by the induction hypothesis that D satisfies the desired conditions. We may therefore assume that E is non-empty.

Assume that T_1 is non-empty. Then the word $Z^{-1}EF$ is cancellable, from which it follows by the non-emptiness of Z and in view of (55) that $F^{-1}HEF$ is cancellable. But, since E is non-empty, this is only possible if H is empty, that is, if $A^{rn}ZA_1^{-rn}Z^{-1}$ is uncancellable. But it then follows from (54) that the word $ZA_1^{-rn}Z^{-1}A^{rn}$ is cancellable. For this case, we look at the uncancellable word $T_2D_2^jT_2^{-1}$, which is equal to the word $A^{-rn}T_1^{-1}D_1^jT_1A^{rn}$, where D_2 is a cyclic shift of D_1. From (54) there follows

$$[ZA_1^{-rn}Z^{-1}, A^{rn}]_0 \,\overline{\overline{\circ}}\, T_2D_2^jT_2^{-1} ,$$

which is analogous to (54). Since $A^{rn}ZA^{-rn}Z^{-1}$ is uncancellable by hypothesis, we obtain, as in the foregoing, that the word T_2 must be empty. Thus one of the words T_1 and T_2 must be empty.

By the obvious analogy, we can restrict the proof to the case where T_1 is empty. Then relation (56) has the form

$$EFA_1^{-rn+3}Z^{-1} \stackrel{\circ}{=} D_1^j. \tag{57}$$

It is clear from (55) that the word $A^{r(n-3)/2}$ is either a start of E or an end of Z^{-1}. We then have $\partial(D_1) > \partial(A^{r(n-3)})$ by 2.12, and this leads to $l(A) = 1$, since $n \geqslant 1003$ and $l(D_1) < 6$. By the obvious symmetry, $l(B) = 1$ as well.

Suppose, for example, that $A \stackrel{\circ}{=} a^t$. Then $A_1 \stackrel{\circ}{=} a^t$ and $E \stackrel{\circ}{=} a^i$ for some $|i| \leqslant |trn|$. Condition (57) can be written in the form

$$a^{trn-i}Z_1a^{-trn}Z_1^{-1}a^i \stackrel{\circ}{=} D_1^j,$$

where $a^{-i}Z_1 \stackrel{\circ}{=} Z$. By 2.12, we have $j = 1$, and since the word D_1 is cyclically uncancellable, it follows that Z_1 starts and ends with b-letters. Since $l(D_1) < 6$, it follows that $l(Z_1) = 1$, that is, $Z \stackrel{\circ}{=} a^{-i}b^k$ for some k. Suppose that $R_1B^{rn}R_1^{-1}$ is the result of cancelling the word $T^{-1}RB^{rn}R^{-1}T$. Since $l(B) = 1$ and the right hand side of relation (53) has the form $a^ib^ka^{-trn}b^{-k}a^{-i}$, then, independent of whether R_1 is empty or not, we have $l([R_1^{-1}, A^{-rn}, R_1^{-1}]_0) = 1$ and $B \stackrel{\circ}{=} b^s$ for some s. Then $R_1 \stackrel{\circ}{=} a^i$ and $k = srn$. Hence

$$D_1 \stackrel{\circ}{=} a^{trn-i}b^{srn}a^{-trn}b^{-srn}a^i,$$

and by hypothesis D is a cyclic shift of D_1. This is what we wanted to prove.

2.15. *For an arbitrary set Π of prime numbers, the group $\Gamma(n, \Pi)$ constructed in* 2.2 *satisfies the identical relation* (1) *if and only if $r \in \Pi$.*

Suppose that $r \in \Pi$. We consider arbitrary elements x and y from $\Gamma(n, \Pi)$. We have $x, y \in \mathscr{B}$. Suppose that $X \in x$ and $X \in \mathscr{A}_\alpha$. By 2.7, there exists a word $S \in \mathscr{A}_{\alpha+2}$ and an elementary period A of rank $\gamma \leqslant \alpha + 1$ such that A^q occurs in some word from $\mathscr{M}_{\alpha-1}$ and

$$X \overset{\alpha}{\sim} [S, A^k, S^{-1}]_\alpha \tag{58}$$

for some $k > 0$. If the elementary period A is distinguished in some rank $< \gamma$, then by IV.2.37 we have $[A]_\gamma^n = 1$, whence by (58) it follows easily that for $\delta = \max(\alpha, \gamma)$, $[X]_\delta^{rn} \overset{\delta}{\sim} A$, that is $x^{rn} = 1$ in $\Gamma(n, \Pi)$. We may therefore assume that A is a non-distinguished elementary period of rank γ. Then, for any $i \geqslant 1$, we have $A^i \in \mathscr{A}_{\gamma+1}$, whence

$$[X]_\alpha^{rn} \overset{\alpha}{\sim} [S, A^{krn}, S^{-1}]_\alpha \tag{59}$$

follows from (58).

Moreover, in view of 2.10, there is a cyclic shift A_1 of the period A and a word TE such that the relations

$$S \overset{\alpha}{\sim} TE$$

and

$$[S, A^{krn}, S^{-1}]_\alpha \overset{\alpha}{\sim} TA_1^{krn}T^{-1} \tag{60}$$

hold, where $TA_1^{krn}T^{-1} \in \mathscr{K}_\alpha$. Clearly, we may assume that $S \in \mathscr{M}_{\alpha+1}$. Then no distinguished elementary $((n+1)/2+43)$-powers of rank $\alpha+1$ occur in T, from which it follows easily that $TA_1^{krn}T^{-1} \in \mathscr{K}_j$ for arbitrary $j > \alpha$, in view of the condition $S \in \mathscr{A}_{\alpha+2}$ and the fact that A_1 is non-distinguished. From this, in view of relation (6), we obtain that $TA_1^{krn}T^{-1} \in \mathscr{A}$ by VI.1.1.

From (59) and (60), it follows that

$$[X]_\alpha^{rn} \overset{\alpha}{\sim} TA_1^{krn}T^{-1}. \tag{61}$$

For an arbitrary word $Y \in y$, we find in an analogous way a word $T_1 \in \mathscr{A}$ and a distinguished period B_1 of rank β such that the relation

$$[Y]_\lambda^{rn} \overset{\lambda}{\sim} T_1B_1^{srn}T_1^{-1} \tag{62}$$

holds for some λ, where $T_1B_1^{srn}T_1^{-1} \in \mathscr{A}$.

Clearly, we may assume that the words $TA_1^{krn}T^{-1}$ and $T_1B_1^{srn}T_1^{-1}$ both belong to the set $\mathscr{A}_\varepsilon$ for some ε, where $\varepsilon \geqslant \alpha$ and $\varepsilon \geqslant \lambda$.

Suppose that

$$Z \overset{\varepsilon}{\sim} [TA_1^{krn}T^{-1}, T_1B_1^{srn}T_1^{-1}, TA_1^{-krn}T^{-1}, T_1B_1^{-srn}T_1^{-1}]_\varepsilon .$$

By II.7.13, we have $Z \in \mathscr{A}_{\varepsilon+3}$. Further, by 2.7 there exists a word $H \in \mathscr{A}$ and an elementary period of some rank $\eta \leqslant \varepsilon+4$ such that

$$Z \overset{\varepsilon+3}{\sim} [H, D^t, H^{-1}]_{\varepsilon+3} \tag{63}$$

for some t.

We set $\mu = \max(\varepsilon+4, \alpha, \lambda)$. Then from (61), (62) and (63) we get

$$[[X]_\mu^{rn}, [Y]_\mu^{rn}, [X^{-1}]_\mu^{rn}, [Y^{-1}]_\mu^{rn}]_\mu \overset{\mu}{\sim} [H, D^t, H^{-1}]_\mu . \tag{64}$$

This means that the period D is distinguished in rank μ. Since $\mu \geqslant \eta$, it then follows from 2.11, II.6.4 and II.2.12 that D is distinguished in rank $\eta-1$. Consequently, D is a distinguished elementary period of rank η and, by IV.2.37, we have $[D]_\eta^n = \Lambda$. Then $[D^t]_\mu^n = \Lambda$, from which we get

$$[[[X]_\mu^{rn}, [Y]_\mu^{rn}, [X^{-1}]_\mu^{rn}, [Y^{-1}]_\mu^{rn}]_\mu]_\mu^n \overset{\mu}{\sim} [H, [D^t]_\mu^n, H^{-1}]_\mu \overset{\mu}{\sim} \Lambda ,$$

by (64). But this means that the elements $x = \{X\}$ and $y = \{Y\}$ satisfy relation (1).

We shall now show that, for an arbitrary prime number k such that $k \notin \Pi$, the identity

$$(x^{kn}y^{kn}x^{-kn}y^{-kn})^n = 1 \tag{65}$$

does not hold in $\Gamma(n, \Pi)$.

We consider the word

$$E \rightleftharpoons (a^{kn}b^{kn}a^{-kn}b^{-kn})^n .$$

It is clear that E is an elementary word of rank 2 with period $a^{kn}b^{kn}a^{-kn}b^{-kn}$. Since $k \notin \Pi$, it follows from 2.14 that no distinguished elementary 9-power of rank 1 occurs in E. By 2.14, the period $a^{kn}b^{kn}a^{-kn}b^{-kn}$ is not distinguished in rank 0, and by 2.11 it is not distinguished in rank 1. Consequently, E is a non-distinguished elementary word of rank 2 with period $a^{kn}b^{kn}a^{-kn}b^{-kn}$. Clearly, no distinguished elementary q-powers of rank $\geqslant 2$ occur in E. Therefore, in view of the assertion mentioned at the end of section 2.4, if we had $E \overset{\alpha}{\sim} \Lambda$ for some α, then we would have $E \underset{\circ}{\equiv} \Lambda$. On the other hand, $\{E\}$ is the result of substituting the values $x = \{a\}$ and $y = \{b\}$ in the left-hand side of (65). Since $\neg(E \sim \Lambda)$, the identity (65) does not hold in $\Gamma(n, \Pi)$. This is what we wanted to prove.

Theorem 2.1 is a direct consequence of assertion 2.15. We adduce some simple corollaries of Theorem 2.1.

2.16. *No infinite subset of the system of identities* (1) *is equivalent to any finite system of group identities.*

Indeed, if some group identity follows from (1), then in deducing it we need only use a finite number of the identities occurring in this system. Consequently, if some infinite subset of (1) were equivalent to some finite system of group identities, then the subset would be equivalent to some finite subsystem of it. But this is impossible, by 2.1.

2.17. *There are continuously many distinct varieties of groups, corresponding to the distinct subsets of the system of identities* (1).

2.18. *There exists a group given by two generators and a recursively enumerable set of identical relations with insoluble word problem, that is, there is no algorithm which can decide for an arbitrary word whether it is equal to* 1 *in the group or not.*

Suppose that Π is a recursively enumerable but undecidable set of prime numbers. We consider the group $\Gamma(\Pi)$, given by two generators and all the identities (1) for $r \in \Pi$ and $n = 1003$. It is clear that every relation in $\Gamma(\Pi)$ is an identical relation. By 2.1, an arbitrary identity (1) holds in $\Gamma(\Pi)$ if and only if $r \in \Pi$. Hence the word problem for $\Gamma(\Pi)$ is undecidable.

In conclusion, we remark that there remains still the very interesting and apparently difficult problem concerning the existence of a group given by a finite number of generators and a finite number of identical relations, and having undecidable word problem.

References

1. Burnside, W.: On an unsettled question in the theory of discontinuous groups, Quart. J. Pure and Appl. Math. **33** (1902), 230–238.
2. Sanov, I. N.: Solution of the Burnside problem for exponent 4, Učen. Zap. Leningrad Univ. **10** (1940), 166–170.
3. Hall, M. Jr.: Solution of the Burnside problem for exponent 6, Proc. Nat. Acad. Sci. USA **43** (1957), 751–753.
4. Novikov, P. S.: On periodic groups, Dokl. Akad. Nauk. SSSR **127**, No. 4 (1959), 749–752.
5. Novikov, P. S., Adian, S. I.: Infinite periodic groups I, II, III, Izv. Akad. Nauk. SSSR Ser. Mat. **32**, No. 1,2,3 (1968), 212–244, 251–524, 709–731.
6. Novikov, P. S., Adian, S. I.: Defining relations and the word problem for free groups of odd exponent, Izv. Akad. Nauk. SSSR, Ser. Mat. **32**, No. 4 (1968), 971–979.
7. Novikov, P. S., Adian, S. I.: Commutative subgroups and the conjugacy problem for free groups of odd exponent, Izv. Akad. Nauk. SSSR, Ser. Mat. **32**, No. 5 (1968), 1176–1190.
8. Adian, S. I.: Infinite irreducible systems of group identities, Izv. Akad. Nauk. SSSR, Ser. Mat. **34**, No. 4 (1970), 715–734.
9. Adian, S. I.: Identical relations in groups, International Congress of Mathematicians, Nice, 1970. Lectures by Soviet Mathematicians, "Nauka", Moscow 1972, 7–13.
   Adjan, S. I.: Identités dans les groupes, Actes Congrès. intern. math. 1970, T. 1 (1971), 263–267.
10. Adian, S. I.: Subgroups of free groups of odd exponent, Trudy Mat. Inst. Steklov Akad. Nauk SSSR **112** (1971), 64–72.
11. Adian, S. I.: Certain torsion-free groups, Izv. Akad. Nauk. SSSR Ser. Mat. **35**, No. 3 (1971), 459–468.
12. Golod, E. S.: Nil-algebras and residually finite groups, Izv. Akad. Nauk SSSR Ser. Mat. **28**, No. 2 (1964), 273–276.
13. Aršon, S. E.: Proof of the existence of *n*-valued infinite asymmetric sequences, Mat. Sb. **2(44)**, No. 4 (1937), 769–779.
14. The Kourovka Notebook (unsolved problems in group theory). Novosibirsk 1969.
15. Neumann, Hanna: Varieties of groups, Ergebnisse der Mathematik und ihrer Grenzgebiete, Band 37. Springer-Verlag New York, Inc., New York, 1967.
    Translation: *Mnogoobrazija grupp*, Izdat. "Mir", Moscow 1969.
16. Ol'šanskii, A. Ju.: On the finite basis problem for laws in groups, Izv. Akad. Nauk. SSSR Ser. Mat. **34**, No. 2 (1970), 376–384.
17. Vaughan-Lee, M. R.: Uncountably many varieties of groups, Bull. London Math. Soc. **2** (1970), 280–286.
18. Baer, R.: Endlichkeitskriterien für Kommutatorgruppen, Math. Ann. **124**, No. 2 (1952), 161–177.
19. Milnor, J.: Problem 5603, Amer. Math. Monthly **75**, No. 6 (1968), 685–686.
20. Milnor, J.: Growth of finitely generated solvable groups, J. Diff. Geometry **2**, No. 4 (1968), 447–449.
21. Wolf, J.: Growth of finitely generated solvable groups and curvature of Riemannian manifolds, J. Diff. Geometry **2**, No. 4 (1968), 421–446.
22. Bass, H.: The degree of polynomial growth of finitely generated nilpotent groups. Proc. London Math. Soc. **25**, No. 4 (1972), 603–614.

23. Kesten, H.: Symmetric random walks on groups, Trans. Amer. Math. Soc. **92**, No. 2 (1959), 336–354.
24. Tartakovskiĭ, V. A.: Solution of the word problem for groups with k-reducible basis for $k > 6$, Izv. Akad. Nauk. SSSR Ser. Mat. **13**, No. 6 (1949), 483–494.
25. Britton, J. L.: The Existence of infinite Burnside groups, Word Problems. North-Holland Publishing Company, Amsterdam-London, 1973, 67–348.
26. Adian, S. I.: Periodic products of groups, Trudy Mat. Inst. Steklov, Akad. Nauk SSSR **142** (1976), 3–21.
27. Kurosh, A. G.: Theory of groups. Izdat. Nauka, 1967.

Subject Index

Together with all the fundamental concepts, the index gives informal, and as far as possible concise references to the majority of the assertions proved, including auxiliary ones. We recommend that the reader become familiar with the index during his first reading of the book. This will promote understanding of the essence of the lemmas, whose formulation is cumbersome.

As well as to page numbers, reference is given to subsections containing the various assertions and concepts. If a reference relates to several successive subsections, the page number indicates the first of them.

Index of Notation

$l_\alpha(E) = l_{\alpha,A}(E)$	the number of segments of the elementary word E of rank α 12, 4.11; 56, 5.1
$\rho_\alpha(V) = \rho_{\alpha,A}(V)$	density of the occurrence V in rank α relative to the original period A 11, 4.8; 56, 5.1
$\rho_{\alpha,A}(X)$	density of the periodic or integral word X of rank α relative to the period A 11, 4.8
$I_\alpha(X, Y)$	the number of kernels $V \in \mathrm{Ker}(\alpha, X)$ not related to $f_\alpha(V; X, Y)$ 173, 2.13
$I_\alpha(P{*}E{*}Q, X, Y)$	the number of kernels V contained in $P{*}E{*}Q$ not related to $f_\alpha(V; X, Y)$ 178, 2.26
$\mu_\alpha(X)$	the maximal numeral of a kernel of rank α that is mutually normalised with some kernel situated to the right and not related to it 192, 3.12

Other Notation

$\rightleftharpoons$	equality by definition 1, Ch. 1
$\overset{\circ}{\equiv}$	letter-for-letter equality of words 1, §1
$\subset$	inclusion of one set in another 1, Ch. 1
$\varnothing$	the empty set 1, Ch. 1
$A^0 = \Lambda$	the empty word 1, §1; 3, §2
A^t	the word in which A is repeated t times 2, §2
A^{-1}	the word inverse to A 2, 1.1
$\mathrm{Bas}(V)$	base of the occurrence V 2, 1.2
$F{*}H$	section of the word FH 67, 6.1
$\{E\}$	the class of words equivalent to E 244, 1.3
$\langle d\rangle$	the subgroup generated by the element d in the given group 269, 1,3; 276, 1.9

Ergebnisse der Mathematik und ihrer Grenzgebiete

A Series of Modern Surveys in Mathematics

---

1. Bachmann: Transfinite Zahlen
2. Miranda: Partial Differential Equations of Elliptic Type
4. Samuel: Méthodes d'algèbre abstraite en géométrie algébrique
5. Dieudonné: La géométrie des groupes classiques
6. Roth: Algebraic Threefolds with Special Regard to Problems of Rationality
7. Ostmann: Additive Zahlentheorie. 1. Teil: Allgemeine Untersuchungen
8. Wittich: Neuere Untersuchungen über eindeutige analytische Funktionen
10. Suzuki: Structure of a Group and the Structure of its Lattice of Subgroups
11. Ostmann: Additive Zahlentheorie. 2. Teil: Spezielle Zahlenmengen
13. Segre: Some Properties of Differentiable Varieties and Transformations
14. Coxeter/Moser: Generators and Relations for Discrete Groups
15. Zeller/Beekmann: Theorie der Limitierungsverfahren
16. Cesari: Asymptotic Behavior and Stability Problems in Ordinary Differential Equations
17. Severi: Il teorema di Riemann-Roch per curve, superficie e varietà questioni collegate
18. Jenkins: Univalent Functions and Conformal Mapping
19. Boas/Buck: Polynomial Expansions of Analytic Functions
20. Bruck: A Survey of Binary Systems
21. Day: Normed Linear Spaces
23. Bergmann: Integral Operators in the Theory of Linear Partial Differential Equations
25. Sikorski: Boolean Algebras
26. Künzi: Quasikonforme Abbildungen
27. Schatten: Norm Ideals of Completely Continuous Operators
28. Noshiro: Cluster Sets
30. Beckenbach/Bellman: Inequalities
31. Wolfowitz: Coding Theorems of Information Theory
32. Constantinescu/Cornea: Ideale Ränder Riemannscher Flächen
33. Conner/Floyd: Differentiable Periodic Maps
34. Mumford: Geometric Invariant Theory
35. Gabriel/Zisman: Calculus of Fractions and Homotopy Theory
36. Putnam: Commutation Properties of Hilbert Space Operators and Related Topics
37. Neumann: Varieties of Groups
38. Boas: Integrability Theorems for Trigonometric Transforms
39. Sz.-Nagy: Spektraldarstellung linearer Transformationen des Hilbertschen Raumes
40. Seligman: Modular Lie Algebras
41. Deuring: Algebren
42. Schütte: Vollständige Systeme modaler und intuitionistischer Logik
43. Smullyan: First-Order Logic
44. Dembowski: Finite Geometries
45. Linnik: Ergodic Properties of Algebraic Fields
46. Krull: Idealtheorie
47. Nachbin: Topology on Spaces of Holomorphic Mappings
48. A. Ionescu Tulcea/C. Ionescu Tulcea: Topics in the Theory of Lifting
49. Hayes/Pauc: Derivation and Martingales
50. Kahane: Séries de Fourier absolument convergentes
51. Behnke/Thullen: Theorie der Funktionen mehrerer komplexer Veränderlichen
52. Wilf: Finite Sections of Some Classical Inequalities
53. Ramis: Sous-ensembles analytiques d'une variété banachique complexe
54. Busemann: Recent Synthetic Differential Geometry
55. Walter: Differential and Integral Inequalities
56. Monna: Analyse non-archimédienne
57. Alfsen: Compact Convex Sets and Boundary Integrals

58. Greco/Salmon: Topics in m-adic Topologies
59. López de Medrano: Involutions on Manifolds
60. Sakai: C^*-Algebras and W^*-Algebras
61. Zariski: Algebraic Surfaces
62. Robinson: Finiteness Conditions and Generalized Soluble Groups, Part 1
63. Robinson: Finiteness Conditions and Generalized Soluble Groups, Part 2
64. Hakim: Topos annelés et schémas relatifs
65. Browder: Surgery on Simply-Connected Manifolds
66. Pietsch: Nuclear Locally Convex Spaces
67. Dellacherie: Capacités et processus stochastiques
68. Raghunathan: Discrete Subgroups of Lie Groups
69. Rourke/Sanderson: Introduction of Piecewise-Linear Topology
70. Kobayashi: Transformation Groups in Differential Geometry
71. Tougeron: Idéaux de fonctions différentiables
72. Gihman/Skorohod: Stochastic Differential Equations
73. Milnor/Husemoller: Symmetric Bilinear Forms
74. Fossum: The Divisor Class Group of a Krull Domain
75. Springer: Jordan Algebras and Algebraic Groups
76. Wehrfritz: Infinite Linear Groups
77. Radjavi/Rosenthal: Invariant Subspaces
78. Bognár: Indefinite Inner Product Spaces
79. Skorohod: Integration in Hilbert Space
80. Bonsall/Duncan: Complete Normed Algebras
81. Crossley/Nerode: Combinatorial Functors
82. Petrov: Sums of Independent Random Variables
83. Walker: The Stone-Čech Compactification
84. Wells/Williams: Embeddings and Extensions in Analysis
85. Hsiang: Cohomology Theory of Topological Transformation Groups
86. Olevskii: Fourier Series with Respect to General Orthogonal Systems
87. Berg/Forst: Potential Theory on Locally Compact Abelian Groups
88. Weil: Elliptic Functions according to Eisenstein and Kronecker
89. Lyndon/Schupp: Combinatorial Group Theory
90. Edwards/Gaudry: Littlewood-Paley and Multiplier Theory
91. Gunning: Riemann Surfaces and Generalized Theta Functions
92. Lindenstrauss/Tzafriri: Classical Banach Spaces I
93. Besse: Manifolds all of whose Geodesics are Closed
94. Heyer: Probability Measures on Locally Compact Groups
95. Adian: The Burnside Problem and Identities in Groups